Fermented Foods and Microbes Related to Health

Fermented Foods and Microbes Related to Health

Hiroshi Kitagaki

MDPI

Basel • Beijing • Wuhan • Barcelona • Belgrade • Novi Sad • Cluj • Manchester

Editor
Hiroshi Kitagaki
Natural Science Educational
Organization
Saga University
Saga city
Japan

Editorial Office
MDPI AG
Grosspeteranlage 5
4052 Basel, Switzerland

This is a reprint of articles from the Special Issue published online in the open access journal *Fermentation* (ISSN 2311-5637) (available at: www.mdpi.com/journal/fermentation/special_issues/fermented_food2).

For citation purposes, cite each article independently as indicated on the article page online and as indicated below:

Lastname, A.A.; Lastname, B.B. Article Title. *Journal Name* **Year**, *Volume Number*, Page Range.

ISBN 978-3-7258-2254-6 (Hbk)
ISBN 978-3-7258-2253-9 (PDF)
doi.org/10.3390/books978-3-7258-2253-9

Contents

fermentation

MDPI

Article

Aspergillus-Derived Cellulase Preparation Exhibits Prebiotic-like Effects on Gut Microbiota in Rats

Yongshou Yang [1,2], Thanutchaporn Kumrungsee [2,*], Norihisa Kato [2,*], Shinji Fukuda [3,4,5], Manabu Kuroda [6] and Shotaro Yamaguchi [6]

1 School of Life Sciences, Anhui University, Hefei 23061, China; 21193@ahu.edu.cn
2 Graduate School of Integrated Sciences for Life, Hiroshima University, Higashi, Hiroshima 739-8528, Japan
3 Institute for Advanced Biosciences, Keio University, Tsuruoka 997-0052, Japan; sfukuda@sfc.keio.ac.jp
4 Transborder Medical Research Center, Faculty of Medicine, University of Tsukuba, Tsukuba 305-8575, Japan
5 Gut Environmental Design Group, Kanagawa Institute of Industrial Science and Technology, Kawasaki 210-0821, Japan
6 Amano Enzyme Inc., Nagoya 460-8630, Japan; manabu_kuroda@amano-enzyme.com (M.K.); shotaro_yamaguchi@amano-enzyme.com (S.Y.)
* Correspondence: kumrung@hiroshima-u.ac.jp (T.K.); nkato@hiroshima-u.ac.jp (N.K.); Tel.: +81-82-424-7980 (T.K.); +81-82-426-0774 (N.K.)

Abstract: *Aspergillus*-derived cellulase, which is involved in the production of *Aspergillus*-fermented foods, has been employed in the food and animal feed industries. However, the effect of dietary *Aspergillus* cellulase on health is poorly understood. Previously, we discovered that supplemental *Aspergillus*-derived protease and lipase preparations had substantial bifidogenic effects on rats fed a high-fat diet. Therefore, this study reports on the effects of a 0.1% dietary *Aspergillus*-derived cellulase preparation (CEL) on the gut microbiota of rats fed a high-fat diet. Gene sequencing analysis of 16S rRNA revealed that CEL treatment markedly affected the microbiota profiles of the cecal contents ($p < 0.05$). Notably, CEL markedly increased the relative abundance (RA) of typical probiotics, such as *Bifidobacterium* and *Lactobacillus*, at the genus level (28- and 5-fold, respectively, $p < 0.05$). Similarly, at the family level, CEL treatment significantly increased the RA of Bifidobacteriaceae and Lactobacillaceae ($p < 0.05$). Furthermore, CEL increased the RA of other genera, such as *Collinsella* and *Enterococcus*, but decreased the RA of *Oscillospira*, *Dorea* and *Coprobacillus* ($p < 0.05$). The effects on these genera are similar to those reported for typical prebiotic oligosaccharides. Overall, this study demonstrates the prebiotic-like effects of dietary CEL by significantly increasing *Bifidobacterium* and *Lactobacillus* abundance.

Keywords: fermentation foods; *Aspergillus*; cellulase; *Bifidobacterium*; *Lactobacillus*; prebiotic; dietary enzyme supplements; 16S rRNA gene sequencing

Citation: Yang, Y.; Kumrungsee, T.; Kato, N.; Fukuda, S.; Kuroda, M.; Yamaguchi, S. *Aspergillus*-Derived Cellulase Preparation Exhibits Prebiotic-like Effects on Gut Microbiota in Rats. *Fermentation* **2022**, *8*, 71. https://doi.org/10.3390/fermentation8020071

Academic Editor: Hiroshi Kitagaki

Received: 19 January 2022
Accepted: 6 February 2022
Published: 8 February 2022

1. Introduction

Aspergillus species, such as *Aspergillus oryzae* and *Aspergillus niger*, have been used in the food fermentation industry in Japan and East Asia. Extracellular hydrolysis enzymes, which are responsible for fermentation during the *Aspergillus*-associated fermentation process, are produced and released into the reaction system [1]. The extracted *Aspergillus* enzymes, such as proteases, lipases, amylases and cellulases, have been used in food processing. Previously, we found that dietary supplementation with an *A. oryzae*-derived protease preparation and purified acid protease caused a bifidogenic effect by striking an elevation in the cecal levels of *Bifidobacterium*, a typical probiotic (beneficial bacteria for host health), in rats fed a high-fat (HF) diet [2,3]. We speculated that the increase in free amino acids (available amino acids) in the gut, induced by supplemental *Aspergillus* proteases, promotes *Bifidobacterium* growth [4]. The effect of an *Aspergillus* protease preparation is similar to that of prebiotics, such as short-chain non-digestible carbohydrates, e.g., fructo-oligosaccharides (FOS) and galacto-oligosaccharides (GOS), which can increase the number of typical probiotics, *Bifidobacterium* and *Lactobacillus* [5]. Prebiotics are well known to be selectively utilized by live probiotics, and promote the growth of probiotics, resulting in health benefits for the host [5]. A subsequent study in our laboratory demonstrated the powerful bifidogenic effect of *Aspergillus* lipase in rats fed an HF diet [6].

Recently, we carried out a preliminary investigation into the bifidogenic effects of several other digestive enzyme preparations derived from *Aspergillus* in rats fed an HF diet. Using quantitative polymerase chain reaction (qPCR) analysis, we found a remarkable increase in the cecal abundance of major probiotics, including *Lactobacillus* and *Bifidobacterium* bacteria, following consumption of an *Aspergillus*-derived cellulase preparation (CEL). Generally, cellulase enzymes, such as β-1,4-endoglucanase, cellobiohydrolase and β-glucosidase, degrade cellulose to β-glucose [7]. Cellulase enzymes have been widely used in the food and animal feed industries to improve nutrient availability and promote antioxidant properties by releasing antioxidants, such as polyphenols [8,9]. Furthermore, there are many applications of cellulase in the food industry, including the tenderization of fruits, clarification of fruit juices, extraction of flavoring materials and essential oils, and improvement in the aroma and taste of food items [10]. In the production of *Aspergillus*-fermented foods, *Aspergillus*-derived cellulase is thought to play an important role in the fermentation process [1,11]. Additionally, *Aspergillus* cellulase is included as a dietary enzyme supplement for gut health [12]. However, to the best of our knowledge, limited information is available regarding how dietary exogenous cellulase modulates the gut environment. In the present study, we hypothesized that dietary supplemental CEL modulates the composition of the gut microbiota. Thus, this study used 16S rRNA gene sequencing analysis to examine the effect of CEL on the gut microbiota in rats fed an HF diet. The study was conducted with rats fed an HF diet, since HF diet-induced colon dysbiosis, inflammation and diseases have been reported to be suppressed by dietary prebiotic oligosaccharides [13]. Herein, we report the first evidence for the prebiotic-like effect of supplemental CEL on gut microbiota in rats.

2. Materials and Methods

2.1. Animals and Diets

Sixteen male Sprague Dawley rats (four weeks old) were purchased from Charles River, Japan. The rats were individually housed in cages in a controlled-temperature environment (23 ± 2 °C), a 12 h light–dark cycle and relative humidity of 50%–60%. After being acclimatized for 7 days, the rats were randomly divided into the following two groups based on their diet: an HF diet (control; Ctrl) [2] or an HF diet mixed with 0.1% (*w/w*) CEL (*A. niger*-derived cellulase preparation, commercial name: Cellulase A "Amano" 3, Amano Enzyme Inc. Nagoya, Japan). The optimum pH was 4.5 (stable at pH 2.0–8.0), and the cellulase activity was 30,000 U/g at pH 4.5. The optimum temperature was 55 °C. CEL had slight activities of protease and lipase equivalent to 3% of the protease activity of *Aspergillus* protease preparation (Protease A "Amano" SD, Amano Enzyme Inc. Nagoya, Japan) used in our study [6] and to less than 0.1% of the lipase activity of *Aspergillus* lipase preparation (Lipase AP12, Amano Enzyme Inc. Nagoya, Japan) used in our study [6]). The HF diet contained 30% beef tallow, 20% casein, 0.3% L-cystine, 1% vitamin mixture (AIN-93), 3.5% mineral mixture (AIN-93G), 5% cellulose, 20% sucrose and 20.2% α-corn starch. During the two-week experimental period, equal amounts of the experimental diets were given daily in food cups (9, 10, 12, 14, and 15 g on days 1, 2–4, 5–7, 8–12, and 13–14, respectively) to prevent differences in food intake. All of the given diet was consumed each day. The rats had ad libitum access to drinking water. The study protocols were approved by the Ethics Committee of Hiroshima University (protocol identity No. C15-12).

2.2. Sample Collection

At the end of the two-week treatment period, the rats were anesthetized (13:00–15:00 h) by inhaling isoflurane in a desiccator to minimize suffering, and then euthanized by decapitation. The cecum was immediately excised, and its contents were removed entirely, weighed, and stored at −80 °C until subsequent analysis of cecal microbiota and organic acids.

2.3. 16S rRNA Gene-Based Microbiome Analysis

Total bacterial DNA in cecal contents was extracted using the QIAamp Stool Mini Kit, according to the manufacturer's instructions. Then, extracted bacterial DNA was quantified using NanoDrop spectrometry (NanoDrop Technology, Wilmington, DL, USA). The V1–V2 region of the 16S rRNA genes was amplified from the DNA isolated from cecal contents using the following bacterial universal primer set: 27F (5′- ACACTCTTTCCCTACACGACGC-TCTTCCGATCTAGRGTTTGATYMTGGCTCAG-3′) and 338R(5′-GTGACTGGAGTTCAGA-CGTGTGCTCTTCCGATCTTGCTGCCTCCCGTAGGAGT-3′). The following library preparation was performed as described previously [14]. Finally, all the barcoded V1–V2 PCR amplicons were sequenced using Illumina MiSeq sequencing technology at a read length of 2 × 300-bp (Illumina, San Diego, CA, USA), based on the manufacturer's instructions.

2.4. Bioinformatics Analysis

Fast Length Adjustment of SHort reads (FLASH, version 1.2.11) [15] was used to assemble the paired-end reads. Assembled reads with an average Q-value < 25 were filtered out using an in-house script. The same numbers of filtered reads were randomly selected from each sample and used for further analysis [6]. The selected reads were then processed using the Quantitative Insights Into Microbial Ecology pipeline (QIIME, version 1.9.1) [6]. The high-quality sequences were clustered into operational taxonomic units (OTUs) at 97% sequence similarity, and OTUs were assigned to the Greengenes database (version 13.8).

2.5. Analyses of Cecal Organic Acids and pH

The concentrations of organic acids in cecal contents were measured by gas chromatography/mass spectrometry as previously described [16]. For the analysis of pH in the cecal contents, 100 mg of freeze-dried cecal contents was mixed with 1 mL of Milli Q water. The pH value of the sample was measured by COMPACT pH Meter B-71X (Horiba Ltd., Kyoto, Japan).

2.6. Data Analysis

Data are expressed as mean ± standard error. Statistical analysis was performed by Welch's t-test. Data separation in the principal coordinate analysis (PCoA) ordination of beta diversity was tested using the ANOSIM statistical test in vegan-R, and p-values were generated based on 999 permutations. Some bacterial taxa data were subjected to linear discriminant analysis effect size (LEfSe) analysis, which uses the two-tailed nonparametric Kruskal–Wallis test to evaluate the significance of differences between taxa. $p < 0.05$ was considered to indicate a statistically significant difference. For the relationship between organic acids and microbiota composition, Pearson's correlation coefficient (r) was calculated, and the resulting correlation matrix was visualized by using R software (version 4.0.2).

2.7. Evaluation of the Risk of Bias in the Methodology

The risk of bias of this study was assessed using the Systematic Review Centre for Laboratory Animal Experimentation (SYRCLE) risk of bias (RoB) tool [17]. Two independent authors (YY and NK) evaluated the following nine items: (1) sequence generation, (2) baseline characteristics, (3) allocation concealment, (4) random housing, (5) intervention blinding, (6) random outcome assessment, (7) outcome blinding, (8) incomplete outcome data and (9) selective outcome reporting. All items were judged as 'yes' (low risk of bias) by two authors (YY and NK) independently.

3. Results

3.1. Food Intake, Body Weight and Cecal Content Weight

The total food intake for the two weeks and the final body weight were unaffected by the dietary treatment (data not shown). The weight of the cecal contents in the CEL group was markedly greater than that in the Ctrl group (5.05 ± 0.31 g and 1.44 ± 0.06 g, respectively; $p < 0.05$).

3.2. Cecal Microbiota

For 16S rRNA gene sequencing-based microbiota analysis, a total of 417,428 high-quality reads were passed through the QIIME filter. Unweighted and weighted UniFrac PCoA and ANOSIM analyses were conducted to compare the microbial community structures (Figure 1A,B). The results of the UniFrac PCoA and ANOSIM analyses indicated that the microbial composition was distinctly separated between the Ctrl and CEL groups in both the unweighted and weighted analyses ($p < 0.05$). However, the different alpha-diversity indices indicated a lower bacterial diversity in the CEL group than in the Ctrl group (Figure 1C,D; $p < 0.05$).

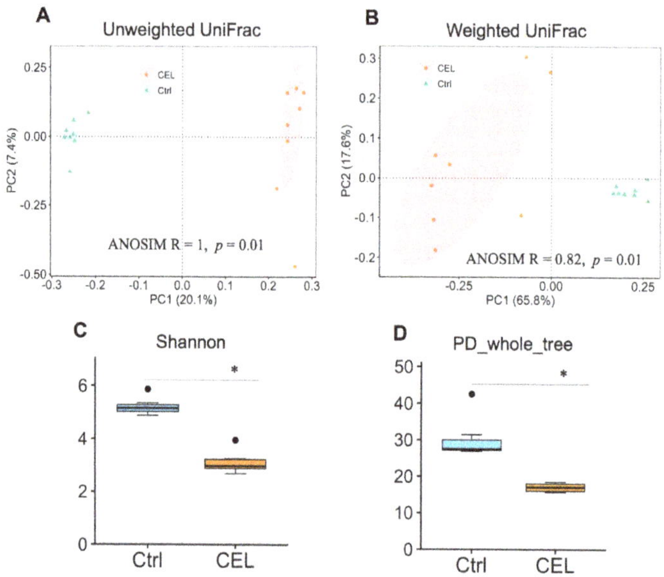

Figure 1. Effects of supplemental CEL on cecal microbiome profiles and alpha diversity. PCoA of unweighted (**A**) and weighted UniFrac (**B**) and PERMANOVA analyses were performed to compare the gut microbiome profiles of the experimental groups. The diversity of the gut microbiota within samples was measured by (**C**) Shannon index and (**D**) PD whole tree. Data are presented as a boxplot with median and min–max whiskers. The dots (•) in the boxplots are outliers. * Significantly different at $p < 0.05$ (Welch's *t*-test).

The LEfSe analysis results indicated that 60 bacterial taxa differed between the Ctrl and CEL groups (Figure 2; $p < 0.05$). This analysis identified that the bacterial species *Collinsella, Lactobacillus, Bifidobacterium, Eggerthella, Enterococcus, Akkermansia, Dehalobacterium, Adlercreutzia, Coprobacillus, Dorea, rc4-4, Oscipillospira, Roseburia, Coprococcus, Allobaculum, Ruminococcus* and *Parabacteroides* varied between the two groups.

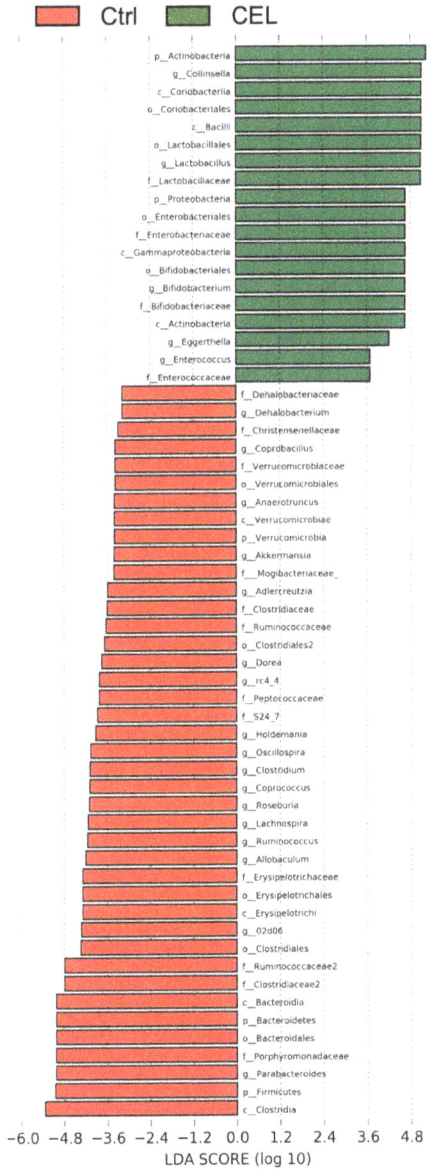

Figure 2. Different taxa between the Ctrl and CEL groups. LEfSe analysis was performed to compare the different taxa between the Ctrl and CEL groups. The two-tailed nonparametric Kruskal–Wallis test was used to evaluate the significance of differences between taxa at $p < 0.05$.

Among the four most abundant phyla, supplemental CEL significantly decreased the RA of Firmicutes and Bacteroidetes but enriched the RA of Actinobacteria and Proteobacteria (Figure 3A; $p < 0.05$). The top nine bacterial taxa are displayed at the family level to address the domain taxa of the microbial groups (Figure 3B); supplemental CEL significantly increased the RA of Bifidobacteriaceae (28-fold), Lactobacillaceae (5-fold), Coribacteriaceae (100-fold) and Enterobacteriaceae (13-fold) ($p < 0.05$). In contrast, it reduced the RA of bacterial taxa, including Lachnospiaceae (1.6-fold), Porphyromonadaceae

(15-fold), Ruminococcaceae (77-fold), Clostridiaceae (12-fold) and Erysipelotrichaceae (12-fold) ($p < 0.05$).

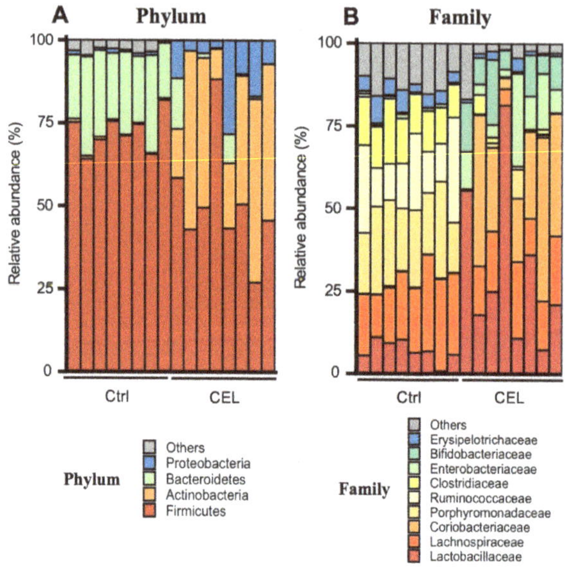

Figure 3. Effects of supplemental CEL on cecal microbiota composition at the phylum (**A**) and family (**B**) levels.

The resulting RA of the genera are shown in Table 1. Supplemental CEL significantly increased the RA of Lactobacillus (5-fold), Collinsella (526-fold), Bifidobacterium (28-fold) and Enterococcus (42-fold) ($p < 0.05$). Meanwhile, CEL significantly decreased the RA of Parabacteroides (15-fold), Allobaculum (8-fold), Oscillospira (13-fold), rc4-4, Dorea, Coprobacillus and Adlercreutzia (Table 1; $p < 0.05$). The RA of Ruminococcus and Roseburia were not significantly affected ($p > 0.05$). The genera with a mean RA less than 0.2% in all the groups were not considered for statistical analysis. The results of the effects of CEL on Lactobacillus and Bifidobacterium levels were similar to those of our preliminary study using qPCR analysis (data not shown).

Table 1. Effects of supplemental CEL on the relative abundance of genera in cecum of rats.

Phylum	Family	Genus	Ctrl	CEL
			(% of total bacteria)	
Firmicute	Lactobacillaceae	*Lactobacillus*	0.698 ± 0.108	3.186 ± 0.833 *
Actinobacteria	Coriobacteriaceae	*Collinsella*	0.005 ± 0.003	2.578 ± 0.601 **
Actinobacteria	Bifidobacteriaceae	*Bifidobacterium*	0.034 ± 0.010	0.965 ± 0.198 **
Firmicute	Enterococaceae	*Enterococcus*	0.02 ± 0.01	0.83 ± 0.15 **
Firmicute	Ruminococacceae	*Ruminococcus*	3.13 ± 1.38	0.01 ± 0.00
Firmicute	Erysipelotrichaceae	*Allobaculum*	3.03 ± 0.68	0.38 ± 0.20 **
Firmicute	Lachnospiraceae	*Roseburia*	2.60 ± 1.34	0.00 ± 0.00
Firmicute	Ruminococaceae	*Oscillospira*	2.32 ± 0.31	0.18 ± 0.08 **
Bacteroidetes	Porphyromonadaceae	*Parabacteroides*	2.19 ± 0.16	0.15 ± 0.01 **
Actinobacteria	Peptococcaceae	*rc4-4*	1.27 ± 0.16	0.00 ± 0.00 **
Firmicute	Lachnospiraceae	*Dorea*	1.12 ± 0.17	0.00 ± 0.00 **
Firmicute	Erysipelotrichaceae	*Coprobacillus*	0.35 ± 0.06	0.00 ± 0.00 **
Actinobacteria	Eggerthellaceae	*Adlercreutzia*	0.21 ± 0.00	0.00 ± 0.00 **

Mean ± SE (n = 8). * $p < 0.05$, ** $p < 0.01$ (Welch's t-test).

3.3. Cecal Organic Acids and pH

Table 2 indicates the concentrations of cecal organic acids. Treatment with CEL significantly increased the concentrations of lactate (219-fold) and total organic acids (3-fold), while it significantly decreased those of acetate (4-fold), propionate (10-fold) and n-butyrate (5-fold) ($p < 0.05$). There was a significant inverse association of lactate with propionate, n-butyrate and acetate levels ($r = -0.91$, $r = -0.82$ and $r = -0.74$, respectively; $p < 0.01$). Figure 4 further indicates the relationship between the levels of organic acids and various bacteria. There was a strong correlation between lactate levels and the RA of the lactate-producing bacteria *Lactobacillus*, *Bifidobacterium* and *Enterococcus*. In general, the RA of genera such as *Oscillospira*, *Dorea* and *Coprobacillus* had a strong positive association with propionate levels, but a strong negative association with lactate levels (Figure 4). The pH in the cecal contents of the CEL group was significantly lower than that in the Ctrl group (5.40 ± 0.10 and 8.29 ± 0.20, respectively; $p < 0.01$). There was a strong inverse association between the pH values and the levels of total organic acids ($r = -0.97$; $p < 0.001$).

Table 2. Effects of supplemental CEL on the levels of organic acids in cecum of rats.

Organic Acids	Ctrl	CEL
	(μmol/g dry wt of cecal contents)	
Acetate	40.3 ± 5.9	10.8 ± 4.0 *
Propionate	11.1 ± 0.0	1.1 ± 0.2 *
n-Butyrate	7.8 ± 0.9	1.5 ± 1.2 *
Lactate	0.9 ± 0.1	188.7 ± 13.1 *
Succinate	13.9 ± 3.1	12.5 ± 3.3
Total organic acids	75.2 ± 8.8	213.9 ± 10.0 *

Mean ± SE (n = 8). * $p < 0.05$ (Welch's *t*-test).

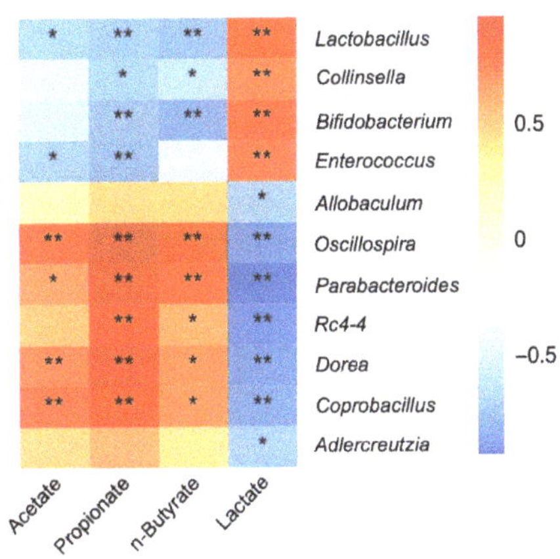

Figure 4. Correlation matrix (correlation coefficient) between the levels of organic acids and the relative abundances of genera. * $p < 0.05$, ** $p < 0.01$.

4. Discussion

4.1. *Bifidobacterium* and *Lactobacillus*

Our previous study revealed that the dietary consumption of 0.1% *Aspergillus* lipase and protease preparations for two weeks markedly increased the RA of *Bifidobacterium* in the cecum, but not *Lactobacillus* [3,6]. Moreover, this study discovered that the consumption of 0.1% *Aspergillus* cellulase preparation significantly increased the RA of both probiotics, such as *Bifidobacterium* and *Lactobacillus*. Hence, our findings suggest a potential role of *Aspergillus* cellulase as a prebiotic-like ingredient for enhancing typical probiotic levels, i.e., bacteria of the genera *Bifidobacterium* and *Lactobacillus*. Because a wide range of plant-derived foods contain cellulose, cellulase preparations may be effective prebiotics for use in the food and animal feed industries. Currently, various prebiotics are well known for providing health benefits by enhancing the abundance of probiotics [18]. To the best of our knowledge, this study, along with previous studies [3,6] on *Aspergillus* protease and lipase, implies that *Aspergillus*-derived cellulase may be a new type of "prebiotic digestive enzyme", as proposed by our recent study [6].

Previously, we discovered that dietary *Aspergillus*-derived acid protease had a strong bifidogenic effect [3], but dietary *Aspergillus*-derived alkaline protease had no effect (Yang et al., unpublished results). In the study, we speculated that the acid protease might be stable under acidic conditions, such as the stomach, and relatively resistant to gut digestive proteases. This might enable the intact acid protease to be delivered to the colon lumen, increasing the number of colonic free amino acids, which are essential for *Bifidobacterium* growth. In this context, we were interested to see that the CEL remains stable under acidic conditions (pH 2.0). We believe that this acid-resistant property might partially contribute to the substantial increase in *Bifidobacterium* and *Lactobacillus*.

Cellulase is responsible for the hydrolytic conversion of cellulose to metabolites, including shorter cello-polysaccharides, cello-oligosaccharides (COS), cellobiose and beta-glucose [7,8]. According to research, dietary supplemental COS significantly enhances the abundance of *Lactobacillus* bacteria in pig jejunal contents [19]. Furthermore, a recent in vitro study reported that COS treatment significantly enhanced the growth of *Lactobacillus* bacteria [20]. Thus, the enrichment of *Lactobacillus* by CEL may be, at least partially, mediated through mechanisms involving COS. However, neither of these studies indicated any effect of COS on the abundance of *Bifidobacterium*. In this study, CEL treatment markedly increased the RA of *Bifidobacterium* (a 28-fold increase). Accordingly, COS cannot account for the strong bifidogenic effect of CEL. Therefore, further studies are necessary to elucidate the mechanisms underlying the bifidogenic effect of supplemental *Aspergillus* cellulase.

4.2. Other Genera

This study further indicated that CEL markedly increased the RA of *Collinsella* and *Enterococcus*, but decreased the RA of seven genera, including *Parabacteroides, Allobaculum, Oscillospira, rc4-4, Dorea, Coprobacillus* and *Adlercreutzia*. There is very limited information about the effects of prebiotic oligosaccharides on *Parabacteroides, Allobaculum, rc4-4* and *Adlercreutzia*, as well as their roles in gut health. Therefore, the implications of modulating *Collinsella, Enterococcus, Oscillospira, Dorea* and *Coprobacillus* by CEL are discussed below.

Accumulating evidence indicates that the treatment of rats with inulin and oligosaccharides enhances *Collinsella* and *Bifidobacterium* abundance in the guts [21]. A recent study also revealed that *Aspergillus* protease and lipase preparations significantly increased the RA of *Collinsella* [6]. These findings are similar to our current results, indicating a marked increase in the level of *Collinsella*, as induced by CEL. *Collinsella* species might be beneficial to health; their enhanced abundance following dietary supplementation with oligofructose-enriched inulin in obese women is associated with an improved profile of hippurate, a microbial co-metabolite, indicating a healthier phenotype [22]. Furthermore, *Collinsella* exists at lower abundance in patients with inflammatory bowel disease or chronic pancreatitis than in healthy controls [23]. A study by Saalman et al. [24] suggested the potential use of this genus in treating inflammatory bowel disease. Overall, this study suggests that the significant increase in the abundance of *Collinsella* by CEL is beneficial to health; however, further studies are necessary to validate this position.

Additionally, the current study indicates a higher abundance of *Enterococcus* species in rats fed CEL. *Enterococcus faecalis* improves host health [25,26] and is clinically relevant for the treatment of chronic recurrent bronchitis [27]. Some *Enterococcus* species are employed as probiotics and in the production of feed additives to prevent diarrhea in animals [24]. Interestingly, several *Enterococcus* species isolated from food possess antioxidant activities [28]. Studies have shown that prebiotic oligosaccharides enhance the abundance of

Enterococcus species in mice and in perioperative colorectal cancer patients [29,30]. In addition, our previous study showed a significant increase in the RA of *Enterococcus* in rats fed *Aspergillus* protease and lipase preparations [6]. The higher abundance of commensal *Enterococcus* in rats fed CEL might be beneficial to the rats' health. However, *Enterococcus* species are a leading cause of hospital-associated bacteremia, endocarditis and urinary tract infections [31]. Therefore, further studies are necessary to determine the implications of increased *Enterococcus* in rats administered CEL.

It is worth noting that the CEL treatment significantly decreased the abundance of bacteria from the following genera: *Oscillospira*, *Dorea* and *Coprobacillus*. These findings agree with previous research on typical prebiotic oligosaccharides [32–36], and *Aspergillus* protease and lipase preparations (6). The current information on the roles of *Oscillospira*, *Dorea* and *Coprobacillus* in gut health is limited. Therefore, the implications of their modulation remain unclear.

4.3. Bacterial Diversity

In this study, contrary to expectations, CEL treatment significantly lowered bacterial diversity compared to the control. Microbial diversity is considered beneficial for community stability and host health [37,38]. However, this may not always be the case, and assumptions of increased diversity could be oversimplified for complicated interactive mechanisms in health and disease [39]. We believe that the reduced bacterial diversity in the CEL group might be associated with the depletion of several bacterial species, including *Parabacteroides*, *Allobaculum*, *Oscillospira*, *Dorea* and *Coprobacillus* (Figure 2, Table 1).

4.4. Organic Acids

Furthermore, in this study, it is interesting that CEL markedly increased cecal lactate levels, which were significantly associated with the modulation of the RA of lactate-producing bacteria, such as *Lactobacillus*, *Bifidobacterium* and *Enterococcus*. Meanwhile, CEL decreased the levels of other organic acids, such as acetate, propionate and butyrate. Notably, there was a significant inverse relationship between lactate and propionate levels. Propionate is produced microbially from lactate in the human colon [40]. Thus, CEL may reduce the metabolic conversion of lactate into propionate. Lactate has previously been studied in vitro for its free radical scavenging and antioxidant properties [41]. According to recent studies, lactate exhibits an inflammatory or anti-inflammatory role depending on its effects on immune cells and disease types [42]. Therefore, the implications of lactate accumulation in the CEL group remain unexplored. Additionally, CEL increased the total organic acids and lowered the pH. Interestingly, there was a strong inverse association between total organic acids and pH. The increased organic acids by CEL may cause the lower pH.

4.5. Limitations of This Study

One limitation of this study was that the cellulase preparation was crude and unpurified, despite having high cellulase activity. Therefore, factors related to cellulase preparation, besides the cellulase itself, may be responsible for modulating the gut microbiota. Since the activities of protease and lipase in CEL were slight (Section 2.1: Animals and Diets), the possibility that the protease and lipase in CEL modulate the gut microbiota was neglected. Further research is necessary to determine the effects of purified *Aspergillus*-derived cellulase on intestinal microbiota, as well as the relationship between cellulase activity and gut microbiota modulation. Although the cellulase preparation used here is crude, the preparation is actually used for the improvement of food digestion and food production. Therefore, the finding of the prebiotic-like effect of CEL is of great significance in terms of application.

5. Conclusions

Our results indicate that CEL treatment increased the RA of typical probiotics, such as *Bifidobacterium* and *Lactobacillus*. CEL also modulated the RA of other genera, including *Collinsella, Enterococcus, Oscillospira, Dorea* and *Coprobacillus*, as reported for typical prebiotic oligosaccharides. These findings suggest a potential role for *Aspergillus* cellulase as a prebiotic digestive enzyme in the food and animal feed industries, in addition to the established benefits to food digestion. This study may also help to elucidate the health benefits of *Aspergillus*-fermented foods and dietary enzyme supplements containing *Aspergillus* cellulase. Interestingly, the modulations of the genera *Bifidobacterium, Collinsella, Enterococcus, Oscillospira, Dorea* and *Coprobacillus* are similar to those reported for *Aspergillus* protease and lipase preparations [6]. Thus, the colonic digestion of carbohydrates, proteins and lipids may have a similar impact on these genera. Currently, our group is conducting metabolomics studies to elucidate the mechanisms through which *Aspergillus* cellulase modulates microbiota, as well as the effects of *Aspergillus* cellulase on gut health and diseases.

Author Contributions: Conceptualization, investigation, methodology, validation, writing—original draft: Y.Y., T.K. and N.K.; formal analysis, software, methodology, visualization: Y.Y., T.K., S.F., M.K. and S.Y.; literature search: Y.Y. and N.K.; project administration, funding acquisition, supervision, writing—review and editing: N.K., T.K., S.F., M.K. and S.Y. All authors have read and agreed to the published version of the manuscript.

Funding: This study was supported, in part, by the grant from the HIRAKU consortium, Hiroshima University (to T.K.). Amano Enzyme Inc. (to N.K.), JSPS KAKENHI (18H04805 to S.F.), AMED-CREST (JP21gm1010009 to S.F.), JST ERATO (JPMJER1902 to S.F.), the Takeda Science Foundation (to S.F.) and the Food Science Institute Foundation (to S.F.).

Institutional Review Board and Statement: The study was conducted according to the guidelines of the Declaration of Helsinki and approved by the Ethics Committee of Hiroshima University (ethical approval No. C15-12), approved on 16 June 2015.

Informed Consent Statement: Not applicable.

Data Availability Statement: The data of the 16S rRNA gene sequences of gut microbiota presented in this study are available from the DDBJ database (http://getentry.ddbj.nig.ac.jp/) (accessed on 8 October 2021) under accession number DRA012837. Further inquiries can be directed to the corresponding authors.

Conflicts of Interest: This study received funding from Amano Enzyme Inc. The funders had no role in the design of the study; in the collection, analyses, or interpretation of data; in the writing of the manuscript, or in the decision to publish the results. All authors declare no other competing interests in the current study.

References

1. Ichishima, E. Enzymology of the national microorganisms of Japan in a historical context. *J. Biotechnol. Biochem.* **2019**, *2*, 11–18.
2. Yang, Y.; Sitanggang, N.V.; Kato, N.; Inoue, J.; Murakami, T.; Watanabe, T.; Iguchi, T.; Okazaki, Y. Beneficial effects of protease preparations derived from Aspergillus on the colonic luminal environment in rats consuming a high-fat diet. *Biomed. Rep.* **2015**, *3*, 715–720. [CrossRef] [PubMed]
3. Yang, Y.; Iwamoto, A.; Kumrungsee, T.; Okazaki, Y.; Kuroda, M.; Yamaguchi, S.; Kato, N. Consumption of an acid protease derived from Aspergillus oryzae causes bifidogenic effect in rats. *Nutr. Res.* **2017**, *44*, 60–66. [CrossRef] [PubMed]
4. Yang, Y.; Kumrungsee, T.; Kuroda, M.; Yamaguchi, S.; Kato, N. Feeding Aspergillus protease preparation combined with adequate protein diet to rats increases levels of cecum gut-protective amino acids, partially linked to Bifidobacterium and Lactobacillus. *Biosci. Biotech. Biochem.* **2019**, *83*, 1901–1911. [CrossRef] [PubMed]
5. Gibson, G.R.; Hutkins, R.; Sanders, M.E.; Prescott, S.L.; Reimer, R.A.; Salminen, S.J.; Scott, K.; Stanton, C.; Swanson, K.S.; Cani, P.D.; et al. Expert consensus document: The International Scientific Association for Probiotics and Prebiotics (ISAPP) consensus statement on the definition and scope of prebiotics. *Nat. Rev. Gastroenterol. Hepatol.* **2017**, *14*, 491–502. [CrossRef]
6. Yang, Y.; Kumrungsee, T.; Kato, N.; Fukuda, F.; Kuroda, M.; Yamaguchi, S. Supplemental Aspergillus lipase and protease preparations display powerful bifidogenic effects and modulate the gut microbiota community of rats. *Fermentation* **2021**, *7*, 294.
7. Jecu, L. Solid state fermentation of agricultural wastes for endoglucanase production. *Ind. Crop Prod.* **2000**, *11*, 1–5. [CrossRef]

8. Kumar, V.A.; Kurup, R.S.C.; Snishamol, C.; Prabhu, G.N. Role of Cellulases in Food, Feed, and Beverage Industries. In *Green Bio-processes. Energy, Environment, and Sustainability*; Parameswaran, B., Varjani, S., Raveendran, S., Eds.; Springer: Singapore, 2019; pp. 323–343.
9. Sharma, A.; Tewari, R.; Rana, S.S.; Soni, R.; Soni, S.K. Cellulases: Classification, Methods of Determination and Industrial Applications. *Appl. Biochem. Biotechnol.* **2016**, *179*, 1346–1380. [CrossRef]
10. Ejaz, U.; Sohail, M.; Ghanemi, A. Cellulases: From Bioactivity to a Variety of Industrial Applications. *Biomimetics* **2021**, *6*, 44. [CrossRef]
11. Yamane, Y.; Fukuda, H.; Mikami, S.; Kizaki, Y.; Wakabayasi, S. Properties of Cellulase Produced by Aspergillus oryzae and its Effect on Sake Moromi Fermentation. *Nippon Jyozo Kyokaishi (J. Brew. Soc. Japan)* **2004**, *99*, 84–92. (In Japanese)
12. Text Sison, G. Digestive Enzymes: Uses, Common Brands, and Safety Information. Available online: https://www.singlecare.com/blog/digestive-enzymes/ (accessed on 12 November 2021).
13. Van Hul, M.; Karnik, K.; Canene-Adams, K.; De Souza, M.; Abbeele, P.V.D.; Marzorati, M.; Delzenne, N.M.; Everard, A.; Cani, P.D. Comparison of the effects of soluble corn fiber and fructooligosaccharides on metabolism, inflammation, and gut microbiome of high-fat diet-fed mice. *Am J. Physiol. Endocrinol. Metab.* **2020**, *319*, E779–E791. [CrossRef]
14. Ishii, C.; Nakanishi, Y.; Murakami, S.; Nozu, R.; Ueno, M.; Hioki, K.; Aw, W.; Hirayama, A.; Soga, T.; Ito, M.; et al. A metabologenomic approach reveals changes in the intestinal environment of mice fed on American diet. *Int. J. Mol. Sci.* **2018**, *19*, 4079. [CrossRef]
15. Magoč, T.; Salzberg, S.L. FLASH: Fast length adjustment of short reads to improve genome assemblies. *Bioinformatics* **2011**, *27*, 2957–2963. [CrossRef]
16. Zhang, S.; Wang, H.; Zhu, M. A sensitive GC/MS detection method for analyzing microbial metabolites short chain fatty acids in fecal and serum samples. *Talanta* **2019**, *196*, 249–254. [CrossRef]
17. Hooijmans, C.R.; Rovers, M.M.; BM de Vries, R.; Leenaars, M.; Ritskes-Hoitinga, M.; Langendam, M.W. SYRCLE's risk of bias tool for animal studies. *BMC Med. Res. Meth.* **2014**, *14*, 43. [CrossRef]
18. Tsai, Y.-L.; Lin, T.-L.; Chang, C.-J.; Wu, T.-R.; Lai, W.-F.; Lu, C.-C.; Lai, H.-C. Probiotics, prebiotics and amelioration of diseases. *J. Biomed. Sci.* **2019**, *26*, 3. [CrossRef]
19. Jiao, L.F.; Song, Z.H.; Ke, Y.L.; Xiao, K.; Hu, C.H.; Shi, B. Cello-oligosaccharide influences intestinal microflora, mucosal architecture and nutrient transport in weaned pigs. *Anim. Feed Sci. Technol.* **2014**, *195*, 85–91. [CrossRef]
20. Zhong, C.; Ukowitz, C.; Domig, K.J.; Nidetzky, B. Short-Chain Cello-oligosaccharides: Intensification and Scale-up of Their Enzymatic Production and Selective Growth Promotion among Probiotic Bacteria. *J. Agric. Food Chem.* **2020**, *68*, 8557–8567. [CrossRef]
21. Klancic, T.; Laforest-Lapointe, I.; Choo, A.; Nettleton, J.E.; Chleilat, F.; Tuplin, E.W.N.; Alukic, E.; Cho, N.A.; Nicolucci, A.C.; Arrieta, M.; et al. Prebiotic Oligofructose Prevents Antibiotic-Induced Obesity Risk and Improves Metabolic and Gut Microbiota Profiles in Rat Dams and Offspring. *Mol. Nutr. Food Res.* **2020**, *64*, e2000288. [CrossRef]
22. Dewulf, E.M.; Cani, P.; Claus, S.; Fuentes, S.; Puylaert, P.G.B.; Neyrinck, A.; Bindels, L.B.; De Vos, W.M.; Gibson, G.R.; Thissen, J.-P.; et al. Insight into the prebiotic concept: Lessons from an exploratory, double blind intervention study with inulin-type fructans in obese women. *Gut* **2013**, *62*, 1112–1121. [CrossRef]
23. Kassinen, A.; Krogius-Kurikka, L.; Mäkivuokko, H.; Rinttilä, T.; Paulin, L.; Corander, J.; Malinen, E.; Apajalahti, J.; Palva, A. The Fecal Microbiota of Irritable Bowel Syndrome Patients Differs Significantly From That of Healthy Subjects. *Gastroenterology* **2007**, *133*, 24–33. [CrossRef]
24. Saalman, R.; Alderberth, I.; Wold, A.; Sjoberg, F. Use of Collinsella for Treatment of Inflammatory Bowel Disease. U.S. Patent 15/510, 245, 1 February 2018.
25. Hanchi, H.; Mottawea, W.; Sebei, K.; Hammami, R. The Genus Enterococcus: Between Probiotic Potential and Safety Concerns–An Update. *Front. Microbiol.* **2018**, *9*, 1791. [CrossRef]
26. Fugaban, J.I.F.; Holzapfel, W.H.; Todorov, S.D. Probiotic potential and safety assessment of bacteriocinogenic Enterococcus faecium strains with antibacterial activity against Listeria and vancomycin-resistant enterococci. *Curr. Res. Microb. Sci.* **2021**, *2*, 100070. [CrossRef]
27. Habermann, W.; Zimmermann, K.; Skarabis, H.; Kunze, R.; Rusch, V. The effect of a bacterial immunostimulant (human Enterococcus faecalis bacteria) on the occurrence of relapse in patients with chronic bronchitis. *Arzneimittelforschung* **2001**, *51*, 931–937. [PubMed]
28. Pieniz, S.; Andreazza, R.; Anghinoni, T.; Camargo, F.; Brandelli, A. Probiotic potential, antimicrobial and antioxidant activities of Enterococcus durans strain LAB18s. *Food Control* **2014**, *37*, 251–256. [CrossRef]
29. Mao, B.; Li, D.; Zhao, J.; Liu, X.; Gu, Z.; Chen, Y.Q.; Zhang, H.; Chen, W. Metagenomic insights into the effects of fructo-oligosaccharides (FOS) on the composition of fecal microbiota in mice. *J. Agric. Food Chem.* **2015**, *63*, 856–863. [CrossRef] [PubMed]
30. Xie, X.; He, Y.; Li, H.; Yu, D.; Na, L.; Sun, T.; Zhang, D.; Shi, X.; Xia, Y.; Jiang, T.; et al. Effects of prebiotics on immunologic indicators and intestinal microbiota structure in perioperative colorectal cancer patients. *Nutrition* **2019**, *61*, 132–142. [CrossRef]
31. Wang, L.; Pan, M.; Li, D.; Yin, Y.; Jiang, T.; Fang, S.; Wang, G.; Zhao, J.; Zhang, H.; Chen, W. Metagenomic insights into the effects of oligosaccharides on the microbial composition of cecal contents in constipated mice. *J. Funct. Foods* **2017**, *38*, 486–496. [CrossRef]

32. Dubin, K.; Pamer, E.G. Enterococci and Their Interactions with the Intestinal Microbiome. *Microbiol. Spectrum* **2014**, *5*, 1–24.
33. Fei, Y.; Wang, Y.; Pang, Y.; Wang, W.; Zhu, D.; Xie, M.; Lan, S.; Wang, Z. Xylooligosaccharide Modulates Gut Microbiota and Alleviates Colonic Inflammation Caused by High Fat Diet Induced Obesity. *Front. Physiol.* **2020**, *10*, 1601. [CrossRef]
34. Garcia-Mazcorro, J.F.; Barcenas-Walls, J.R.; Suchodolski, J.S.; Steiner, J.M. Molecular assessment of the fecal microbiota in healthy cats and dogs before and during supplementation with fructo-oligosaccharides (FOS) and inulin using high-throughput 454-pyrosequencing. *PeerJ* **2017**, *5*, e3184. [CrossRef]
35. Everard, A.; Lazarevic, V.; Gaïa, N.; Johansson, M.; Ståhlman, M.; Bäckhed, F.; Delzenne, N.; Schrenzel, J.; Francois, P.; Cani, P. Microbiome of prebiotic-treated mice reveals novel targets involved in host response during obesity. *ISME J.* **2014**, *8*, 2116–2130. [CrossRef]
36. Alexander, C.; Cross, T.-W.L.; Devendran, S.; Neumer, F.; Theis, S.; Ridlon, J.M.; Suchodolski, J.S.; de Godoy, M.R.C.; Swanson, K.S. Effects of prebiotic inulin-type fructans on blood metabolite and hormone concentrations and faecal microbiota and metabolites in overweight dogs. *Br. J. Nutr.* **2018**, *120*, 711–720. [CrossRef]
37. Audebert, C.; Even, G.; Cian, A.; Loywick, A.; Merlin, S.; Viscogliosi, E.; Chabe, M.; Blastocystis Investigation Group. Colonization with the Enteric Protozoa Blastocystis is Associated with Increased Diversity of Human Gut Bacterial Microbiota. *Sci..Rep.* **2016**, *6*, 25255. [CrossRef]
38. Sjöberg, F.; Barkman, C.; Nookaew, I.; Östman, S.; Adlerberth, I.; Saalman, R.; Wold, A.E. Low-complexity microbiota in the duodenum of children with newly diagnosed ulcerative colitis. *PLoS ONE* **2017**, *12*, e0186178.
39. Shade, A. Diversity is the question, not the answer. *ISME J.* **2017**, *11*, 1–6. [CrossRef]
40. Flint, H.J.; Duncan, S.H.; Scott, K.P.; Louis, P. Links between diet, gut microbiota composition and gut metabolism. *Proc. Nutr. Soc.* **2015**, *74*, 13–22. [CrossRef]
41. Groussard, C.; Morel, I.; Chevanne, M.; Monnier, M.; Cillard, J.; Delamarche, A. Free radical scavenging and antioxidant effects of lactate ion: An in vitro study. *J. Appl. Physiol.* **2000**, *89*, 169–175. [CrossRef]
42. Manoharan, I.; Prasad, P.D.; Thangaraju, M.; Manicassamy, S. Lactate-Dependent Regulation of Immune Responses by Dendritic Cells and Macrophages. *Front. Immunol.* **2021**, *12*, 691134. [CrossRef]

fermentation

MDPI

Article

Effects of Fermented Vegetable Consumption on Human Gut Microbiome Diversity—A Pilot Study

Kerstin Thriene [1,†], Sina S. Hansen [1,†], Nadine Binder [2] and Karin B. Michels [1,*]

1 Institute for Prevention and Cancer Epidemiology, Faculty of Medicine and Medical Center, University of Freiburg, 79110 Freiburg, Germany; kerstin.thriene@uniklinik-freiburg.de (K.T.); sina.s.ullrich@gmail.com (S.S.H.)
2 Institute of General Practice/Family Medicine, Faculty of Medicine and Medical Center, University of Freiburg, 79110 Freiburg, Germany; nadine.binder@uniklinik-freiburg.de
* Correspondence: karin.michels@uniklinik-freiburg.de
† These authors contributed equally to this work.

Abstract: The interest in fermented food products has been increasing in recent years. Details about their microbial composition and the effects of their consumption on the human gut microbiome are of particular interest. However, evidence regarding their potential to increase gut microbial diversity, a measure likely associated with health, is lacking. To address this, we analyzed the microbial composition of commercially available fermented vegetables using 16S rRNA sequencing. We also conducted a pilot study to assess the feasibility of studying the effects of regular consumption of fermented vegetables on the gut microbiome. Six healthy male volunteers participated in a randomized crossover trial, with two two-week intervention phases. Volunteers consumed 150 g/d of either sauerkraut or a variety of six different commercially available fermented vegetables. This study is registered at the German Clinical Trials Register (DRKS-ID: DRKS00014840). *Lactobacillales* was the dominant family in all fermented vegetables studied. However, the alpha diversity, richness and evenness of the microbiota differed substantially among the different products. The number of species per product varied between 20 and 95. After consumption of both sauerkraut and the selection of fermented vegetables, we observed a slight increase in alpha diversity. Specifically, the amount of the genus *Prevotella* decreased while the amount of *Bacteroides* increased after both interventions. However, these initial observations need to be confirmed in larger studies. This pilot study demonstrates the feasibility of this type of research.

Keywords: fermented food; human gut microbiota; dietary intervention; crossover study; pilot study

Citation: Thriene, K.; Hansen, S.S.; Binder, N.; Michels, K.B. Effects of Fermented Vegetable Consumption on Human Gut Microbiome Diversity —A Pilot Study. *Fermentation* **2022**, *8*, 118. https://doi.org/10.3390/fermentation8030118

Academic Editor: Hiroshi Kitagaki

Received: 31 January 2022
Accepted: 4 March 2022
Published: 8 March 2022

1. Introduction

Fermented foods have been part of the human diet for centuries and have been associated with various health benefits including longevity, reduced risk of metabolic and immune-mediated disease and overall health [1,2]. Daily consumption of sauerkraut over a period of six weeks has been reported to induce changes in gut microbial composition, accompanied by improved gastrointestinal symptoms in patients with irritable bowel syndrome [3]. In addition to prolonging shelf-life, fermentation also increases the nutritional value of food products—potentially beneficial metabolites including vitamins, bioactive peptides and phytochemicals are synthesized during the fermentation process [4]. For vegetables, lactic acid fermentation is a common type of fermentation. This process is for example used for the production of sauerkraut or kimchi. During the fermentation process, lactic acid bacteria metabolize carbohydrates present in the raw vegetables into lactic acid. This leads to a drop in pH, which prevents the growth of undesirable microbes such as molds and therefore prevents mal-fermentation [5]. Lactic acid fermentation can be induced by adding salt at a concentration of approximately 2% (brine) to the shredded vegetables and creating an anaerobic environment, which favors the growth of lactic acid bacteria. Sauerkraut and other fermented vegetables are generally produced by adding a brine solution. While the bacterial community present in sauerkraut has been reported to establish itself within approximately four weeks of incubation and to remain stable during the fermentation process [6], long-term survival of microbes is not guaranteed during shelf-life.

With the growing recognition of the importance of the gut microbiome to human health and physiology, fermented foods have experienced a surge in popularity [1,7]. The composition of the human microbiota is affected by various factors including lifestyle, age, and diet [8–10]. While a dysbiotic microbiome has been associated with a number of diseases such as inflammatory bowel disease, asthma, obesity, and depression, its role in the development of these disorders remains unclear [11–14]. However, these observations suggest that the gut microbiome may be a promising target for maintaining or improving human health. Fermented foods are often promoted as beneficial for gut health; however, the effect of daily consumption of fermented vegetables on the gut microbiome has not been sufficiently explored. Therefore, we investigated the bacterial content of various commercially available fermented vegetables and conducted a pilot study to evaluate the feasibility of conducting a feeding study with these products.

2. Materials and Methods

2.1. Analysis of Fermented Vegetables

We explored the bacterial composition of six commercially available fermented vegetables: beetroot with goji berries, carrot with ginger, cauliflower with curcuma, sauerkraut with cranberries (referred to as "pink sauerkraut"), daikon kimchi (all from complete-organics GmbH, Munich, Germany), and traditional sauerkraut (Eden Frischkost- (L+)-Sauerkraut, Radolfszell, Germany). We selected these products, because fermented vegetables, in particular sauerkraut, have a very high abundant microbial diversity compared to other fermented foods. To maintain consistency of microbial content as much as possible throughout our study, we chose commercially available products. The fermented vegetables we selected were the only products available that were non-pasteurized and contained live bacteria, according to the manufacturers' information. We tested the number of live lactic acid bacteria (LAB) in some batches of the fermented vegetables by cultivation on LAB-selective agar to ensure that the products contained live LAB.

According to the manufacturer's information, the traditional sauerkraut was produced by addition of starter cultures while the other fermented vegetables were produced using wild fermentation. A list of the ingredients can be found in Appendix A Table A1. Distributors were unaware of the purpose of our purchases.

2.2. 16 S Ribosomal RNA Gene Sequencing

A sample of each fermented vegetable product was sent to the ZIEL Core Facility Microbiome of the Technical University Munich, Germany, for sequencing. The samples were processed according to the protocol by Reitmeier et al. (2020) [15].

2.3. Pilot Intervention Study

2.3.1. Study Design

To assess the feasibility of a feeding study using fermented vegetables and to assess their possible effects on the composition of the human gut microbiome, a pilot dietary intervention study was conducted. This study was advertised internally at the University Hospital Freiburg. As this study was a pilot study, we included six participants to test the feasibility of conducting a large-scale study, the participants' acceptance and preference of the different fermented vegetables, and early tendencies in potentially expectable effects on the gut microbiome. We restricted this pilot study to male volunteers because men are generally more difficult to recruit for feeding studies, so that we could evaluate their compliance and the feasibility in a longer-term feeding study. We conducted a randomized crossover trial, comprising eight weeks in total with one two-week washout phase, followed by two two-week intervention phases, which were separated by a second two-week washout phase (Figure 1). Exclusion criteria included a history of chronic diseases, acute or chronic gastrointestinal symptoms, one or more episodes of strong diarrhea within the past two months, regular intake of oral probiotic supplements or any antibiotics within the past two months, severe dietary restraints, plans to change diet within the next two months or consumption of more than two standard drinks of alcohol per day.

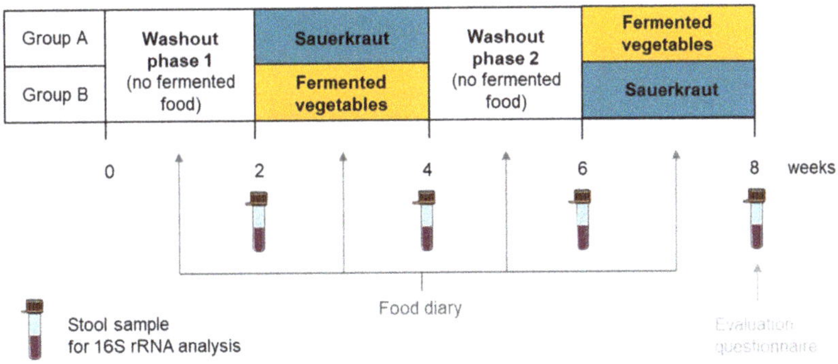

Figure 1. Study design of the randomized crossover trial, with two two-week intervention phases, each of which was preceded by a two-week washout phase. Participants consumed 150 g/d of either sauerkraut or a variety of six different fermented vegetables. Stool samples were collected after week 2, 4, 6 and 8 of this study.

The intervention consisted of consumption of 150 g/day of either (i) traditional sauerkraut or (ii) a selection of six different fermented vegetables: fermented beetroot, carrot, cauliflower, kimchi, pink sauerkraut, and traditional sauerkraut (as described under Section 2.1). Each product was assigned to be consumed on a certain day of the week (with traditional sauerkraut twice a week). In both parts of this study, the daily servings had to be consumed as 75 g portions twice a day. Participants were asked to refrain from consumption of other fermented foods, such as yogurt, kefir, or additional fermented vegetables throughout the entire study period. Stool samples were collected the day before and the day after each intervention period to establish the stool sampling process and pipeline for 16S rRNA analysis. Compliance was assessed by three-day food diaries that participants kept during each study period. Participants were asked to keep diaries of adverse reactions to the intervention. At the end of this study, participants filled in a questionnaire to evaluate the feasibility of a long-term study with a comparable design, as well as their preferences regarding the intervention. This study was approved by the ethics committee of the Albert-Ludwigs-University Freiburg.

2.3.2. Stool Sample Collection

Stool samples were collected by participants at home in two containers ((i) native, (ii) with 96% ethanol). Participants were instructed to keep the samples in the refrigerator after collection and to return them as soon as possible (within 24 h) to our lab. There, they were aliquoted and stored at −80 °C until subsequent analysis.

2.3.3. Fecal Microbial DNA Isolation

Stool samples were analyzed at the Laboratory of B. Grimbacher, University Medical Center Freiburg, Freiburg, Germany. Microbial DNA was extracted from a 2 mL aliquot using QIAamp DNA Stool Mini Kit (Qiagen, Hilden, Germany) following the manufacturer's instructions with modifications, similarly as described before [16–18]. Temperature of stool lysis was increased from 70 (suggested temperature in the protocol) to 95 °C for the lysis of bacteria that are known to be difficult to lyse, e.g., for Gram-positive bacteria. In a later step, 400 μL supernatant (instead of suggested volume 200 μL) was pipetted into

15 μL of proteinase K to which 400 μL AL buffer (instead of suggested volume of 200 μL) was added followed by thorough mixing and incubation at 70 °C for 10 min. Afterwards, a spin column was loaded twice with 400 μL of the lysate.

2.3.4. 16 S Ribosomal RNA Gene Sequencing

Variable (V) region 3 and 4 amplicons of 16S rRNA gene were sequenced following 16S metagenomic sequencing library protocol by Illumina [19]. 16S Amplicon PCR Forward Primer 5′ TCGTCGGCAGCGTCAGATGTGTATAAGAGACAGCCTACGGGNGGCWGCAG and 16S Amplicon PCR Reverse Primer 5′ GTCTCGTGGGCTCGGAGATGTGTATAAGAGACAGGACTACHVGGGTATCTAATCC were chosen as specific primers for the regions of interest of the 16S rRNA gene (V3 and V4). This reaction was set up with 2.5 μL of 5 ng/μL concentrated DNA, 5 μL 16S Amplicon PCR Forward Primer, 5 μL 16S Amplicon PCR Reverse Primer and 12.5 μL 2 × KAPA HiFi HotStart ReadyMix (final volume: 25 μL). PCR products were purified using magnetic beads (Beckman Coulter, Agencourt AMPure XP-Kit). The Index PCR was set up with 5 μL PCR product, 5 μL Nextera Index Primer 1.5 μL Nextera Index Primer 2 (different combination of primers for every sample), 25 μL 2 × KAPA HiFi HotStart ReadyMix and 10 μL nuclease-free water (final volume: 50 μL). After a second PCR clean up, the size of the PCR product was visualized by gel electrophoresis. In addition, loading on a Tape Station using the D1000 High-Sensitivity Reagents from Agilent Technologies allowed quantification of the PCR product and preparation of final library by equimolar pooling of 96 samples. Library was denatured with NaOH, diluted, mixed with 5% PhiX control and finally loaded on MiSeq for high-throughput sequencing (2 × 300 cycle V3 kit).

2.4. Bioinformatics and Data Analysis

All preprocessing and downstream analyses were conducted in R (v.4.1.0), an open-source free software environment for statistical computing and graphics [20]. Raw amplicon sequences were processed following the Divisive Amplicon Denoising Algorithm 2 (DADA2) workflow [21] with the DADA2 R software package (v. 1.22.0). DADA2 is able to classify sequences in the most strain-specific manner possible. Sample composition is inferred by partitioning the amplicon reads into partitions that match the error model. DADA2 derives exact amplicon sequence variants (ASVs) from the amplicon data and resolves biological differences of only 1 or 2 nucleotides. This results in an error-corrected table of the absolute abundances of the ASVs in each sample. Specifically, we followed all standard processing steps suggested by the DADA2 workflow including quality filtering, dereplication, learning the dataset-specific error model, ASV inference, chimera removal and taxonomic assignment. Within the denoise-paired function, sequences were truncated at position 260 (fermented vegetables) and 250 (stool) due to a drop-off of the quality score. Forward and reverse readings were merged, and chimeras were removed. Taxonomic assignments were obtained with the pre-trained Naïve Bayes classifier using the Silva 138.1 prokaryotic SSU taxonomic database, which specifically acknowledges the new taxonomy of Lactobacillus as characterized in [22]. A phyloseq object was created by means of the phyloseq R-package (v. 1.38.0) [23]. The species richness was determined by counting the number of different species present (observed) and alpha diversity was analyzed using Shannon diversity (richness and evenness). Beta diversity was assessed by computing weighted Bray–Curtis distances to compare microbial communities based on relative abundance. Non-metric multidimensional scaling (NMDS) was performed to compare beta diversity among groups. Relative abundances as well as alpha diversity measures were calculated using the R-package microbiome (v. 1.16.0). For computing the beta diversity, we used the R-package phyloseq. Graphics were generated using the ggplot2 R-package (v. 3.3.5). For exploring community composition, a genus-level dataset was created by agglomerating the data at the genus level, where genera that were less than 5% abundant in any of the samples (fermented vegetables) or had a relative abundance of less than 0.1% in half of the samples (stool) were aggregated into the category "Other".

Due to the fact that this is a pilot study with a small number of participants, we focused on descriptive analyses and have deliberately refrained from any significance testing.

3. Results

3.1. Bacterial Composition of Commercially Available Fermented Vegetables

Fermented Vegetable Microbiome

Lactic acid bacteria (*Lactobacillales*) were detected in all analyzed fermented vegetables, among which *Lactiplantibacillus* was the dominant genus in most fermented vegetables displaying 38% of the total genera in a grouped analysis with all samples, followed by *Lacticaseibacillus* and *Latilactobacillus* as well as *Pediococcus* (Figure 2A).

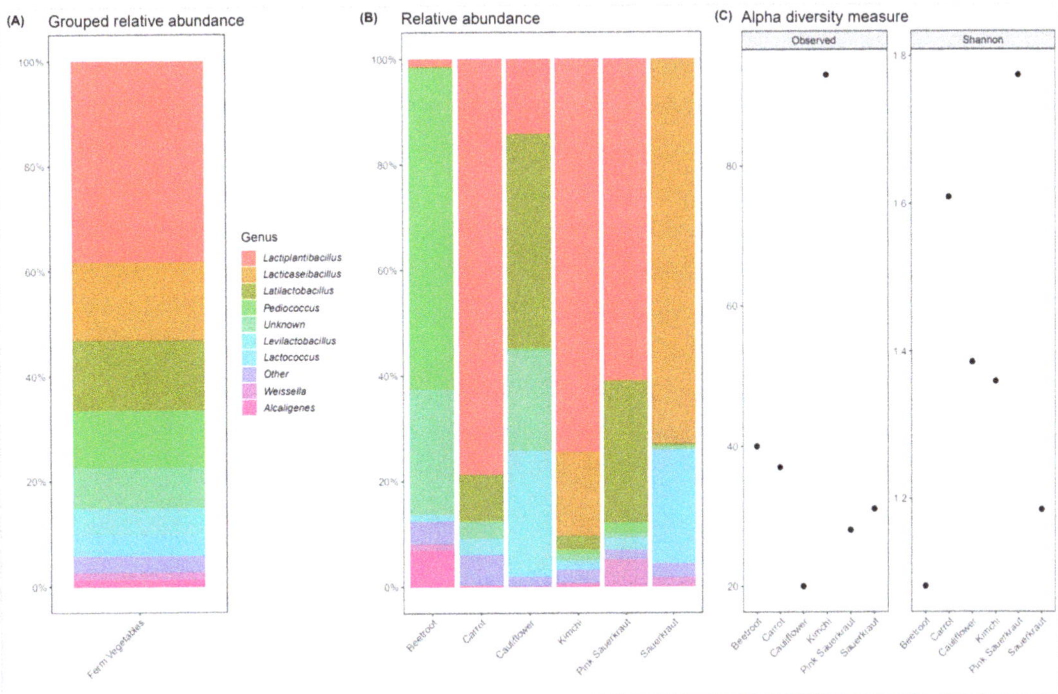

Figure 2. Bacterial composition of several commercially available fermented vegetables. Samples were analyzed by 16S rRNA sequencing. (**A**) Grouped relative abundance of the fermented vegetables—the most abundant genera are displayed in the chart, genera that were abundant less than 5% in any of the samples were aggregated into one category ("Other"), and genera with no taxonomic annotation are categorized as "unknown"; (**B**) read counts (abundance) of the main bacteria species found in the fermented vegetables; (**C**) alpha diversity of bacterial taxa found in the fermented vegetables.

The relative abundance of the detected genera varied between the different fermented vegetables (Figure 2B). *Lactiplantibacillus* was the dominant genus in carrots, kimchi and pink sauerkraut. In fermented beetroot, *Pediococcus* was the most dominant genus, while in the traditional sauerkraut, *Lacticaseibacillus* was most dominant, followed by *Lactococcus*. In cauliflower, *Latilactobacillus* followed by *Levilactobacillus* were the most abundant genera.

In general, alpha diversity, richness and evenness of the fermented vegetable microbiota differed substantially among the different products (Figure 2C). The number of different species (ASVs) ranged between 20 and 95 in the samples analyzed (observed richness). According to the Shannon index, pink sauerkraut was the most diverse sample while fermented beetroot and traditional sauerkraut displayed the lowest diversity.

3.2. Effects of Fermented Vegetable Consumption on the Human Gut Microbiome

Six healthy male volunteers (age: 25.5 ± 2.9 yrs, BMI: 24.3 ± 1.2k g/m^2) participated in this feasibility study. Compliance was very high; all participants completed all interventions. In total, 96% of the food and symptoms diaries were returned as requested. According to the food diaries, participants adhered to the intervention protocol. Four participants reported side effects, which were mainly related to gastrointestinal functions, such as flatulence and isolated episodes of diarrhea. Four participants declared their willingness to participate in a longer feeding study of this type. Three participants preferred the consumption of a selection of fermented vegetables over sauerkraut consumption. All participants returned stool samples at all requested time points. Samples stabilized in ethanol were provided by five participants, whereas native samples were provided by four participants. Further analysis was conducted using the ethanol samples for all participants except one, from whom only native samples were available.

Gut Microbiome Analysis

An increase in relative abundance was detected within the genera *Bacteroides* and *Ruminococcus* (Figure 3A). The genera *Prevotella* and *Faecalibacterium* displayed a decrease after both intervention cycles.

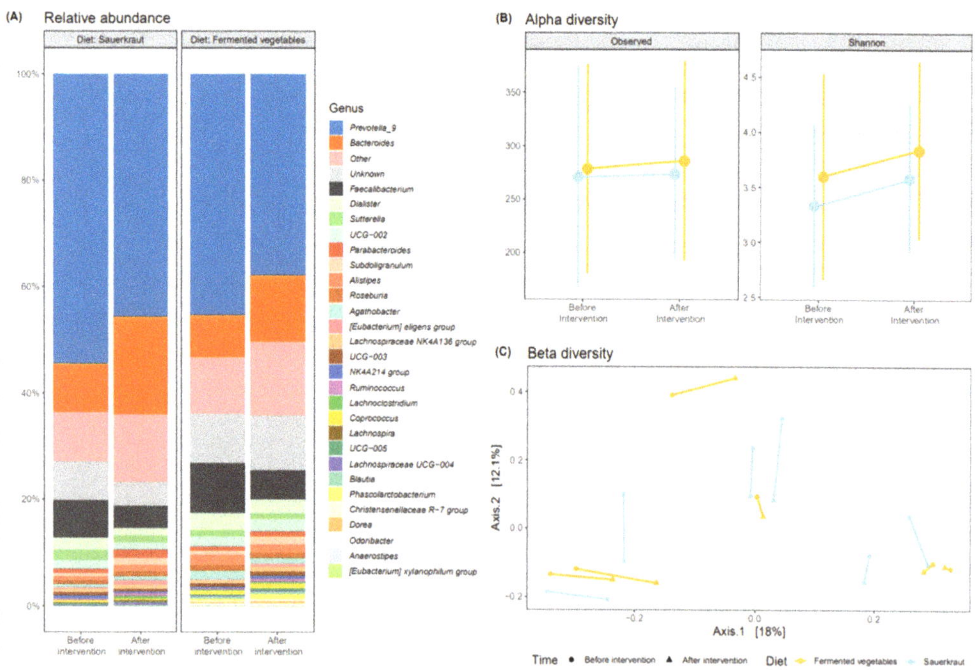

Figure 3. Human gut microbiome changes after consumption of fermented vegetables. (**A**) Grouped relative abundance of the main phyla found in the stool samples—the most abundant genera are displayed in the chart, genera that had a relative abundance of more than 0.1% in half of the samples were aggregated into one category ("Other"), and genera with no taxonomic annotation are categorized as "unknown"; (**B**) alpha diversity measures representing the diversity of bacterial taxa before and after intervention; (**C**) beta diversity of bacterial taxa found in the stool samples displaying each participant.

No clear change in alpha diversity in response to either intervention was apparent according to the observed index. A slight increase in alpha diversity (the Shannon index) was observed after consumption of both sauerkraut (pre intervention: 3.31 ± 0.74, post intervention: 3.58 ± 0.68) and the selection of fermented vegetables (pre: 3.60 ± 0.93, post: 3.84 ± 0.81) (Figure 3B). Bray–Curtis dissimilarity revealed no clustering of the samples according to dietary intake and time point. For most participants, baseline and intervention samples did not differ very much, pointing towards an individualized and heterogeneous microbial composition (Figure 3C).

4. Discussion

In this study, the microbial composition of commercially available fermented vegetables was analyzed using 16S rRNA sequencing. The effect of consumption of these vegetables on the composition of the human gut microbiome was also examined through a pilot human intervention study.

We analyzed the bacterial composition of commercially available fermented vegetables. As expected, the lactic acid bacteria family was the most represented in all fermented vegetables assessed. The genus *Lactiplantibacillus* was the most common in half of the fermented vegetables. Species of the *Lactiplantibacillus* genus are found in various fermented vegetables, especially sauerkraut and kimchi, and produce lactic acid [24]. The genera *Lacticaseibacillus* and *Latilactobacillus* are widely found in fermented foods with some species exhibiting probiotic properties. Among other genera, the genus *Pediococcus* is important for its preservative properties. This genus converts the sugars in fresh vegetables into lactic acid, which causes a drop in pH and prevents the food from spoiling [25]. The second most prominent genus detected in the sauerkraut was *Lactococcus*, which is made up of homofermentative species that produce lactic acid through fermentation of glucose. Many *Lactococcus* species are widely used to produce fermented dairy products. Another genus that was detected in the fermented vegetables, but primarily in the two sauerkraut types, was *Weissella*. Species of this genus are found in various fermented foods, many of them producing exopolysaccharides that are involved in adhesion to surfaces and influence the viscosity and structure of the respective fermented food [26].

The commercially available fermented vegetables that were analyzed in this study contain a variety of different microbial components, of which many species have been described to display probiotic effects. According to Hill et al., probiotics are defined as live microorganisms that are intended to have health benefits when administered in adequate amounts [27]. However, the exact mechanisms that explain the effects each species has on the human host are still not fully understood. For example, it has remained unclear whether the live bacteria consumed adhere to the intestinal mucosa and persistently populate the human gut or whether these products have to be consumed on a regular basis to enrich the host gut flora. In comparison, fermented milk products also contain probiotic bacteria, but display a less complex microbial composition. Traditional yogurt contains *Lactobacillus delbrueckii* subsp. *bulgaricus* and *Streptococcus thermophiles*. Often other LAB or bacteria from the genera *Bifidobacterium* are added to promote probiotic effects [28,29]. Other fermented dairy products such as kefir, a slightly carbonated viscous fermented milk drink, contain a more complex community of bacteria and yeasts when traditionally produced. However, in industrial production, instead of spontaneous fermentation or fermentation with kefir grains, standardized starter cultures are often used to achieve consistent product quality and suppress undesirable fermentation processes, resulting in a less complex microbial composition [30–32]. Our feasibility study demonstrated a high compliance, with no major side effects, indicating that longer-term studies investigating daily effects of fermented vegetables on health are feasible. Our pilot study was not powered to reveal any but substantial shifts in the gut microbiome. Hence, larger studies, potentially with longer intervention periods, are required to draw firm conclusions regarding the general population. Nevertheless, our study suggests a minor increase in gut microbiome diversity in healthy, young volunteers. The biggest change on the genera level was observed in the two genera *Prevotella* and *Bacteroides*. The relative abundance of *Bacteroides* nearly doubled while the relative abundance of *Prevotella* decreased in a corresponding manner. Both genera are closely related and belong to the phylum *Bacteroidota*. An antagonistic relation has been proposed in the literature describing some *Prevotella* species as being associated with chronic inflammatory conditions [33].

As reflected in the beta diversity, the gut microbiome composition differed vastly between volunteers. This was to be expected, as it has been established that the composition of the gut microbiome differs between individuals and is influenced by a great number of parameters such as diet, environment, and living situation [34]. All participants were allowed to follow their individual habitual diet in addition to the interventions. Additionally, the intervention period was relatively short. For more profound and sustained effects, fermented vegetables may have to be consumed longer in order to enable establishment of the potentially induced changes. Moreover, the small sample size fosters an intra-individual variation that exceeds the inter-individual variation in gut microbial composition. The study of the impact of fermented vegetables is also limited by difficulties in standardizing the intervention, since bacterial composition will vary even in a standardized production environment.

Our pilot study suggests that an intervention with 150 g of different fermented vegetables per day is feasible. This is consistent with findings from other studies such as by Nielsen et al. including an intervention with 75 g sauerkraut per day [3] or a study by Han et al., who investigated effects of the consumption of 180 g kimchi per day in a study with obese women [35].

5. Conclusions

Fermented vegetables contain a vast variety of bacteria that differ between respective products. Therefore, a broader exposure of a microbial variety can already be achieved by consumption of a variety of different fermented vegetables. Our study suggests that intervention studies with fermented vegetables are feasible. In addition, it provides a first look at the effects of the consumption of fermented vegetables on the human gut microbiome, suggesting an increase in microbial diversity in the human gut. This study lays a foundation for further research providing insights into the health functions of fermented vegetables.

Author Contributions: Conceptualization, S.S.H. and K.B.M.; methodology, K.T., S.S.H., N.B. and K.B.M.; software, N.B.; formal analysis, N.B.; data curation, N.B. and S.S.H., writing—original draft preparation, K.T., S.S.H. and N.B.; writing—review and editing, K.T., S.S.H., N.B. and K.B.M.; visualization, N.B.; supervision, K.B.M.; project administration, S.S.H., funding acquisition, S.S.H. All authors have read and agreed to the published version of the manuscript.

Funding: This research was funded by Forschungskommission, Medical Faculty, Albert-Ludwigs-University Freiburg, grant number: ULL1174/18.

Institutional Review Board Statement: This study was conducted in accordance with the Declaration of Helsinki, and approved by the Ethics Committee of Albert-Ludwigs-University Freiburg (protocol code 232/18, date of approval: 25 September 2018).

Informed Consent Statement: Informed consent was obtained from all subjects involved in this study.

Data Availability Statement: The data presented in this study are available on request from the corresponding author.

Acknowledgments: We thank the team of the ZIEL Core Facility Microbiome, TUM, Freising, Germany, for sequencing the fermented vegetables. Additionally, we thank the team of the Next Generation Sequencing Core Facility (NGS-CF),University Medical Center Freiburg, Germany, for sequencing the stool samples. We acknowledge support by the Open Access Publication Fund of the University of Freiburg.

Conflicts of Interest: The authors declare no conflict of interest.

Appendix A

Table A1. Compositional details of the fermented vegetables used in this study.

Type of Fermented Vegetable	Ingredients	Manufacturer
Cauliflower	Cauliflower (96%), natural sea salt, fresh turmeric, spices (turmeric, cumin)	completeorganics GmbH, Munich, Germany
Pink Sauerkraut	White cabbage (72%), red cabbage, natural sea salt, cranberries (dried)	completeorganics GmbH, Munich, Germany
Carrots	Carrots (96%), natural sea salt, ginger (fresh)	completeorganics GmbH, Munich, Germany
Beet Root	Beetroot (93%), horseradish, natural sea salt, goji berries (dried)	completeorganics GmbH, Munich, Germany
Daikon Kimchi	Radish (84%), carrots, red peppers, natural sea salt, spices (ginger, chili)	completeorganics GmbH, Munich, Germany
Traditional Sauerkraut	White cabbage, sea salt, lactic acid bacteria	Eden Frischkost-(L+)-Sauerkraut, Radolfszell, Germany

References

1. Marco, M.L.; Heeney, D.; Binda, S.; Cifelli, C.J.; Cotter, P.D.; Foligne, B.; Gänzle, M.; Kort, R.; Pasin, G.; Pihlanto, A.; et al. Health benefits of fermented foods: Microbiota and beyond. *Curr. Opin. Biotechnol.* **2017**, *44*, 94–102. [CrossRef] [PubMed]
2. Dimidi, E.; Cox, S.R.; Rossi, M.; Whelan, K. Fermented Foods: Definitions and Characteristics, Impact on the Gut Microbiota and Effects on Gastrointestinal Health and Disease. *Nutrients* **2019**, *11*, 1806. [CrossRef] [PubMed]
3. Nielsen, E.S.; Garnås, E.; Jensen, K.J.; Hansen, L.H.; Olsen, P.S.; Ritz, C.; Krych, L.; Nielsen, D.S. Lacto-fermented sauerkraut improves symptoms in IBS patients independent of product pasteurisation—A pilot study. *Food Funct.* **2018**, *9*, 5323–5335. [CrossRef] [PubMed]
4. Septembre-Malaterre, A.; Remize, F.; Poucheret, P. Fruits and vegetables, as a source of nutritional compounds and phytochemicals: Changes in bioactive compounds during lactic fermentation. *Food Res. Int.* **2018**, *104*, 86–99. [CrossRef]
5. Jung, J.Y.; Lee, S.H.; Kim, J.M.; Park, M.S.; Bae, J.-W.; Hahn, Y.; Madsen, E.L.; Jeon, C.O. Metagenomic Analysis of Kimchi, a Traditional Korean Fermented Food. *Appl. Environ. Microbiol.* **2011**, *77*, 2264–2274. [CrossRef]
6. Zabat, M.A.; Sano, W.H.; Wurster, J.I.; Cabral, D.J.; Belenky, P. Microbial Community Analysis of Sauerkraut Fermentation Reveals a Stable and Rapidly Established Community. *Foods* **2018**, *7*, 77. [CrossRef]
7. Rezac, S.; Kok, C.R.; Heermann, M.; Hutkins, R. Fermented Foods as a Dietary Source of Live Organisms. *Front. Microbiol.* **2018**, *9*, 1785. [CrossRef]
8. Zmora, N.; Suez, J.; Elinav, E. You are what you eat: Diet, health and the gut microbiota. *Nat. Rev. Gastroenterol. Hepatol.* **2019**, *16*, 35–56. [CrossRef]
9. Müller, M.; Canfora, E.E.; Blaak, E.E. Gastrointestinal Transit Time, Glucose Homeostasis and Metabolic Health: Modulation by Dietary Fibers. *Nutrients* **2018**, *10*, 275. [CrossRef]
10. Roager, H.M.; Hansen, L.B.S.; Bahl, M.I.; Frandsen, H.L.; Carvalho, V.; Gøbel, R.J.; Dalgaard, M.D.; Plichta, D.R.; Sparholt, M.H.; Vestergaard, H.; et al. Colonic transit time is related to bacterial metabolism and mucosal turnover in the gut. *Nat. Microbiol.* **2016**, *1*, 16093. [CrossRef]
11. Koopman, M.; El Aidy, S. Depressed gut? The microbiota-diet-inflammation trialogue in depression. *Curr. Opin. Psychiatry* **2017**, *30*, 369–377. [CrossRef]
12. Gevers, D.; Kugathasan, S.; Denson, L.A.; Vázquez-Baeza, Y.; Van Treuren, W.; Ren, B.; Schwager, E.; Knights, D.; Song, S.J.; Yassour, M.; et al. The Treatment-Naive Microbiome in New-Onset Crohn's Disease. *Cell Host Microbe* **2014**, *15*, 382–392. [CrossRef]
13. Arrieta, M.-C.; Stiemsma, L.T.; Dimitriu, P.A.; Thorson, L.; Russell, S.; Yurist-Doutsch, S.; Kuzeljevic, B.; Gold, M.J.; Britton, H.M.; Lefebvre, D.L.; et al. Early infancy microbial and metabolic alterations affect risk of childhood asthma. *Sci. Transl. Med.* **2015**, *7*, 307ra152. [CrossRef]
14. Greenblum, S.; Turnbaugh, P.J.; Borenstein, E. Metagenomic systems biology of the human gut microbiome reveals topological shifts associated with obesity and inflammatory bowel disease. *Proc. Natl. Acad. Sci. USA* **2011**, *109*, 594–599. [CrossRef]
15. Reitmeier, S.; Kiessling, S.; Neuhaus, K.; Haller, D. Comparing Circadian Rhythmicity in the Human Gut Microbiome. *STAR Protoc.* **2020**, *1*, 100148. [CrossRef]
16. Haring, E.; Andrieux, G.; Uhl, F.M.; Krausz, M.; Proietti, M.; Sauer, B.; Esser, P.R.; Martin, S.F.; Pfeifer, D.; Schmitt-Graeff, A.; et al. Therapeutic targeting of endoplasmic reticulum stress in acute graft-versus-host disease. *Haematologica* **2021**, *4*, 2782–2858. [CrossRef]

17. Norona, J.; Apostolova, P.; Schmidt, D.; Ihlemann, R.; Reischmann, N.; Taylor, G.; Köhler, N.; De Heer, J.; Heeg, S.; Andrieux, G.; et al. Glucagon-like peptide 2 for intestinal stem cell and Paneth cell repair during graft-versus-host disease in mice and humans. *Blood* **2020**, *136*, 1442–1455. [CrossRef]
18. Haring, E.; Uhl, F.M.; Andrieux, G.; Proietti, M.; Bulashevska, A.; Sauer, B.; Braun, L.M.; Gomez, E.D.V.; Esser, P.R.; Martin, S.F.; et al. Bile acids regulate intestinal antigen presentation and reduce graft-versus-host disease without impairing the graft-versus-leukemia effect. *Haematologica* **2020**, *106*, 2131–2146. [CrossRef]
19. Kozich, J.J.; Westcott, S.L.; Baxter, N.T.; Highlander, S.K.; Schloss, P.D. Development of a dual-index sequencing strategy and curation pipeline for analyzing amplicon sequence data on the MiSeq Illumina sequencing platform. *Appl. Environ. Microbiol.* **2013**, *79*, 5112–5120. [CrossRef]
20. R Core Team. *R: A Language and Environment for Statistical Computing*; R Foundation for Statistical Computing: Vienna, Austria, 2020.
21. Callahan, B.J.; Mcmurdie, P.J.; Rosen, M.J.; Han, A.W.; Johnson, A.J.A.; Holmes, S.P. DADA$_2$: High-resolution sample inference from Illumina amplicon data. *Nat. Methods* **2016**, *13*, 581–583. [CrossRef]
22. Zheng, J.; Wittouck, S.; Salvetti, E.; Franz, C.M.A.P.; Harris, H.M.B.; Mattarelli, P.; O'Toole, P.W.; Pot, B.; Vandamme, P.; Walter, J.; et al. A taxonomic note on the genus Lactobacillus: Description of 23 novel genera, emended description of the genus Lactobacillus Beijerinck 1901, and union of Lactobacillaceae and Leuconostocaceae. *Int. J. Syst. Evol. Microbiol.* **2020**, *70*, 2782–2858. [CrossRef]
23. McMurdie, P.J.; Holmes, S. phyloseq: An R package for reproducible interactive analysis and graphics of microbiome census data. *PLoS ONE* **2013**, *8*, e61217. [CrossRef]
24. Gustaw, K.; Niedźwiedź, I.; Rachwał, K.; Polak-Berecka, M. New Insight into Bacterial Interaction with the Matrix of Plant-Based Fermented Foods. *Foods* **2021**, *10*, 1603. [CrossRef]
25. Porto, M.C.W.; Kuniyoshi, T.M.; Azevedo, P.O.S.; Vitolo, M.; Oliveira, R.P.S. *Pediococcus* spp.: An important genus of lactic acid bacteria and pediocin producers. *Biotechnol. Adv.* **2017**, *35*, 361–374. [CrossRef]
26. Kavitake, D.; Devi, P.B.; Shetty, P.H. Overview of exopolysaccharides produced by Weissella genus—A review. *Int. J. Biol. Macromol.* **2020**, *164*, 2964–2973. [CrossRef]
27. Hill, C.; Guarner, F.; Reid, G.; Gibson, G.R.; Merenstein, D.J.; Pot, B.; Morelli, L.; Canani, R.B.; Flint, H.J.; Salminen, S.; et al. Expert consensus document: The International Scientific Association for Probiotics and Prebiotics consensus statement on the scope and appropriate use of the term probiotic. *Nat. Rev. Gastroenterol. Hepatol.* **2014**, *11*, 506–514. [CrossRef]
28. Volokh, O.; Klimenko, N.; Berezhnaya, Y.; Tyakht, A.; Nesterova, P.; Popenko, A.; Alexeev, D. Human Gut Microbiome Response Induced by Fermented Dairy Product Intake in Healthy Volunteers. *Nutrients* **2019**, *11*, 547. [CrossRef]
29. Stiemsma, L.T.; Nakamura, R.E.; Nguyen, J.G.; Michels, K.B. Does Consumption of Fermented Foods Modify the Human Gut Microbiota? *J. Nutr.* **2020**, *150*, 1680–1692. [CrossRef]
30. Kwak, H.; Park, S.; Kim, D. Biostabilization of Kefir with a Nonlactose-Fermenting Yeast. *J. Dairy Sci.* **1996**, *79*, 937–942. [CrossRef]
31. Marsh, A.J.; O'Sullivan, O.; Hill, C.; Ross, R.; Cotter, P.D. Sequencing-Based Analysis of the Bacterial and Fungal Composition of Kefir Grains and Milks from Multiple Sources. *PLoS ONE* **2013**, *8*, e69371. [CrossRef]
32. Bourrie, B.C.T.; Willing, B.P.; Cotter, P.D. The Microbiota and Health Promoting Characteristics of the Fermented Beverage Kefir. *Front. Microbiol.* **2016**, *7*, 647. [CrossRef]
33. Ley, R.E. *Prevotella* in the gut: Choose carefully. *Nat. Rev. Gastroenterol. Hepatol.* **2016**, *13*, 69–70. [CrossRef] [PubMed]
34. Conlon, M.A.; Bird, A.R. The Impact of Diet and Lifestyle on Gut Microbiota and Human Health. *Nutrients* **2014**, *7*, 17–44. [CrossRef] [PubMed]
35. Han, K.; Bose, S.; Wang, J.-H.; Kim, B.-S.; Kim, M.J.; Kim, E.-J.; Kim, H. Contrasting effects of fresh and fermented kimchi consumption on gut microbiota composition and gene expression related to metabolic syndrome in obese Korean women. *Mol. Nutr. Food Res.* **2015**, *59*, 1004–1008. [CrossRef] [PubMed]

fermentation

Article

Antioxidant Effect via Bioconversion of Isoflavonoid in *Astragalus membranaceus* Fermented by *Lactiplantibacillus plantarum* MG5276 In Vitro and In Vivo

Ji Yeon Lee , Hye Min Park and Chang-Ho Kang *

MEDIOGEN, Co., Ltd., Biovalley 1-ro, Jecheon-si 27159, Korea; ljy341@naver.com (J.Y.L.); hyeminpark95@naver.com (H.M.P.)
* Correspondence: changho-kang@naver.com; Tel.: +82-43-842-1940

Abstract: In this study, the antioxidant mechanism of *Astragalus membranaceus* fermented by *Lactiplantibacillus plantarum* MG5276 (MG5276F-AM) was evaluated in HepG2 cells and in an animal model. HPLC analysis was performed to confirm the bioconversion of the bioactive compounds in *A. membranaceus* by fermentation. Calycosin and formononetin, which were not detected before fermentation (NF-AM), were detected after fermentation (MG5276F-AM), and its glycoside was not observed in MG5276F-AM. In HepG2 cells, MG5276F-AM alleviated H_2O_2-induced oxidative stress by mediating lipid peroxidation and glutathione levels, and upregulated antioxidant enzymes including catalase (CAT), superoxide dismutase (SOD), and glutathione peroxidase (GPx). In the *t*BHP-injected mouse model, administration of MG5276F-AM reduced hepatic aspartate transaminase, alanine transaminase, and lipid peroxidation. MG5276F-AM also modulated antioxidant enzymes as well as HepG2 cells. Thus, fermentation of *A. membranaceus* with *L. plantarum* MG5276 elevated the isoflavonoid aglycone by hydrolysis of its glycosides, and this bioconversion enhanced antioxidant activity both in vitro and in vivo.

Keywords: antioxidant; formononetin; calycosin; HepG2 cells; t-BHP

Citation: Lee, J.Y.; Park, H.M.; Kang, C.-H. Antioxidant Effect via Bioconversion of Isoflavonoid in *Astragalus membranaceus* Fermented by *Lactiplantibacillus plantarum* MG5276 In Vitro and In Vivo. *Fermentation* **2022**, *8*, 34. https://doi.org/10.3390/fermentation8010034

Academic Editor: Hiroshi Kitagaki

Received: 26 November 2021
Accepted: 13 January 2022
Published: 16 January 2022

1. Introduction

Antioxidants prevent oxidative stress produced by a variety of environmental conditions, including alcohol consumption, smoking, and drug addiction [1]. Chronic oxidative stress is caused by the accumulation of reactive oxygen species (ROS), including hydrogen peroxide (H_2O_2), hydroxyl radicals (OH^-), and free radicals (O^-) in organelles. ROS induce cancer, aging, and inflammation by causing cell membrane degradation, lipid oxidation, protein denaturation, and DNA damage [2]. To protect against oxidative stress, organelles have enzymatic and non-enzymatic antioxidant defense systems [3]. Oxidative stress accumulated in cells is relieved by antioxidant enzymes such as SOD, CAT, and GPx. Free radicals are converted to H_2O_2 by SOD and degraded to H_2O by CAT and/or GPx to relieve oxidative stress [4]. What is more, non-enzymatic defense system against oxidative stress involves flavonoids such as flavanoid, flavonols, isoflavonoids, and phenolic acid, which are abundant in foods, plants and beverages [5]. The balance between the activity and intracellular levels of antioxidants is important for human health [4].

Bioconversion is the conversion of organic materials from natural product by biological processes [6]. In the food industry, bioconversion is used to produce feed and energy using food waste, and to improve bioactivity and bioavailability [7,8]. In particular, lactic acid bacteria, which are probiotics, are widely used in the functional food industry [9]. Among the lactic acid bacteria used for bioconversion, *Lactiplantibacillus plantarum* has been used to convert flavonoid glycosides in *Cudrania tricuspidata* leaves into its aglycone, and to convert ginsenoside Rb2 and Rb3 in red ginseng to ginsenoside Rd [10,11]. It has also been reported that immune system regulation, whitening, and antioxidant activity of various natural products are enhanced through bioconversion using *L. plantarum* [6,12].

Roots of *Astragalus membranaceus* (*A. membranaceus*) have been grown in Korea, China, and Mongolia [13]. *A. membranaceus* has been used traditionally for the treatment of cancer, hyperglycemia, and immune modulating-related diseases, and contains mainly polysaccharides, astragaloside, and various flavonoids [14]. In a previous study, *A. membranaceus* fermented by *L. plantarum* MG5276 (MG5276F-AM) had antioxidant and anti-inflammatory effects through decreasing nitro oxide (NO) production and gene expression of inducible nitric oxide synthase (iNOS) and cyclooxygenase-2 (COX-2) in RAW264.7 cells [15]. As a further study, the bioconversion of isoflavonoid glycosides calycosin-7-*O*-glucoside and formononetin-7-*O*-glucoside in *A. membranaceus* by fermentation was confirmed through HPLC analysis, and the antioxidant efficacy of MG5276F-AM was verified in cells and animal models.

2. Materials and Methods

2.1. Preparation and Fermentation of Astragulus Membranaceus Extract

The fermentation method was performed as previously described [15]. Briefly, *L. plantarum* MG5276, isolated from fermented foods, was cultured in ManRogosaSharpe (MRS; Difco, Sparks, MD, USA) broth at 37 °C for 24 h. The roots of *A. membranaceus* were purchased from the Chungbuk Herb Farming Association (Jecheon, Korea). *A. membranaceus* (100 g) was extracted in hot water (*w*:*v* = 1:2, 60 °C) for 15 h. The *A. membranaceus* extract was filtered using Whatman No. 2 filter paper (GE Healthcare Bio-Science, PA, USA). Then, 2% (*v*/*v*) of *L. plantarum* MG5276 (OD_{600} = 1.0, 10^8–10^9 CFU/mL) was inoculated and fermented for *A. membranaceus* extract at 37 °C for 15 h. After fermentation, the supernatant was filtered and freeze-dried for subsequent experiments.

2.2. High Perfomance Liquid Chromatography (HPLC) Analysis

2.2.1. Pre-treatment of Samples for Analysis

Previously study reported that isoflavonoids from AM was contained in ethyl acetate fraction [16,17]. Therefore, the isoflavonoid-rich sample for HPLC analysis was conducted by ethyl acetate fraction. NF-AM and MG5276F-AM (1 g, each) were dissolved in water then fractionated with ethyl acetate. Fractions of each sample were dissolved in methanol (50 mg/mL each) following concentration under reduced pressure.

2.2.2. Analysis of Isoflavonoids by HPLC

For qualitative and quantitative analysis of isoflavonoids, HPLC (Shimadzu, Kyoto, Japan) was performed using a reverse-phase system. Calycosin, formononetin, calycosin-7-*O*-glucoside (calycosin-G), and formononetin-7-*O*-glucoside (formononetin-G), used as a standard, were from Sigma-Aldrich (St. Louis, MO, USA). A Kinetex EVO C18 (4.6 × 250 mm, 5 µm; Phenomenex, Torrance, CA, USA) column was used with a mobile phase consisting of 0.1% acetic acid in water (A, *v*/*v*) and 0.1% acetic acid in acetonitrile (B, *v*/*v*). The injection volume was 10 µL, the flow rate was 1.0 mL/min, and UV detection was conducted at 260 nm. The elution program was modified from Park's program [13]. The elution program was as follows (B%): 10–15%, 10 min; 15–20%, 15 min; 20–30%, 25 min; 30–40%, 45 min; 40–100%, 55 min; 100%, 65 min.

2.3. Cell Culture

HepG2 cells were obtained from the KTCC (Seoul, Korea). HepG2 cells were cultured in minimum essential medium (MEM) supplemented with 10% fetal bovine serum (FBS) and 1% penicillin-streptomycin (Gibco BRL, Burlington, ON, Canada) at 37 °C in a 5% CO_2 incubator. The medium was replaced every 2 days for subculture.

2.4. Cell Viability

Cell viability was measured using the 3-(4,5-Dimethylthiazol-2-yl)-2,5-Diphenyltetrazolium bromide (MTT) assay [18]. Briefly, cells (4×10^4 cells/well) were seeded in 96 well plates incubated at 37 °C in 5% CO_2 for 24 h. After aspirating the supernatant, 100 µL of NF-AM and MG5276F-AM (25 and 50 µg/mL in serum free MEM) was treated in each well. After 24 h, H_2O_2 (1 mM) with samples was added to the each well for 24 h. After aspirating the supernatant, 100 µL of MTT solution (0.1 mg/mL in serum free MEM) was added to each well, incubated for 2 h. Formazan produced by MTT solution was dissolved by DMSO (150 µL) and absorbance at 550 nm was measured using an Epoch2 microplate reader (Biotek, Winooski, VT, USA).

2.5. Quantitative Real-Time Polymerase Chain Reaction (qRT-PCR)

Total RNA was isolated from hepatocytes using NuceloZOL (MACHEREY-NAGEL GmbH & Co., KG, Dueren, Germany), and mRNA (1 µg) was reverse-transcribed into cDNA using Reverse Transcriptase Premix (Intron, Seongnam-si, Korea). qRT-PCR was performed using iQ™ SYBR® Green Supermix (Bio-Rad, Hercules, CA, USA) with the gene primers listed in Table 1. Relative gene expression was normalized to glyceraldehyde 3-phosphate dehydrogenase (GAPDH). Expression levels were analyzed using the $2^{-\Delta\Delta CT}$ method [19].

Table 1. Primer sequences for qRT-PCR.

Gene [1]		Primer Sequence (5′→3′)	Product Size (bp)
CAT	Forward	GAACTGTCCCTACCGTGCTCGA	156
	Reverse	CCAGAATATTGGATGCTGTGCTCCAGG	
SOD	Forward	AATGGACCAGTGAAGGTGTGGGG	186
	Reverse	CACATTGCCCAAGTCTCCAACATGC	
GPX	Forward	CGGCCCAGTCGGTGTATGC	122
	Reverse	CGTGGTGCCTCAGAGGGAC	
GAPDH	Forward	ACCCACTCCTCCACCTTTG	178
	Reverse	CTCTTGTGCTCTTGCTGGG	

[1] qRT-PCR, quantitative real-time polymerase chain reaction; *CAT*, catalase; *SOD*, superoxide dismutase; *GPX*, glutathione peroxide; *GAPDH*, glyceraldehyde 3-phosphate dehydrogenase.

2.6. Animal Treatments

All animal experimental protocols used in this study were approved by the Institutional Animal Care and Use Committee (IACUC) of the Korea Institute of NDIC (P214057, Seoul, Korea, 21 October 2021). Five-week-old female C57BL/6 mice were used in this study (Orient Bio., Seongnam-si, Korea). The mice were housed in standard polycarbonate cages (n = 4–5 per cage) at a controlled temperature of 21 ± 2 °C and humidity of 50 ± 20% under a 12-h light/dark cycle and given free access to standard rodent chow (Research Diets, Inc., New Brunswick, NJ, USA) and water. All solutions used for the treatments were freshly dissolved on the experimental days in physiological saline (Dai Han Pharm. Co., Ltd., Seoul, Korea).

After a week of acclimatization, the mice were randomly divided into four groups (n = 7 per group) as follows: (1) normal group (saline), (2) *tert*-butyl hydroperoxide (*t*-BHP) in saline-injected group, (3) *t*-BHP + MG5276F-AM (100 mg/kg), and (4) *t*-BHP + MG5276F-AM (300 mg/kg). The extracts were administered orally for 14 days with a feeding needle catheter, and the mice received an intraperitoneal (i.p.) injection of *t*-BHP (0.5 mmol/kg) after 3 h from the last feed [20]. Following euthanasia, the liver tissues were dissected and stored at −80 °C for biochemical parameter analysis.

2.7. Biochemical Parameters of Serum and Liver Tissues

Serum alanine aminotransferase (ALT) and aspartate aminotransferase (AST) levels were evaluated using a chemistry analyzer (AU480, Beckman Coulter, CA, USA). The liver tissue was lysed using RIPA buffer (Biosesang, Seongnam-si, Korea) containing protease inhibitor (Gendepot, Katy, TX, USA) and homogenized using TissueLyser II (Qiagen, Hilden, Germany) with stainless steel beads (Qiagen). Malondialdehyde (MDA) (LS Bio, Seattle, DC, USA), glutathione (GSH, Abcam, Cambridge, UK), SOD (Abcam), CAT (Abcam), and GPx (Abcam) were evaluated using ELISA kits. The absorbance was measured using a SpectraMax M2 microplate reader (Molecular Devices, San Jose, CA, USA).

2.8. Statistical Analysis

Data are expressed as mean ± standard error of the mean (SEM) and analyzed by multiple comparisons one-way analysis of variance (ANOVA) using Duncan's multiple range test, IBM SPSS Statistics 21 software program (SPSS Inc., Chicago, IL, USA). The results were considered statistically significant if p values were less than 0.05.

3. Results and Discussion

3.1. Analysis Method Vaildation of Isoflavonoids Aglycones and Its Glycosides

The analytical method used for quantification of calycosin, formononetin, calycosin-G, and formononetin-G using HPLC was verified by the limit of detection (LOD) using the slope, and limit of quantification (LOQ) using the standard deviation obtained through linearity (Table 2). The retention times of isoflavonoid aglycones (calycosin and formononetin) and its glycosides (calycosin-G and formononetin-G) were 15.74, 30.36, 64.76, and 44.71 min, respectively. The calibration curves of isoflavonoids and their glycosides were Y = 57,122X − 19,253, Y = 46,662X − 9324, Y = 55,047X − 22,903, and Y = 52,371X − 22,302, respectively. In addition, LOD and LOQ were determined to be 1.66, 2.47, 2.93, and 3.43 µg/mL and 5.02, 7.49, 8.87, and 10.40 µg/mL, respectively.

Table 2. Calibration curves, the LOD and LOQ values of standards for quantification in NF-AM and MG5276F-AM (n = 3).

Standards [1]	Rt (min)	Calibration Curve [2]	R [2,3]	LOD (µg/mL)	LOQ (µg/mL)
Calycosin-G	15.74	Y = 57,122X − 19,253	1.00	1.66	5.02
Formononetin-G	30.36	Y = 46,662X − 9324	1.00	2.47	7.49
Calycosin	34.76	Y = 55,047X − 22,903	1.00	2.93	8.87
Formononetin	44.71	Y = 52,371X − 22,302	1.00	3.43	10.40

[1] Calycosin-G, calycosin-7-*O*-glucoside; Formononetin-G, formononetin-7-*O*-glucoside. [2] Where the Y and X are the peak area and concentration of the analyses (µg/mL), respectively. [3] Correlation coefficients for three data points in the calibration curve. Rt, retention time; LOD, limit of detection; LOQ, limit of quantification.

Linearity should be evaluated through the y-intercept and the slope of the regression line with a correlation coefficient of 0.999 or higher after calculating the calibration curve with the analyzed area values [21]. Since the correlation coefficient calculated through the regression curves of all standards was 0.999 or more, it could be quantified using this analysis method.

3.2. Quantification of Isoflavonoids Aglycones and Its Glycosides Changed through the Fermentation Process of *Astragalus membranaceus* Extracts

A. membranaceus extract contains various compounds, including dihydroflavones, isoflavonoids, flavonoids, pterocarpans, phenolic acids, and saponins [14]. Among these compounds, isoflavonoids, such as calycosin and formononetin, are the main components that reduce oxidative stress [22]. In addition, the glycoside in the various extracts can be converted by using the lactobacilli produced through the fermentation process [23]. To confirm the isoflavonoid change in *A. membranaceus* extract through fermentation, it was analyzed using the method described in Section 3.1.

The chromatograms of the isoflavonoid aglycones (calycosin and formononetin) and its glycosides (calycosin-G and formononetin-G) matched with UV spectrum, in NF-AM and MG5276F-AM, are shown in Figure 1. Retention time each isoflavonoid is indicated in

Table 1. The chromatograms confirmed that the isoflavonoid glycoside that existed before fermentation (Figure 1A) disappeared after fermentation, and isoflavonoid aglycones were generated (Figure 1B).

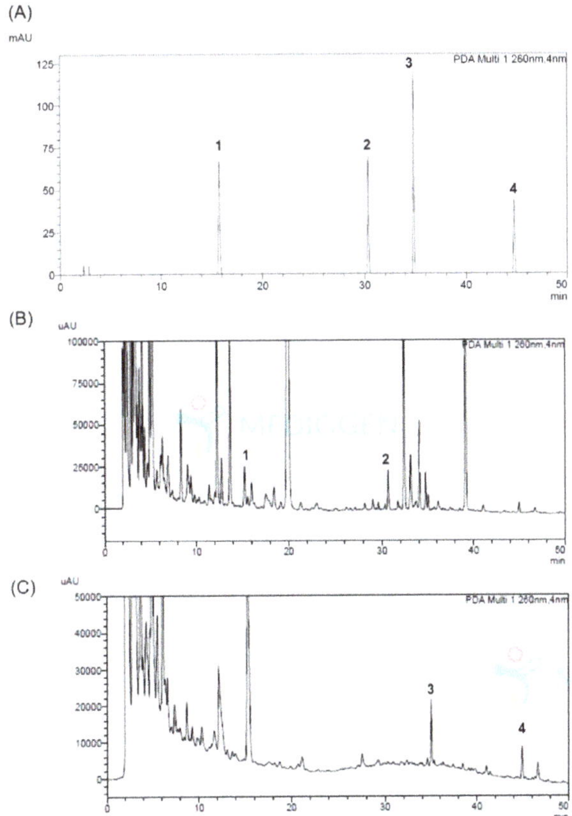

Figure 1. The chromatograms of (**A**) standards (20 ppm, each), (**B**) NF-AM (50 mg/mL), and (**C**) MG5276F-AM (50 mg/mL) by HPLC analysis. 1, calycosin-7-*O*-glucoside; 2, formonetin-7-*O*-glucoside; 3, calycosin; 4, formononetin.

The contents of isoflavonoid aglycones and their glycosides are listed in Table 3. In NF-AM, only the isoflavonoids glycosides, calycosin-G (0.91 ± 0.01 mg/100 g dried weight) and formononetin-G (0.58 ± 0.00 mg/100 g dried weight), were analyzed. In MG5276F-AM, only isoflavonoids aglycones, calycosin (0.88 ± 0.02 mg/100 g dried weight) and formononetin (0.87 ± 0.00 mg/100 g dried weight), were analyzed.

Table 3. Contents of isoflavonoids aglycones and its glycosides in NF-AM and MG5276F-AM (*n* = 3).

Sample [1]	Content (mg/100 g Dried Weight)			
	Calycosin	Formononetin	Calycosin-G	Formononetin-G
NF-AM	n.d. [2]	n.d.	0.91 ± 0.01	0.58 ± 0.00
MG5276F-AM	0.88 ± 0.02	0.87 ± 0.00	n.d.	n.d.

[1] Calycosin-G, calycosin-7-*O*-glucoside; Formononetin-G, formononetin-7-O-glucoside; NF-AM, *A. membranaceus*, not fermented; MG5276F-AM, *A. membranaceus* fermented with *L. plantarum* MG5276. [2] n.d. not detected.

L. plantarum contains various enzymes, including phosphohydrolase, α-galactosidase, α-glucosidase, acid phosphatase, and β-glucosidase [24]. Calycosin and formononetin are bound to 7-*O*-glucoside, which can be degraded by the β-glucosidase produced by *L. plantarum* [25]. In addition, many studies have been conducted using lactic acid bacteria for the bioconversion of glycosides to aglycones of bioactive compounds [23,26,27]. In our previously study, we only confirmed that calycosin content was increased by MG5276 fermentation [15]. However, in this study, we conducted quantitative and qualitative analyzes of glycosides and other isoflavonoids to determine why the calycosin content was increased. Our results confirmed that glycosides were decomposed to produce aglycones when the *A. membranaceus* extract was fermented. However, the formononetin content after fermentation was higher than that of formononetin-7-*O*-glucoside before fermentation. These results show that formononetin is synthesized by other enzymes and compounds other than β-glucosidase via the biosynthetic pathway [28]. Isoflavonoid glycosides have a very low absorption rate in the body; therefore, a bioconversion process that converts them into an isoflavonoid aglycone is urgently needed to compensate for this [27]. Therefore, conversion to aglycone is important for increasing bioactivity by increasing absorption in the body. Overall, fermentation of *A. membranaceus* extract with *L. plantarum* MG5276 suggests that bioactivity and bioavailability can be enhanced by increasing the isoflavonoid aglycone.

3.3. *Cytoprotective Effect of NF-AM and MG5276F-AM in H_2O_2-Induced HepG2 Cells*

To establish a concentration of NF-AM and MG5276F-AM that is not toxic to cells, viability was measured in HepG2 cells (Figure 2A). Based on these results, only NF-AM and MG5276F-AM at concentrations up to 50 μg/mL were used in subsequent experiments. The cytoprotective effects of NF-AM and MG5276F-AM were investigated in H_2O_2-induced HepG2 cells (Figure 2B). Treatment with H_2O_2 (0.1 to 1 mM) on HepG2 cells reduced cell viability in a concentration-dependent manner. Cell morphology and toxicity were observed at a concentration of 1 mM H_2O_2 (Figure S1). As a result of treat with NF-AM and MG5276F-AM and 1 mM H_2O_2 in HepG2 cells, the viability of MG5276F-AM was significantly higher than that of NF-AM.

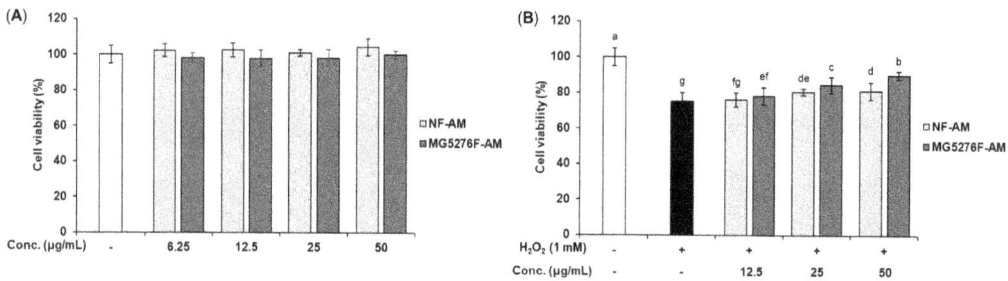

Figure 2. Cell viability of NF-AM and MG5276F-AM on hepatocytes. (**A**) HepG2 cells were treated with NF-AM and MG5276F-AM (6.25, 12.5, 25, and 50 μg/mL) for 24 h. (**B**) Viability of H_2O_2-induced HepG2 cells pre-treated with or without NF-AM and MG5276F-AM (6.25, 12.5, 25, and 50 μg/mL). Values are presented as mean ± SEM (n = 3). Means with the different letters indicate significantly differences determined by Duncan's multiple range tests ($p < 0.05$).

Oxidative stress factors, including H_2O_2, cause cytotoxicity by damaging cells and DNA [1]. It has been reported that 1 mM H_2O_2 increases cell toxicity and ROS production in HepG2 [29,30]. These reports are consistent with our study showing that H_2O_2 causes toxicity at a concentration of 1 mM. In addition, fermentation of grape skin can protect HepG2 cells from H_2O_2 damage via bioconversion to produce compounds with antioxidant activity [31]. Similarly, in our study, MG4276F-AM showed higher cell protection against oxidative damage than NF-AM because of bioconversion through fermentation.

3.4. Effect of NF-AM and MG5276F-AM on Antioxidant Enzymes in H_2O_2-Induced HepG2 Cells

To confirm that fermenting *A. membranaceus* can enhance the activity of intracellular antioxidant enzymes, mRNA expression levels of CAT, SOD, and GPx were measured in H_2O_2-induced HepG2 cells (Figure 3). When HepG2 cells were treated with only 1 mM H_2O_2, the mRNA expression levels of CAT, SOD, and GPx were significantly reduced. However, NF-AM and MG5276F-AM increased in a concentration-dependent manner the mRNA expression levels of CAT, SOD, and GPx in H_2O_2-induced HepG2 cells; in particular, MG5276F-AM (50 μg/mL) restored the mRNA expression levels of CAT, SOD, and GPx in the control.

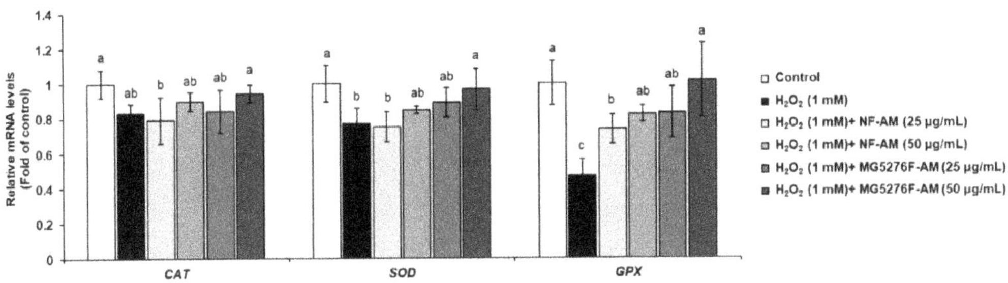

Figure 3. Antioxidant enzyme levels of CAT, SOD, and GPx in H_2O_2-induced HepG2 cells treated with NF-AM and MG5276F-AM (25 and 50 μg/mL). Values are presented as mean ± SEM (n = 3). Means with the different letters indicate significantly differences determined by Duncan's multiple range tests ($p < 0.05$).

CAT, SOD, and GPx are known to protect cells from oxidative damage caused by H_2O_2 by indirectly and/or directly reducing free radicals [5]. Calycosin, an isoflavonoid mainly found in the roots of *A. membranaceus*, elevates intracellular antioxidant enzymes by binding to the cell membrane [32]. In addition, formononetin, which is the main isoflavonoid of *A. membranaceus*, has a protective effect against H_2O_2-induced oxidative damage in HepG2 cells [33]. Thus, it was confirmed that the cytoprotective effect of MG5276F-AM on H_2O_2-treated HepG2 cells was due to the reduction of oxidative damage by isoflavonoid conferred by fermentation, which increased the mRNA expression level of antioxidant enzymes. In our previously study, MG5276F-AM decreased NO production and inflammatory related mRNA expression, iNOS, and COX-2 [15]. NO produced by the expression of iNOS and COX-2 is the reactive nitrogen species that causes oxidative stress together with ROS [34]. Thus, reduction of factors associated with inflammation by MG5276F-AM participate in expression of antioxidant enzymes that defend against oxidative stress.

3.5. Effect of NF-AM and MG5276F-AM on AST and ALT in t-BHP-Injected Mice

AST and ALT induce toxicity in the liver by causing oxidative stress [35]. To determine the effect of MG5276F-AM on oxidative damage by *t*-BHP injection, serum AST and ALT levels were measured (Figure 4). *t*-BHP injection is ultimately toxic because of the response to factors related to oxidative stress [36]. Thus, an increase in serum AST and ALT levels compared to the non-injection group indicated that oxidative stress in the liver was normally induced by *t*-BHP [35]. The serum AST and ALT levels were increased in the *t*-BHP-injected group, by 2.11- and 2.22-fold, respectively, compared to the non-injection group. In the MG5276F-AM group, both serum AST and ALT levels were significantly reduced at all concentrations.

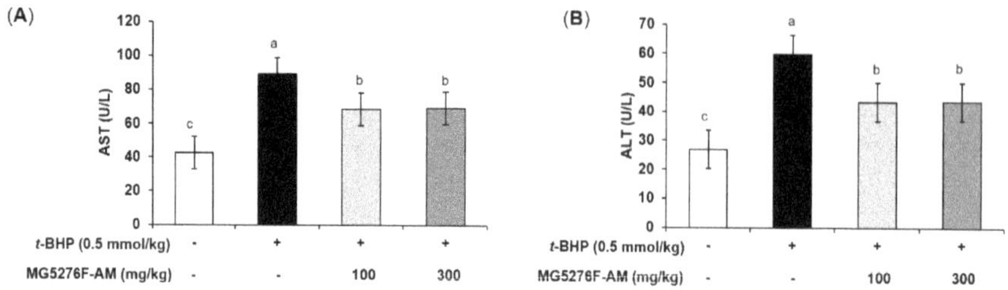

Figure 4. The serum AST (**A**), and ALT (**B**) levels of MG5276F-AM (100 and 300 mg/kg) in *t*-BHP injected mice. Values are presented as mean ± SEM ($n = 7$). Means with the different letters indicate significantly differences determined by Duncan's multiple range tests ($p < 0.05$).

3.6. *Effect of NF-AM and MG5276F-AM on Antioxidant Parameters in t-BHP-Injected Mice*

The *t*-BHP-injected animal model is mainly used to observe the antioxidant effects of many compounds and extracts. Lipid peroxidation (MDA levels) and GSH in serum are used as early-stage biomarkers of oxidative stress [35]. To confirm the antioxidant effect of MG5276F-AM, MDA and GSH levels were evaluated (Figure 5A,B). Figure 5A shows that MG5276F-AM inhibited MDA levels (0.72-fold of the *t*-BHP-injected group at 300 mg/kg) as much as the normal group compared to mice injected with *t*-BHP (1.52-fold of non-injection group). With respect to glutathione levels, *t*-BHP-injected mice had 0.92 times the normal control; however, MG5276F-AM slightly increased glutathione levels (Figure 5B). In addition, it was confirmed that CAT, SOD, and GPx were affected by MG5276F-AM, as in vitro (Figure 5C–E). MG5276F-AM restored the levels of CAT and GPx lowered by *t*-BHP injection to the same extent as in the normal group, but no change in SOD levels was observed in any group.

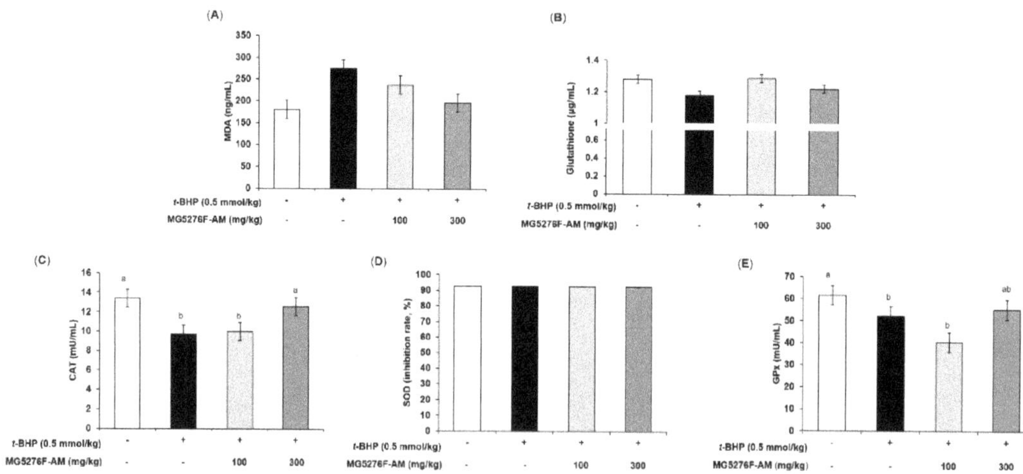

Figure 5. Effect of MG5276F-(100 and 300 mg/kg) on (**A**) MDA, (**B**) glutathione levels, (**C**) CAT, (**D**) SOD, and (**E**) GPx in *t*-BHP injected mice. Values are presented as mean ± SEM ($n = 7$). Means with the different letters indicate significantly differences determined by Duncan's multiple range test ($p < 0.05$). MG5276F-AM, *A. membranaceus* fermented with *L. plantarum* MG5276.

t-BHP induces lipid peroxidation through iron-dependent reactions. In addition, *t*-BHP is oxidized to *t*-butoxyl radicals, and endogenous GSH is involved and plays a major role in inducing oxidative stress [36]. It was reported that fermented soybean extract showed antioxidant effects via recovery of lipid peroxidation and glutathione levels in *t*-BHP injected rats, and increased antioxidant enzymes in HepG2 cells and rats [37]. In our study, MG5276F-AM alleviated lipid peroxidation and glutathione levels. Additionally, CAT and GPx were increased by MG5276F-AM as in HepG2 cells. Taken together, MG5276F-AM is valuable as a functional food with antioxidant effects in both HepG2 cells and *t*-BHP-injected mouse models induced by oxidative stress, via the increase of aglycone through bioconversion of isoflavonoids by fermentation.

4. Conclusions

This study investigated the ability of *L. plantarum* MG5276, which abundantly produces β-glucosidase, to enhance the antioxidant activity of *A. membranaceus* by bioconversion of isoflavone glycosides to aglycone. Fermentation using *L. plantarum* MG5276 of *A. membranaceus* achieved biological conversion of isoflavonoid glycosides, calycosin-G, and formononetic-G, into aglycones, calycosin, and formononetin. In addition, *A. membranaceus* fermented by *L. plantarum* MG5276 showed antioxidant efficacy against both H_2O_2-induced oxidative stress in HepG2 cells and in animal models injected with *t*-BHP. In conclusion, *A. membranaceus* fermented by *L. plantarum* MG5276 can have value as a functional food with enhanced antioxidant activity.

Supplementary Materials: The following are available online at https://www.mdpi.com/article/10.3390/fermentation8010034/s1, Figure S1: Effect of H_2O_2 on HepG2 cells viability.

Author Contributions: Conceptualization, C.-H.K.; methodology, J.Y.L.; validation, J.Y.L. and H.M.P.; formal analysis, J.Y.L. and H.M.P.; investigation, J.Y.L.; resources, C.-H.K.; writing—original draft preparation, J.Y.L.; writing—review and editing, J.Y.L. and C.-H.K.; supervision, C.-H.K. All authors have read and agreed to the published version of the manuscript.

Funding: This research received no external funding.

Institutional Review Board Statement: The animal study protocol was approved by the Institutional Ethics Committee of NDIC (P214057).

Informed Consent Statement: Not applicable.

Data Availability Statement: Not applicable.

Acknowledgments: This research was supported by the Encouragement program for the healthcare natural products industry (Chungcheongbuk-do, Jecheon-si).

Conflicts of Interest: The authors declare no conflict of interest.

References

1. Young, I.; Woodside, J. Antioxidants in health and disease. *J. Clin. Pathol.* **2001**, *54*, 176–186. [CrossRef]
2. Yashin, A.; Yashin, Y.; Xia, X.; Nemzer, B. Antioxidant activity of spices and their impact on human health: A review. *Antioxidants* **2017**, *6*, 70. [CrossRef] [PubMed]
3. Chaves, N.; Santiago, A.; Alías, J.C. Quantification of the antioxidant activity of plant extracts: Analysis of sensitivity and hierarchization based on the method used. *Antioxidants* **2020**, *9*, 76. [CrossRef] [PubMed]
4. Matés, J.M.; Pérez-Gómez, C.; De Castro, I.N. Antioxidant enzymes and human diseases. *Clin. Biochem.* **1999**, *32*, 595–603. [CrossRef]
5. Adwas, A.A.; Elsayed, A.; Azab, A.; Quwaydir, F. Oxidative stress and antioxidant mechanisms in human body. *J. Appl. Biotechnol.* **2019**, *6*, 43–47.
6. Kim, J.G.; Dong, X.; Park, S.H.; Bayazid, A.B.; Jeoung, S.A.; Lim, B.O. Bioconversion of black rice and blueberry regulate immunity system through regulation of MAPKs, NF-kB in RAW264. 7 macrophage cells. *Food Agr. Immunol.* **2021**, *32*, 471–481. [CrossRef]
7. Czekała, W. Concept of IN-OIL project based on bioconversion of by-products from food processing industry. *J. Ecol. Eng.* **2017**, *18*, 180–185. [CrossRef]
8. Eom, S.J.; Kim, K.-T.; Paik, H.-D. Microbial bioconversion of ginsenosides in *Panax ginseng* and their improved bioactivities. *Food Rev. Int.* **2018**, *34*, 698–712. [CrossRef]
9. Lee, N.-K.; Paik, H.-D. Bioconversion using lactic acid bacteria: Ginsenosides, GABA, and phenolic compounds. *J. Microbiol. Biotechnol.* **2017**, *27*, 869–877. [CrossRef]
10. Lee, Y.; Oh, J.; Jeong, Y.-S. *Lactobacillus plantarum*-mediated conversion of flavonoid glycosides into flavonols, quercetin, and kaempferol in *Cudrania tricuspidata* leaves. *Food Sci. Biotechnol.* **2015**, *24*, 1817–1821. [CrossRef]

11. Jung, J.; Jang, H.J.; Eom, S.J.; Choi, N.S.; Lee, N.-K.; Paik, H.-D. Fermentation of red ginseng extract by the probiotic *Lactobacillus plantarum* KCCM 11613P: Ginsenoside conversion and antioxidant effects. *J. Ginseng Res.* **2019**, *43*, 20–26. [CrossRef]
12. Kim, B.-H.; Jang, J.-O.; Lee, J.-H.; Park, Y.; Kim, J.-G.; Yoon, Y.-C.; Jeong, S.J.; Kwon, G.-S.; Lee, J.-B. Increased anti-oxidative activity and whitening effects of a Saposhnikovia extract following bioconversion fermentation using *Lactobacillus plantarum* BHN-LAB 33. *J. Life Sci.* **2019**, *29*, 1208–1217.
13. Park, J.-Y. Protective Activity of Roasted *Astragalus membranaceus* against UVB-Induced Damage in Human Dermal Fibroblasts and Artificial Skin Tissues. Master's Thesis, Chungbuk National University, Cheongju, Korea, 2020.
14. Li, Y.; Guo, S.; Zhu, Y.; Yan, H.; Qian, D.-W.; Wang, H.-Q.; Yu, J.-Q.; Duan, J.-A. Comparative analysis of twenty-five compounds in different parts of *Astragalus membranaceus* var. mongholicus and *Astragalus membranaceus* by UPLC-MS/MS. *J. Pharm. Anal.* **2019**, *9*, 392–399. [CrossRef] [PubMed]
15. Park, H.M.; Lee, J.Y.; Kim, M.Y.; Kang, C.-H.; Hwang, H.S. Anti-oxidative and anti-inflammatory activities of *Astragalus membranaceus* fermented by *Lactiplantibacillus plantarum* on LPS-Induced RAW 264.7 cells. *Fermentation* **2021**, *7*, 252. [CrossRef]
16. Xu, X.; Li, F.; Zhang, X.; Li, P.; Zhang, X.; Wu, Z.; Li, D. In vitro synergistic antioxidant activity and identification of antioxidant components from *Astragalus membranaceus* and *Paeonia lactiflora*. *PLoS ONE* **2014**, *9*, e96780. [CrossRef] [PubMed]
17. Zhang, Y.; Hu, G.; Lin, H.C.; Hong, S.J.; Deng, Y.H.; Tang, J.Y.; Seto, S.W.; Kwan, Y.W.; Waye, M.M.Y.; Wang, Y.T. Radix astragali extract promotes angiogenesis involving vascular endothelial growth factor receptor-related phosphatidylinositol 3-kinase/Akt-dependent pathway in human endothelial cells. *Phytother. Res* **2009**, *23*, 1205–1213. [CrossRef] [PubMed]
18. Van Meerloo, J.; Kaspers, G.J.; Cloos, J. Cell sensitivity assays: The MTT assay. In *Cancer Cell Culture*; Humana Press: Totowa, NJ, USA, 2011; pp. 237–245.
19. Livak, K.J.; Schmittgen, T.D. Analysis of relative gene expression data using real-time quantitative PCR and the $2^{-\Delta\Delta CT}$ method. *Methods* **2001**, *25*, 402–408. [CrossRef] [PubMed]
20. Kim, D.H.; Kim, M.; Oh, S.B.; Lee, K.M.; Kim, S.M.; Nho, C.W.; Yoon, W.B.; Kang, K.; Pan, C.H. The protective effect of antioxidant enriched fractions from colored potatoes against hepatotoxic oxidative stress in cultured hepatocytes and mice. *J. Food Biochem.* **2017**, *41*, e12315. [CrossRef]
21. Part 1: Validation of analytical procedures: Text and methodology. In *Ich Harmonised Tripartite Guideline*; Somatek Inc.: San Diego, CA, USA, 2005; p. 5.
22. Toda, S.; Shirataki, Y. Inhibitory effects of isoflavones in roots of *Astragalus membranaceus* Bunge (Astragali Radix) on lipid peroxidation by reactive oxygen species. *Phytother. Res.* **1998**, *12*, 59–61. [CrossRef]
23. Ávila, M.; Hidalgo, M.; Sánchez-Moreno, C.; Pelaez, C.; Requena, T.; de Pascual-Teresa, S. Bioconversion of anthocyanin glycosides by Bifidobacteria and Lactobacillus. *Food Res. Int.* **2009**, *42*, 1453–1461. [CrossRef]
24. Georgieva, R.; Iliev, I.; Haertlé, T.; Chobert, J.-M.; Ivanova, I.; Danova, S. Technological properties of candidate probiotic *Lactobacillus plantarum* strains. *Int. Dairy J.* **2009**, *19*, 696–702. [CrossRef]
25. Yuksekdag, Z.; Cinar Acar, B.; Aslim, B.; Tukenmez, U. β-Glucosidase activity and bioconversion of isoflavone glycosides to aglycones by potential probiotic bacteria. *Int. J. Food Prop.* **2017**, *20*, S2878–S2886. [CrossRef]
26. Rekha, C.; Vijayalakshmi, G. Bioconversion of isoflavone glycosides to aglycones, mineral bioavailability and vitamin B complex in fermented soymilk by probiotic bacteria and yeast. *J. Appl. Microbiol.* **2010**, *109*, 1198–1208. [CrossRef] [PubMed]
27. Kim, B.-H.; Jang, J.-O.; Lee, J.-H.; Park, Y.-E.; Kim, J.-G.; Yoon, Y.-C.; Jeong, S.J.; Kwon, G.-S.; Lee, J.-B. Bioconversion of glycosides isoflavones to aglycone isoflavones by *Lactobacillus rhamnosus* BHN-LAB 76 under anaerobic conditions. *Korean J. Food Preserv.* **2019**, *26*, 148–156. [CrossRef]
28. Pan, H.; Fang, C.; Zhou, T.; Wang, Q.; Chen, J. Accumulation of calycosin and its 7-*O*-β-D-glucoside and related gene expression in seedlings of *Astragalus membranaceus* Bge. var. mongholicus (Bge.) Hsiao induced by low temperature stress. *Plant Cell Rep.* **2007**, *26*, 1111–1120. [CrossRef] [PubMed]
29. Bak, M.-J.; Jeong, W.-S.; Kim, K.-B. Detoxifying effect of fermented black ginseng on H_2O_2-induced oxidative stress in HepG2 cells. *Int. J. Mol. Med.* **2014**, *34*, 1516–1522. [CrossRef] [PubMed]
30. Ryu, H.M.; Kim, Y.J.; Oh, E.J.; Oh, S.H.; Choi, J.Y.; Cho, J.H.; Kim, C.D.; Park, S.H.; Kim, Y.L. Hypoxanthine induces cholesterol accumulation and incites atherosclerosis in apolipoprotein E-deficient mice and cells. *J. Cell Mol. Med.* **2016**, *20*, 2160–2172. [CrossRef]
31. Liu, J.; Tan, F.; Liu, X.; Yi, R.; Zhao, X. Grape skin fermentation by *Lactobacillus fermentum* CQPC04 has anti-oxidative effects on human embryonic kidney cells and apoptosis-promoting effects on human hepatoma cells. *RSC Adv.* **2020**, *10*, 4607–4620. [CrossRef]
32. Gao, J.; Liu, Z.J.; Chen, T.; Zhao, D. Pharmaceutical properties of calycosin, the major bioactive isoflavonoid in the dry root extract of Radix astragali. *Pharm. Biol.* **2014**, *52*, 1217–1222. [CrossRef]
33. Wang, F.; Zhao, S.; Li, F.; Zhang, B.; Qu, Y.; Sun, T.; Luo, T.; Li, D. Investigation of antioxidant interactions between Radix Astragali and *Cimicifuga foetida* and identification of synergistic antioxidant compounds. *PLoS ONE* **2014**, *9*, e87221. [CrossRef]
34. Sekhar, S.; Sampath-Kumara, K.K.; Niranjana, S.R.; Prakash, H.S. Attenuation of reactive oxygen/nitrogen species with suppression of inducible nitric oxide synthase expression in RAW 264.7 macrophages by bark extract of Buchanania lanzan. *Pharmacogn. Mag.* **2015**, *11*, 283. [PubMed]
35. Oh, J.M.; Jung, Y.S.; Jeon, B.S.; Yoon, B.I.; Lee, K.S.; Kim, B.H.; Oh, S.J.; Kim, S.K. Evaluation of hepatotoxicity and oxidative stress in rats treated with *tert*-butyl hydroperoxide. *Food Chem. Toxicol.* **2012**, *50*, 1215–1221. [CrossRef] [PubMed]

36. Martín, C.; Martínez, R.; Navarro, R.; Ruiz-Sanz, J.I.; Lacort, M.; Ruiz-Larrea, M.B. *tert*-Butyl hydroperoxide-induced lipid signaling in hepatocytes: Involvement of glutathione and free radicals. *Biochem. Pharmacol.* **2001**, *62*, 705–712. [CrossRef]
37. Kim, E.Y.; Hong, K.-B.; Suh, H.J.; Choi, H.-S. Protective effects of germinated and fermented soybean extract against *tert*-butyl hydroperoxide-induced hepatotoxicity in HepG2 cells and in rats. *Food Funct.* **2015**, *6*, 3512–3521. [CrossRef]

fermentation

MDPI

Article

Supplemental *Aspergillus* Lipase and Protease Preparations Display Powerful Bifidogenic Effects and Modulate the Gut Microbiota Community of Rats

Yongshou Yang [1,2], Thanutchaporn Kumrungsee [2,*], Norihisa Kato [2,*], Shinji Fukuda [3,4,5], Manabu Kuroda [6] and Shotaro Yamaguchi [6]

1 School of Life Sciences, Anhui University, Hefei 230036, China; yangyongshou@g.ecc.u-tokyo.ac.jp
2 Graduate School of Integrated Sciences for Life, Hiroshima University, Higashi-Hiroshima 739-8528, Japan
3 Institute for Advanced Biosciences, Keio University, Tsuruoka 997-0052, Japan; sfukuda@sfc.keio.ac.jp
4 Transborder Medical Research Center, Faculty of Medicine, University of Tsukuba, Tsukuba 305-8575, Japan
5 Gut Environmental Design Group, Kanagawa Institute of Industrial Science and Technology, Kawasaki 210-0821, Japan
6 Amano Enzyme Inc., Nagoya 460-8630, Japan; manabu_kuroda@amano-enzyme.com (M.K.); shotaro_yamaguchi@amano-enzyme.com (S.Y.)
* Correspondence: kumrung@hiroshima-u.ac.jp (T.K.); nkato@hiroshima-u.ac.jp (N.K.); Tel.: +81-82-424-7980 (T.K.); +81-82-426-0774 (N.K.)

Abstract: *Aspergillus*-derived protease and lipase, which are involved in the production of *Aspergillus*-fermented foods, are consumed as digestive enzyme supplements. A marked bifidogenic effect of supplemental *Aspergillus* protease preparation (AP) in rats fed with a high-fat diet was identified. This study was conducted to examine whether the consumption of *Aspergillus*-derived lipase exerts similar bifidogenic effect. Rats were fed diets supplemented with either an *Aspergillus*-derived lipase preparation (AL) or AP at 0.1% for two weeks. 16S rRNA gene sequencing analysis indicated that supplemental AL and AP markedly influenced cecal microbial community. At the phylum level, treatment with AL and AP resulted in a lower relative abundance of Firmicutes and Bacteroidetes, but a higher relative abundance of Actinobacteria and Proteobacteria than the control rats ($p < 0.05$). At the genus level, AL and AP remarkedly elevated the relative abundances of *Bifidobacterium*, *Collinsella*, and *Enterococcus*, but significantly reduced those of *Oscillospira*, *Dorea*, and *Coprobacillus* ($p < 0.05$). These modulations were similar to those reported by several studies with typical prebiotic oligosaccharides. Notably, the bifidogenic effect of AL was much greater than that of AP. Our results show that the two different *Aspergillus*-derived preparations, AL and AP, have strong bifidogenic effects and can change the microbiota's composition.

Keywords: *Aspergillus*; lipase; protease; bifidogenic effect; prebiotics; 16S rRNA gene sequencing

check for updates

Citation: Yang, Y.; Kumrungsee, T.; Kato, N.; Fukuda, S.; Kuroda, M.; Yamaguchi, S. Supplemental *Aspergillus* Lipase and Protease Preparations Display Powerful Bifidogenic Effects and Modulate the Gut Microbiota Community of Rats. *Fermentation* **2021**, *7*, 294. https://doi.org/10.3390/fermentation7040294

Academic Editor: Hiroshi Kitagaki

Received: 4 November 2021
Accepted: 29 November 2021
Published: 1 December 2021

1. Introduction

The gut microbiota has recently received increasing attention due to its important roles in host metabolism, physiology, and immune system development. Among many environmental factors, diet is considered a key factor influencing the composition and function of the gut microbiota [1]. For instance, a western diet, characterized by high-fat and low dietary fiber, affects the gut microbiota's composition and causes gut bacterial dysbiosis. Furthermore, an altered gut microbiota is associated with system inflammation and higher incidence of chronic diseases, such as inflammatory bowel disease (IBD), cardiovascular disease, and diabetes [2]. However, increasing evidence indicates that dietary intervention and particular prebiotics supplementation, can reverse high-fat-induced gut dysbiosis. Short-chain nondigestible carbohydrates, such as fructooligosaccharides (FOS, an inulin-type fructan), and galactooligosaccharides (GOS) are well-known prebiotics that can increase populations of certain beneficial bacteria, typically *Bifidobacterium* and *Lactobacillus* species [3]. By modulating imbalances in the gut microbiota, prebiotics have been shown to promote gut health and decrease the risk of bowel and systemic diseases [4,5].

The fungi genus *Aspergillus* contains many species found in various ecological niches. Some nonpathogenic species, such as *Aspergillus oryzae* and *A. niger*, have a long history of application in food fermentation industries in Japan. During *Aspergillus*-associated fermentation, extracellular hydrolytic enzymes, such as proteases, α-amylases, and lipases, are released into the reaction system. They are responsible for the fermentation and release of umami and flavor substances [6,7]. Extracted *Aspergillus* enzymes, such as proteases and lipases, have also been used in food processing [7]. Commercially available digestive supplements containing proteases, α-amylases, and lipases derived from microbial species have been manufactured to treat insufficient food digestion [8]. A previous study of ours found that dietary supplementation with *A. oryzae*-derived protease preparation (AP) and the purified acid protease markedly elevated cecal *Bifidobacterium* spp. levels in rats fed a high-fat (HF) diet [9–11]. Cecal levels of organic acids and fecal levels of immunoglobulin A (IgA) and mucins were also increased by AP consumption, whereas pH in the cecal contents was significantly decreased [9]. The increased abundance of *Bifidobacterium* spp. in the presence of *Aspergillus* protease has been proposed to be mediated by the increased availability of free amino acids following hydrolysis of undigested proteins in the large intestine [11]. Recently, preliminary investigations into the bifidogenic effects of several digestive enzyme preparations for lipids and carbohydrates derived from *Aspergillus* spp. have been completed. Results of the quantitative polymerase chain reaction (qPCR) analysis indicated a striking increase in cecal abundance of *Bifidobacterium* spp. in rats following treatment with *Aspergillus*-derived lipase preparation (AL). To our knowledge, there is limited information on how dietary exogenous lipase supplementation modulates the gut environment. It was hypothesized that dietary supplemental AL would exert bifidogenic effects in a manner similar to that of the AP. Thus, this study examines the effect of AL on the gut microbiota and compares those with the effects of AP in rats fed a HF diet. The study was conducted using rats fed with HF diet since HF diet-induced colon dysbiosis, inflammation, and diseases have been reported to be suppressed by dietary prebiotic oligosaccharides [12].

2. Materials and Methods

2.1. Animals and Diets

Twenty-three male Sprague Dawley rats (four weeks old) were purchased from Charles River Japan. The rats were individually housed in cages in a controlled temperature environment (23 ± 2 °C), 12 h light-dark cycle, and relative humidity of 50–60%. The experimental animals were acclimatized for 7 days and then randomly divided into the following three groups based on their diet: HF diet (Ctrl: Control, Table S1; eight rats) [9]; HF diet + 0.1%(*w*/*w*) *A. niger*-derived lipase preparation (AL: Lipase AP12, lipase activity at pH 6.0, 60,000 U/g; Amano Enzyme Inc. Nagoya, Japan; eight rats), and HF diet + 0.1%(*w*/*w*) *A. oryzae*-derived protease preparation (AP: Protease A "Amano" SD, protease activity at pH 6.0, 100,000 U/g; Amano Enzyme Inc. Nagoya, Japan; seven rats). For two weeks, equal amounts of the experimental diets were given daily in food cups at 18:00 h (9, 10, 12, 14, and 15 g on days 1, 2–4, 5–7, 8–12, and 13–14, respectively) to prevent differences in food intake. All diets were consumed daily. The rats had ad libitum access to fresh water. The study protocols (protocol identical No. C15-12) were approved by the Ethics Committee of the Hiroshima University.

2.2. Sample Collection

At the end of the two-week treatment period, the rats were anesthetized (13:00–15:00 h) by inhalation of isoflurane in a desiccator to minimize suffering, and then euthanized by decapitation. The cecum was immediately excised, and its contents were completely removed weighed, and stored at −80 °C until subsequent analysis of cecal microbiota and organic acids.

2.3. 16S rRNA Gene-Based Microbiome Analysis

Total bacterial DNA in cecal contents was extracted using the QIAamp Stool Mini Kit according to the manufacturer's instructions (QIAGEN, Hilden, Germany). Then, extracted bacterial DNA was quantified using NanoDrop spectrometry (NanoDrop Technology, Wilmington, DL, USA). The V1–V2 region of the 16S rRNA genes was amplified from the DNA isolated from cecal contents using the following bacterial universal primer set: 27F (5′-ACACTCTTTCCCTACACGACGCTCTTCCGATCTAGRGTTTGATYMTGGCTCAG-3′) and 338R(5′-GTGACTGGAGTTCAGACGTGTGCTCTTCCGATCTTGCTGCCTCCCGTAG GAGT-3′). The following library preparation was performed as described previously [13]. Finally, all the barcoded V1–V2 PCR amplicons were sequenced using Illumina MiSeq sequencing technology at a read length of 2 × 300 bp based on the manufacturer's instructions (Illumina, San Diego, CA, USA).

2.4. Bioinformatics Analysis

First, fast-length adjustment of short reads (v.1.2.11) [14] was used to assemble the paired-end reads. Assembled reads with an average *Q*-value < 25 were filtered out using an in-house script. The same numbers of filtered reads were randomly selected from each sample and used for further analysis. The selected reads were then processed using QIIME pipeline (ver. 1.9.1). The high-quality sequences were clustered into operational taxonomic units (OTUs) at 97% sequence similarity, and OTUs were assigned to the Greengenes database.

2.5. Analysis of Cecal Organic Acids

The concentrations of organic acids in cecal contents were measured by gas chromatography/mass spectrometry as previously described [15].

2.6. Data Analysis

Data are expressed as mean ± standard error or boxplots with minimum, maximum and median. Statistical analysis was performed using one-way analysis of variance (ANOVA) and Tukey–Kramer HSD test (Figures 1 and 3–5). The data separation in the principal coordinate analysis ordination of beta-diversity was conducted by the PERMANOVA permutation-based statistical test in vegan package of R, and *p*-values were generated based on 999 permutations (Figure 1). Some bacterial taxa data were subjected to linear discriminant analysis effect size (LEfSe) analysis, which uses the two-tailed nonparametric Kruskal–Wallis test to evaluate the significance of differences between taxa (Figure 2). $p < 0.05$ was considered to indicate a statistically significant different. For the relationship between organic acids and microbiota composition, Pearson's correlation coefficient (r) was calculated, and the resulting correlation matrix was visualized by using R software (version 4.0.2) (Figure S1).

2.7. Evaluation of the Risk of Bias in the Methodology

The risk of bias of this study was assessed using the Systematic Review Centre for Laboratory Animal Experimentation (SYRCLE) risk of bias (RoB) tool [16]. Two independent authors (YY and NK) evaluated the following nine items: (1) sequence generation, (2) baseline characteristics, (3) allocation concealment, (4) random housing, (5) intervention blinding, (6) random outcome assessment, (7) outcome blinding, (8) incomplete outcome data, and (9) selective outcome reporting. All items were judged as 'yes' (low risk of bias) by two authors (YY and NK) independently.

3. Results

3.1. Food intake, Body Weight, and Cecal Content Weight

Total food intake for the two-week experimental period and final body weight were unaffected by dietary treatment (data not shown, $p > 0.05$). The AP and AL groups significantly increased in the weight of cecal contents compared to the Ctrl group (4.72 ± 0.39 g, 3.53 ± 0.36 g, and 1.44 ± 0.06 g, respectively, $p < 0.05$, Tukey–Kramer HSD test).

3.2. Cecal Microbiota

For microbiota analysis by 16S rRNA gene sequencing, a total of 790,645 high-quality reads were passed using the QIIME filter. Unweighted and weighted UniFrac PCoA and PERMANOVA analyses were conducted to compare the microbial structure. The results of UniFrac PCoA and PERMANOVA analyses indicated that the microbial composition differed between the Ctrl and AL groups and between the Ctrl and AP groups (in both unweighted and weighted) (Figure 1A,B, $p < 0.05$). However, the different alpha-diversity indices indicated a higher bacterial diversity in the Ctrl group than in the AL and AP groups (Figure 1C–E, $p < 0.05$).

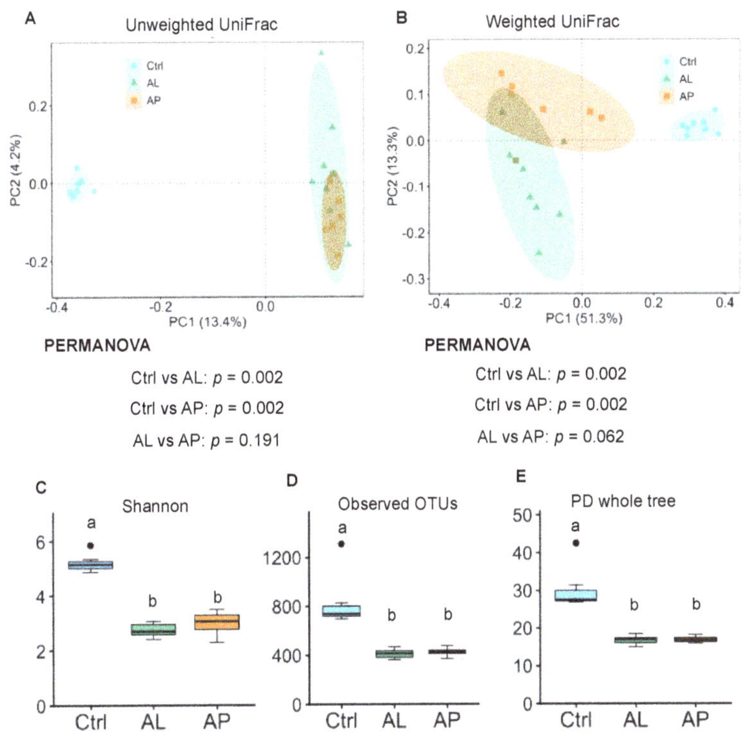

Figure 1. Effects of dietary treatment with AL and AP on cecal microbiome profiles and alpha-diversity. 16S rRNA gene-based microbiome analysis was used to evaluate the effects. PCoA of unweighted (**A**) and weighted UniFrac (**B**) and PERMANOVA analysis were performed to compare the gut microbiome profiles of the experimental groups. The diversity of the gut microbiota within samples was measured by (**C**) Shannon index, (**D**) observed OTUs, and (**E**) PD whole tree. Superscript with different letters indicate significantly difference at $p < 0.05$ (Tukey–Kramer HSD test).

LEfSe analysis results indicated that 45 bacterial taxa differed between the Ctrl and AL groups (Figure 2A, $p < 0.05$), whereas 39 taxa varied between the Ctrl and AP groups

(Figure 2B, $p < 0.05$). This analysis identified the phylum Proteobacteria, Actinobacteria, Bacteroidetes, and Firmicute differed between the Ctrl and AL groups and between the Ctrl and AP groups. This analysis further identified bacterial species *Collinsella, Lactobacillus, Bifidobacterium, Eggerthella, Enterococcus, Dehalobacterium, Coprobacillus, Dorea, Akkermansia, Adlercreutzia, rc4_4, Oscipillospira, Coprococcus, Allobaculum, Ruminococcus, Holdemania, Roseburia*, and *Parabacteroides* varied between the Ctrl and AL groups and between the Ctrl and AP groups.

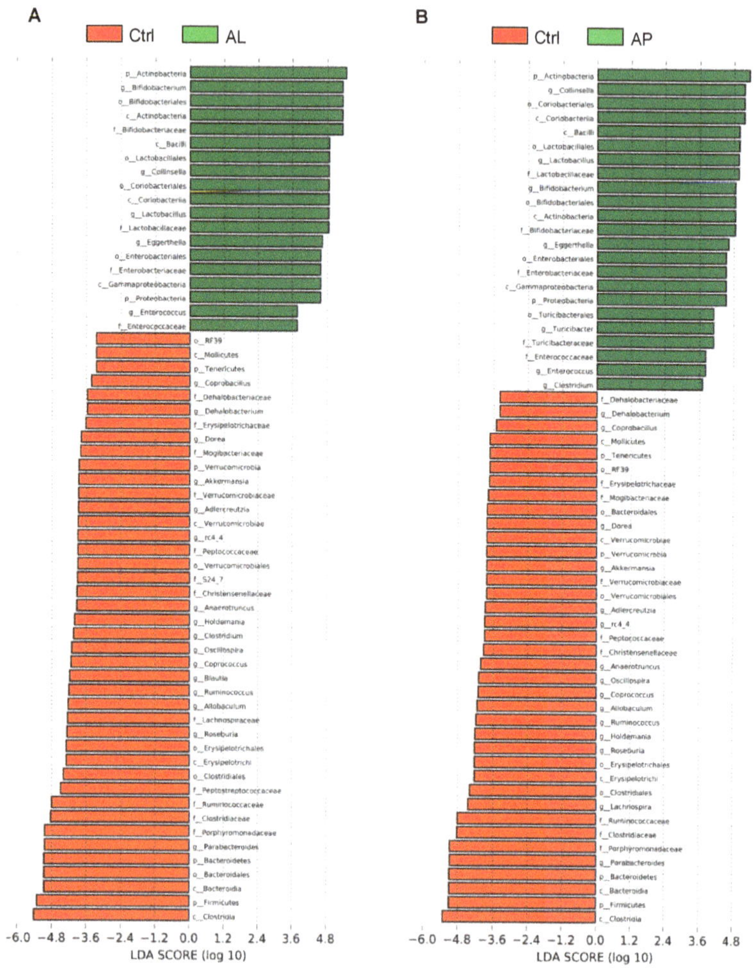

Figure 2. Comparison of the different taxa between the Ctrl and AL groups (**A**), and the Ctrl and AP groups (**B**) through LEfSe analysis. The two-tailed nonparametric Kruskal–Wallis test showed the significant differences in the taxa between the Ctrl group and the AL and AP groups ($p < 0.05$).

At the phylum level (Figure 3A), the relative abundances of Bacteroidetes and Firmicutes were significantly decreased by treatment with AL and AP, while the relative abundances of Proteobacteria and Actinobacteria were markedly increased, compared to Ctrl ($p < 0.05$, Tukey–Kramer HSD test). There was no difference in the phylum levels such as Bacteroides, Firmicute, Proteobaceria, and Actinobacteria between the AL and AP groups ($p > 0.05$). These changes were shown at the family level (Figure 3B), with significant decreases in the relative abundances of Clostridiaceae (phylum: Firmicute), Ruminococcaceae (phylum: Firmicute) and Porphyromonadaceae (phylum: Bacteroides), and significant increases in the relative abundances of Enterobacteriaceae (phylum: Proteobacteria), Coriobacteriaceae (phylum: Actinobacteria) and Bifidobacteriaceae (phylum: Actinobacteria) ($p < 0.05$, Tukey–Kramer HSD test). The levels of Lachnospiraceae (phylum: Firmicute) in the AL group were significantly lower than those in other groups ($p < 0.05$). The levels of Lactobacillaceae were unaffected ($p > 0.05$).

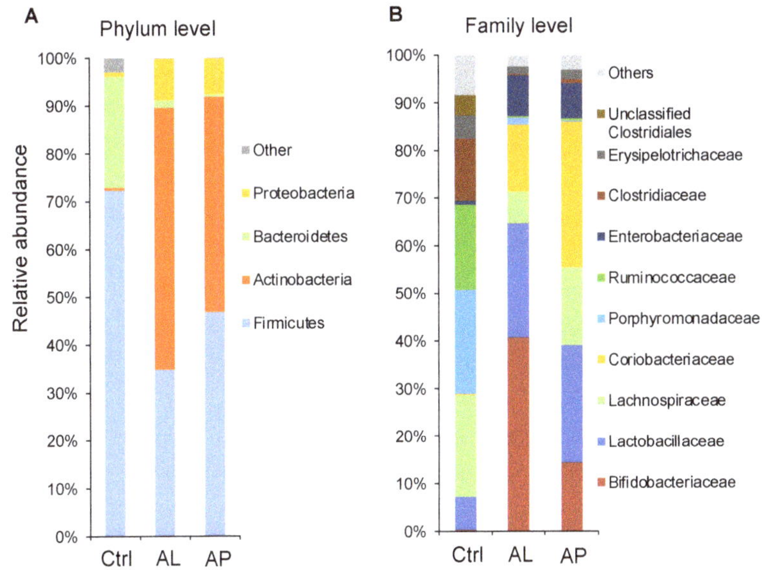

Figure 3. Effects of dietary AL and AP on cecal microbiota composition at the phylum (**A**) and family (**B**) levels.

At the genus level (Figure 4), AL and AP treatments strikingly increased the relative abundances of *Bifidobacterium* (family: Bifidobacteriaceae) by 120- and 43-fold on average, respectively, and *Collinsella* (family: Coriobacteriaceae) by 279- and 609-fold on average, respectively ($p < 0.05$). The levels of *Enterococcus* (family: Enteroccaceae) were also significantly increased by dietary AL and AP (50- and 51-fold on average, respectively, $p < 0.05$) (Figure 4). The relative abundances of *Parabacteroides* (family: Porphyromonadaceae), *Allobaculum* (family: Erysipelotrichaceae), *Oscillospira* (family: Ruminococcaceae), and *Dorea* (family: Lachnospiraceae) were significantly lower in the AL and AP groups ($p < 0.05$). *Coprobacillus* (family: Erysipelotrichaceae) was detected in all the Ctrl rats, but not detectable in the AL and AP rats. A significant increase in the *Bifidobacterium* counts by AL and AP was also observed when expressed as the copy numbers per g wet weight of cecal contents and per total cecal contents (qPCR analysis, data not shown). The relative abundances of *Lactobacillus* species were not affected by dietary treatment ($p > 0.05$). The relative abundances of *Bifidobacterium* in the AL group were much higher than those in the AP group ($p < 0.05$), while those of *Collinsella* in the AL group were significantly lower than those in the AP group ($p < 0.05$) (Figure 4). Treatment with AL and AP had no significant effect on other genus levels such as *Eggerthella*, *Mucispirillum*, *Staphylococcus*, *Lactococcus*, *Turicibacter*, *Clostridium*, *Blautia*, *Coprococcus*, *Roseburia*, *Ruminococcus*, *rc4-4*, *Anaerotruncus*, and *Eubacterium* (data not shown). The genera with a mean relative abundance < 0.2% in all groups were not considered for statistical analysis.

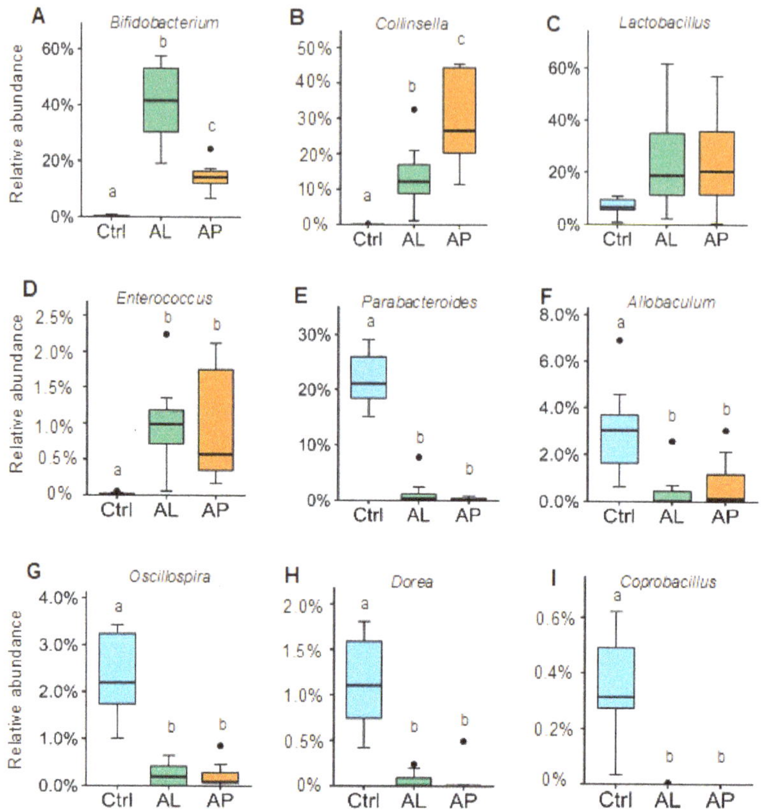

Figure 4. Effects of dietary AL and AP on cecal microbiota composition (**A**): Bifidobacterium, (**B**): Collinsella, (**C**): Lactobacillus, (**D**): Enteroccus, (**E**): Parabacteroides, (**F**): Allobaculum, (**G**): Oscillospira, (**H**): Dorea, and (**I**): Coprobacillus at the genus levels. Data are presented as a boxplot with median and min–max whiskers. The dots (•) in the boxplots are outliers. Superscript with different letters indicate significantly difference at $p < 0.05$ (Tukey–Kramer HSD test).

3.3. Cecal Organic Acids

Figure 5 indicates the concentrations of cecal organic acids. Treatment with AL and AP markedly increased the concentrations of lactate (164- and 144-fold on average, respectively), but significantly decreased those of acetate (2.0- and 2.1-fold on average, respectively, propionate (50- and 25-fold on average, respectively), and butyrate (10- and 5.4-fold on average, respectively) ($p < 0.05$). Figure S1 further indicates the relationship between the levels of organic acids and various bacteria. The levels of lactate significantly correlated with the relative abundances of lactate-producing bacteria *Bifidobacterium* (r = 0.688, $p < 0.001$) and *Enterococcus* (r = 0.473, $p < 0.05$). In general, the relative abundances of the genera such as *Parabacteroides*, *Dorea*, *Oscillospira*, *Akkermansia*, and *Coprobacillus* had a strong positive association with propionate levels, but a strong negative association with lactate levels (Figure S1).

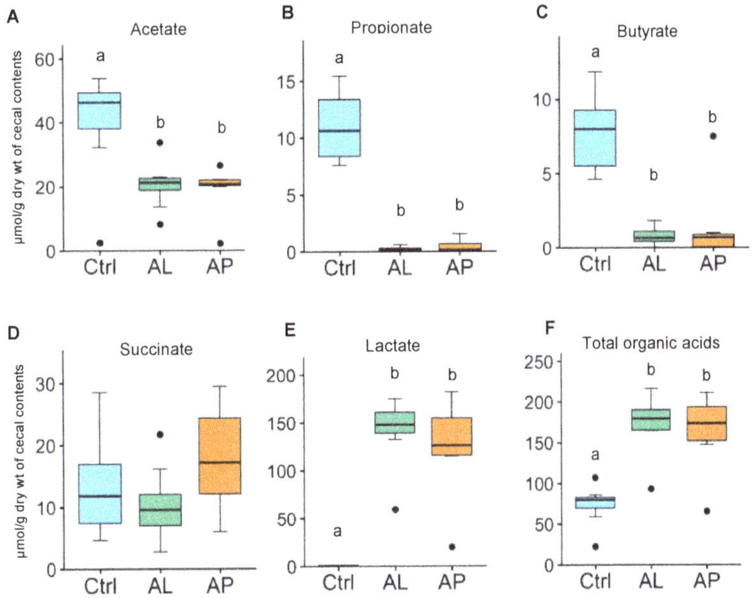

Figure 5. Effects of dietary AL and AP on cecal levels of organic acids (**A**): acetate, (**B**): propionate, (**C**): butyrate, (**D**): succinate, (**E**): lactate, and (**F**): total organic acids. Data are presented as a boxplot with median and min–max whiskers. The dots (•) in the boxplots are outliers. Superscript with different letters indicate significantly difference at $p < 0.05$ (Tukey–Kramer HSD test).

4. Discussion

4.1. Bifidobacterium

This study provided the first evidence of the strong bifidogenic effects of AL in addition to AP. Intriguingly, the bifidogenic effect of AL was much stronger than that of AP. These results imply that the bifidogenic effect is not specific for *Aspergillus* protease, and also might be possible for other hydrolases, such as *Aspergillus* lipase. *Aspergillus* lipase and protease preparations used in this study are often included in commercially available digestive enzyme supplements to improve food digestion. Accordingly, the information on the effect of the enzyme preparations on gut health may be of practical relevance, and may also help in the understanding of the health benefits of *Aspergillus*-fermented foods containing *Aspergillus* lipases and proteases. Currently, several prebiotics are well-known to confer health benefits through increasing probiotics [3,4]. To our knowledge, this study, together with previous studies [9–11], implies that the *Aspergillus*-derived lipase and protease may be considered as a new type of prebiotics, and introduce a new concept of "prebiotic digestive enzymes".

Our previous studies [10,11] found consumption of AP and purified acid protease obtained from *A. oryzae* markedly elevated cecal *Bifidobacterium* abundance. In contrast, the bifidogenic effect was absent in rats fed inactivated acid protease or inactivated AP, implying the involvement of protease activity in the mechanism of prebiotic activity. Additionally, treatment with AP increased cecal free amino acids, which correlated well with modulation of the relative abundance of *Bifidobacterium* [11]. Thus, we speculated that the increased levels of free amino acids (available amino acids) seen following consumption of AP promoted the growth of *Bifidobacterium* species. It is interesting to investigate whether an increased lipase activity in the gut arising from the supplemental AL could promote triglyceride hydrolysis to free long-chain fatty acids, which might increase the growth of *Bifidobacterium* species. The probiotic effects of *Bifidobacterium* are well-studied, imparting positive health benefits to the human host and preventing several disorders, such as gastrointestinal disorders, allergy, liver diseases, obesity, metabolic syndrome, cardiovascular diseases, cancers, autoimmune diseases, and brain diseases [3–5]. Due to their potential health-promoting properties, *Bifidobacterium* is incorporated into various foods as active ingredients. Many studies suggest that consumption of prebiotics, such as inulin, FOS, and GOS selectively increases the population of *Bifidobacterium*, a bifidogenic effect similar to that caused by AL and AP in this study [3–5].

There is growing evidence that gut microbial metabolites are key factors influencing host health. Treating rodents with probiotic bacteria of the genera *Bifidobacterium* and *Lactobacillus* increased gut γ-aminobutyric acid (GABA) levels, decreasing depression-like behavior and visceral pain [17,18]. Our previous study also showed that AP supplementation markedly increased gut microbial metabolites of amino acids, such as GABA and taurine, which are putative gut-protectors [11]. *Bifidobacterium* can metabolize linoleic acid into conjugated linoleic acids (CLAs), exerting several putative health-promoting activities, including anticarcinogenic, antiadipogenic, and antidiabetogenic, anti-inflammatory, and antioxidant actions [19]. A recent study has demonstrated that treatment with inulin stimulated *Bifidobacterium*-mediated the conversion of linoleic acid to CLAs in a human intestinal model [20]. From these findings, we believe that the intake of the *Aspergillus*-derived enzymes, such as AL and AP, is a novel approach to promote the hydrolysis of undigested macronutrients (proteins and lipids) in the intestine to nutrients usable by the gut microbiota, resulting in an elevated abundance of *Bifidobacterium* and bioactive microbial metabolites. Thus, a metabolomics study is necessary to evaluate the impact of supplemental AL and AP on gut microbial metabolites.

4.2. *Collinsella*

There is accumulating evidence that treatment with inulin and oligosaccharides increases the abundances of *Collinsella* and *Bifidobacterium* in the guts of rats [21]. In this study, treatment with AL and AP increased the levels of *Collinsella*, showing that AL and AP are similar to prebiotic oligosaccharides concerning their effect on *Collinsella*. *Collinsella* species appear to be beneficial to health; their increased abundance following dietary supplementation with oligofructose-enriched inulin in obese women is associated with an improved profile of a microbial co-metabolite, hippurate, indicating a healthier phenotype [22]. *Collinsella* exists in lower concentrations in patients with IBD or chronic pancreatitis than healthy controls [23]. A study by Saalman et al. [24] suggested the potential usefulness of this genus for treating IBD. These studies suggest that *Collinsella* is beneficial; however, abundance of *Collinsella* is reportedly positively associated with some disorders, such as type 2 diabetes and rheumatoid arthritis [25,26]. Chen et al. showed that *Collinsella* contributes to pro-inflammatory cytokine production and increased gut permeability in arthritis model mice and Caco-2 cell lines, implying potential adverse effects on some diseases [25]. Altogether, studies regarding the role of *Collinsella* species in host health are as yet inconclusive.

4.3. *Enterococcus*

This study indicated a higher abundance of *Enterococcus* in rats fed AL and AP. *Enterococcus faecalis* exerts positive effects on host health [27] and has clinical relevance for the treatment of chronic recurrent bronchitis [28]. Some members of *Enterococcus* are used as probiotics and in the production of feed additives to prevent diarrhea or to improve growth in animals [27]. Interestingly, several *Enterococcus* spp. from food exert antioxidant activity [29]. Studies have indicated that prebiotic oligosaccharides increase the abundance of *Enterococcus* species in mice and perioperative colorectal cancer patients [30,31]. In this study, both enzyme preparations acted similarly to prebiotic oligosaccharides concerning their effect on *Enterococcus*. However, *Enterococcus* species are a leading cause of hospital-associated bacteremia, endocarditis, and urinary tract infections [32]; they exist in patients with IBD [33]. Despite their pathogenic potential, commensal enterococci display low levels of virulence because they have been safely used for decades as probiotics in humans and farm animals [27]. Further research is required to carefully evaluate the implication of the increased abundances of individual members of *Enterococcus* seen in rats receiving the enzyme preparations used in this study.

4.4. *Oscillospira, Dorea, and Coprobacillus*

It is interesting that treatment with AL and AP markedly decreased the relative abundance of *Oscillospira*, *Dorea*, and *Coprobacillus*. Similarly, there is accumulating evidence that consuming various prebiotic oligosaccharides decreases gut abundance of *Oscillospira* [34,35], *Dorea* [35,36], and *Coprobacillus* [37,38] in mice, rats, and dogs. Hitherto, the impact of these genera on host health may depend on the disease type, and further studies are necessary to understand the implication of lower abundance of the genera by AL and AP.

4.5. *Parabacteroides and Allobaculum*

In this study, the relative abundances of *Parabacteroides* and *Allobaculum* were also significantly decreased by AL and AP. So far, studies on the prebiotic oligosaccharides on gut levels of these genera are limited, and controversial. Presently, the implication of lower abundance of *Parabacteroides* by AL and AP is unknown, and remains to be explored.

4.6. *Bacterial Diversity*

Unexpectedly, in this study, treatment with AL and AP significantly lowered bacterial diversity compared to control. Microbial diversity is considered beneficial for community stability and host health [39,40]. However, this may not always be the case, and assumptions of increased diversity could be oversimplified for complicated interactive mechanisms of health and disease [41]. We believe the reduced bacterial diversity in the AL and AP groups might relate to the depletion of many bacterial species, including *Parabacteroides*, *Allobaculum*, *Oscillospira*, *Dorea*, and *Coprobacillus* (Figures 2 and 4).

4.7. *Organic Acids*

In this study, it is interesting that AL and AP strikingly increased cecal lactate levels, which were significantly associated with the bifidogenic effects. Meanwhile, the levels of other organic acids, such as acetate, propionate, and butyrate, were decreased by AL and AP. Notably, there was a strong inverse association between lactate and propionate levels. Propionate is microbially produced from lactate in the human colon [42]. Accordingly, the metabolic conversion of lactate into propionate might be lowered in the AL and AP groups. Previous in vitro study indicated free radical scavenging and antioxidant effects of lactate [43]. Recent studies have suggested that lactate exhibits an inflammatory or anti-inflammatory role depending on its effects on immune cell types and disease types [44]. Thus, the implication of lactate accumulation in the AL and AP groups remains to be explored.

4.8. *Limitations of this Study*

Study limitations should be highlighted. In this study, both AL and AP used were crude enzyme preparations and contained minor components other than lipase and protease. Thus, other factors, except for lipase and protease in these preparations, might have contributed to the effects observed in the gut microbiota. Further research is recommended to investigate the impact of highly purified lipase and protease obtained from *Aspergillus* species on intestinal microbiota. In addition, this study used wild-type rats, but not disease-model rats. Thus, the effects of *Aspergillus* lipase and protease on disease-model animals remains to be elucidated.

5. Conclusions

This study is the first to show the powerful bifidogenic effect of AL, an impact that, interestingly, was far stronger than that of AP. Additionally, this study specifically revealed the modulations of the relative abundances of other species of the genera *Collinsella, Enterococcus, Oscillospira, Dorea*, and *Coprobacillus* by the enzyme preparations. These modulations were similar to those reported in several studies with typical prebiotic oligosaccharides. The results might provide novel insights into the impacts of *Aspergillus* protease and lipase, and of *Aspergillus*-fermented foods containing *Aspergillus* proteases and lipases on gut health. Additionally, our studies might introduce a new concept of "prebiotic digestive enzymes" based on the findings of the impacts of dietary *Aspergillus* lipase and protease on gut *Bifidobacterium*. Our findings raise a fundamental question of how exogenous digestive enzymes of fats and proteins lead to their bifidogenic effects. Further studies will address this question. Metabolomics study in our group is in progress to investigate the gut metabolites derived from dietary fats and proteins in rats fed AL and AP to understand the impacts of AL and AP on the gut health.

Supplementary Materials: The following are available online at https://www.mdpi.com/article/10.3390/fermentation7040294/s1. Table S1. Composition of basal diet. Figure S1. Correlation matrix (correlation coefficient, r) between cecal levels of organic acids and microbiota composition. The deep color bars indicate correlation strength. * $p < 0.05$, ** $p < 0.01$.

Author Contributions: Conceptualization, Investigation, Methodology, Validation, Writing—original draft: Y.Y., T.K. and N.K.; Formal analysis, Software, Methodology, Visualization: Y.Y., T.K., S.F., M.K. and S.Y.; literature search: Y.Y., N.K.; Project administration, Funding acquisition, Supervision, Writing—review and editing: N.K., T.K., S.F., M.K. and S.Y. All authors have read and agreed to the published version of the manuscript.

Funding: This study was supported in part by the grant from the HIRAKU consortium, Hiroshima University (to T.K.). Amano Enzyme Inc. (to N.K.), JSPS KAKENHI (18H04805 to S.F.), AMED-CREST (JP20gm1010009 to S.F.), JST ERATO (JPMJER1902 to S.F.), the Takeda Science Foundation (to S.F.), the Food Science Institute Foundation (to S.F.), and the Program for the Advancement of Research in Core Projects under Keio University's Longevity Initiative (to S.F.).

Institutional Review Board and Statement: The study was conducted according to the guidelines of the Declaration of Helsinki, and approved by the Ethics Committee of the Hiroshima University (Ethical approval No. C15-12), approved on 16 June 2015.

Data Availability Statement: The data of the 16S rRNA gene sequences of gut microbiota presented in this study can be available from the DDBJ database (http://getentry.ddbj.nig.ac.jp/) (accessed 21 January 2021) under accession number DRA011374. Further inquiries can be directed to the corresponding authors.

Conflicts of Interest: This study received funding from Amano Enzyme Inc. The funders had no role in the design of the study, in the collection, analyses, or interpretation of data, in the writing of the manuscript, or in the decision to publish the results. All authors declare no other competing interests in the current study.

References

1. Hasan, N.; Yang, H. Factors affecting the composition of the gut microbiota, and its modulation. *PeerJ* **2019**, *7*, e7502. [CrossRef] [PubMed]
2. Conlon, M.A.; Bird, A.R. The impact of diet and lifestyle on gut microbiota and human health. *Nutrients* **2015**, *7*, 17–44. [CrossRef]
3. Patel, S.; Goyal, A. The current trends and future perspectives of prebiotics research: A review. *3 Biotech* **2012**, *2*, 115–125. [CrossRef]
4. Tsai, Y.L.; Lin, T.L.; Chang, C.J.; Wu, T.R.; Lai, W.F.; Lu, C.C. Probiotics, prebiotics and amelioration of diseases. *J. Biomed. Sci.* **2019**, *26*, 3. [CrossRef]
5. Stavropoulou, E.; Bezirtzoglou, E. Probiotics in Medicine: A Long Debate. *Front. Immunol.* **2020**, *11*, 2192. [CrossRef] [PubMed]
6. Ichishima, E. Enzymology of the national microorganisms of Japan in a historical context. *J. Biochem. Biotechnol.* **2019**, *2*, 11–18.
7. Takenaka, S.; Nakabayashi, R.; Ogawa, C.; Kimura, Y.; Yokota, S.; Doi, M. Characterization of surface *Aspergillus* community involved in traditional fermentation and ripening of katsuobushi. *Int. J. Food Microbiol.* **2020**, *327*, 108654. [CrossRef]
8. Ianiro, G.; Pecere, S.; Giorgio, V.; Gasbarrini, A.; Cammarota, G. Digestive Enzyme Supplementation in Gastrointestinal Diseases. *Curr. Drug Metabol.* **2016**, *17*, 187–193. [CrossRef]

9. Yang, Y.; Sitanggang, N.V.; Kato, N.; Inoue, J.; Murakami, T.; Watanabe, T.; Iguchi, T.; Okazaki, Y. Beneficial effects of protease preparations derived from *Aspergillus* on the colonic luminal environment in rats consuming a high-fat diet. *Biomed. Rep.* **2015**, *3*, 715–720. [CrossRef]
10. Yang, Y.; Iwamoto, A.; Kumrungsee, T.; Okazaki, Y.; Kuroda, M.; Yamaguchi, S.; Katoa, N. Consumption of an acid protease derived from *Aspergillus oryzae* causes bifidogenic effect in rats. *Nutr. Res.* **2017**, *44*, 60–66. [CrossRef]
11. Yang, Y.; Kumrungsee, T.; Kuroda, M.; Yamaguchi, S.; Kato, N. Feeding *Aspergillus* protease preparation combined with adequate protein diet to rats increases levels of cecum gut-protective amino acids, partially linked to *Bifidobacterium* and *Lactobacillus*. *Biosci. Biotech. Biochem.* **2019**, *83*, 1901–1911. [CrossRef] [PubMed]
12. Van Hul, M.; Karnik, K.; Canene-Adams, K.; De Souza, M.; Abbeele, P.V.D.; Marzorati, M.; Delzenne, N.M.; Everard, A.; Cani, P.D. Comparison of the effects of soluble corn fiber and fructooligosaccharides on metabolism, inflammation, and gut microbiome of high-fat diet-fed mice. *Am. J. Physiol. Endocrinol. Metab.* **2020**, *319*, E779–E791. [CrossRef]
13. Ishii, C.; Nakanishi, Y.; Murakami, S.; Nozu, R.; Ueno, M.; Hioki, K.; Aw, W.; Hirayama, A.; Soga, T.; Ito, M.; et al. A metabologenomic approach reveals changes in the intestinal environment of mice fed on American diet. *Int. J. Mol. Sci.* **2018**, *19*, 4079. [CrossRef] [PubMed]
14. Magoč, T.; Salzberg, S.L. FLASH: Fast length adjustment of short reads to improve genome assemblies. *Bioinformatics* **2011**, *27*, 2957–2963. [CrossRef]
15. Zhang, S.; Wang, H.; Zhu, M. A sensitive GC/MS detection method for analyzing microbial metabolites short chain fatty acids in fecal and serum samples. *Talanta* **2019**, *196*, 249–254. [CrossRef] [PubMed]
16. Hooijmans, C.R.; Rovers, M.M.; BM de Vries, R.; Leenaars, M.; Ritskes-Hoitinga, M.; Langendam, M.W. SYRCLE's risk of bias tool for animal studies. *BMC Med. Res. Meth.* **2014**, *14*, 43. [CrossRef]
17. Patterson, E.; Ryan, P.M.; Wiley, N.; Carafa, I.; Sherwin, E.; Moloney, G.; Franciosi, E.; Mandal, R.; Wishart, D.S.; Tuohy, K.; et al. Gamma-aminobutyric acid-producing lactobacilli positively affect metabolism and depressive-like behaviour in a mouse model of metabolic syndrome. *Sci. Rep.* **2019**, *9*, 16323. [CrossRef]
18. Pokusaeva, K.; Johnson, C.; Luk, B.; Uribe, G.; Fu, Y.; Oezguen, N.; Matsunami, R.K.; Lugo, M.; Major, A.; Mori-Akiyama, Y.; et al. GABA-producing *Bifidobacterium dentium* modulates visceral sensitivity in the intestine. *Neurogastroenterol. Motil.* **2017**, *29*, e12904. [CrossRef]
19. Basak, S.; Duttaroy, A.K. Conjugated Linoleic Acid and Its Beneficial Effects in Obesity, Cardiovascular Disease, and Cancer. *Nutrients* **2020**, *12*, 1913. [CrossRef]
20. Carafa, I.; Masuero, D.; Vrhovsek, U.; Bittante, G.; Franciosi, E.; Tuohy, K.M. Production of conjugated linoleic acid (CLA): Effect of inulin on microbial composition and CLA concentration in a human intestinal model. *Proc. Nutr. Soc.* **2020**, *79 (OCE2)*, E628. [CrossRef]
21. Klancic, T.; Laforest-Lapointe, I.; Choo, A.; Nettleton, J.E.; Chleilat, F.; Noye Tuplin, E.W.; Alukic, E.; Cho, N.A.; Nicolucci, A.C.; Arrieta, M.; et al. Prebiotic Oligofructose Prevents Antibiotic-Induced Obesity Risk and Improves Metabolic and Gut Microbiota Profiles in Rat Dams and Offspring. *Mol. Nutr. Food Res.* **2020**, *64*, e2000288. [CrossRef]
22. Dewulf, E.M.; Cani, P.D.; Claus, S.P.; Fuentes, S.; Puylaert, P.G.B.; Neyrinck, A.M.; Bindels, L.B.; de Vos, W.M.; Gibson, G.R.; Thissen, J.; et al. Insight into the prebiotic concept: Lessons from an exploratory, double blind intervention study with inulin-type fructans in obese women. *Gut* **2013**, *62*, 1112–1121. [CrossRef] [PubMed]
23. Kassinen, A.; Krogius-Kurikka, L.; Mäkivuokko, H.; Rinttilä, T.; Paulin, L.; Corander, J.; Malinen, E.; Apajalahti, J.; Palva, A. The Fecal Microbiota of Irritable Bowel Syndrome Patients Differs Significantly From That of Healthy Subjects. *Gastroenterology* **2007**, *133*, 24–33. [CrossRef] [PubMed]
24. Saalman, R.; Alderberth, I.; Wold, A.; Sjoberg, F. Use of *Collinsella* for Treatment of Inflammatory Bowel Disease. U.S. Patent No. 2018/0028579 A1, 1 February 2018.
25. Chen, J.; Wright, K.; Davis, J.M.; Jeraldo, P.; Marietta, E.V.; Murray, J.; Nelson, H.; Matteson, E.L.; Taneja, V. An expansion of rare lineage intestinal microbes characterizes rheumatoid arthritis. *Genome Med.* **2016**, *8*, 43. [CrossRef] [PubMed]
26. Shah, V.; Lambeth, S.M.; Carson, T.; Lowe, J.; Ramaraj, T.; Leff, J.W.; Luo, L.; Bell, J.C.; Shah, V.O. Composition Diversity and Abundance of Gut Microbiome in Prediabetes and Type 2 Diabetes. *J. Diabetes Obes.* **2015**, *2*, 1–7. [CrossRef]
27. Hanchi, H.; Mottawea, W.; Sebei, K.; Hammami, R. The Genus Enterococcus: Between Probiotic Potential and Safety Concerns–An Update. *Front. Microbiol.* **2018**, *9*, 1791. [CrossRef]
28. Habermann, W.; Zimmermann, K.; Skarabis, H.; Kunze, R.; Rusch, V. The effect of a bacterial immunostimulant (human Enterococcus faecalis bacteria) on the occurrence of relapse in patients with chronic bronchtis. *Arzneimittelforschung* **2001**, *51*, 931–937.
29. Pieniz, S.; Andreazza, R.; Anghinoni, T.; Camargo, F.; Brandelli, A. Probiotic potential, antimicrobial and antioxidant activities of *Enterococcus durans* strain LAB18s. *Food Control* **2014**, *37*, 251–256. [CrossRef]
30. Mao, B.Y.; Li, D.Y.; Zhao, J.X.; Liu, X.M.; Gu, Z.N.; Chen, Y.Q.; Zhang, H.; Chen, W. Metagenomic insights into the effects of fructo-oligosaccharides (FOS) on the composition of fecal microbiota in mice. *J. Agric. Food Chem.* **2015**, *63*, 856–863. [CrossRef] [PubMed]
31. Xie, X.; He, Y.; Li, H.; Yu, D.; Na, L.; Sun, T.; Zhang, D.; Shi, X.; Xia, Y.; Jiang, T.; et al. Effects of prebiotics on immunologic indicators and intestinal microbiota structure in perioperative colorectal cancer patients. *Nutrition* **2019**, *61*, 132–142. [CrossRef]
32. Dubin, K.; Pamer, E.G. Enterococci and Their Interactions with the Intestinal Microbiome. *Microbiol. Spectrum* **2014**, *5*, 5–6.

33. Lo Presti, A.; Zorzi, F.; Del Chierico, F.; Altomare, A.; Cocca, S.; Avola, A.; de Biasio, F.; Russo, A.; Cella, E.; Reddel, S.; et al. Fecal and Mucosal Microbiota Profiling in Irritable Bowel Syndrome and Inflammatory Bowel Disease. *Front. Microbiol.* **2019**, *10*, 1655. [CrossRef]
34. Fei, Y.; Wang, Y.; Pang, Y.; Wang, W.; Zhu, D.; Xie, M.; Lan, S.; Wang, Z. Xylooligosaccharide Modulates Gut Microbiota and Alleviates Colonic Inflammation Caused by High Fat Diet Induced Obesity. *Front. Physiol.* **2020**, *10*, 1601. [CrossRef] [PubMed]
35. Wang, L.; Pan, M.; Li, D.; Yin, Y.; Jiang, T.; Fang, S.; Wang, G.; Zhao, J.; Zhang, H.; Chen, W. Metagenomic insights into the effects of oligosaccharides on the microbial composition of cecal contents in constipated mice. *J. Funct. Foods* **2017**, *38*, 486–496. [CrossRef]
36. Garcia-Mazcorro, J.F.; Barcenas-Walls, J.R.; Suchodolski, J.S.; Steiner, J.M. Molecular assessment of the fecal microbiota in healthy cats and dogs before and during supplementation with fructo-oligosaccharides (FOS) and inulin using high-throughput 454-pyrosequencing. *PeerJ* **2017**, *5*, e3184. [CrossRef] [PubMed]
37. Alexander, C.; Cross, T.-W.L.; Devendran, S.; Neumer, F.; Theis, S.; Ridlon, J.M.; Suchodolski, J.S.; de Godoy, M.R.C.; Swanson, K.S. Effects of prebiotic inulin-type fructans on blood metabolite and hormone concentrations and faecal microbiota and metabolites in overweight dogs. *Br. J. Nutr.* **2018**, *120*, 711–720. [CrossRef] [PubMed]
38. Everard, A.; Lazarevic, V.; Gaïa, N.; Johansson, M.; Ståhlman, M.; Bäckhed, F.; Delzenne, N.; Schrenzel, J.; Francois, P.; Cani, P. Microbiome of prebiotic-treated mice reveals novel targets involved in host response during obesity. *ISME J.* **2014**, *8*, 2116–2130. [CrossRef] [PubMed]
39. Audebert, C.; Even, G.; Cian, A.; Loywick, A.; Merlin, S.; Viscogliosi, E.; Chabe, M.; Blastocystis Investigation Group. Colonization with the Enteric Protozoa Blastocystis is Associated with Increased Diversity of Human Gut Bacterial Microbiota. *Sci. Rep.* **2016**, *6*, 25255. [CrossRef]
40. Sjöberg, F.; Barkman, C.; Nookaew, I.; Östman, S.; Adlerberth, I.; Saalman, R.; Wold, A.E. Low-complexity microbiota in the duodenum of children with newly diagnosed ulcerative colitis. *PLoS ONE* **2017**, *12*, e0186178. [CrossRef]
41. Shade, A. Diversity is the question, not the answer. *ISME J.* **2017**, *11*, 1–6. [CrossRef]
42. Flint, H.J.; Duncan, S.H.; Scott, K.P.; Louis, P. Links between diet, gut microbiota composition and gut metabolism. *Proc. Nutr. Soc.* **2015**, *74*, 13–22. [CrossRef] [PubMed]
43. Groussard, C.; Morel, I.; Chevanne, M.; Monnier, M.; Cillard, J.; Delamarche, A. Free radical scavenging and antioxidant effects of lactate ion: An in vitro study. *J. Appl. Physiol.* **2000**, *89*, 169–175. [CrossRef] [PubMed]
44. Manoharan, I.; Prasad, P.D.; Thangaraju, M.; Manicassamy, S. Lactate-Dependent Regulation of Immune Responses by Dendritic Cells and Macrophages. *Front. Immunol.* **2021**, *12*, 691134. [CrossRef] [PubMed]

fermentation

MDPI

Article

Influence of the Ratio of Sheep and Cow Milk on the Composition and Yield Efficiency of Lećevački Cheese

Siniša Matutinović [1], Ante Rako [2], Milna Tudor Kalit [3,*] and Samir Kalit [3]

[1] Department of Production, Meggle Dairy d.o.o., Bihać, Grabeška 34, 77000 Bihać, Bosnia and Herzegovina; sinisa.matutinovic@gmail.com
[2] Department of Applied Sciences, Institute for Adriatic Crops and Karst Reclamation, Put Duilova 11, 21000 Split, Croatia; Ante.Rako@krs.hr
[3] Department of Dairy Science, University of Zagreb Faculty of Agriculture, Svetošimunska c. 25, 10000 Zagreb, Croatia; skalit@agr.hr
* Correspondence: mtudor@agr.hr; Tel.: +38-51-239-3647

Abstract: Lećevački cheese is a traditional Croatian hard cheese, which belongs to the group of hard Mediterranean cheeses produced from mixed milk (cow and sheep). The aim of this research was to determine the influence of different ratios and compositions of sheep milk on the composition and yield of Lećevački cheese. A total of 15 batches of Lećevački cheese were selected containing different ratios of sheep and cow milk from the regular production of a dairy plant. The ratio of sheep milk was as follows: up to 39%, from 40 to 44%, and from 45 to 50%. For each ratio, five batches were randomly selected. A higher ratio of sheep milk caused a noticeable increase in fat, protein, lactose, and total solids content, while the content of solids-not-fat significantly ($p < 0.05$) increased. A similar trend was found for casein content ($p < 0.1$). The highest ratio of sheep milk in mixed milk increased ($p < 0.05$) the protein content by almost 1%. However, the results showed that it is not reasonable to increase the sheep milk ratio in mixed milk above 44% (v/v) because it causes a higher ($p < 0.01$) moisture content in the cheese, as well as a lower fat content ($p < 0.01$) and fat recovery ($p = 0.07$) during the manufacturing of Lećevački cheese.

Keywords: mixed milk; cow milk; sheep milk; cheese composition; cheese yield; Lećevački cheese; Mediterranean

Citation: Matutinović, S.; Rako, A.; Tudor Kalit, M.; Kalit, S. Influence of the Ratio of Sheep and Cow Milk on the Composition and Yield Efficiency of Lećevački Cheese. *Fermentation* **2021**, *7*, 274. https://doi.org/10.3390/fermentation7040274

Academic Editor: Hiroshi Kitagaki

Received: 27 October 2021
Accepted: 17 November 2021
Published: 24 November 2021

1. Introduction

Milk and dairy products produced from cows, sheep, goats, and buffalos are an important part of Mediterranean nutrition [1,2]. Moreover, the production of cheese from the mixed milk of different dairy animals is a common practice in Mediterranean countries [3]. There are many examples of cheeses produced from mixed milk, such as Teleme, Graviera, Kasseri, Kopanisti, and Kefalotiri from Greece [4–6]; Casciotta d'Urbino, Murazzano, Fossa, Robiola di Roccaverano, and Castelmagno from Italy [7–10]; Cabrales, Mahon, Hispanico, and Iberico from Spain [11,12]; Halloumi from Cyprus [13]; Picante, cheeses with Appéllation d'origine protégée, Amarelo, and Rabaçal from Portugal [14,15]; and Lećevački cheese from Croatia [16,17]. Park et al. [18] postulated that the composition of cow milk changes minimally over the year, taking into consideration that cows' lactation is not influenced by the season [19], which is not the case with sheep and goat milk. Therefore, it is difficult to maintain the standard quality of cheese produced from mixed milk [20].

The composition of the cheese produced from mixed milk strongly depends on different changeable milk parameters and on the composition of the milk that is used in its production, especially the protein and casein content [21]. Moreover, fat content and mineral composition, especially calcium, also influence the composition [18] and yield [22] of produced cheese. Cheese yield efficiency is the most important item in cheesemaking, and its improvement is one of the most important tasks in the countries where milk is processed into cheese [23]. It can be defined as the weight of cheese in relation to the weight of milk used and can be presented as the value of efficiency of the recovery of fat and protein from milk into the cheese [24]. All formulas for the prediction of yield include the milk composition [25]. Cheese yield is lower if it is produced from milk with lower contents of fat, protein, and solids-not-fat [26]. Kalit et al. [27] reported that increasing the protein in milk by adding skim milk powder increases actual and adjusted cheese yield. Temizkan et al. [28] reported that the yield of cheese produced from sheep milk is higher than that produced from cow milk. Consequently, the yield of cheese produced from mixed milk with a higher ratio of sheep milk has to be higher. However, there has not been enough research to confirm that, especially considering the strong influence of season on the composition of sheep milk used in cheese production, which is typical for the Mediterranean area [29]. Different mixtures of sheep and cow milk drastically change the ratio of casein to fat as well as the average size of casein micelles and fat globules in mixed milk, which could also affect cheese yields.

Lećevački cheese is a pastoral and traditional Croatian cheese that belongs to the group of hard Mediterranean cheeses. In contrast to the islands of the Adriatic sea and coastal regions where limited vegetation makes only sheep breeding possible, the environment of the Lećevica region, located in the hinterlands of Dalmatia, is equally suitable for breeding sheep and cows. Such an environment dictated that in the past, this cheese was produced on family farms in the area of Lećevica from full-fat sheep milk, cow milk, or mixed milk, depending on the season and herd composition (the number of sheep and cows that were possessed by farmers). Today, it is produced from mixed sheep and cow milk [16,17]. The composition of mixed milk depends on many different variabilities of milk, including the composition and physico-chemical properties of different milk types used in the production of cheese. The composition and yield of mixed milk cheese directly depend on the ratio of the mixture of milk, as well as the composition of the milk used in its production. The composition of the milk of the indigenous breed, whose milk is used for Lećevački cheese making, is strongly affected by the flock, year, and season [29], so it is difficult to maintain a consistent quality of mixed milk cheeses. Therefore, the aim of this research was to determine the influence of different ratios and compositions of sheep milk on the composition and yield of Lećevački cheese.

2. Materials and Methods

2.1. Sampling of Milk

For research purposes, 15 batches of Lećevački cheese were selected containing different ratios of sheep and cow milk from the regular production from the Mils dairy plant in Split, Croatia. The ratio of sheep milk was as follows: up to 39%, from 40 to 44%, and from 45 to 50%. For each ratio, five batches were randomly selected. Bulk tank-mixed milk samples were taken after mixing sheep and cow milk and before processing into Lećevački cheese. Batches were selected randomly, considering the season. Sheep and cow milk used in the production of Lećevački cheese were collected from the Dalmatian hinterland between March and August. Sheep for milk production were bred semi-extensively, which included pasture during the vegetation and milking period. In addition, 150 g of wheat or barley were given to sheep during lactation. In contrast, the cows were bred intensively and fed with a concentrate according to their milk production. Similar to sheep, cows were outdoors during the vegetation period. Winter feeding included hay and some silage.

2.2. Milk Analysis

To determine the chemical composition, mixed milk samples were analyzed for the content of fat, protein, lactose, total solids, and total solids-not-fat with a mid-infrared spectrometry method using a MilkoScan FT-120 instrument [30]. The casein content of milk was detected using the direct method [31]. The value of pH was determined using the potentiometric method with a Schott CG 842 pH meter.

2.3. Making of Lećevački Cheese and Cheese Sampling

Following pasteurization, the sheep and cow milk were weighed and then mixed in a 600 L cheese vat. The process flow of the making of Lećevački cheese is presented in Table 1.

Table 1. Process flow of making of Lećevački cheese.

Step	Cheesemaking Parameters
Pasteurization of milk	Temperature: 74 ± 2 °C/30 s
Cooling of milk	Temperature: 30–32 °C
Addition of $CaCl_2$	Dosage: 0.02%
Addition of lysozyme	Dosage: 0.0025%
Addition of starter culture *	Dosage: 0.1%
Addition of rennet **	Dosage: 0.0016%
Coagulation	Duration: 30–40 min
Curd cutting	Size of wheat
Curd grain heating	Temperature: 41–42 °C
Drying of curd grains	Duration: 35 min
Cheese forming	Duration: 30 min
Pressing	Duration/pressure: 30 min/100 kPa Duration/pressure: 30 min/200 kPa Duration/pressure: 1 h/300 kPa
Salting (brine)	Salt concentration: 19–20 °Be Temperature: 14–16 °C pH value: 5.1 ± 0.1 Duration: 24 h ± 2 h
Drying of the cheese	Temperature: 14–16 °C RAH: 70% Duration: 24 h
Ripening	Temperature: 14–16 °C RAH: 75–85% Duration: 90 days

* Homofermentative strains composed of *Streptococcus thermophilus*, *Lactobacillus helveticus*, and *Lactococcus lactis*; ** 100% chymosin. RAH = Relative air humidity.

During ripening, the cheese was kept on wooden shelves, turned upside-down, and washed every week. For analysis, 60 wheels of Lećevački cheese were produced from a total of 15 selected batches. Each batch of Lećevački cheese was weighed after 90 days of ripening.

2.4. Composition Analysis of the Cheese

The composition analysis of the cheese included the determination of fat content using the Van Gulik method [32], protein content using the block-digestion method [33], total solids content using the drying method [34], pH value using the potentiometric method using the Schott CG 842 instrument, lactic acid content using the titratable method [35], and salt content using the Volhard method [36].

2.5. Determination of Cheese Yield Efficiency

The cheese yield efficiency was determined by the weighing of the cheeses produced from the previously weighed mixed milk, and it was calculated as kg of milk used for the production of one kg of Lećevački cheese. Predictive cheese yield formulas were used according to the Van Slyke cheese yield equation [37]:

$$Van\ Slyke\ Cheese\ Yield = \frac{[(RF \times \%\ Fat\ in\ milk) + (RC \times Casein\ in\ milk)] \times RS}{(100 - \%\ Moisture\ of\ cheese)}$$

$$RF = \frac{(\%\ fat\ in\ cheese \times cheese\ wt)}{(\%\ fat\ in\ milk \times milk\ wt)}$$

$$RC = \frac{(\%\ casein\ in\ cheese \times cheese\ wt)}{(\%\ casein\ in\ milk \times milk\ wt)}$$

where *RF* is the fat recovered in cheese, *RC* is the casein recovered in cheese, and *RS* represents the other milk solids and added salt recovered in cheese. *RS* values for each cheese were calculated by substituting the *RF* and *RC* values into the following equation:

$$RS = \frac{RF \times \%\ Fat\ in\ milk}{[(RF \times \%\ Fat\ in\ milk) + (RC \times \%\ Casein\ in\ milk) \times FDM]}$$

where *FDM* is fat in dry matter.

2.6. Statistical Analysis

Statistical analysis was done using the statistical program SAS (SAS Institute Inc., 2001, SAS/STAT User's Guide, Version 8.2., Cary, NC, USA). The influence of different ratios of sheep and cow milk on the chemical composition of cheese and cheese yield efficiency was studied by analysis of variance (PROC GLM). To determine the different mean values of the groups where statistical differences were found ($p < 0.05$), a Tukey–Kramer post hoc test was used. The obtained results were expressed as least square mean values ± standard error (LSM ± SE).

3. Results and Discussion

3.1. Mixed Milk Composition

The mixed milk composition used for the production of Lećevački cheese is presented in Table 2. A higher ratio of sheep milk in the mixed milk caused a noticeable, but not significant, increase in the content of fat, protein, lactose, and total solids, while the content of solids-not-fat significantly ($p < 0.05$) increased as a consequence of a higher content of sheep milk in the mixed milk. A similar trend was found for casein content (Table 2, $p < 0.1$). This is in accordance with Bonczar et al. [38], who found that the content of total solids, fat, and protein increased in mixed milk with an equal ratio of sheep and cow milk for the production of Bundz cheese. Similar results were obtained by Vyletelova-Klimešova et al. [39], who found that the addition of sheep milk to cow and goat milk increases the ratio of total solids, fat, protein, and casein in mixed milk. Mixing different ratios of sheep and goat milk for the production of Picante cheese influenced the content of fat and casein in the mixed milk [40]. The authors found that the content of casein was the lowest in the mixed milk with the lowest ratio of sheep milk. A similar effect was noticed when buffalo milk was mixed with cow milk for the production of Mozzarella cheese [21]. Authors found that the ratio of total solids, fat, and protein increased in mixed milk in comparison to cow milk. The pH value of the mixed milk was not changed as a consequence of different ratios of sheep and cow milk, which is in line with Bonczar et al. [38] and Vyletelova-Klimešova et al. [39].

Table 2. Influence of the ratio of sheep and cow milk on the composition of mixed milk used for making Lećevački cheese (least square mean ± standard error).

Component	Ratio of Sheep Milk and Cow Milk (% *v*/*v*)			Level of Significance
	35–39 $n = 5$	40–44 $n = 5$	45–50 $n = 5$	
Fat (g/100 g)	4.43 ± 0.18	4.80 ± 0.20	4.79 ± 0.22	NS
Protein (g/100 g)	4.06 ± 0.09	4.26 ± 0.10	4.37 ± 0.12	NS
Lactose (g/100 g)	4.24 ± 0.05	4.31 ± 0.05	4.30 ± 0.06	NS
Total solids (g/100 g)	13.68 ± 0.25	14.26 ± 0.27	14.36 ± 0.30	NS
SNF (g/100 g)	$9.16^{b} \pm 0.09$	$9.43^{ab} \pm 0.10$	$9.57^{a} \pm 0.11$	$p < 0.05$
Casein (g/100 g)	$3.20^{b} \pm 0.07$	$3.40^{ab} \pm 0.08$	$3.47^{a} \pm 0.08$	$p < 0.1$
pH	6.65 ± 0.37	6.61 ± 0.37	6.64 ± 0.42	NS

NS = Not significant; SNF = solids-not-fat. Means within the same row marked with different letters differ significantly.

3.2. *Lećevački Cheese Composition*

The higher content of sheep milk in the mixed milk significantly ($p < 0.01$) decreased the ratio of the total solids and fat content of Lećevački cheese (Table 3), while the ratio of moisture in total solids-not-fat, salt content, pH value, and lactic acid content were not changed as a consequence of increasing the ratio of sheep milk in the mixed milk. The highest ratio of sheep milk in the mixed milk (45–50% (v/v)) ($p < 0.05$) increased the protein content of Lećevački cheese by almost 1% (Table 3). Freitas et al. [40] found that a higher ratio of sheep milk in mixed milk with goat milk causes a higher retention of moisture in hard cheese. This is a consequence of a noticeably higher casein content in mixed milk with a higher sheep milk content, which was the case in our study as well ($p < 0.1$). The higher capacity of water-binding of the casein matrix system of the cheese is a consequence of the higher casein content in the milk used for the production of this cheese. A similar result was found in the production of Bundz cheese when half of the cow milk was substituted with sheep milk [38]. Gobbetti et al. [7] also found that Fossa (pit) cheese produced from a mix of cow and sheep milk (70:30%) contained less moisture in comparison to Fossa (pit) cheese produced exclusively from cow milk. Authors postulated that lower total solids content in such cheese is a consequence of the lower fat content of the cheese, which was the case in our work as well. Aminifar et al. [41] found that moisture content in Lighvan cheese decreased when at least half of the cow milk was replaced with sheep milk. The same was found by Ponce de Leon-Gonzales et al. [42] in researching the influence of the mixing of sheep milk with cow milk on the composition of reduced fat Muenster-type cheese. They found that cheese produced from mixed milk has a lower moisture and higher protein content. Elgaml et al. [43] found that mixing cow and goat milk in different ratios (from 25% to 50%) changed the moisture content of Halloumi cheese. However, Niro et al. [44] did not find any influence of mixing of sheep milk (18%) with cow milk (82%) on the content of total solids of the cheese produced from mixed milk. The content of moisture in total solids-not-fat did not change significantly as a consequence of different ratios of sheep and cow milk.

Table 3. Influence of the ratio of sheep and cow milk on the composition of Lećevački cheese (least square mean ± standard error).

Component	Ratio of Sheep Milk and Cow Milk (% v/v)			Level of Significance
	35–39 $n = 5$	40–44 $n = 5$	45–50 $n = 5$	
Total solids (g/100 g)	61.02 [a] ± 0.27	60.65 [a] ± 0.30	59.22 [b] ± 0.34	$p < 0.01$
Fat (g/100 g)	31.02 [a] ± 0.42	30.88 [a] ± 0.46	28.34 [b] ± 0.47	$p < 0.01$
Protein (g/100 g)	24.34 [b] ± 0.21	24.31 [b] ± 0.23	25.2 [a] ± 0.26	$p < 0.05$
MNFS (g/100 g)	56.51 ± 0.43	56.93 ± 0.46	56.91 ± 0.50	NS
Salt (g/100 g)	1.41 ± 0.04	1.40 ± 0.05	1.39 ± 0.05	NS
pH	5.09 ± 0.03	5.04 ± 0.03	5.08 ± 0.03	NS
Lactic acid (%)	1.46 ± 0.04	1.40 ± 0.04	1.40 ± 0.04	NS

NS = Not significant; MNFS = moisture in non-fat substance. Means within the same row marked with different letters differ significantly.

The fat content of Lećevački cheese produced from mixed milk with the highest ratio of sheep milk (45–50%) was significantly ($p < 0.01$) lower, which is in accordance with Gobbetti et al. [7]. The authors found that Fossa hard cheese produced from mixed milk (30% sheep and 70% cow milk) had less fat content (32.7%) in comparison to Fossa cheese produced exclusively from cow milk. Fossa cheese produced from pure cow milk had a fat content of 38.8%. This could be a consequence of the fact that the fat globules of cow milk are larger than those of sheep milk. Larger fat globules are better entrapped within the cheese casein matrix in comparison to smaller fat globules. Bonczar et al. [38] determined that cheese produced from mixed sheep and cow milk (50:50%) contained less fat in comparison to cheese produced only from cow milk, but more fat than cheese produced only from sheep milk. Similar to the result of this research, Freitas et al. [40] found that cheese produced from mixed milk with a higher ratio of sheep milk in comparison to goat milk contained less fat. The fat and protein of milk represent the greatest part of the total solids of the cheese produced from that milk, so the higher content of protein in mixed milk could cause the cheese to have a lower fat content as was the case in our research (Table 3), especially taking into consideration that the higher sheep milk ratio in the mixed milk caused a lowering of the total solids in Lećevački cheese. In contrast, Ponce de Leon-Gonzales et al. [42] did not find any differences in the fat content of Muenster cheese when 20% of cow milk was replaced with sheep milk, nor did Niro et al. [44] in Caciocavallo cheese produced with 18% sheep milk and 82% cow milk. Moreover, Freitas and Malcata [14] made cheese from mixed goat and sheep milk in different ratios (20:80% and 40:60%) and did not find any differences in the fat content of these cheeses, while Halloumi cheese produced from mixed goat and cow milk with 25% and 50% of goat milk had a higher fat content [43]. In contrast to our research, Vyletelova-Klimešova et al. [39] found that cheese produced from mixed milk (cow, sheep, and goat milk) contained more fat.

This investigation showed that mixed milk with 40–44% of sheep milk did not influence the protein content of Lećevački cheese in comparison to the ratio of 35–39%, which is in accordance with Niro et al. [44], who found that replacing 18% of cow milk with sheep milk did not change the protein content of Caciocavallo cheese. However, further increasing the ratio of sheep milk in the mixed milk (45–50%) caused a significant ($p < 0.05$) increase in the protein content of Lećevački cheese. Gobbetti et al. [7], as well as Ponce de Leon Gonzales et al. [42], concluded that hard cheese produced with mixed milk with a higher ratio of sheep milk contains a higher level of protein, which is a result of the fact that sheep milk has a higher protein content in comparison to cow milk. The obtained results are in line with those previously reported [38,39,43]. In accordance with our research, Fenelon et al. [45] and Madsen and Ardo [46] reported that a decrease of fat in cheese causes moisture and proteins to increase.

Our investigation showed that different ratios of sheep and cow milk did not significantly influence the salt content of the cheese (Table 3), which is in line with the results of previous reports [6,28,42,47]. In contrast, Aminifar et al. [41] reported that the type of milk used in the production of Linghvan cheese, which ripened in 12% brine for three months, influenced the salt content of the cheese. They reported that cheese produced from sheep milk contained less salt than cheese produced from mixed milk (50:50%), or from pure cow milk. Moreover, Elgaml et al. [43] found that different ratios of goat and cow milk (25:75% and 50:50%) influenced the salt content of the cheese, which they connected with the different moisture content in these cheeses. Cheese with a higher moisture content absorbs more salt by diffusion [48]. In spite of the fact that a significantly ($p < 0.01$) different content of moisture was found among Lećevački cheeses produced with different ratios of sheep and cow milk, these differences did not influence the intensity of the absorption of salt during the brining of Lećevački cheese in this research.

This research showed that different ratios of sheep and cow milk did not have an influence on the pH value of Lećevački cheese (Table 3). Regardless of the type of milk used for cheesemaking, three main factors directly influence the pH value of cheese: the fermentation of lactose, accumulation of amino acids during ripening of the cheese [49], and fat content of the cheese [45,50]. No differences were found in the concentration of lactic acid as an indicator of the fermentation of lactose among the cheeses produced with different ratios of sheep and cow milk (Table 3). Moreover, there were no differences in the concentration of amino acids, expressed as the ratio of trichloroacetic acid-soluble nitrogen to the total nitrogen, of Lećevački cheese made from the different ratios of sheep and cow milk (data are not published). Similar results were obtained by Ponce de Leon-Gonzales et al. [42], Bonczar et al. [38], and Aminifar et al. [41], who also did not find any differences in the pH value of the cheeses produced with different ratios of sheep and cow milk. Moreover, Freitas et al. [40] did not find any differences in pH value as a consequence of mixing sheep and goat milk. However, Niro et al. [44] reported that the addition of sheep milk to cow milk (18:82%) decreases pH value as a consequence of the higher production of lactic acid in the manufacturing of pasta filata cheeses. When the ratio of goat milk is increased in a mixture with cow milk (from 25% to 50%), the pH value decreases [43]. Some authors reported that full-fat cheeses have a higher pH value in comparison to corresponding reduced fat cheese [45]. In spite of significant ($p < 0.01$) differences in the fat content of Lećevački cheese produced from the different ratios of sheep and cow milk, these differences were not sufficient to influence the pH value of the Lećevački cheese in our experiment.

3.3. Cheese Yield Efficiency

A higher ratio of sheep milk in mixed milk caused a noticeable tendency ($p = 0.11$) to increase actual yield (Table 4). There could be two reasons for this trend: a higher SNF ($p < 0.05$) and a higher casein ($p < 0.1$) content of the mixed milk with a higher ratio of sheep milk (Table 2). This was the reason for less mixed milk with a higher ratio of sheep milk being required for the making of one kg of Lećevački cheese. Additionally, Lećevački cheese produced with a higher ratio of sheep milk had a higher moisture content (Table 3). It is well known that cheese that retains more moisture during its production has a higher cheese yield [51]. On average, from 100 kg of mixed milk, 13.54 kg of Lećevački cheese was made. Zedan et al. [21] reported that the cheese yield of fresh Mozzarella produced from mixed buffalo and cow milk is higher (11.66 kg/100 kg of milk) in comparison to the cheese produced from pure cow milk (10.66 kg/100 kg of milk). They explained that the differences are due to higher total solids, which means a higher fat and protein content of buffalo milk in comparison to cow milk. Authors cited El-Zoughby (1998), who reported that cheese yields of Mozzarella produced from different types of milk vary greatly. For buffalo milk it was 11.62 kg/100 kg of milk, for cow milk it was 8.95 kg/100 kg of milk, for goat milk it was 7.93 kg/100 kg of milk, for mixed buffalo and cow milk it was 10.50 kg/100 kg of milk, for mixed buffalo and goat milk it was 10.35 kg/100 kg of milk, for mixed cow and goat milk it was 8.37 kg/100 kg of milk, and for mixed buffalo, cow, and goat milk it was 9.15 kg/100 kg of milk. Similar results were obtained by Economides et al. [13], who found there to be a higher yield of Halloumi cheese produced from mixed sheep and goat milk in comparison to the cheese produced from cow milk. Temizkan et al. [28] determined that actual and adjusted yields were higher for the Kashar cheese produced from sheep milk (13.77 kg/100 kg of milk and 12.31 kg/100 kg of milk, respectively) in comparison to Kashar cheese produced from cow milk (8.04 kg/100 kg of milk and 7.53 kg/100 kg of milk, respectively). Wendorff [52] reported that the cheese yield of cheese produced from sheep milk varies between 16.1 kg for Manchego cheese and 21.9 kg for blue cheese, while Kalit [24] reported that the actual yield of semi-hard Tounj cheese produced from raw cow milk was 11.23 kg/100 kg of milk. Merćep et al. [53] reported that the actual yield of semi-hard cow Trappist cheese varies between 9.25 kg and 9.49 kg/100 kg of milk. Tuta [26] determined that the actual cheese yield of semi-hard Trappist cheese varies between 8.78 and 9.53 kg/100 kg of milk, whereas the yield of PDO Paški cheese, which is produced exclusively from the milk of the Croatian Paška sheep breed, varies between 18.92 and 19.60 kg/100 kg of milk, depending on the type of rennet used [54]. Finally, Faccia et al. [55] found that the yield of the Fior di Latte cheese produced from mixed sheep and goat milk is higher than the corresponding cheese produced from cow milk. Moreover, from Table 4 it can be seen that Van Slyke cheese yield increased as a consequence of increasing the ratio of sheep milk up to 44%, while further increasing the ratio of sheep milk in mixed milk did not cause a further increase of the Van Slyke cheese yield. This could be associated with the trend of the decreasing recovery of fat, which dropped significantly ($p = 0.07$) from 0.76 (35–39% of sheep milk) to 0.66 in mixed milk with 45–50% of sheep milk (Table 4). Casein recovery was stable regardless of the ratio of sheep milk in mixed milk. Milani and Wendorff [56] determined that fat recovery varied from 0.83 to 0.84, while casein recovery varied from 0.94 to 0.96 in the making of hard sheep cheeses, while fat and protein recovery in the production of cow semi-hard Tounj cheese were 0.85 and 0.77, respectively [24]. Our results are noticeably lower than those obtained by other authors, probably due to the fact that the fat globules of cow milk are larger in diameter than those of sheep milk [57], which could be the reason that these fat globules were not efficiently retained in the casein matrix of the mixed milk with a higher sheep milk ratio (due to the different casein composition of cow and sheep milk). The other reason could be the fact that the Lećevački cheese for this research was produced in the regular production from the Mils dairy plant in Split (Croatia) and not in fully controlled experimental conditions.

Table 4. Influence of the ratio of sheep and cow milk on the yield efficiency of Lećevački cheese (least square mean ± standard error).

Component	Ratio of Sheep Milk and Cow Milk (% *v/v*)			Level of Significance
	35–39 $n = 5$	40–44 $n = 5$	45–50 $n = 5$	
Actual yield *	7.48 ± 0.23	7.38 ± 0.26	7.25 ± 0.29	$p = 0.11$
Van Slyke cheese yield	10.63 ± 0.33	10.96 ± 0.35	10.74 ± 0.37	NS
Fat recovery	0.76 ± 0.03	0.74 ± 0.04	0.66 ± 0.05	$p = 0.07$
Casein recovery	0.82 ± 0.04	0.83 ± 0.05	0.82 ± 0.05	NS

NS = Not significant. * kg of milk required for the production of 1 kg of cheese.

4. Conclusions

In spite of the fact that different mixtures of sheep and cow milk could drastically change mixed milk composition, these changes only have a slight influence on hard cheese composition such as in the case of Lećevački cheese. This could allow cheesemakers to choose from a wide range of mixtures of sheep and cow milk depending on the season due to the fact that the quantity of sheep milk produced in Mediterranean areas such as the coast of Croatia is strongly dependent on the season. However, our results showed that it is not reasonable to increase the sheep milk ratio in mixed milk above 44% (*v*/*v*) because it causes higher moisture content in the cheese, as well as lower fat content and fat recovery during the manufacturing of Lećevački cheese.

Author Contributions: Conceptualization and methodology, S.K.; investigation, S.M.; formal analysis, S.M. and M.T.K.; data curation, A.R.; writing—original draft preparation, S.M. and S.K.; writing—review and editing, M.T.K. and A.R.; supervision, S.K., A.R. and M.T.K. All authors have read and agreed to the published version of the manuscript.

Funding: This work was supported by a grant from the Split-Dalmatia County, Agriculture Development Program, Agreement Number 402-01/11-05/31.

Institutional Review Board Statement: Not applicable.

Informed Consent Statement: Not applicable.

Data Availability Statement: Not applicable.

Conflicts of Interest: The authors declare no conflict of interest.

References

1. Tornambe, G.; Di Grigoli, A.; Alicata, M.L.; De Pasquale, C.; Bonnano, A. Comparing quality characteristics of "Caciocavallo Palermitano" cheese from traditional and intensive production systems. In Proceedings of the 15th Meeting of the FAO-CIHEAM Mountain Pastures Network, Les Diablerets, Vaud, Switzerland, 7–9 October 2009; pp. 153–156.
2. Selvaggi, M.; Laudadio, V.; Dario, C.; Tufarelli, V. Investigating the genetic polymorphism of sheep milk proteins: A useful tool for dairy production. *J. Sci. Food Agric.* **2014**, *94*, 3090–3099. [CrossRef] [PubMed]
3. Hayaloglu, A.; Ozer, B. Cheeses of Turkey: 2. Varieties ripened under brine. *Dairy Sci. Technol.* **2008**, *88*, 225–244. [CrossRef]
4. Alichanidis, E.; Polychroniadou, A. Characteristics of major traditional regional cheese varieties of East-Mediterranean countries: A review. *Dairy Sci. Technol.* **2008**, *88*, 495–510. [CrossRef]
5. Nega, A.; Moatsou, G. Proteolysis and related enzymatic activities in ten Greek cheese varieties. *Dairy Sci. Technol.* **2012**, *92*, 57–73. [CrossRef]
6. Pappa, E.C.; Kandarakis, I.; Zefridis, G.K.; Antifantakis, E.M.; Robertson, J.A.; Mills, E.N.C. Teleme white-brined cheese from sheep or goat milk. *Int. Dairy J.* **2012**, *24*, 83–86.

7. Gobbetti, M.; Folkertsma, B.; Fox, P.F.; Corsetti, A.; Smacchi, E.; De Angelis, M.; Rossi, J.; Kilcawley, K.; Cortini, M. Microbiology and biochemistry of Fossa (pit) cheese. *Int. Dairy J.* **1999**, *9*, 763–773. [CrossRef]
8. Gobbetti, M. Extra-hard varieties. In *Cheese: Chemistry, Physics and Microbiology*, 3rd ed.; Fox, P.F., McSweeney, P.L.C., Cogan, T.M., Guinee, T.P., Eds.; Elsevier Academic Press: London, UK, 2004; Volume 2, pp. 51–70.
9. Pirisi, A.; Comunian, R.; Urgeghe, P.P.; Scintu, M.F. Sheep's and goat's dairy products in Italy: Technological, chemical, microbiological, and sensory aspects. *Small Rumin. Res.* **2011**, *101*, 102–112. [CrossRef]
10. Bertolino, M.; Dolci, P.; Giordano, M.; Rolla, L.; Zeppa, G. Evolution of chemico-physical characteristics during manufacture and ripening of Castelmano PDO cheese in wintertime. *Food Chem.* **2011**, *129*, 1001–1011. [CrossRef]
11. Medina, M.; Nuñez, M. Cheeses made from ewe's and goat's milk. In *Cheese: Chemistry, Physics and Microbiology*, 3rd ed.; Fox, P.F., McSweeney, P.L.C., Cogan, T.M., Guinee, T.P., Eds.; Elsevier Academic Press: London, UK, 2004; Volume 2, pp. 279–299.
12. Martinez, S.; Franco, I.; Carballo, J. Spanish goat and sheep milk cheeses. *Small Rumin. Res.* **2011**, *101*, 41–54. [CrossRef]
13. Economides, S.; Georghiades, E.; Mavrogenis, A.P. The effect of different milks on the yield and chemical composition of Halloumi cheese. *Tech. Bull. Agric. Res. Inst.* **1987**, *90*, 2–7.
14. Freitas, A.C.; Malcata, F.X. Influence of milk type, coagulant, salting procedure and ripening time on the final characteristics of Picante cheese. *Int. Dairy J.* **1996**, *6*, 1099–1116. [CrossRef]
15. Freitas, A.C.; Malcata, F.X. Microbiology and biochemistry of cheeses with Appélation d'Origine Proteggée and manufactured in the Iberian peninsula from ovine and caprine milks. *J. Dairy Sci.* **2000**, *83*, 584–602. [CrossRef]
16. Bubić, M. Contribution to the Knowledge of Quality and Production of Lećevački Cheese. Master's Thesis, Faculty of Veterinary, University of Zagreb, Zagreb, Croatia, 1981.
17. Matutinović, S.; Rako, A.; Kalit, S.; Havranek, J. The importance of traditional cheeses with a special focus on Lećevački cheese. *Mljekarstvo* **2007**, *57*, 49–65.
18. Park, Y.W.; Juarez, M.; Ramos, M.; Haenlein, G.F.W. Physico-chemical characteristics of goat and sheep milk. *Small Rumin. Res.* **2007**, *68*, 88–113. [CrossRef]
19. Teter, A.; Kędzierska-Matysek, M.; Barłowska, J.; Król, J.; Brodziak, A. Nutritional value and coagulation properties of milk from local cow breeds, including the selected macro- and micronutrients and trace elements. *Mljekarstvo* **2020**, *70*, 210–220. [CrossRef]
20. Campos, G.; Robles, L.; Alonso, R.; Nuñez, M.; Picon, A. Microbial dynamics during the ripening of a mixed cow and goat milk cheese manufactured using frozen goat milk curd. *J. Dairy Sci.* **2011**, *94*, 4766–4776. [CrossRef] [PubMed]
21. Zedan, I.A.; Abou-Shaloue, Z.; Zaky, S.M. Quality evaluation of Mozzarella cheese from different milk types. *Alex. Sci. Exch. J.* **2014**, *35*, 162–177.
22. Fox, P.F.; Cogan, T.M. Factors that affect the quality of cheese. In *Cheese: Chemistry, Physics and Microbiology*, 3rd ed.; Fox, P.F., McSweeney, P.L.C., Cogan, T.M., Guinee, T.P., Eds.; Elsevier Academic Press: London, UK, 2004; Volume 1, pp. 583–608.
23. Pretto, D.; De Marchi, M.; Penasa, M.; Cassandro, M. Effect of milk composition and coagulation traits on Grana Padano cheese yield under field conditions. *J. Dairy Res.* **2013**, *80*, 1–5. [CrossRef]
24. Kalit, S. Biochemical Changed of Tounj Cheese during Ripening. Ph.D. Thesis, Faculty of Agriculture, University of Zagreb, Zagreb, Croatia, 2003.
25. Katz, G.; Merin, U.; Bezman, D.; Lavie, S.; Lemberskly-Kuzin, L.; Leitner, G. Real-time evaluation of individual cow milk for higher cheese-milk quality with increased cheese yield. *J. Dairy Sci.* **2015**, *99*, 4178–4187. [CrossRef] [PubMed]
26. Tuta, N. Production of Trappist-Style Semi-Hard Cheese. Bachelor's Thesis, Department for Chemistry and Technology, University Department for Professional Studies, University of Split, Split, Croatia, 2011.
27. Kalit, S.; Tudor Kalit, M.; Dolenčić Špehar, I.; Salajpal, K.; Samaržija, D.; Anušić, J.; Rako, A. The Influence of Milk Standardization on Chemical Composition, Fat and Protein Recovery, Yield and Sensory Properties of Croatian PGI Lički Škripavac Cheese. *Foods* **2021**, *10*, 690. [CrossRef] [PubMed]
28. Temizkan, R.; Yasar, K.; Hayaloglu, A.A. Changes during ripening in chemical composition, proteolysis, volatile composition and texture in Kashar cheese made using raw bovine, ovine or caprine milk. *Int. J. Food Sci. Technol.* **2014**, *49*, 2643–2649. [CrossRef]
29. Matutinović, S.; Kalit, S.; Salajpal, K.; Vrdoljak, J. Effects of flock, year and season on the quality of milk from an indigenous breed in the sub-Mediterranean area. *Small Rumin. Res.* **2011**, *100*, 159–163. [CrossRef]
30. International Organization for Standards. *ISO 9622. Milk and Liquid Milk Products—Guidelines for the Application on Mid-Infrared Spectrometry*; International Organization for Standards: Geneva, Switzerland, 2013.
31. International Organization for Standards. *ISO 17997-2. Milk—Determination of Casein-Nitrogen Content—Part 1: Direct Method*; International Organization for Standards: Geneva, Switzerland, 2013.
32. International Organization for Standards. *ISO 3433. Cheese—Determination of Fat Content—Van Gulik Method*; International Organization for Standards: Geneva, Switzerland, 2008.
33. International Organization for Standards. *ISO 8968-2. Milk—Determination of Nitrogen Content—Part 2: Block-digestion method (Macro method)*; International Organization for Standards: Geneva, Switzerland, 2001.
34. International Organization for Standards. *ISO 5534. Cheese and Processed Cheese—Determination of the Total Solids Content (Reference Method)*; International Organization for Standards: Geneva, Switzerland, 2004.
35. Association of Official Analytical Chemists (AOAC). *Acidity of Milk, Titrimetric Method. Official Method*; Association of Official Analytical Chemists: Washington, DC, USA, 2000.

36. Association of Official Analytical Chemists (AOAC). *Chloride (Total) in Cheese Volhard Method. Official Method*; Association of Official Analytical Chemists: Washington, DC, USA, 2006.
37. Wendorff, W.L.; Kalit, S. Processing of sheep milk. In *Handbook of Milk of Non-Bovine Mammals*, 2nd ed.; Park, W.Y., Haenlein, G.F.W., Wendorf, W.L., Eds.; Wiley Blackwell: West Sussex, UK, 2017; pp. 222–260.
38. Bonczar, G.; Regula-Sardat, A.; Pustkowiak, H.; Zebrovska, A. Effect of mixing of ewe's and cow's milk on Bundz cheese properties. *Zywnošć Nauka Technol. Jakošć* **2009**, *16*, 96–106.
39. Vyletelova-Klimešova, M.; Hanuš, O.; Horaček, J.; Vorlova, L.; Nemečkova, I.; Nejeschlebova, L.; Kopecky, J. Characteristic and quality and food safety of regional cheese produced from mixed milk. *Acta Univ. Agric. Silvic* **2014**, *62*, 1171–1182. [CrossRef]
40. Freitas, A.C.; Fresno, J.M.; Prieto, B.; Malcata, F.X.; Carballo, J. Effects of ripening time and combination of ovine and caprine milks on proteolysis of Picante cheese. *Food Chem.* **1997**, *60*, 219–229. [CrossRef]
41. Aminifar, M.; Hamedi, M.; Emamdjomeh, Z.; Mehdinia, A. The effect of ovine and bovine milk on the textural properties of Linghvan cheese during ripening. *Int. J. Dairy Technol.* **2013**, *66*, 45–53. [CrossRef]
42. Ponce de Leon-Gonzalez, L.; Wendorf, W.L.; Ingham, B.H.; Thomas, D.L.; Jeaggi, J.J.; Houck, K.B. Influence of ovine milk in mixture with bovine milk on the quality of reduced fat Muenster-type cheese. *J. Dairy Sci.* **2002**, *85*, 36–42. [CrossRef]
43. Elgaml, N.B.; Moussa, M.A.M.; Saleh, A.E. Comparison of the properties of Halloumi cheese made from goat milk, cow milk and their mixture. *J. Sustain. Agric. Sci.* **2017**, *43*, 77–87. [CrossRef]
44. Niro, S.; Fratianni, A.; Tremonte, P.; Sorrentino, E.; Tipaldi, L.; Panfili, G.; Coppola, R. Innovative Caciocavallo cheeses made from a mixture of cow milk with ewe or goat milk. *J. Dairy Sci.* **2014**, *97*, 1296–1304. [CrossRef]
45. Fenelon, M.A.; Guinee, T.P. Primary proteolysis and textural changes during ripening in Cheddar cheeses manufactured to different fat contents. *Int. Dairy J.* **2000**, *10*, 151–158. [CrossRef]
46. Madsen, J.S.; Ardo, Y. Exploratory study of proteolysis, rheology and sensory properties of Danbo cheese with different fat contents. *Int. Dairy J.* **2001**, *11*, 423–431. [CrossRef]
47. Freitas, A.C.; Sousa, M.I.; Malcata, F.X. Effect of ripening time and the combination of ewe and goat milk on the microflora of Picante cheese. *Ital. J. Anim. Sci.* **1995**, *7*, 361–377.
48. Scott, R. *Cheesemaking Practice*; Robinson, R.K., Wilbey, R.A., Eds.; Kluwer Academic/Plenum Publishers: New York, NY, USA, 1998; ISBN 9781461558194.
49. Fox, P.F. Cheese. An overview. In *Cheese: Hemistry, Physics and Microbiology*, 3rd ed.; Fox, P.F., McSweeney, P.L.C., Cogan, T.M., Guinee, T.P., Eds.; Elsevier Academic Press: London, UK, 2004; Volume 2, pp. 1–18.
50. Van Hekken, D.L.; Park, Y.W.; Tunick, M.H. Effect of reducing fat content on the proteolytic and reological properties of Cheddar-like caprine milk cheese. *Small Rumin. Res.* **2013**, *110*, 46–51. [CrossRef]
51. Vacca, G.M.; Stocco, G.; Dettori, M.L.; Summer, A.; Cipolat-Gotet, C.; Bittante, G.; Pazzola, M. Cheese yield, cheesemaking efficiency, and daily production of 6 breeds of goats. *J. Dairy Sci.* **2018**, *101*, 7817–7832. [CrossRef]
52. Wendorff, W.L. Milk composition and cheese yield. In Proceedings of the 8th Great Lakes Dairy Sheep Symposium, Ithaca, NY, USA, 7–9 November 2002; pp. 104–117.
53. Mercep, A.; Kirin, S.; Zdolec, N.; Cvrtila Fleck, Ž.; Filipović, I.; Njari, B.; Mitak, M.; Kozačinski, L. Quality of trapist cheese from Croatian dairy plant. *Mljekarstvo* **2010**, *60*, 288–298.
54. Oštarić, F.; Antunac, N.; Prpić, Z.; Mikulec, N. Influence of rennet on the quality of Paški cheese. *Mljekarstvo* **2015**, *65*, 101–110.
55. Faccia, M.; Trani, A.; Gambacorta, G.; Loizzo, P.; Cassone, A.; Caponio, F. Production technology and characterization of Fior di latte cheeses made from sheep and goats milks. *J. Dairy Sci.* **2015**, *98*, 402–1410. [CrossRef]
56. Milani, F.X.; Wendorff, W.L. Goat and sheep milk products in the United States. *Small Rumin. Res.* **2011**, *101*, 134–139. [CrossRef]
57. Havranek, J.; Kalit, S.; Antunac, N.; Samaržija, D. *Cheesemaking*; Croatian Dairy Union: Zagreb, Croatia, 2014; ISBN 9789537472085.

fermentation

Article

The Effect of Microbial Transglutaminase on the Viscosity and Protein Network of Kefir Made from Cow, Goat, or Donkey Milk

Lívia Darnay [1,*], Adrienn Tóth [2], Barbara Csehi [2], Anna Szepessy [1], Martin Horváth [2], Klára Pásztor-Huszár [2] and Péter Laczay [1]

[1] Department of Food Hygiene, University of Veterinary Medicine Budapest, István u. 2, H-1078 Budapest, Hungary; szepessy.anna@univet.hu (A.S.); laczay.peter@univet.hu (P.L.)
[2] Department of Livestocks Product and Food Preservation Technology, Institute of Food Science and Technology, Hungarian University of Agriculture and Life Sciences, Ménesi út 43-45, H-1118 Budapest, Hungary; toth.adrienn@uni-mate.hu (A.T.); csehi.barbara@uni-mate.hu (B.C.); horvath.martin@univet.hu (M.H.); Pasztorne.Huszar.Klara@uni-mate.hu (K.P.-H.)
* Correspondence: darnay.livia@univet.hu

Abstract: In this study, we aim to decrease the fermentation time and to produce low-fat set-type kefir with adequate textural properties using microbial transglutaminase without inactivation. In addition, we reveal the effect of microbial transglutaminase, during and after fermentation, on kefir made with cow, goat, or donkey milk, which is a novel approach. Fermentation is followed by continuous pH and viscosity measurements; the final product is characterized by dry matter content, whey separation, protein pattern, and viscosity parameters, as well as gel firmness. The results show that already 0.5 U/g protein dosage of MTGase decreases pH levels independent of milk type, but MTGase does not influence the kinetics of fermentation. Apparent viscosity could be measured from different stages of fermentation depending on milk type (cow milk, 6 h; goat milk, 8 h; and donkey milk, 9 h). The final product characteristics show that the higher the casein ratio of the applied milk, the better the viscosity and gel firmness of the kefir due to the high reaction affinity of MTGase.

Keywords: set-type kefir; microbial transglutaminase; viscosity; fermentation

Citation: Darnay, L.; Tóth, A.; Csehi, B.; Szepessy, A.; Horváth, M.; Pásztor-Huszár, K.; Laczay, P. The Effect of Microbial Transglutaminase on the Viscosity and Protein Network of Kefir Made from Cow, Goat, or Donkey Milk. *Fermentation* **2021**, *7*, 214. https://doi.org/10.3390/fermentation7040214

Academic Editor: Hiroshi Kitagaki

Received: 30 August 2021
Accepted: 29 September 2021
Published: 1 October 2021

1. Introduction

To reach optimal taste and texture properties, kefir fermentation lasts approximately 18–24 h at 25–27 °C, based on the scientific literature [1,2]. Nowadays, this is considered to be a time-consuming effort; more than two shifts of stable and concise human work are essential. Furthermore, scientists have tended to shift product development to low fat alternatives, which agrees with marketing strategies that try to follow a low-calorie diet due to less physical activity. This tendency is also visible in fermented dairy studies that have demonstrated good textural properties even with low-fat cow milk for yoghurt [3–6] and for kefir [1,7,8]. Dairy studies in the field have applied texture modifiers such as xanthan [9] or, in most cases, microbial transglutaminase [1,2,7,8,10,11], which are well-known, in the food industry, for their protein crosslinking action [12]. However, in most cases, studies have inactivated microbial transglutaminase (MTGase, E.C. 2.3.2.13) at 80–85 °C for 1 min to avoid the development of pudding-like textures [1,3,4,7], which results in extra heat treatment costs and time loss due to the time needed to cool milk back to the inoculation temperature (25–43 °C), according to the fermented dairy product. In addition to these, published studies with kefir fermentation exist only for cow milk at low-fat levels [1,7,8].

Our research aimed to analyze the final low-fat kefir product and the kinetics of fermentation, both without inactivation of MTGase. Furthermore, we investigated the effect of MTGase with cow milk, as well as goat and donkey milk. Our final aim was to find the shortest fermentation time which enabled low-fat kefir products with appropriate textural properties.

2. Materials and Methods

2.1. Kefir Production Procedure

Low-fat set-type kefir was fermented from 3 different UHT heat-treated milk types: cow milk from a large-sized dairy plant (Alföldi Tej Ltd., Székesfehérvár, Hungary); goat milk from a middle-sized dairy plant (Girau, Arborea, Italy), and donkey milk (fat 1.02 ± 0.14, protein 3.45 ± 0.01) provided by a middle-sized dairy plant (Virágoskút Ltd., Balmazújváros, Hungary). The 3 applied milk types were analyzed using a Lactoscan MCC WS (Mikotronic Ltd., Tsentar, Bulgaria) to determine fat, protein, lactose, and solid nonfat (SNF) contents.

The Activa YG (enzyme activity 113 U/g) commercial preparation of MTGase was kindly provided by Barentz Hungary Ltd., the Hungarian distributor of Ajinomoto GmbH (Hamburg, Germany). Three different enzyme dosages (0.5, 1, and 1.5 U/g protein) were used for manufacturing the low-fat set-type kefir, in addition to a control kefir (without enzyme addition). Therefore, altogether, 4 different kefir/milk types were made for one experiment that was repeated three times.

The kefir starter culture, mainly consisting of mesophilic lactic acid bacteria (*Lactococcus lactis, Leuconostoc mesenteroides*) and yeasts (*Debaryomyces hansenii, Candida Colliculosa*), was kindly provided by BiaRia Ltd. (Hungary).

The kefir milk was tempered to 26 °C, and inoculated with 0.8 m/v% kefir starter culture and commercial enzyme preparation (Activa YG), simultaneously. The enzyme was not inactivated, consistent with common practice in the Hungarian food industry, due to the high cost of extra heat treatment. Then, the inoculated and partially enzyme-treated milk types were distributed in 250 mL PP plastic containers and incubated at 26 °C. The fermentation time differed according to milk type: cow milk 10 h, goat and donkey milk 11 h (see also Figure 1a–c). The kinetics of fermentation was followed by pH and rheological measurements. Fermentation was already stopped at pH 4.90 ± 0.05 to avoid over acidification during the time needed to cool down the kefir from 26 to 3 °C. The set-type kefir samples were kept in an industrial refrigerator at 3 ± 1 °C overnight for aging and checked for final pH (cow kefir 4.67 ± 0.05, goat kefir 4.60 ± 0.21, and donkey kefir 4.64 ± 0.10), and then used for further analysis. The analyses of different rheological properties during and after fermentation were done from designated, separate containers. All data presented in this study are the results of 5 independent trials with all milk types.

2.2. Monitoring Lactic Acid Fermentation

The lactic acid fermentation was monitored with simultaneous pH and apparent viscosity measurements at fermentation temperature (26 °C) every 2 h, until a pH of 4.8 was reached. The pH was measured every 2 h with a Testo 206 portable pH meter (Testo AG., Titisee-Neustadt, Germany). The rheological measurements were performed with a Rheomat 115 (Contraves, Switzerland) rotational viscometer, in duplicate. From separate containers, 100 g of each kefir sample was weighed and put into the stainless-steel cylindrical container of the viscosimeter for measurement. The apparent viscosity was calculated according to a previous study [13].

2.3. Dry Matter Content Measurements of the Final Kefir Products

The dry matter (DM) content was determined with an MB 160 moisture analyzer (VWR, Debrecen, Hungary) at 105 °C until constant weight (weight loss < 0.1%/min). All samples were analyzed in triplicate.

2.4. *Syneresis-Whey Separation Measurements*

The set-type kefirs (50 mL) were centrifuged (6000 rpm, 10 min, 10 °C) with a Beckman J2–21 centrifuge and the separated whey was measured. The results of whey loss were calculated to milliliters of whey separated from 100 mL kefir.

2.5. *SDS-PAGE Analysis of the Final Kefir Products*

Handcast gels 4–15% (*w*/*v*) (acrylamide/bis-acrylamide, 83 mm × 73 mm × 1.0 mm) were used for the electrophoresis and a vertical system was applied during the measurement (Bio-Rad mini-Protean Tetra System, Bio-Rad, Hercules, CA, USA). The protein sizes in the samples were monitored using SDS-PAGE. The range of molecular standard (Precision Plus Protein All Blue Standards, Bio-Rad, Hercules, CA, USA) was 250–10 kDa. The sample extracts were prepared by dilution with the sample buffer (2 × Laemmli sample buffer and 2-mercaptoethanol, Bio-Rad, Hercules, CA, USA). In the case of the cow, goat, and donkey samples 10-fold dilutions were used. The mixture was heated at 100 °C for 2 min. From the dilutions, 5 µL of protein solutions were loaded into the wells and run for 45 min at 200 V. The gels were stained for 30 min with 0.2% Coomassie brilliant blue (R-250, Bio-Rad). The stained gel images were captured using a Gel Doc XR+ System (Bio-Rad). The identification and the densities of the bands were quantified by using Quantity One and Image Lab 6.1 software programs (Bio-Rad, Hercules, CA, USA).

2.6. *Gel Firmness Measurements of the Final Kefir Products*

The gel structure of the produced set-type kefir was measured with a TA.XTplus (Stable Micro Systems, Great Britain) texture analyzer. The samples were tempered in their original PP containers to 10 °C, and analyzed with a 35 mm diameter cylinder probe, applying 0.5 mm/sec measurement speed. The gel strength was the force recorded at the maximum penetration depth (10 mm). The evaluation of data was performed with the software of the Texture Exponent 32 instrument.

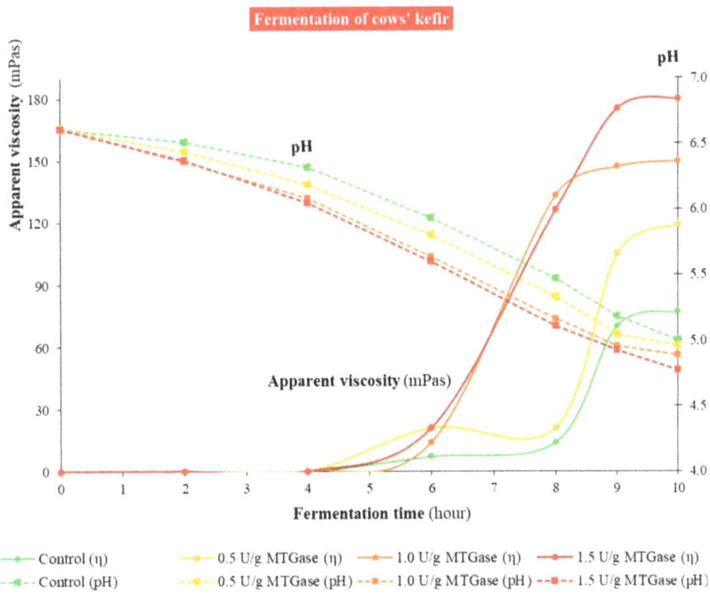

(a)

Figure 1. *Cont.*

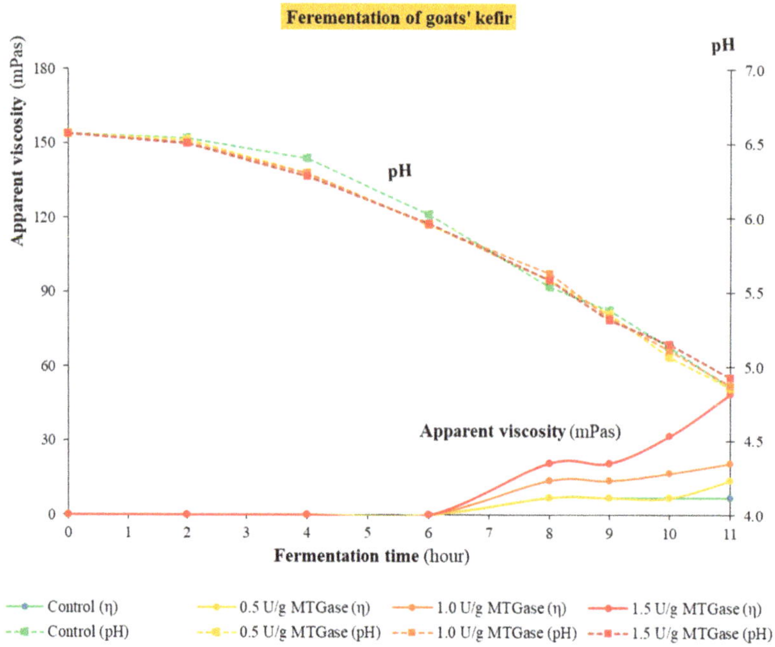

(b)

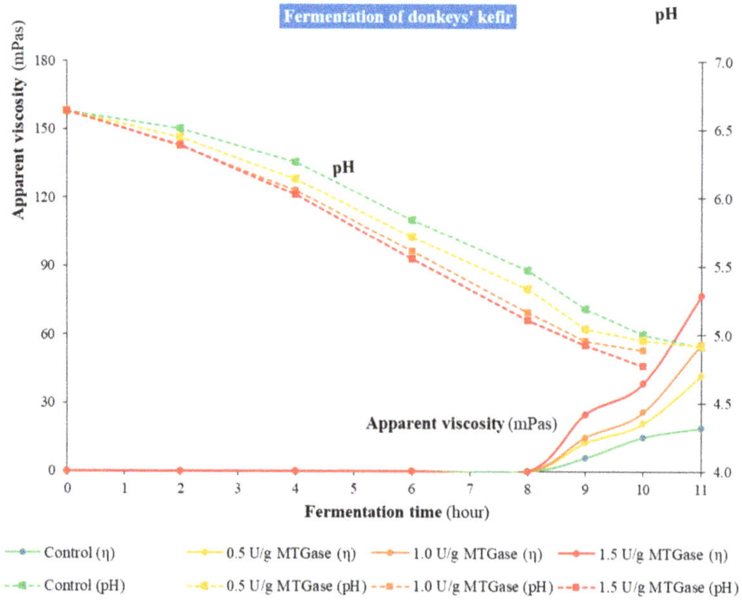

(c)

Figure 1. Kinetics of kefir fermentation: (**a**) cow milk kefir at different MTGase dosages; (**b**) goat milk kefir at different MTGase dosages; (**c**) donkey milk kefir at different MTGase dosages.

2.7. Rheological Measurements of the Final Kefir Products

For the analyses of the rheological properties of the final kefir samples, 15 mL samples from the designated containers were placed in the sample holder of an Anton Paar (France) MCR 92 rheometer. The shear stress values of the samples were investigated using a CC27 system between 10 and 1000 1/s shear rate, similar to [14], at 15 °C. The Herschel–Bulkley model was used to analyze the flow curves (shear rate-shear stress diagrams). This model was used to describe the rheological properties of samples at 15 °C. Most of the R^2 values of the fitted model were higher than 0.99.

2.8. Statistical Analysis

Analysis of variance (ANOVA) along with post hoc tests (Tukey HSD and LSD) were used to compare the variables of pH, dry matter, whey retention, and Herschel–Bulkley model parameters. The effects of enzyme treatment and milk type were analyzed for the mentioned parameters. The data analysis was carried out by SPSS Statistics software 22 (IBM Corp., Armonk, NY, USA).

3. Results

3.1. Analysis and Comparison of the Applied Milk Types for Kefir Production

The chemical composition of the three applied milk types are summarized in Table 1.

Table 1. Chemical composition of the three applied milk types for kefir production.

Chemical Composition	Cow	Goat	Donkey
Fat (%)	1.26 ± 0.17	1.34 ± 0.02	1.02 ± 0.14 *
Protein (%)	3.29 ± 0.08	3.14 ± 0.14	3.45 ± 0.01 *
Lactose (%)	4.90 ± 0.12	4.65 ± 0.18	5.23 ± 0.02 *
Solid nonfat (%)	8.92 ± 0.22	8.45 ± 0.33	9.50 ± 0.03 *
pH	6.64 ± 0.04	6.56 ± 0.02	7.14 ± 0.05 *

Stars indicate significant statistical differences ($p < 0.05$) from the values of the same row.

In this study, we aimed to have standard low-fat milk in all the applied milk types to show the texture enhancement and whey retention ability of MTGase at these reduced fat levels. This is also important due to the growing consumer demand, mainly in the USA, for fermented foods with reduced fat level [15]. Donkey milk was significantly different from cow and goat milk, in all the analyzed parameters. The main difference was that goat milk was albumin milk with a very low-level casein fraction.

3.2. Kinetics of Kefir Fermentation

The fermentation process of kefir was followed by pH and viscosity measurements every 2 h, as shown in Figure 1a–c.

MTGase enzyme lowers the pH values (only until 0.5 U/g protein) of cow and donkey milk, however, goat milk was not affected. The application of MTGase during fermentation caused higher apparent viscosity, than control samples, which was found for all milk types. The statistical analysis revealed that the final pH of the kefir, for all milk types, differed significantly ($p = 0.05$) depending on the enzyme addition. It is recognized that cow milk, as casein milk, has the highest apparent viscosity values, which is reflected by a high substrate affinity to MTGase, as has been reported earlier (Ajinomoto 1998), whereas goat and donkey milk, which are mainly comprised of albumin proteins, have relatively low crosslinking activity with MTGase. The fermentation time needed was the fastest as compared with previous studies that achieved 18 h [1] and 24 h [8] fermentation times, both at 25 °C temperature with cow and soy milk, respectively. This was possible due to the lactic acid bacteria and yeast colonies of the kefir starter culture and the applied dosage (0.8 m/v%).

3.3. Physical-Chemical Quality Characteristics of the Final Kefir Products

The results of the dry matter content for the final kefir products, using three different milk types, are shown on Figure 2.

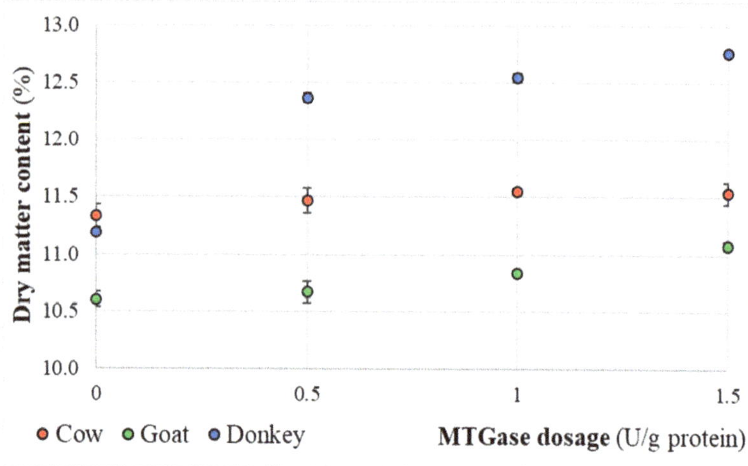

Figure 2. Dry matter content of the set-type kefir products made from different milk types (cow, goat, or donkey milk) and MTGase dosages (0–1.5 U/g protein).

The dry matter contents of the control cow and donkey kefirs were non-significantly different due to their similar protein levels (cow milk protein level 3.29 ± 0.08 and donkey milk protein level 3.45 ± 0.01), but kefir made from goat milk had significantly (p = 0.05) lower dry matter content than that of the other milk types. The use of MTGase resulted in significantly (p = 0.05) higher dry matter content in donkey kefir at all applied enzyme dosages.

The effect of MTGase on whey retention (also known as syneresis) is shown in Figure 3.

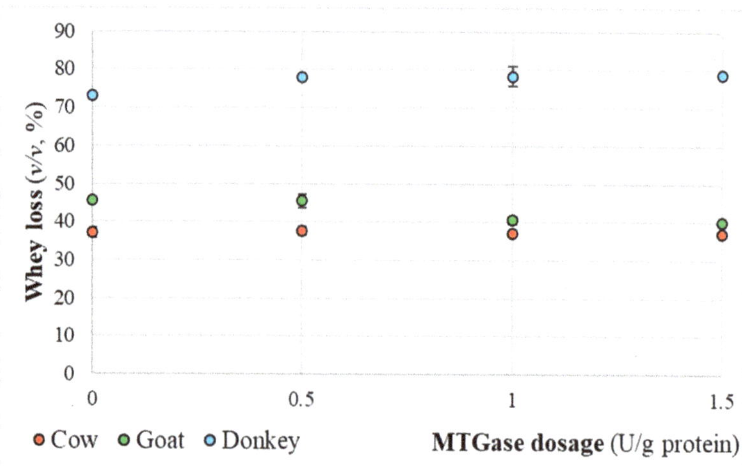

Figure 3. Syneresis of the set-type kefir products made from athe different milk types (cow, goat, or donkey milk) and MTGase dosages (0–1.5 U/g protein).

Microbial transglutaminase is also preferred due to its whey retention ability, which has been proven by a previous kefir study during storage of low-fat cow kefir [7]. However, in our study, syneresis was not significantly ($p = 0.05$) influenced by the use of MTGase in the case of cow kefir. These results were consistent for all enzyme dosages and milk types during six independent trials (two measurements/trial); therefore, it can be concluded that the lower the fat level of the applied milk, the less effective MTGase is, in the case of set-type kefir products. This was also clearly reflected by results of kefir made with donkey milk (fat 1.02 ± 0.014) as compared with cow milk (1.26 ± 0.17) or goat milk (1.34 ± 0.02), both of which resulted in less whey loss (see on Figure 3) due to their higher initial milk fat level.

3.4. Protein Patterns of the Kefirs Based on Gel Electrophoresis

Figure 4 shows the results of the SDS-PAGE analysis of the kefirs made from three different milk types.

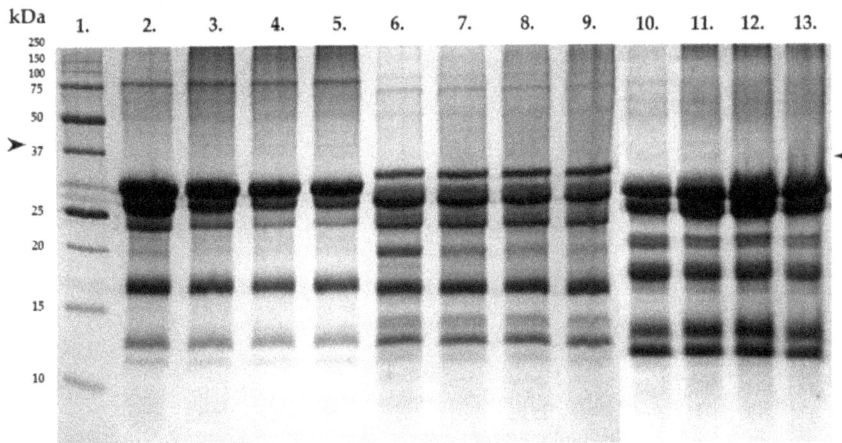

Figure 4. Results of SDS-PAGE of untreated and m-TG treated cow, goat, and donkey kefir: (**1**) Molecular standard; (**2–5**) control (2), cow kefir with 0.5 U/g protein MTGase (3), cow kefir with 1 U/g protein MTGase (4), cow kefir with 1.5 U/g protein MTGase (5); (**6–9**) control (6), goat kefir with 0.5 U/g protein MTGase (7), goat kefir with 1 U/g protein MTGase (8), goat kefir with 1.5 U/g protein MTGase (9); (**10–13**) control (10), donkey kefir with 0.5 U/g protein MTGase (11), donkey kefir with 1 U/g protein MTGase (12), donkey kefir with 1.5 U/g protein MTGase (13).

The proteins contained in the kefir samples prepared with and without m-TG were analyzed by means of SDS-PAGE. The first column is the molecular standard, which makes the identification of proteins possible from 250 to 10 kDa. The whey protein fraction in bovine milk is dominated by β-lactoglobulin (β-lg), which accounts for ~50% of the total whey protein, and α-lactalbumin (α-lac) comprises 20% of the total whey protein [16]. In parallel, we found that proteins were detectable during molecular weight separation at the molecular weights of typical milk proteins. Caseins (αS1-CN, β-CN, and κ-CN) range from 23 to 29 kDa, β-lactoglobulin (β-LG) at ~16 kDa, and α-lactalbumin (α-LA) variants from 11 to 12 kDa. In the case of cow, goat, and donkey kefir samples treated with transglutaminase, we were able to detect clearly visible transglutaminase in some of the examined samples. Bands of the enzyme were detectable at ~38 kDa in the samples. However, it can be seen that the m-TG bands with the highest intensity were observed in the case of cow and donkey kefirs in the 0.5, 1, and 1.5 U/g protein samples.

3.5. Gel Firmness of the Final Kefir Products

Gel firmness or gel strength represents the hardness of kefir curd, and therefore the quality of the set-type fermented dairy products. The results of the set-type kefir gels are presented on Figure 5.

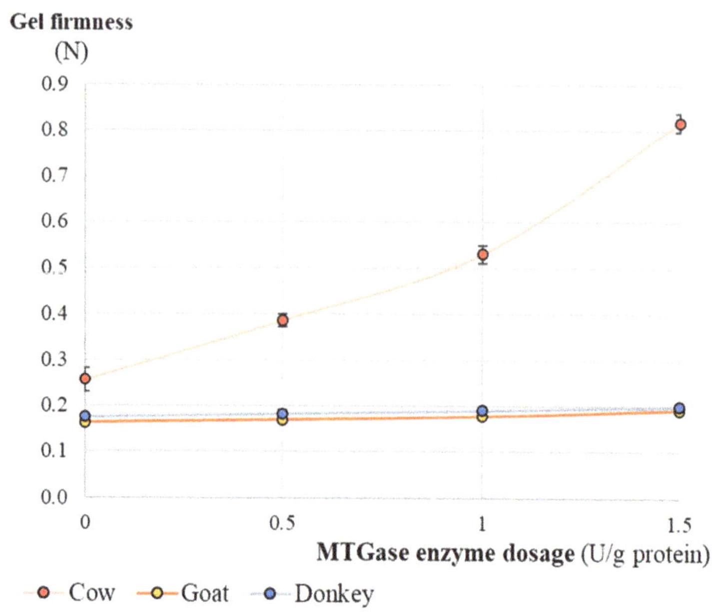

Figure 5. Effect of MTGase on the gel firmness of kefir made from cow, goat, or donkey milk.

According to our study, application of MTGase increased the gel firmness but only in the case of low-fat kefir made from cow milk. This is reflected by the significantly higher apparent viscosity values described and discussed earlier. The obtained results are the averages of six independent trials (six measurement/trial). Probably, the difference in gel firmness is mostly due to the high casein/whey protein ratio of cow milk (casein protein 86% and whey protein 14%), which are highly reactive with MTGase. However, goat milk has a casein/whey protein ratio (81:19) that is similar to that of cow milk [17], but it has a lower concentration of αs1-casein, which is the main substrate of MTGase. Donkey milk is albumin milk with very low casein fraction (casein protein 47% and whey protein 53%) [18] and a subsequent amount of α-lactalbumin that has very low reaction ability with MTGase [19], which is also reflected in the experienced gel firmness values.

3.6. Rheological Characterization of the Final Kefir Products

3.6.1. Rheological Flow Curves

Figure 6 shows the flow curves of the kefir samples made from different milks with different MTGase concentrations. The samples from cow milk with 1 and 1.5 U/g protein MTGase concentrations have a very specific flow behavior in the initial shear rates (between 10 and 40 1/s).

This may be explained by the breaking of kefir curd caused by spinning of the measuring head. A similar flow behavior was reported by [20] in the case of HHP-treated liquid egg white. In the case of kefir made from goat milk, an almost Newtonian fluid behavior is represented in Figure 6b. The higher the MTGase concentration used, the higher the shear stress value measured. The shear stress of the kefir samples made from goat milk are MTGase concentration dependent. The higher the MTGase concentration used, the higher the shear stress value measured. The kefir from goat milk had the lowest shear stress value which showed that its curd was the thinnest as compared with all the investigated samples. In contrast, the samples made from donkey milk had pseudoplastic (shear thickener) flow behavior, which is shown in Figure 6c). The lowest shear stress value was measured for the control sample and the highest shear stress value was measured in the case of 1.5 U/g protein MTGase concentrations. A study by [10] reported similar results in the case of kefirs made from cow milk with different MTGase concentrations. It was found that a higher MTGase concentration increased the viscosity of samples. In our results, this was demonstrated not only in the case of cow milk samples but also in the case of goat and donkey milk samples.

3.6.2. Herschel–Bulkley Model

Table 2 summarizes the calculated τ_0 values of the kefir samples using the Herschel–Bulkley model. Cow milk samples have an increasing tendency with increasing MTGase concentration.

The increasing tendency may be clarified by the higher initial shear stress values measured (when the spinning of the measuring head was breaking the curd). The samples from donkey milk have a similar increasing tendency, although statistically significant only in the case of the 1.5 U/g protein concentration. The τ_0 values of kefir samples made from goat milk show a decreasing tendency with higher MTGase concentration, which is statistically significant, even in the case of 0.5 U/g protein MTGase concentration.

Table 3 represents the consistency values of samples from the calculated Herschel–Bulkley models. The samples made from all milk types show an increasing tendency in calculated consistency, i.e., the higher the enzyme concentration used, the higher the C values calculated.

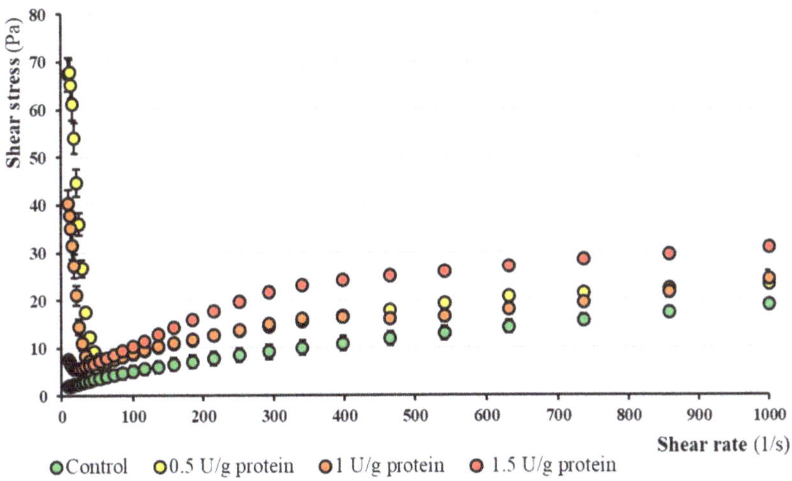

(a)

Figure 6. *Cont.*

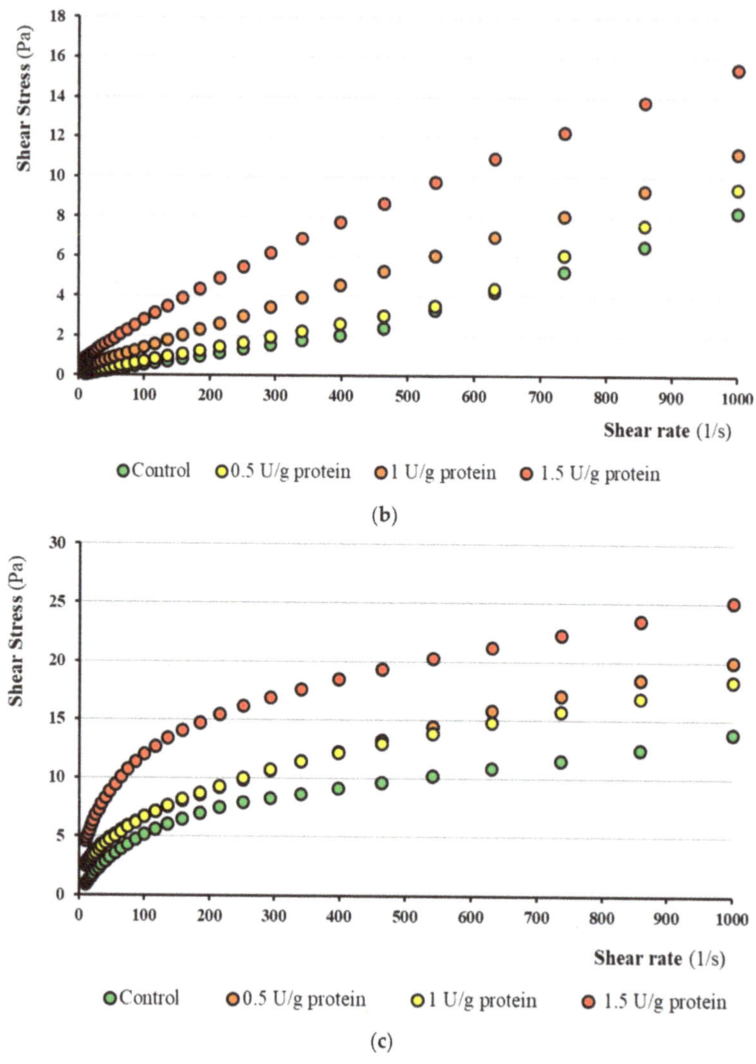

Figure 6. (**a**) Flow curves of the kefir samples made from cow milk; (**b**) flow curves of the kefir samples made from goat milk; (**c**) flow curves of kefir samples made from donkey milk.

Table 2. Calculated τ_0 values of the kefir samples using the Herschel–Bulkley model (significant difference to control ($p = 0.05$), [A] Tukey HSD and [B] LSD).

τ_0 (Pa)	Cow	Goat	Donkey
Control	0.94 ± 0.04	1.44 ± 0.04	0.00 ± 0.00
0.5 U/g protein	0.34 ± 0.01 [AB]	1.42 ± 0.04 [AB]	0.00 ± 0.00
1 U/g protein	13.82 ± 0.52 [AB]	0.40 ± 0.01 [AB]	0.01 ± 0.00
1.5 U/g protein	19.66 ± 1.22 [AB]	0.30 ± 0.01 [AB]	1.36 ± 0.03 [AB]

Table 3. Calculated **C** (consistency) values of the kefir samples using the Herschel–Bulkley model (significant difference to control (p = 0.05), ^{A} Tukey HSD and ^{B} LSD).

C (Pas)	Cow	Goat	Donkey
Control	0.22 ± 0.03	0.00 ± 0.00	0.06 ± 0.00
0.5 U/g protein	1.38 ± 0.01 ^{AB}	0.00 ± 0.00	0.05 ± 0.00 ^{AB}
1 U/g protein	2.69 ± 0.06 ^{AB}	0.01 ± 0.00 ^{B}	0.89 ± 0.02 ^{AB}
1.5 U/g protein	4.27 ± 0.07 ^{AB}	0.06 ± 0.00 ^{AB}	2.31 ± 0.00 ^{AB}

An increasing tendency is visible in shear stress values as well (Figure 6a–c). A slightly increasing tendency is found in the case of samples made from goat milk, but only 1 and 1.5 U/g protein MTGase concentrations were found to be statistically significant. Although the consistency value of donkey kefir was slightly decrease by using 0.5 U/g protein MTGase concentration, 1 and 1.5 U/g protein concentrations increased the consistency as compared with the control sample. The most consistent kefir sample was made from cow milk applying an enzyme concentration of 1.5 U/g protein MTGase, which was well represented by the shear stress values (Figure 6a). Similar to our results, a study by [9] found that the addition of MTGase increased the consistency of kefir drinks made from cow milk. In addition, Persian gum has been shown to improve the consistency and viscosity of kefir samples [2]; therefore, the consistency of kefirs made from goat and donkey milk with MTGase may also be improved by the addition of Persian gum.

4. Discussion

The kinetics of pH decreasing during fermentation was mainly influenced by the milk type applied for the given kefir production. This reflects the importance of selecting an adequate kefir starter culture according to the milk type used for kefir production. In the cases of kefir made from goat or donkey milk, the fermentation lasted 11 h, which was 1 h longer than that from cow milk. The applied kefir culture and dosage resulted in 8–14 h less fermentation time in the case of cow kefir as compared with previous studies, which also highlights the possibility of producing low-fat set-type kefir with less manufacturing time and cost. Gel electrophoresis showed detectable difference in the protein patterns of cow and goat milk treated with MTGase. Whey retention was improved by MTGase according to the fat levels of the applied milk types, which was the most obvious in donkey kefir. The rheological measurements revealed the importance of the casein ratio in the milk used for kefir fermentation, since cow milk with a high casein content could be crosslinked with MTGase, and therefore reached a high viscosity level during the fermentation as well as in the final product. Our study revealed that MTGase is only effective if low-fat set-type kefir is made from cows' milk. In the case of goat and donkey milk, other texture modifiers are also highly recommended at low-fat level.

Author Contributions: Conceptualization, L.D.; methodology, L.D., A.T., and B.C.; formal analysis, L.D., A.T., B.C., A.S., and M.H.; investigation, L.D., A.T., B.C., A.S. and M.H.; resources, K.P.-H., and P.L.; writing—original draft preparation and review and editing, L.D., A.T., and B.C. All authors have read and agreed to the published version of the manuscript.

Funding: This research received no external funding.

Institutional Review Board Statement: Not applicable.

Informed Consent Statement: Not applicable.

Acknowledgments: We would like to express our profound gratitude to Barentz Hungary Ltd. for providing us with the commercial microbial transglutaminase enzyme preparation used in this study. We wish to thank BiaRia Hungary for providing us with the applied kefir starter culture.

Conflicts of Interest: The authors declare no conflict of interest.

References

1. Temiz, H.; Dağyıldız, K. Effects of Microbial Transglutaminase on Physicochemical, Microbial and Sensorial Properties of Kefir Produced by Using Mixture Cow's and Soymilk. *Korean J. Food Sci. Anim. Resour.* **2017**, *37*, 606–616. [CrossRef] [PubMed]
2. Beirami-Serizkani, F.; Hojjati, M.; Jooyandeh, H. The effect of microbial transglutaminase enzyme and Persian gum on the characteristics of traditional kefir drink. *Int. Dairy J.* **2021**, *112*, 104843. [CrossRef]
3. Bönisch, M.P.; Huss, M.; Weitl, K.; Kulozik, U. Transglutaminase cross-linking of milk proteins and impact on yoghurt gel properties. *Int. Dairy J.* **2007**, *17*, 1360–1371. [CrossRef]
4. Guyot, C.; Kulozik, U. Effect of transglutaminase-treated milk powders on the properties of skim milk yoghurt. *Int. Dairy J.* **2011**, *21*, 628–635. [CrossRef]
5. Romeih, E.A.; Abdel-Hamid, M.; Awad, A.A. The addition of buttermilk powder and transglutaminase improves textural and organoleptic properties of fat-free buffalo yoghurt. *Dairy Sci. Technol.* **2014**, *94*, 297–309. [CrossRef]
6. Pakseresht, S.; Mazaheri Tehrani, M.; Razavi, S.M.A. Optimization of low-fat set-type yoghurt: Effect of altered whey protein to casein ratio, fat content and microbial transglutaminase on rheological and sensorial properties. *J. Food Sci. Technol.* **2017**, *54*, 2351–2360. [CrossRef] [PubMed]
7. Wróblewska, B.; Kołakowski, P.; Pawlikowska, K.; Troszyńska, A.; Kaliszewska, A. Influence of the addition of transglutaminase on the immunoreactivit of milk proteins and sensory quality of kefir. *Food Hydrocoll.* **2009**, *23*, 2434–2445. [CrossRef]
8. Beirami-Serizkani, F.; Hojjati, M.; Jooyandeh, H. Influence of the addition of transglutaminase on the immunoreactivity of milk proteins and sensory quality of kefir. In Proceedings of the 4th International Student Bioltechnology Congress, Tehran, Iran, 25–27 February 2019; Volume 23, pp. 238–239.
9. Sabooni, P.; Pourahmad, R.; Adeli, H.R.M. Improvement of viability of probiotic bacteria, organoleptic qualities and physical characteristics in kefir using transglutaminase and xanthan. *Acta Sci. Pol. Technol. Aliment.* **2018**, *17*, 141–148. [CrossRef]
10. Setyawardani, T.; Sumarmono, J.; Rahardjo, A.H.D.; Widayaka, K.; Santosa, R.S. Texture Profile and Sensory Characteristics of Kefir with Colostrum Addition. *IOP Conf. Ser. Earth Environ. Sci.* **2019**, *372*, 012036. [CrossRef]
11. Baranowska, M.; Staniewski, B.; Bohdziewicz, K. Effect of transglutaminase on selected properties of kefir. *Milchwissenschaft* **2012**, *67*, 424–427.
12. Kieliszek, M.; Misiewicz, A. Microbial transglutaminase and its application in the food industry. *A Review. Folia Microbiol.* **2014**, *59*, 241–250. [CrossRef] [PubMed]
13. Darnay, L.; Len, A.; Koncz, Á.; Friedrich, L. Small angle neutron scattering study of nanostructural changes in microbial transglutaminase-treated low-fat yogurt during fermentation. *Food Sci. Biotechnol.* **2015**, *24*, 2125–2128. [CrossRef]
14. Hidas, K.I.; Németh, C.; Le Nguyen, P.F.; Visy, A.; Tóth, A.; Barkó, A.; Friedrich, L.; Nagy, A.; Nyulas-Zeke, I.C. Effect of cryogenic freezing on the rheological and calorimetric properties of pasteurized liquid egg yolk. *Czech J. Food Sci.* **2021**, *2021. 39*, 181–188. [CrossRef]
15. Astrup, A.; Rica, N.; Geiker, W.; Magkos, F. Effects of Full-Fat and Fermented Dairy Products on Cardiometabolic Disease: Food Is More Than the Sum of Its Parts. *Adv. Nutr.* **2019**, *10*, 924S–930S. [CrossRef] [PubMed]
16. Crowley, S.V.; Dowling, A.P.; Caldeo, V.; Kelly, A.L.; O'Mahony, J.A. Impact of α-lactalbumin:β-lactoglobulin ratio on the heat stability of model infant milk formula protein systems. *Food Chem.* **2016**, *194*, 184–190. [CrossRef] [PubMed]
17. Mati, A.; Senoussi-Ghezali, C.; Si Ahmed Zennia, S.; Almi-Sebbane, D.; El-Hatmi, H.; Girardet, J.-M. Dromedary camel milk proteins, a source of peptides having biological activities–A review. *Int. Dairy J.* **2017**, *73*, 25–37. [CrossRef]
18. Vincenzetti, S.; Pucciarelli, S.; Polzonetti, V.; Polidori, P. Role of proteins and of some bioactive peptides on the nutritional quality of donkey milk and their impact on human health. *Beverages* **2017**, *3*, 34. [CrossRef]
19. AJINOMOTO: ACTIVA General Information–Transglutaminase Basics. 1998. Available online: http://docplayer.net/20711824-Activa-general-information.html (accessed on 22 August 2021).
20. Németh, C.; Tóth, A.; Hidas, K.I.; Surányi, J.; Friedrich, L. HHP treatment of liquid egg products. *J. Phys. Conf. Ser.* **2020**, *1609*, 012012. [CrossRef]

fermentation

MDPI

Article

Probiotic and Antioxidant Properties of Lactic Acid Bacteria Isolated from Indigenous Fermented Tea Leaves (Miang) of North Thailand and Promising Application in Synbiotic Formulation

Kridsada Unban [1], Wirunya Chaichana [2], Sasitorn Baipong [2], Aliyu Dantani Abdullahi [3], Apinun Kanpiengjai [4], Kalidas Shetty [5] and Chartchai Khanongnuch [1,6,7,*]

1 Division of Biotechnology, School of Agro-Industry, Faculty of Agro-Industry, Chiang Mai University, Chiang Mai 50100, Thailand; kridsada_u@cmu.ac.th
2 Division of Food Science and Technology, School of Agro-Industry, Faculty of Agro-Industry, Chiang Mai University, Chiang Mai 50100, Thailand; wirunya_ch@hotmail.com (W.C.); sasitorn.bai@cmu.ac.th (S.B.)
3 Interdisciplinary Program in Biotechnology, The Graduate School, Chiang Mai University, Chiang Mai 50200, Thailand; abdullahialiyud.ada@gmail.com
4 Division of Biochemistry and Biochemical Technology, Department of Chemistry, Faculty of Science, Chiang Mai University, Chiang Mai 50200, Thailand; apinun.k@cmu.ac.th
5 Department of Plant Sciences, Global Institute of Food Security and International Agriculture (GIFSIA), North Dakota State University, Fargo, ND 58108, USA; kalidas.shetty@ndsu.edu
6 Research Center for Multidisciplinary Approaches to Miang, Chiang Mai University, Chiang Mai 50200, Thailand
7 Science and Technology Research Institute, Chiang Mai University, Chiang Mai 50200, Thailand
* Correspondence: chartchai.k@cmu.ac.th or ck_biot@yahoo.com; Tel.: +66-53-948261

Citation: Unban, K.; Chaichana, W.; Baipong, S.; Abdullahi, A.D.; Kanpiengjai, A.; Shetty, K.; Khanongnuch, C. Probiotic and Antioxidant Properties of Lactic Acid Bacteria Isolated from Indigenous Fermented Tea Leaves (Miang) of North Thailand and Promising Application in Synbiotic Formulation. *Fermentation* **2021**, *7*, 195. https://doi.org/10.3390/fermentation7030195

Academic Editor: Hiroshi Kitagaki

Received: 26 August 2021
Accepted: 12 September 2021
Published: 16 September 2021

Abstract: Miang, a traditional fermented tea from Northern Thailand, potentially hosts beneficial probiotic bacteria. A total of 133 isolates of lactic acid bacteria (LAB) isolated from Miang were evaluated for probiotic potential. Among them, 5 strains showed high tolerance to bile and acidic conditions and were selected for further evaluation. All selected strains showed inhibitory activity against human pathogens, including *Bacillus cereus*, *Staphylococcus aureus*, and *Salmonella* ser. Typhimurium. Nucleotide sequences analysis of the 16S rRNA gene revealed that 3 isolates were identified as *Lactobacillus pentosus;* the remaining were *L. plantarum* and *Pediococcus pentosaceus*, respectively. All 5 strains showed a high survival rate of more than 90% when exposed to simulated gastrointestinal conditions and were also susceptible to antibiotics such as erythromycin, tetracycline, and gentamycin, and resistant to vancomycin, streptomycin, and polymycin. In addition, the selected isolates exhibited different degrees of cell surface hydrophobicity (58.3–92.9%) and auto-aggregation (38.9–46.0%). The antioxidant activity reflected in DPPH scavenging activities of viable cells and their cell-free culture supernatants (CFCS) were also found in selected LAB isolates. Moreover, selected LAB isolates showed ability to grow on commercial prebiotics (GOS, FOS or XOS). The preliminary study of spray-drying using cyclodextrin as thermoprotectant suggested that all strains can be designed as a powdered formulation. *L. pentosus* A14-6 was the best strain, with high tolerance against simulated gastrointestinal conditions, high cell surface hydrophobicity, effective response to tested commercial oligosaccharides, especially XOS, and the highest cell antioxidant properties. *L. pentosus* A14-6 was therefore targeted for further applications in food and synbiotic applications.

Keywords: probiotic potential; fermented tea; Miang; synbiotic; antioxidant activity

1. Introduction

Probiotics are live microorganisms which exert beneficial effects on host health when consumed in adequate amounts [1]. Some lactic acid bacteria (LAB) have been widely used as probiotics in humans and animals, and the strains most used as probiotics belong to either the genus *Bifidobacterium* or *Lactobacillus* [2,3]. The approval process for probiotics requires certain imperative characteristics such as the resistance to bile and low pH, antibiotic susceptibilities, and antimicrobial activity [2]. Additionally, other properties are also beneficial for probiotic cultures, such as desirable technological, sensorial, and safety features [4]. Moreover, the probiotic culture should be well-adapted to fermented dairy product environments (i.e., presence of curing salts, acidity, and temperature) in order to compete with the endogenous microbiota and grow to levels that enable the display of health-promoting effects [5]. Although dairy products are the most commonly used food vehicles for probiotic delivery, probiotics can also be included in different fermented or unfermented foods [6]. However, the classical obstacle of living probiotic applications in various food vehicles is the survival of probiotic microbes. The vehicle food products require specific temperatures to preserve and sustain the survival rate of probiotics [7]. Besides the direct application of living probiotic LAB using food vehicles, the applications of LAB in the form of either nutraceuticals or food supplements are also gaining interest [8]. The commercial product formulation of probiotics combined with specific growth-promoting carbon sources called prebiotics is also well accepted among consumers and physicians [9,10]. Some oligosaccharides are commercially produced for use as prebiotics, including xylooligosaccharides (XOS), fructooligosaccharide (FOS), and galactooligosaccharides (GOS) [11].

Traditional fermented products constitute an alternative and readily available delivery matrix for LAB starter cultures with attractive functional characteristics particularly with additive probiotic properties [12]. Miang is a traditional fermented food product made of tea leaves (*Camellia sinensis* var. *assamica*) which is commonly produced and consumed in northern Thailand and neighboring countries [13]. The manufacturing process of Miang includes many steps following the inherited protocol depending on local communities, and the most important step of Miang production process is the natural fermentation for a few weeks or up to one year without the use of any preservatives [14,15]. LAB is a key group of microorganisms having an important role in Miang fermentation. Various strains of LAB, including *Lactobacillus* sp., *Pediococcus* sp. and an *Enterococcus* sp., have been isolated from Miang [16–20]. Recent studies exploring the microbial community during Miang fermentation by non-filamentous growth-based fermentation (NFP process) [21], and filamentous growth-based fermentation (FFP process) [22] have confirmed the important role of LAB and their diversity in Miang samples. This indicates the relevance of Miang as potential source of probiotic bacteria.

Therefore, the objectives of this study were to screen for beneficial probiotics from LAB derived from Miang tea fermentation process and to investigate their characteristics and capabilities for use as human probiotics. This study also provides the rationale for targeting specific LAB strains and aligning them to commercial prebiotics to advance commercial synbiotic products as functional foods or nutraceuticals.

2. Materials and Methods

2.1. Bacterial Strains and Culture Conditions

The LAB strains isolated from Miang samples were collected from different locations in Upper-Northern Thailand including Chiang Mai, Chiang Rai, Nan, and Phrae, as described in a previous study [16]. A total of 133 LAB were maintained in de Man, Rogosa, and Sharpe (MRS) broth (HiMedia, Mumbai, India) containing 30% (*v*/*v*) glycerol and stored at −80 °C. Four pathogenic bacteria were used as antimicrobial activity indicators including *Bacillus cereus* TISTR 747, *Salmonella* ser. Typhimurium TISTR 1472, *Staphylococcus aureus* TISTR 746, and Lactobacillus acidophilus TISTR 2365 was used as the reference probiotic strains. To prepare seed inoculum, 1 mL aliquot of LAB was added to 10 mL MRS broth vials and statically incubated at 37 °C for 12 h. The LAB isolates were spread on MRS agar supplemented with 125 ppm bromocresol purple. Plates were incubated for 48 h at 37 °C.

A single colony of LAB was inoculated into 5 mL of the MRS broth and further incubated at 37 °C under static conditions for 18 h.

2.2. Screening and Selection of Acid and Bile Salt Tolerant LAB as the Potent Probiotic

All LAB isolates were tested for their tolerance to acidic conditions and bile salts. Tolerance to acidic conditions was determined using the method of Argyri, et al. [23], with some modification. Briefly, an overnight culture of LAB strains in MRS broth at 37 °C was harvested by centrifugation at 8000× *g* for 10 min at 4 °C, then washed twice with phosphate-buffered saline (PBS), pH 7.2. The washed cell pellets were resuspended in PBS to approximately 10^8 CFU/mL. A total 0.1 mL of LAB cell suspension was transferred into 10 mL of PBS pH 2.0 adjusted by 1.0 M hydrochloric acid, and cells suspended with PBS (pH 7.2) were used as control. All mixtures were incubated at 37 °C for 3 h. A viable cell count was determined by plating on MRS agar and incubated at 37 °C for 12 h. The viable cell count was expressed as the log value of colony-forming units per mL (logCFU/mL).

The bile salts tolerance of LAB isolates was determined according to the method of García-Hernández, et al. [24]. Briefly, 0.1 mL of LAB cell suspension was inoculated into 10 mL of PBS supplemented with 0.3% (*w*/*v*) bile salts (Oxgall, Merck, Germany) and PBS without bile salts served as the control; all cultures were incubated at 37 °C for 3 h. The viable cell count was determined, and the survival rate was calculated as follows:

$$\text{Survival (\%)} = [\text{final (logCFU/mL)}/\text{control (logCFU/mL)}] \times 100 \qquad (1)$$

2.3. Identification and Characterization of Lactic Acid Bacteria

The selected LAB isolates showing the acid and bile salt tolerance properties were identified based on morphological characteristics and 16S rDNA sequence analysis. The genomic DNA was extracted using a Wizard Genomic DNA purification kit (Promega Corp., Madison, WI, USA) following the manufacturer's protocol. Each genomic DNA obtained from the pure culture was used as a template with the primers 27F (5′-AGAGTTTGATCCTGGCTCAG-3′) and 1525R (5′-AAGGAGGTGWTCCARCC-3′) for 16S rRNA gene amplification. The PCR products were purified by GF-1 PCR clean-up gel extraction kits (Vivantis, Malaysia), visualized by electrophoresis on 1.0% (*w*/*v*) agarose gels, and were sent for sequencing service at a sequencing service provider (1st BASE Laboratory Company, Singapore). The sequenced 16S rRNA gene was employed to search the closest sequences using basic local alignment search tool (BLAST) available at the National Center for Biotechnology Information (NCBI) GenBank databases (http://www.ncbi.nlm.nih.gov (accessed on 28 August 2021)). A multiple sequence alignment was performed, and the phylogenetic tree was constructed by the neighbor-joining method using MEGA (Molecular Evolution Genetic Analysis) software, version 4.0 [25]. In order to identify the LAB isolates of the *L. plantarum* group, the isolates were subjected to *recA* gene analysis using species-specific PCR. A multiplex PCR assay was performed with the *recA* gene-based primers paraF, pentF, planF, and pREV, the annealing temperature was 56 °C [26]. The expected sizes of the amplicons were 318 bp for *L. plantarum*, 218 bp for *L. pentosus*, and 107 bp for *L. paraplantarum*. The identified sequences were submitted to NCBI GenBank with the following accession number: MW564014-MW564018.

2.4. Antimicrobial Activity against Pathogens

The selected LAB isolates were investigated for their antagonistic activity in cell-free culture supernatant (CFCS) against a variety of gastrointestinal pathogenic bacteria, including *B. cereus*, *S.* Typhimurium, and *S. aureus* by the agar well diffusion method. Briefly, an overnight culture of pathogenic bacteria (approximately 10^6–10^8 CFU/mL) was gently swabbed on the surface of NA plate. Sterile filter paper discs (8.0 mm diameter) containing 20 μL of the unneutralized CFCS and neutralized CFCS (neutralized to pH 7 by addition of 5 N NaOH) obtained from the MRS culture broth of the LAB isolates were placed on the surface of the swabbed agar plates. The presence of growth inhibition was observed from the appearance of clear zone around the well after plates were incubated at 37 °C for 24 h.

2.5. Resistance to Simulated Gastrointestinal Conditions

Gastrointestinal tolerance was determined as described by Sriphannam, et al. [27], with some modification. The selected LAB isolates and reference probiotic strains (*L. acidophilus*) were inoculated in MRS broth, incubated at 37 °C for 18 h, and harvested by centrifugation at 8000× *g* with 4 °C for 10 min. After washing twice with sterile PBS, the cell pellets were resuspended in electrolyte solution (6.2 g/L NaCl, 2.2 g/L KCl, 0.22 g/L $CaCl_2$, and 1.2 g/L $NaHCO_3$) for an initial viable cell count of 10^8 CFU/mL. One milliliter aliquot was aseptically removed, serially diluted in 0.85% (*w*/*v*) NaCl, and spread on MRS agar to determine the number for the surviving bacterial count (CFU/mL) at time 0. To simulate the dilution and possible hydrolysis reaction of bacteria in the human oral cavity, 5 mL of cell suspension in electrolyte solution was mixed with the same volume of sterile electrolyte solution supplemented with lysozyme (Wako Pure Chemical Industries Ltd., Japan) to obtain a final concentration of 100 ppm, and the sample was then incubated for 5 min at 37 °C. The mixture solution was subsequently diluted 3:5 with an artificial gastric fluid consisting of 0.3% (*w*/*v*) pepsin (Fluka Biochemika, Germany) in the electrolyte solution, with the pH adjusted to 2.5. After 1 h of incubation at 37 °C, a sample was taken to measure the viable cells on an MRS agar plate. For simulation of the conditions in the small intestine, the remaining volume was then diluted 1:4 using an artificial duodenal secretion (6.4 g/L $NaHCO_3$, 0.24 g/L KCl, 1.28 g/L NaCl, 0.5% (*w*/*v*) bile salts, and 0.1% (*w*/*v*) pancreatin) (Fluka Biochemika, Germany) at pH 7.2 and was incubated at 37 °C. At 2 and 3 h of the incubation time, samples were taken for measurement of viable cells by plate count technique on MRS agar plates. The viable cell count was determined and the survival rate was calculated.

2.6. Auto-Aggregation

An auto-aggregation study was carried based on the Xu, et al. [28] method, with some modification. A total of 10 mL of LAB cell suspension in PBS (10^8 CFU/mL) was vortexed homogeneously for 10 s, and 1 mL of cell suspension was measured at 600 nm ($A_{initial}$). After being incubated at 37 °C without disturbing for 2 h, the absorbance of the upper fraction (1 mL) was measured (A_{final}). The auto-aggregation percentage was calculated using the following equation:

$$\text{Auto-aggregation (\%)} = [(A_{initial} - A_{final})/A_{initial}] \times 100 \quad (2)$$

where $A_{initial}$ and A_{final} are the absorbance at 0 and 2 h, respectively.

2.7. Cell Surface Hydrophobicity

The cell surface hydrophobicity of the selected LAB was determined in terms of the bacterial cell adhesion to solvents based on the ability of cells to bind to hydrocarbons, according to the methodology described by García-Hernández, et al. [24]. Chloroform was chosen as a nonpolar solvent because it reflects cell surface hydrophobicity and hydrophilicity. Briefly, an overnight culture of selected LAB was harvested by centrifugation at 8000× *g* with 4 °C for 10 min, washed twice with PBS buffer (pH 7.2), and resuspended with PBS to an absorbance of 1.0 at 600 nm ($A_{initial}$). An equal volume of chloroform (BDH Chemicals, Ltd., Poole, England) was added and mixed by vortex mixer for 5 min. After 1 h of incubation at 37 °C, the optical density of aqueous phase was measured at 600 nm (A_{final}). Isolates with cell surface hydrophobicity above 50% were considered hydrophobic. Cell surface hydrophobicity was calculated using the following equation:

$$\text{Cell surface hydrophobicity (\%)} = [(A_{initial} - A_{final})/A_{initial}] \times 100 \quad (3)$$

2.8. Antibiotic Resistance

The LAB strains were tested for antibiotic susceptibilities by the disc diffusion method. Antibiotic discs including erythromycin (15 μg/disc), tetracycline (30 μg/disc), gentamycin (10 μg/disc), kanamycin (30 μg/disc), vancomycin (30 μg/disc), polymycin (30 μg/disc), and streptomycin (10 μg/disc) were used [29]. Briefly, 50 μL of overnight-grown LAB culture on MRS broth (approximately 10^7–10^8 CFU/mL) was spread on the MRS agar plate and the antibiotic discs (HiMedia, Mumbai, India) were placed on it. The plates were incubated at 37 °C for 24 h and the inhibition zone diameters were measured. The results were expressed as sensitive, S; intermediate, I; or resistant, R as described by the Clinical and Laboratory Standard Institute [30].

2.9. Hemolytic Activity Test

The hemolysis assay of the LAB strains was evaluated using Columbia blood agar (Himedia, Mumbai, India) supplemented with 5% (*v*/*v*) sheep blood according to Angmo, et al. [31], with some modification. The strains were streaked on a blood agar plate and incubated at 37 °C for 48 h. The characteristics of hemolysis, shown as clear zones around colonies on blood agar, were recorded and classified as hemolytic (β-hemolysis); green-hued zones around colonies (α-hemolysis) and no clear zones around colonies (γ-hemolysis) were considered non-hemolytic.

2.10. DPPH Free Radical Scavenging Ability

The overnight culture of LAB strains in MRS broth was centrifuged at 8000× *g* for 10 min at 4 °C to separate the CFCS and cells. The CFCS was harvested and filtered through 0.2 μm Acrodisc® syringe filters (Pall Netherlands B.V., Medemblik, The Netherlands). The cell pellets were resuspended in PBS to reach a concentration of 10^8 CFU/mL and served as intact cells. The radical scavenging capacities of LAB strains were evaluated using 2,2-diphenyl-1-picrylhydrazyl hydrate (DPPH) (Sigma-Aldrich, St. Louis, MO, USA) free radical assays [32]. The assay was carried out by mixing 800 μL of freshly prepared 0.2 mM of DPPH solution in 80% methanol with 400 μL of CFCS or intact cells and vortexed for 30 s. The mixture was left at room temperature in the dark for 30 min. Scavenging ability was measured by decrease in absorbance at 517 nm. Uninoculated MRS broth and PBS were used as control samples for the DPPH scavenging measurement of CFCS and intact cells, respectively. The percentage scavenging potential was estimated using the following equation:

$$\text{DPPH scavenging capacity (\%)} = (A_{control} - A_{sample})/A_{control} \times 100 \quad (4)$$

2.11. Prebiotic Utilization

The ability to utilize different prebiotics of LAB strains was evaluated and presented as a prebiotic score as described by Kondepudi, et al. [33]. Commercial FOS, GOS, and XOS purchased from Wako Pure Chemical Industries (Tokyo, Japan) were used as the carbon sources for LAB cultivation. An overnight culture of LAB strains in MRS broth at 37 °C was harvested by centrifugation (8000× *g* for 10 min at 4 °C) and the cell pellets were washed twice with PBS buffer. The washed cell pellets were resuspended in PBS. The inoculum size of the bacterial viable cell (approximately 10^8 CFU) was transferred into 10 mL of modified MRS medium containing 10 g/L of individual prebiotics (GOS, FOS, or XOS) as a sole carbon source and statically incubated at 37 °C for 24 h. The modified MRS broth containing 10 g/L of glucose as the sole carbon source was used as positive control. Prebiotic utilization was determined by measuring the viable cells (logCFU/mL) at 24 h. The prebiotic score is the highest growth achieved by a strain in the presence of MRS supplemented with prebiotics relative to their growth in glucose, which was considered 100%. The prebiotic score was calculated as below:

$$\text{Prebiotic score (\%)} = (A/B) \times 100 \quad (5)$$

where A and B are the mean viable cell values (logCFU/mL) of a strain grown in the presence of each prebiotic (GOS, FOS, or XOS) and glucose, respectively, after 24 h cultivation.

2.12. Viability of Probiotic Strains after Spray-Drying

An overnight culture of LAB strains in MRS broth at 37 °C was harvested by centrifugation at 8000× *g* for 10 min at 4 °C, then washed twice with PBS buffer. The washed cell pellets were resuspended in maltodextrin (15%, *w*/*v*) to approximately 10^7 CFU/mL. A laboratory scale spray dryer (model B-290 Buchi mini spray dryer, Flawil, Switzerland) was used to process samples at a constant air inlet temperature of 180 °C, and the flow rate of the drying air was set at 40 m^3/h, leading to an outlet temperature of around 80–85 °C, in order to obtain powders with less than 5% moisture. Cell viability was tested before and after the spray-drying procedure. The survival percentage was calculated as follows:

$$\text{Survival rate (\%)} = (N/N_0) \times 100 \quad (6)$$

where N represents the number of viable cells per gram of dry matter after drying, and N_0 is the number of viable cells per gram of dry matter in the bacterial suspension before drying [34].

2.13. Statistical Analysis

All the experiments were performed in triplicate, and the results were calculated as the mean ± standard deviation (SD). One-way analysis of variance (ANOVA) was used to study significant differences between means with significance level $p < 0.05$ using SPSS statistical software, version 20.0 (SPSS Inc., Chicago, IL, USA).

3. Results and Discussion

3.1. Screening and Selection of Acid and Bile Salt Tolerance LAB as Potential Probiotic

Tolerance to low pH and bile salts are generally considered an essential assessment criteria for probiotic strains to exert their beneficial effects on the gastrointestinal tract environment [35]. The pH of the human stomach rises from 1–2 to 4–5 after the ingestion of food, therefore the majority of in vitro assays are designed to select for strains that can tolerate the extreme low pH range from 2–3 [36]. This assay was designed to exclude some strains that may actually possess probiotic properties but were unable to tolerate extreme low acidity. Therefore, a total of 133 LAB isolates obtained from Miang were investigated for their acid tolerance at pH 2.0. The results determined that only 16 LAB isolates showed the ability to tolerate pH 2.0 for 3 h of incubation with a survival rate of more than 80%, which represented 12% of the total LAB isolates (Table 1). Isolates A9-2 and A14-6 were the most tolerant at pH 2.0, with survival rates of 95%. Acidic conditions have a large effect on the survival rates of *Lactobacillus*. According to Mathara, et al. [37], the percentage of LAB strains isolated from traditional fermented dairy products with a favorable resistance at pH 2.0 was 22.2% of the overall strains. Almost none of the 47 *Lactobacillus* strains isolated from ripened Parmigiano–Reggiano cheese could survive at pH 2.0 [38].

According to Goldin and Gorbach [39], tolerance to bile salt concentration of 0.15–0.3% is recommended for probiotics, since it is in the range of the physiological conditions found in the gastrointestinal tract. Moreover, Jose, et al. [40] reported that the maximum concentration of bile salt that can be found in an average healthy person is 0.3%. Therefore, 0.3% bile was used in this study, and all 16 acid-tolerant LAB isolates demonstrated bile resistance ability, as noted in the results presented in Table 1. The results revealed that 6 of 16 isolates were sensitive to 0.3% bile salt (less than 50%), whereas the others had tolerance in the range of 65 to 92%. Only 5 of 16 isolates showed the ability to tolerate bile salt at concentration of 0.3% with survival rates over 80% during 3 h incubation. On evaluating the combined tolerance to acidic conditions and bile salt, which is an important criterion for effective probiotics selection, as mentioned previously, all five isolates including A9-2, A14-6, A26-8, CMY9, and CMY46 were selected for further identification and probiotic characterization.

Table 1. Viability and survival percentages of sixteen LAB isolates after being incubated at pH 2.0 and 0.3% (*w*/*v*) bile salt at 37 °C for 3 h.

Isolates	Viable Cell (logCFU/mL)		Survival Rate (%)	Viable Cell (logCFU/mL)		Survival Rate (%)
	pH 7.2	pH 2.0		Without Bile	0.3% Bile	
A5-1	8.57 ± 0.12	6.89 ± 0.03	80.4 ± 0.4^{i}	8.40 ± 0.05	4.15 ± 0.06	49.4 ± 0.5^{h}
A9-2	8.23 ± 0.01	7.82 ± 0.08	95.0 ± 0.6^{a}	8.55 ± 0.07	7.85 ± 0.08	91.8 ± 1.4^{a}
A10-1	8.44 ± 0.04	6.95 ± 0.12	82.3 ± 1.7^{hi}	8.23 ± 0.12	6.32 ± 0.04	76.8 ± 0.6^{e}
A13-5	8.24 ± 0.04	7.46 ± 0.10	90.5 ± 0.8^{bcd}	8.45 ± 0.04	3.82 ± 0.07	45.2 ± 0.8^{ij}
A14-2	8.11 ± 0.12	6.71 ± 0.11	82.7 ± 0.9^{ghi}	8.06 ± 0.07	3.87 ± 0.05	48.0 ± 1.3^{hi}
A14-6	8.68 ± 0.09	8.26 ± 0.07	95.2 ± 0.5^{a}	8.52 ± 0.09	7.75 ± 0.21	91.0 ± 1.8^{a}
A21-4	8.67 ± 0.06	7.75 ± 0.02	89.4 ± 1.4^{cde}	8.15 ± 0.16	5.28 ± 0.02	64.8 ± 0.9^{g}
A26-8	8.45 ± 0.17	7.84 ± 0.04	92.8 ± 0.3^{ab}	8.66 ± 0.02	7.45 ± 0.06	86.0 ± 1.5^{b}
A27-3	8.16 ± 0.10	7.31 ± 0.06	89.6 ± 0.5^{cde}	8.43 ± 0.12	6.61 ± 0.08	78.4 ± 1.6^{de}
A29-1	8.71 ± 0.12	7.33 ± 0.03	84.2 ± 0.7^{gh}	8.89 ± 0.13	6.33 ± 0.05	71.2 ± 0.7^{f}
CMY1	8.10 ± 0.09	7.14 ± 0.04	88.1 ± 0.6^{de}	8.77 ± 0.07	3.48 ± 0.17	39.7 ± 0.5^{k}
CMY9	8.14 ± 0.07	7.42 ± 0.09	91.2 ± 0.4^{bc}	8.45 ± 0.13	7.15 ± 0.03	84.6 ± 1.1^{bc}
CMY12	8.27 ± 0.07	7.21 ± 0.15	87.2 ± 0.6^{ef}	8.62 ± 0.04	3.75 ± 0.12	43.5 ± 0.9^{j}
CMY34	8.62 ± 0.15	7.36 ± 0.08	85.4 ± 1.2^{fg}	8.28 ± 0.04	3.46 ± 0.05	41.8 ± 1.6^{jk}
CMY46	8.57 ± 0.02	7.85 ± 0.06	91.6 ± 1.5^{cd}	8.41 ± 0.06	6.89 ± 0.07	81.9 ± 0.8^{cd}
CMY47	8.46 ± 0.08	7.58 ± 0.10	89.6 ± 0.6^{de}	8.32 ± 0.08	6.52 ± 0.13	78.4 ± 1.9^{de}

Note: Means in column with different superscripts are statistically different at $p < 0.05$.

3.2. Identification of LAB

Five selected isolates that were determined to be Gram-positive by forming a yellow clear zone on MRS agar were presumptively considered LAB. Most of the colonies were round, smooth, convex surface, off-white or yellow colonies. The cell morphology after 24 h incubation of A9-2, A14-6, A26-8, and CMY46 were observed as a rod shape, while the CMY9 was cocci. The nucleotide sequences of 16S rRNA genes of all five selected LAB isolates were also determined, with the purpose of classifying the species of the selected isolates. The full-length 16S rRNA genes were sequenced and compared in similarity with the bacterial 16S rRNA gene in the NCBI database; these results are shown in Table 2. The isolated CMY9 showed more than 99.9% similarity to *Pediococcus pentosaceus*, and the other isolates (A9-2, A14-6, A26-8, and CMY46) showed more than 98% similarity to the *Lactobacillus plantarum* group including *L. pentosus* and *L. plantarum*. Based on high similarity of up to 99% of 16S rRNA gene sequences between *L. pentosus* and *L. plantarum*, it was difficult to differentiate all of selected strains. Confirmation of *recA* gene analysis found that isolate A9-2 was identified to be *L. plantarum*, whereas the isolates A14-6, A26-8, and CMY46 were identified to be *L. pentosus*. The phylogenetic tree of all five LAB isolates is shown in Figure 1. Our results supported the previous findings which reported that the most common species of lactobacilli isolates found in fermented fruits and vegetables including fermented cucumbers, fermented olives, and fermented tea leaves were *L. pentosus*, *L. plantarum*, *L. vaccinostercus*, *L. thaitandensis*, *L. camellia*, and *P. siamensis* [16,17,19,32,41,42].

Table 2. Molecular identification of LAB isolates by 16S rRNA gene sequence analysis.

Isolates	Closest Species	Similarity (%)	Length (bp)	Accession Number	Note
A9-2	*Lactobacillus plantarum* ATCC 14917	99.8	1486	MW564014	*recA* gene confirmed
A14-6	*Lactobacillus pentosus* DSM 20314	99.9	1467	MW564015	*recA* gene confirmed
A26-8	*Lactobacillus pentosus* DSM 20314	99.9	1476	MW564016	*recA* gene confirmed
CMY9	*Pediococcus pentosaceus* DSM 20336	99.9	1476	MW564017	-
CMY46	*Lactobacillus pentosus* DSM 20314	99.5	1472	MW564018	*recA* gene confirmed

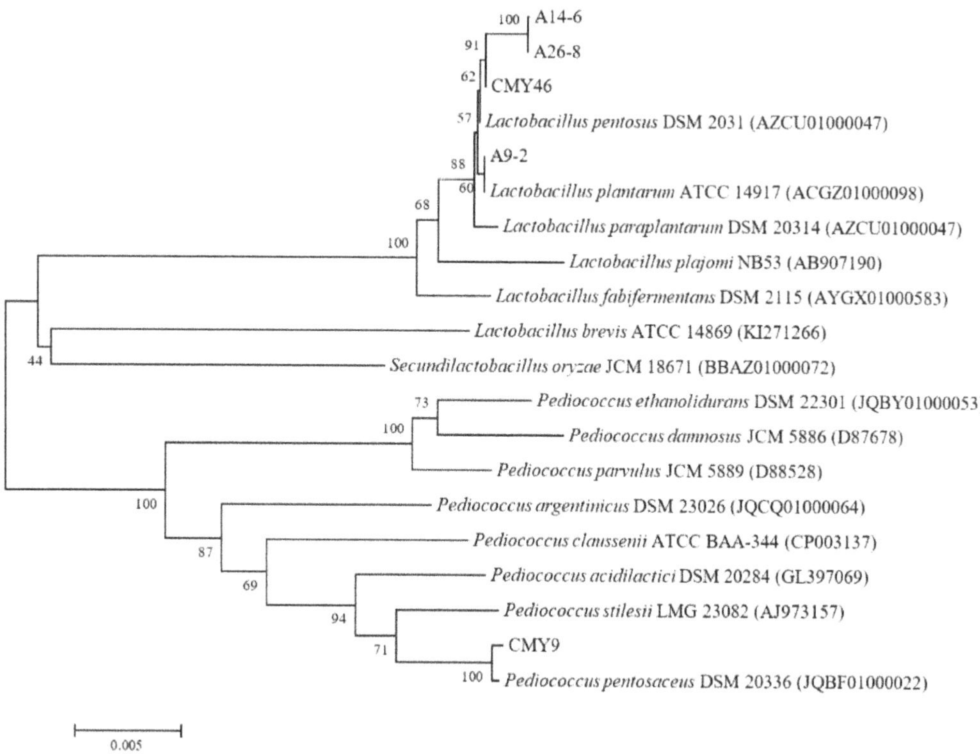

Figure 1. Phylogenetic tree of five selected LAB isolates based on 16S rRNA gene sequence analysis and other related species.

3.3. Antimicrobial Activity against Pathogens

The selected LAB isolates were evaluated for antimicrobial potential against the indicator microorganisms *B. cereus*, *S. aureus*, and *S.* Typhimurium, by the disc diffusion method. An unneutralized CFCS from LAB isolates and a reference strain (*L. acidophilus*) showed a clear zone of inhibition against all indicator microorganisms tested. The results revealed that all LAB isolates exhibited average inhibition zone of 10.4–14.4 mm. *Lactobacillus pentosus* A14-6 was the most effective in inhibiting target pathogens, with 14.4 ± 0.5 and 13.3 ± 0.3 mm clear zones against *S. aureus* and *S.* Typhimurium, respectively, and showed an inhibition zone better than *L. acidophilus* against both pathogens (13.8 ± 0.3 and 12.0 ± 0.4 mm), as shown in Figure 2A–C. In contrast, neutralized CFCS of all selected isolates did not show clear zone formation (Figure 2D–F). Generally, an unneutralized CFCS always shows an inhibitory effect because of the acidic pH, which is mostly not favorable for the growth of most pathogenic bacteria, whereas the inhibitory properties found in neutralized CFCS are commonly caused by bacteriocin or bacteriocin-like metabolites [27,43]. The study by Sankar, et al. [44] found that *L. plantarum* showed antibacterial activity against *S. aureus*, *Enterococcus fecalis*, *Escherichia coli*, and *Listeria monocytogenes*. Among these, the highest growth inhibition recorded was against *S. aureus*, and minimum activity was observed against *L. monocytogenes*.

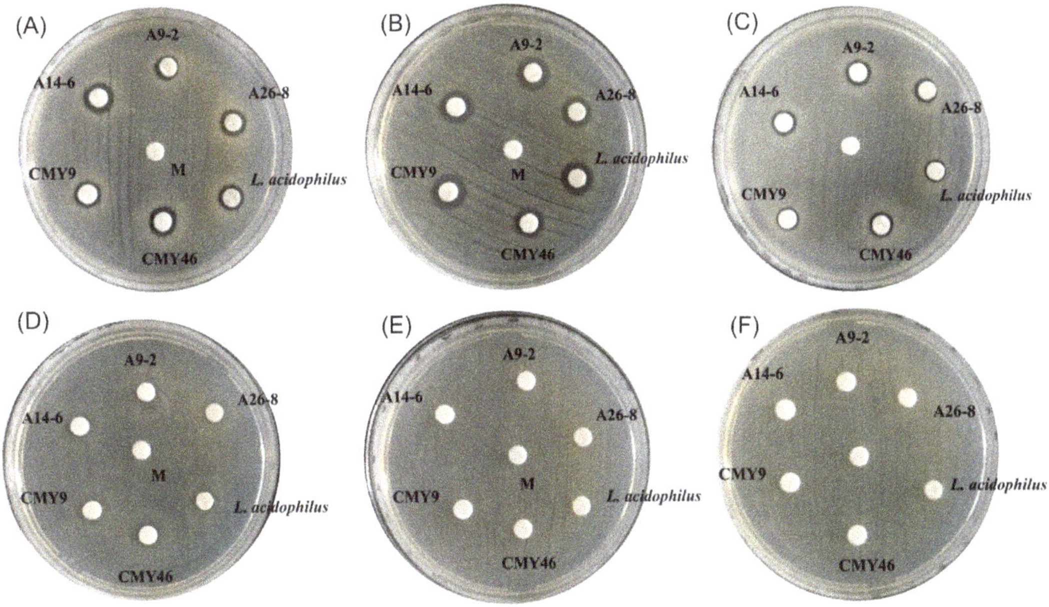

Figure 2. Disc diffusion assay of cell-free culture supernatants (CFCS), obtained from *L. plantarum* A9-2, *L. pentosus* A14-6, *L. pentosus* A26-8, *P. pentosaceus* CMY9, *L. pentosus* CMY46, and *L. acidophilus* culture in MRS broth (M) uninoculated MRS broth, against *Bacillus cereus* (**A**,**C**), *Staphylococcus aureus* (**B**,**E**), *Salmonella* Typhimurium (**C**,**F**). (**A–C**): unneutralized CSCF, (**D–F**): neutralized CFCS (pH 7).

3.4. Survival in the Simulated Gastrointestinal Conditions

An in vitro model for the evaluation of survival in simulated gastrointestinal conditions was used for the investigation of five LAB isolates along with the probiotic reference strain, *L. acidophilus*. The effects of artificial saliva, gastric, and duodenal juices on the viability of LAB isolates are presented in Figure 3. The statistical comparison of the viability for each species at the end of the treatments revealed that the surviving capability of LAB isolates against the artificial gastrointestinal conditions was clearly comparable to the *L. acidophilus* probiotic strain [45]. All LAB isolates showed tolerance and a good survival rate of more than 90% after being tested under the stress of simulated gastrointestinal conditions. The LAB candidates showed higher surviving capability against in vitro gastrointestinal conditions than the *L. acidophilus* probiotic strain. Another consideration is that tea leaves contain phenolic compounds, particularly tannins and other tea phenolics, which are considered to inhibit microbial growth [14,16,46] and the microorganisms living and compatible in Miang substrate might have mechanisms allowing them to tolerate the tea tannins. In addition, the cell wall of *L. plantarum* and *L. pentosus* contain the meso-diaminopimelic acid (mDAP) peptidoglycan, which means they can survive in the tannin-rich substrate [14,16,18] and therefore potentially have greater survival capability against in vitro gastrointestinal conditions than the *L. acidophilus* probiotic strain.

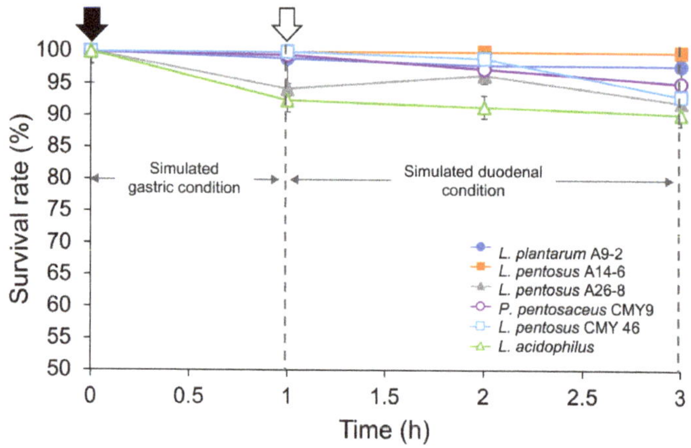

Figure 3. Survival of *L. plantarum* A9-2, *L. pentosus* A14-6, *L. pentosus* A26-8, *P. pentosaceus* CMY9, *L. pentosus* CMY46, and *L. acidophilus* under simulated gastrointestinal conditions. Black and white arrows indicate addition of simulated gastric juice and simulated duodenal juice, respectively.

3.5. Resistance to Antibiotics

Antibiotics are major antimicrobial agents utilized to fight bacterial pathogens. However, antibiotic resistance and its transfer to pathogens in the natural environment can cause significant danger and suffering for many people with pathogen infections [47]. Therefore, it is desirable that probiotics are sensitive to commonly prescribed antibiotics at low concentrations. The results of antibiotic susceptibilities (Table 3) showed all five LAB candidates were sensitive to erythromycin and tetracycline, and were moderately susceptible to gentamycin. On the other hand, all strains were resistant to kanamycin, vancomycin, and polymycin. The vancomycin resistance of the selected probiotic LAB in this experiment supports the fact that the majority of the lactobacilli were intrinsically resistant to glycopeptide [2]. The susceptibility and resistance of LAB against various antibiotics is variable depending on the species; for example, LAB isolated from infant feces are resistant to kanamycin and streptomycin [48]. Thus, the resistance mechanisms observed among these strains are probably inherent or intrinsic to their species and could therefore not be attributed to the acquisition of resistance genes [49]. It has been reported that probiotic strains should be susceptible to at least two clinically relevant antibiotics [50]. All selected LAB strains in this experiment were susceptible to erythromycin and tetracycline, and this meets the good probiotics criteria in term of antibiotic-resistant properties. Furthermore, all LAB strains showed negative hemolytic activity, which is the supporting property for their in vivo safety.

Table 3. Susceptibility of LAB isolates to antibiotics.

Isolates	Antibiotics						
	VA	K	CN	S	TE	E	PB
L. plantarum A9-2	R	I	S	R	S	S	R
L. pentosus A14-6	R	R	I	R	S	S	R
L. pentosus A26-8	R	R	I	R	S	S	R
P. pentosaceus CMY9	R	R	I	R	S	S	R
L. pentosus CMY46	R	R	I	R	S	S	R

Note: VA: vancomycin (30 μg/disc), K: kanamycin (30 μg/disc), CN: gentamycin (10 μg/disc), S: streptomycin (10 μg/disc), TE: tetracycline (30 μg/disc), E: erythromycin (15 μg/disc) PB: polymycin (30 μg/disc). Zone of clearing: various antibiotic resistant (R), moderately susceptible (I), and susceptible (S).

3.6. Hydrophobicity and Auto-Aggregation

Hydrophobicity and auto-aggregation assays were employed as indirect screening tools to test and select the adhesion potentiality of probiotic bacteria to host intestinal mucosa [51]. The cell surface hydrophobicities of the selected LAB strains are presented in Figure 4A. All isolates showed a high percentage of hydrophobicity towards chloroform which ranged from 58.3 to 92.9% after a 1 h incubation period. The highest hydrophobicity percentage was found with *L. pentosus* A14-6. Generally, hydrophobicity above 40% is desired for a probiotic strain [52], and all selected probiotic LAB in this study therefore qualified in terms of cell hydrophobicity properties. Previous studies have reported on the high degree of variation of the hydrophobicity property among probiotic LAB [31,53,54]. *L. plantarum* and *L. casei* originally isolated from Ladakh fermented foods have shown cell surface hydrophobicities ranging from 5–74% with *n*-hexadecane [31]. The high hydrophobicity property of probiotic strains represents higher interaction with the epithelium cells of gastrointestinal tract, which indicate the better exclusion of pathogens [55]. Differences in cell surface hydrophobicity are caused from variations in the levels of cell surface protein expression of the species [56].

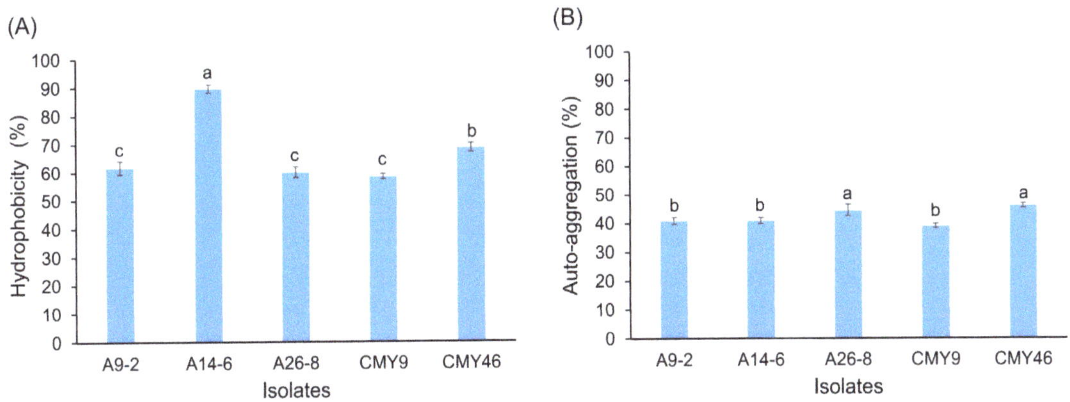

Figure 4. Cell surface hydrophobicity (**A**), and auto-aggregation (**B**) of *L. plantarum* A9-2, *L. pentosus* A14-6, *L. pentosus* A26-8, *P. pentosaceus* CMY9, and *L. pentosus* CMY46. Different letters (a–c) indicate significant differences of the values ($p < 0.05$).

Auto-aggregation of probiotic LAB is believed to align with adhesion of the LAB to the intestinal epithelium. The percentages of auto-aggregation of all selected LABs were found to be in the range of 38.9 to 46.0% after 2 h incubation (Figure 4B), which indicated moderate capability to colonize host intestinal cells, while the auto-aggregative values of S. Typhimurium and *S. aureus* were 33.7% and 15.8%, respectively, after incubation at 37 °C for 2 h [28]. The higher auto-aggregation properties are reported to increase in relation to LAB colonization in the gastrointestinal tract. Auto-aggregation values are strain-specific and the wide range of auto-aggregation abilities (1.63–80.50%) have been reported for *L. brevis, L. plantarum, L. acidophilus, L. curvatus, L. sake, L. fermentum*, and *P. pentosaceus* isolated from fermented foods [31].

3.7. Radical Scavenging Activities of LAB

Antioxidant activities of CFCS and intact cells of the five selected LAB isolates were evaluated by the DPPH free radical scavenging assay method and their antioxidant capacities are presented in Figure 5A. CFCS of all probiotic LAB strains showed higher radical scavenging activities compared to their intact cells. However, the intact cells of *L. pantarum* A9-2, *L. pentosus* A14-6, and *L. pentosus* A26-8 showed higher DPPH scavenging activity than *P. pentosaceus* CMY9 and *L. pentosus* CMY46. Recently, LAB such as *Lactobacillus* sp. and *Bifidobacterium* sp. revealed significant antioxidant characteristics, which a study has suggested means that the antioxidant properties mainly depend on the type of probiotic bacteria used [57]. Moreover, the antioxidant activity of *L. plantarum* derived from pickles, tea, sauerkraut, fermented dairy products, fermented beverages, and the feces of healthy infants have also been reported [32,58,59]. The mechanisms of the antioxidant activity of LAB intact cells were suggested to be responsible due to metal ion chelation, enzymes such as antioxidases from probiotics, antioxidant compounds produced by probiotic cells, or countering radicals generated in the intestinal tract [60,61]. However, the oxidation-resistant ability of probiotics and their mechanisms are not completely clear.

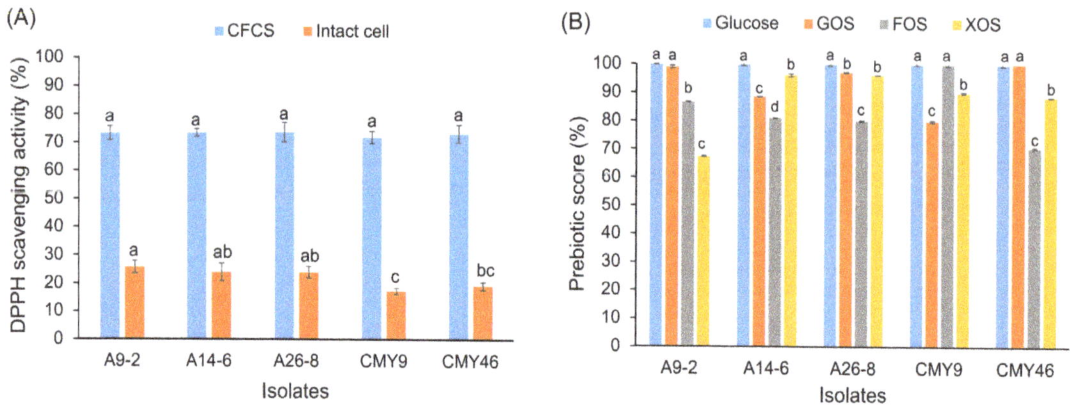

Figure 5. DPPH scavenging activity (**A**) and prebiotic scores (**B**) of *L. plantarum* A9-2, *L. pentosus* A14-6, *L. pentosus* A26-8, *P. pentosaceus* CMY9, and *L. pentosus* CMY46. Different letters (a–c) indicate significant differences of the values ($p < 0.05$).

3.8. Prebiotic Utilization and Prebiotic Scores

The synbiotic concept was created to overcome the difficulties of growing of probiotics in the gastrointestinal tract. Synbiotics have beneficial synergistic effects, greater than those observed for the individual administration of only prebiotics or probiotics [62]. Therefore, the study of prebiotic utilization by the selected LAB was explored. The growths of all LAB strains in the presence of various prebiotics, including GOS, XOS, and FOS, in comparison to glucose (positive control) are presented in Figure 5B. *L. plantarum* A9-2, *L. pentosus* A26-8, and *L. pentosus* CMY46 showed the maximum growth with GOS, meanwhile, the strains *L. pentosus* A14-6 and *P. pentosaceus* CMY9 showed their maximum growth with XOS and FOS, respectively. The results from this study indicate that all selected probiotic LAB had potential for application in the design of formulation of synbiotics with various types of commercial prebiotics. However, the improved specificity of probiotic LAB and prebiotic types may lead to the highest efficiency of formulated synbiotic products. Therefore, a determination of specific properties between prebiotics and probiotics is the most appropriate approach to be further fine tuned for commercial applications.

3.9. Viability of Probiotic Strains during Spray-Drying

The effect of spray-drying on the viability of the LAB strains is shown in Table 4. The viability of *L. pantarum* A9-2, *L. pentosus* A14-6, *L. pentosus* A26-8, *P. pentosaceus* CMY9, and *L. pentosus* CMY46 decreased after spray-drying by 2.01, 2.03, 2.10, 1.69 and 1.88 log, respectively. The survival rates of all strains were up to 75% after spray-drying. Even with reduction of microbial count, this preliminary study of spray-drying indicated the feasibility of applying these LAB strains at industrial levels and can be starting point for improving final microbial counts. The improved viability and efficacy of probiotics after exposure to the spray-drying process may be improved by the use of other thermoprotectants such as different sugars, skim milk, and whey protein. Previous reports mentioned that the conditions of the spray-drying process influence the quality and efficacy of spray-dried probiotic powder [63–66]. Therefore, the optimization of the spray-drying process is required before applying these selected LAB strains in a commercial scale.

Table 4. Survival of five selected LAB isolates after spray-drying with maltodextrin as the thermoprotectant.

Isolates	Viable Cell (logCFU/g)		Survival Rate (%)
	Before Spray-Drying	After Spray-Drying	
L. plantarum A9-2	8.42 ± 0.02	6.41 ± 0.07	76.1 ± 0.65 [bc]
L. pentosus A14-6	8.61 ± 0.06	6.58 ± 0.11	76.4 ± 0.74 [bc]
L. pentosus A26-8	8.49 ± 0.11	6.39 ± 0.05	75.3 ± 0.38 [c]
P. pentosaceus CMY9	8.55 ± 0.07	6.86 ± 0.04	80.2 ± 0.18 [a]
L. pentosus CMY46	8.36 ± 0.09	6.48 ± 0.05	77.5 ± 0.23 [b]

Note: Means in column with different superscripts are statistically different at $p < 0.05$.

4. Conclusions

The results obtained from this study indicate that Miang, traditional fermented tea leaves, can serve as a beneficial source of potential probiotic candidates. Five selected LAB strains originally isolated from Miang samples clearly demonstrated survival under simulated gastrointestinal conditions, which indicates their beneficial capabilities could be applied as probiotics. These selected LAB strains also showed specificity for being utilized in combinations of commercial prebiotics and tolerance against spray-drying process. Among the five strains, *L. pentosus* A14-6 in particular had the most promising probiotic potential, with cellular antioxidative characteristics that support its feasibility for application in various food products or as targeted towards the development of synbiotic nutraceuticals. However, the safety and functional properties of these strains should be further evaluated to confirm their health-beneficial properties in in vivo models.

Author Contributions: Conceptualization, K.U., W.C., S.B. and C.K.; methodology and formal analysis, W.C., K.U., A.D.A., S.B. and A.K.; investigation, W.C., K.U., S.B. and C.K.; writing—original draft preparation, K.U., W.C., S.B. and C.K.; writing—review and editing, K.U., S.B., A.K., K.S. and C.K.; supervision, C.K. All authors have read and agreed to the published version of the manuscript.

Funding: The authors are grateful to Chiang Mai University for financial support via post-doctoral fellowship.

Institutional Review Board Statement: Not applicable.

Informed Consent Statement: Not applicable.

Data Availability Statement: Not applicable.

Acknowledgments: The authors are grateful to Chiang Mai University for financial support via postdoctoral fellowship and Faculty of Agro-Industry for research facilities. This work was partially supported by the National Science Technology and Innovation Policy Office (STI) and the Agricultural Research Development Agency (ARDA).

Conflicts of Interest: The authors declare no conflict of interest.

References

1. Hill, C.; Guarner, F.; Reid, G.; Gibson, G.R.; Merenstein, D.J.; Pot, B.; Morelli, L.; Canani, R.B.; Flint, H.J.; Salminen, S. Expert consensus document: The International Scientific Association for Probiotics and Prebiotics consensus statement on the scope and appropriate use of the term probiotic. *Nat. Rev. Gastroenterol. Hepatol.* **2014**, *11*, 506. [CrossRef]
2. Lee, K.W.; Shim, J.M.; Park, S.-K.; Heo, H.-J.; Kim, H.-J.; Ham, K.-S.; Kim, J.H. Isolation of lactic acid bacteria with probiotic potentials from kimchi, traditional Korean fermented vegetable. *LWT-Food Sci. Technol.* **2016**, *71*, 130–137. [CrossRef]
3. Vinderola, G.; Cespedes, M.; Mateolli, D.; Cardenas, P.; Lescano, M.; Aimaretti, N.; Reinheimer, J. Changes in gastric resistance of *Lactobacillus casei* in flavoured commercial fermented milks during refrigerated storage. *Int. J. Dairy Technol.* **2011**, *64*, 269–275. [CrossRef]
4. Byakika, S.; Mukisa, I.M.; Byaruhanga, Y.B.; Muyanja, C. A review of criteria and methods for evaluating the probiotic potential of microorganisms. *Food Rev. Int.* **2019**, *35*, 427–466. [CrossRef]

5. Rubio, R.; Jofré, A.; Martín, B.; Aymerich, T.; Garriga, M. Characterization of lactic acid bacteria isolated from infant faeces as potential probiotic starter cultures for fermented sausages. *Food Microbiol.* **2014**, *38*, 303–311. [CrossRef]
6. Forssten, S.D.; Sindelar, C.W.; Ouwehand, A.C. Probiotics from an industrial perspective. *Anaerobe* **2011**, *17*, 410–413. [CrossRef]
7. Tripathi, M.K.; Giri, S.K. Probiotic functional foods: Survival of probiotics during processing and storage. *J. Funct. Foods* **2014**, *9*, 225–241. [CrossRef]
8. Reque, P.M.; Brandelli, A. Encapsulation of probiotics and nutraceuticals: Applications in functional food industry. *Trends Food Sci. Technol.* **2021**, *114*, 1–10. [CrossRef]
9. Sartor, R.B. Therapeutic manipulation of the enteric microflora in inflammatory bowel diseases: Antibiotics, probiotics, and prebiotics. *Gastroenterology* **2004**, *126*, 1620–1633. [CrossRef] [PubMed]
10. Vallianou, N.; Stratigou, T.; Christodoulatos, G.S.; Tsigalou, C.; Dalamaga, M. Probiotics, prebiotics, synbiotics, postbiotics, and obesity: Current evidence, controversies, and perspectives. *Curr. Obes. Rep.* **2020**, *9*, 1–14. [CrossRef] [PubMed]
11. Pandey, K.R.; Naik, S.R.; Vakil, B.V. Probiotics, prebiotics and synbiotics—A review. *J. Food Sci. Technol.* **2015**, *52*, 7577–7587. [CrossRef] [PubMed]
12. Patel, A.; Prajapati, J.; Holst, O.; Ljungh, A. Determining probiotic potential of exopolysaccharide producing lactic acid bacteria isolated from vegetables and traditional Indian fermented food products. *Food Biosci.* **2014**, *5*, 27–33. [CrossRef]
13. Khanongnuch, C.; Unban, K.; Kanpiengjai, A.; Saenjum, C. Recent research advances and ethno-botanical history of miang, a traditional fermented tea (*Camellia sinensis* var. *assamica*) of Northern Thailand. *J. Ethn. Foods* **2017**, *4*, 135–144. [CrossRef]
14. Kanpiengjai, A.; Chui-Chai, N.; Chaikaew, S.; Khanongnuch, C. Distribution of tannin-tolerant yeasts isolated from Miang, a traditional fermented tea leaf (*Camellia sinensis* var. *assamica*) in northern Thailand. *Int. J. Food Microbiol.* **2016**, *238*, 121–131. [CrossRef]
15. Unban, K.; Khatthongngam, N.; Shetty, K.; Khanongnuch, C. Nutritional biotransformation in traditional fermented tea (Miang) from north Thailand and its impact on antioxidant and antimicrobial activities. *J. Food Sci. Technol.* **2019**, *56*, 2687–2699. [CrossRef] [PubMed]
16. Chaikaew, S.; Baipong, S.; Sone, T.; Kanpiengjai, A.; Chui-chai, N.; Asano, K.; Khanongnuch, C. Diversity of lactic acid bacteria from Miang, a traditional fermented tea leaf in northern Thailand and their tannin-tolerant ability in tea extract. *J. Microbiol.* **2017**, *55*, 720–729. [CrossRef]
17. Gharaei-Fathabad, E.; Eslamifar, M. Isolation and applications of one strain of *Lactobacillus paraplantarum* from tea leaves (*Camellia sinensis*). *Am. J. Food Technol.* **2011**, *6*, 429–434. [CrossRef]
18. Okada, S.; Takahashi, N.; Ohara, N.; Uchimura, T.; Kozaki, M. Microorganisms involving in the fermentation of Japanese fermented tea leaves, 2: Microorganisms in fermentation of Goishi-cha, Japanese fermented tea leaves. *J. Jpn. Soc. Food. Sci. Technol.* **1996**, *43*, 1019–1027. [CrossRef]
19. Sukontasing, S.; Tanasupawat, S.; Moonmangmee, S.; Lee, J.S.; Suzuki, K.I. *Enterococcus camelliae* sp. nov., isolated from fermented tea leaves in Thailand. *Int. J. Syst. Evol. Microbiol.* **2007**, *57*, 2151–2154. [CrossRef]
20. Tanasupawat, S.; Pakdeeto, A.; Thawai, C.; Yukphan, P.; Okada, S. Identification of lactic acid bacteria from fermented tea leaves (miang) in Thailand and proposals of *Lactobacillus thailandensis* sp. nov., *Lactobacillus camelliae* sp. nov., and *Pediococcus siamensis* sp. nov. *J. Gen. Appl. Microbiol.* **2007**, *53*, 7–15. [CrossRef]
21. Unban, K.; Khatthongngam, N.; Pattananandecha, T.; Saenjum, C.; Shetty, K.; Khanongnuch, C. Microbial community dynamics during the non-filamentous fungi growth-based fermentation process of Miang, a traditional fermented tea of north Thailand and their product characterizations. *Front. Microbiol.* **2020**, *11*, 1–15. [CrossRef] [PubMed]
22. Kodchasee, P.; Nain, K.; Abdullahi, A.D.; Unban, K.; Saenjum, C.; Shetty, K.; Khanongnuch, C. Microbial dynamics-linked properties and functional metabolites during Miang fermentation using the filamentous fungi growth-based process. *Food Biosci.* **2021**, *41*, 100998. [CrossRef]
23. Argyri, A.A.; Zoumpopoulou, G.; Karatzas, K.A.G.; Tsakalidou, E.; Nychas, G.J.E.; Panagou, E.Z.; Tassou, C.C. Selection of potential probiotic lactic acid bacteria from fermented olives by in vitro tests. *Food Microbiol.* **2013**, *33*, 282–291. [CrossRef] [PubMed]
24. García-Hernández, Y.; Pérez-Sánchez, T.; Boucourt, R.; Balcázar, J.L.; Nicoli, J.R.; Moreira-Silva, J.; Rodríguez, Z.; Fuertes, H.; Nuñez, O.; Albelo, N. Isolation, characterization and evaluation of probiotic lactic acid bacteria for potential use in animal production. *Res. Vet. Sci.* **2016**, *108*, 125–132. [CrossRef]
25. Tamura, K.; Dudley, J.; Nei, M.; Kumar, S. MEGA4: Molecular evolutionary genetics analysis (MEGA) software version 4.0. *Mol. Biol. Evol.* **2007**, *24*, 1596–1599. [CrossRef]
26. Torriani, S.; Felis, G.E.; Dellaglio, F. Differentiation of *Lactobacillus plantarum*, *L. pentosus*, and *L. paraplantarum* by *recA* gene sequence analysis and multiplex PCR assay with *recA* gene-derived primers. *Appl. Environ. Microbiol.* **2001**, *67*, 3450–3454. [CrossRef]
27. Sriphannam, W.; Lumyong, S.; Niumsap, P.; Ashida, H.; Yamamoto, K.; Khanongnuch, C. A selected probiotic strain of *Lactobacillus fermentum* CM33 isolated from breast-fed infants as a potential source of β-galactosidase for prebiotic oligosaccharide synthesis. *J. Microbiol.* **2012**, *50*, 119–126. [CrossRef]
28. Xu, H.; Jeong, H.; Lee, H.; Ahn, J. Assessment of cell surface properties and adhesion potential of selected probiotic strains. *Lett. Appl. Microbiol.* **2009**, *49*, 434–442. [CrossRef] [PubMed]
29. Charteris, W.P.; Kelly, P.M.; Morelli, L.; Collins, J.K. Antibiotic susceptibility of potentially probiotic *Lactobacillus* species. *J. Food Prot.* **1998**, *61*, 1636–1643. [CrossRef]

30. CLSI. *Performance Standards for Antimicrobial Susceptibility Testing; Twenty-Fifth Informational Supplement*; Clinical and Laboratory Standards Institute: Wayne, PA, USA, 2012.
31. Angmo, K.; Kumari, A.; Bhalla, T.C. Probiotic characterization of lactic acid bacteria isolated from fermented foods and beverage of Ladakh. *LWT-Food Sci. Technol.* **2016**, *66*, 428–435. [CrossRef]
32. Cao, Z.; Pan, H.; Li, S.; Shi, C.; Wang, S.; Wang, F.; Ye, P.; Jia, J.; Ge, C.; Lin, Q. In vitro evaluation of probiotic potential of lactic acid Bacteria isolated from Yunnan De'ang pickled tea. *Probiotics Antimicrob. Proteins* **2019**, *11*, 103–112. [CrossRef] [PubMed]
33. Kondepudi, K.K.; Ambalam, P.; Nilsson, I.; Wadstr, T. Prebiotic-non-digestible oligosaccharides preference of probiotic bifidobacteria and antimicrobial activity against *Clostridium difficile*. *Anaerobe* **2012**, *18*, 489–497. [CrossRef]
34. Zhu, Z.; Luan, C.; Zhang, H.; Zhang, L.; Hao, Y. Effects of spray drying on *Lactobacillus plantarum* BM-1 viability, resistance to simulated gastrointestinal digestion, and storage stability. *Dry. Technol.* **2016**, *34*, 177–184. [CrossRef]
35. Quinto, E.J.; Jiménez, P.; Caro, I.; Tejero, J.; Mateo, J.; Girbés, T. Probiotic lactic acid bacteria: A review. *Food Nutr. Sci.* **2014**, *5*, 1765. [CrossRef]
36. Papadimitriou, K.; Zoumpopoulou, G.; Foligné, B.; Alexandraki, V.; Kazou, M.; Pot, B.; Tsakalidou, E. Discovering probiotic microorganisms: *In vitro, in vivo*, genetic and omics approaches. *Front. Microbiol.* **2015**, *6*, 58. [CrossRef] [PubMed]
37. Mathara, J.M.; Schillinger, U.; Kutima, P.M.; Mbugua, S.K.; Guigas, C.; Franz, C.; Holzapfel, W.H. Functional properties of *Lactobacillus plantarum* strains isolated from Maasai traditional fermented milk products in Kenya. *Curr. Microbiol.* **2008**, *56*, 315–321. [CrossRef] [PubMed]
38. Solieri, L.; Bianchi, A.; Mottolese, G.; Lemmetti, F.; Giudici, P. Tailoring the probiotic potential of non-starter *Lactobacillus* strains from ripened Parmigiano Reggiano cheese by in vitro screening and principal component analysis. *Food Microbiol.* **2014**, *38*, 240–249. [CrossRef]
39. Goldin, B.; Gorbach, S. Probiotics for humans. In *Probiotics, the Scientific Basis*; Fuller, R., Ed.; Chapman and Hall: London, UK, 1992; pp. 355–376.
40. Jose, N.M.; Bunt, C.R.; Hussain, M.A. Comparison of microbiological and probiotic characteristics of lactobacilli isolates from dairy food products and animal rumen contents. *Microorganisms* **2015**, *3*, 198–212. [CrossRef]
41. Klayraung, S.; Viernstein, H.; Sirithunyalug, J.; Okonogi, S. Probiotic properties of Lactobacilli isolated from Thai traditional food. *Sci. Pharm.* **2008**, *76*, 485–504. [CrossRef]
42. Tanasupawat, S.; Ezaki, T.; Suzuki, K.I.; Okada, S.; Komagata, K.; Kozaki, M. Characterization and identification of *Lactobacillus pentosus* and *Lactobacillus plantarum* strains from fermented foods in Thailand. *J. Gen. Appl. Microbiol.* **1992**, *38*, 121–134. [CrossRef]
43. Gao, Z.; Daliri, E.B.-M.; Wang, J.; Liu, D.; Chen, S.; Ye, X.; Ding, T. Inhibitory effect of lactic acid bacteria on foodborne pathogens: A review. *J. Food Prot.* **2019**, *82*, 441–453. [CrossRef]
44. Sankar, N.R.; Priyanka, V.D.; Reddy, P.S.; Rajanikanth, P.; Kumar, V.K.; Indira, M. Purification and characterization of bacteriocin produced by *Lactobacillus plantarum* isolated from cow milk. *Int. J. Microbiol. Res.* **2012**, *3*, 133–137.
45. Kechagia, M.; Basoulis, D.; Konstantopoulou, S.; Dimitriadi, D.; Gyftopoulou, K.; Skarmoutsou, N.; Fakiri, E.M. Health benefits of probiotics: A review. *ISRN Nutr.* **2013**, *2013*, 481651. [CrossRef]
46. Unban, K.; Kodchasee, P.; Shetty, K.; Khanongnuch, C. Tannin-tolerant and extracellular tannase producing *Bacillus* isolated from traditional fermented tea leaves and their probiotic functional properties. *Foods* **2020**, *9*, 490. [CrossRef] [PubMed]
47. Mathur, S.; Singh, R. Antibiotic resistance in food lactic acid bacteria-a review. *Int. J. Food Microbiol.* **2005**, *105*, 281–295. [CrossRef] [PubMed]
48. Salminen, S.; von Wright, A.; Morelli, L.; Marteau, P.; Brassart, D.; de Vos, W.M.; Fondén, R.; Saxelin, M.; Collins, K.; Mogensen, G. Demonstration of safety of probiotics-a review. *Int. J. Food Microbiol.* **1998**, *44*, 93–106. [CrossRef]
49. Manini, F.; Casiraghi, M.; Poutanen, K.; Brasca, M.; Erba, D.; Plumed-Ferrer, C. Characterization of lactic acid bacteria isolated from wheat bran sourdough. *LWT-Food Sci. Technol.* **2016**, *66*, 275–283. [CrossRef]
50. Borriello, S.; Hammes, W.; Holzapfel, W.; Marteau, P.; Schrezenmeir, J.; Vaara, M.; Valtonen, V. Safety of probiotics that contain lactobacilli or bifidobacteria. *Clin. Infect. Dis.* **2003**, *36*, 775–780. [CrossRef]
51. Menezes, A.G.T.; Ramos, C.L.; Cenzi, G.; Melo, D.S.; Dias, D.R.; Schwan, R.F. Probiotic potential, antioxidant activity, and phytase production of indigenous yeasts isolated from indigenous fermented foods. *Probiotics Antimicrob. Proteins* **2020**, *12*, 280–288. [CrossRef]
52. Del Re, B.; Sgorbati, B.; Miglioli, M.; Palenzona, D. Adhesion, autoaggregation and hydrophobicity of 13 strains of *Bifidobacterium longum*. *Lett. Appl. Microbiol.* **2000**, *31*, 438–442. [CrossRef]
53. dos Santos Leandro, E.; Ginani, V.C.; de Alencar, E.R.; Pereira, O.G.; Rose, E.C.P.; do Vale, H.M.M.; Pratesi, R.; Hecht, M.M.; Cavalcanti, M.H.; Tavares, C.S.O. Isolation, identification, and screening of lactic acid bacteria with probiotic potential in silage of different species of forage plants, cocoa beans, and artisanal salami. *Probiotics Antimicrob. Proteins* **2020**, *13*, 173–186. [CrossRef]
54. Vinderola, C.G.; Reinheimer, J.A. Lactic acid starter and probiotic bacteria: A comparative *"in vitro"* study of probiotic characteristics and biological barrier resistance. *Food Res. Int.* **2003**, *36*, 895–904. [CrossRef]
55. Missotten, J.; Goris, J.; Michiels, J.; Van Coillie, E.; Herman, L.; De Smet, S.; Dierick, N.; Heyndrickx, M. Screening of isolated lactic acid bacteria as potential beneficial strains for fermented liquid pig feed production. *Anim. Feed Sci. Technol.* **2009**, *150*, 122–138. [CrossRef]
56. De Vries, M.C.; Vaughan, E.E.; Kleerebezem, M.; de Vos, W.M. *Lactobacillus plantarum*-survival, functional and potential probiotic properties in the human intestinal tract. *Int. Dairy J.* **2006**, *16*, 1018–1028. [CrossRef]

57. Ozcan, T.; Yilmaz-Ersan, L.; Akpinar-Bayizit, A.; Delikanli, B. Antioxidant properties of probiotic fermented milk supplemented with chestnut flour (*Castanea sativa* Mill). *J. Food Process. Preserv.* **2017**, *41*, e13156. [CrossRef]
58. Das, D.; Goyal, A. Antioxidant activity and γ-aminobutyric acid (GABA) producing ability of probiotic *Lactobacillus plantarum* DM5 isolated from Marcha of Sikkim. *LWT-Food Sci. Technol.* **2015**, *61*, 263–268. [CrossRef]
59. Li, S.; Zhao, Y.; Zhang, L.; Zhang, X.; Huang, L.; Li, D.; Niu, C.; Yang, Z.; Wang, Q. Antioxidant activity of *Lactobacillus plantarum* strains isolated from traditional Chinese fermented foods. *Food Chem.* **2012**, *135*, 1914–1919. [CrossRef]
60. Amaretti, A.; Di Nunzio, M.; Pompei, A.; Raimondi, S.; Rossi, M.; Bordoni, A. Antioxidant properties of potentially probiotic bacteria: In vitro and in vivo activities. *Appl. Microbiol. Biotechnol.* **2013**, *97*, 809–817. [CrossRef] [PubMed]
61. Wang, Y.; Wu, Y.; Wang, Y.; Xu, H.; Mei, X.; Yu, D.; Wang, Y.; Li, W. Antioxidant properties of probiotic bacteria. *Nutrients* **2017**, *9*, 521. [CrossRef]
62. Batista, V.L.; Da Silva, T.F.; De Jesus, L.C.L.; Dias Coelho-Rocha, N.; Barroso, F.A.L.; Tavares, L.M.; Azevedo, V.A.D.C.; Mancha-Agresti, P.D.C.; Drumond, M.M. Probiotics, prebiotics, synbiotics, and paraprobiotics as a therapeutic alternative for intestinal mucositis running head: Alternative treatment for intestinal mucositis. *Front. Microbiol.* **2020**, *11*, 544490. [CrossRef] [PubMed]
63. De Castro-Cislaghi, F.P.; Carina Dos Reis, E.S.; Fritzen-Freire, C.B.; Lorenz, J.G.; Sant'Anna, E.S. *Bifidobacterium* Bb-12 microencapsulated by spray drying with whey: Survival under simulated gastrointestinal conditions, tolerance to NaCl, and viability during storage. *J. Food Eng.* **2012**, *113*, 186–193. [CrossRef]
64. Markowiak, P.; Śliżewska, K. Effects of probiotics, prebiotics, and synbiotics on human health. *Nutrients* **2017**, *9*, 1021. [CrossRef] [PubMed]
65. Reddy, K.B.P.K.; Madhu, A.N.; Prapulla, S.G. Comparative survival and evaluation of functional probiotic properties of spray-dried lactic acid bacteria. *Int. J. Dairy Technol.* **2009**, *62*, 240–248. [CrossRef]
66. Santivarangkna, C.; Higl, B.; Foerst, P. Protection mechanisms of sugars during different stages of preparation process of dried lactic acid starter cultures. *Food Microbiol.* **2008**, *25*, 429–441. [CrossRef]

fermentation

Article

Anti-Osteoporotic Activity of *Pueraria lobata* Fermented with *Lactobacillus paracasei* JS1 by Regulation of Osteoblast Differentiation and Protection against Bone Loss in Ovariectomized Mice

Seon Yu Kim [1,†], Hee-Ju Lee [1,†], Taehyun Kim [1,†], Yeong-Geun Lee [1], Jeong Eun Kwon [1,*] and Se Chan Kang [1,2,*]

1 Department of Biotechnology, Kyung Hee University, Yongin 17104, Korea; sykim@nmr.kr (S.Y.K.); leeheeju815@gmail.com (H.-J.L.); silsoo96@naver.com (T.K.); lyg629@nate.com (Y.-G.L.)
2 Mbiometherapeutics Co., Ltd., Yongin 16942, Korea
* Correspondence: jjung1169@khu.ac.kr (J.E.K.); sckang@khu.ac.kr (S.C.K.); Tel.: +82-31-201-5636 (J.E.K.); +82-31-201-2687 (S.C.K.)
† These three authors contributed equally to this work.

Abstract: Osteoporosis is the most common bone disease associated with low bone mineral density. It is the process of bone loss and is most commonly caused by decreased estrogen production in women, particularly after menopause. *Pueraria lobata,* which contains various metabolites, especially isoflavone, is widely known as regulator for bone mineral contents. In this study, the effects of the *P. lobata* extract (PE) with or without fermentation with *Lactobacillus paracasei* JS1 (FPE) on osteoporosis were investigated in vitro and in vivo. The effects of PE and FPE on human osteoblastic MG63 cells, RAW 264.7 cells, and ovariectomized (OVX)-induced model mice were analyzed at various ratios. We found that FPE increased calcium deposition and inhibited bone resorption by in vitro assay. Furthermore, treatment with PE and FPE has significantly restored destroyed trabecular bone in the OVX-induced bone loss mouse model. Overall, FPE demonstrated bioactivity to prevent bone loss by decreasing bone turnover.

Keywords: fermentation; *Lactobacillus*; *Pueraria lobata*; osteoblast; ovariectomized-induced model

Citation: Kim, S.Y.; Lee, H.-J.; Kim, T.; Lee, Y.-G.; Kwon, J.E.; Kang, S.C. Anti-Osteoporotic Activity of *Pueraria lobata* Fermented with *Lactobacillus paracasei* JS1 by Regulation of Osteoblast Differentiation and Protection against Bone Loss in Ovariectomized Mice. *Fermentation* **2021**, *7*, 186. https://doi.org/10.3390/fermentation7030186

Academic Editor: Hiroshi Kitagaki

Received: 30 July 2021
Accepted: 6 September 2021
Published: 9 September 2021

1. Introduction

Isoflavones found in soybeans are nonsteroidal, phytoestrogenic, and anti-oxidative compounds with potential roles in the prevention of chronic diseases, including hormone-dependent diseases, cardiovascular diseases, breast and prostate cancer, osteoporosis, and postmenopausal symptoms [1]. For example, the efficacy of isoflavone derivatives such as dihydrodaidzein (DHD) and equol (EQ) have been evaluated against various hormone-dependent diseases due to their strong binding affinity for estrogen receptors.

Osteoporosis is a bone disease characterized by the destruction of bone tissue, loss of bone mass, and an increased risk of bone fractures [2–4]. Osteoporosis can be classified as primary or secondary depending on its cause. Primary osteoporosis induced by menopause causes a rapid decrease in estrogen levels; leads to the activation of osteoclasts, the cells involved in bone resorption; and imparts the inactivation of osteoblasts, the cells involved in bone formation [5]. When the balance between osteoblasts and osteoclasts is lost, bone resorption is promoted, bone turnover rate is increased, and bone mineral density is lowered, leading to increased fracture and osteoporosis [6]. Secondary osteoporosis is usually caused by other medical conditions or treatments that interfere with the attainment of peak bone mass and may cause bone loss. These medical conditions include genetic diseases, serious kidney failure, rheumatoid arthritis, chronic obstructive pulmonary diseases, endocrine disorders, nutritional imbalance, inflammatory disease, and drug abuse [7–9].

Treatment of osteoporosis generally involves anti-resorptive and anabolic medications in combination with calcium and vitamin D supplementation. With respect to anti-resorptive drugs, bisphosphonates and calcitonin are the most commonly used for the treatment of postmenopausal osteoporosis [10,11]. Regarding anabolic drugs, human parathyroid hormone analogs promote osteoblast formation [12,13]. Selective estrogen receptor modulators (SERMs) improve bone mineral density (BMD) via the mechanisms of hormone replacement therapy; however, side effects have been reported, including increased risk of stroke and venous thromboembolism [12]. Aside from conventional drug therapy, alternative treatments such as traditional medicine are also used to mitigate osteoporosis. Natural products, especially herbal medicine, have received growing attention in the prevention of osteoporosis. For instance, flavonoids derived from *Epimedium brevicornum* Maxim have been proven to improve estrogen deficiency-induced osteoporosis in ovariectomized rats via direct osteoblast stimulation and inhibition of bone resorption, leading to an overall anabolic effect on periosteum and trabecula [10].

Likewise, isoflavones have been demonstrated as valuable for attenuating menopause-induced osteoporotic bone loss by regulating bone-related factors [14]. *Pueraria lobata*, which is known for its main bioactive components, isoflavones, is a perennial vine native to Asia and primarily subtropical and temperate regions of China, Japan, and Korea. The starch extracted from its roots is used in cooking and herbal medicines [15,16] in a wide variety of applications. Recent research has focused on its antidipsotropic activity for the treatment of alcohol-related health effects [17]. Furthermore, *P. lobata* has demonstrated antioxidant, anti-diabetic, and anti-inflammatory activity [18–21]. With strong applicability to the treatment of osteoporosis, *P. lobata* has been reported to regulate receptor activator of nuclear factor kappa-B ligand (RANKL)-mediated osteoclastogenesis [22], proliferation, differentiation, and mineralization [23]. To enhance the efficacy of *P. lobata*, bioconversion with *Lactobacillus paracasei* JS1 was performed.

Lactobacillus paracasei JS1 is a gram-positive, non-spore-forming microorganism and lactic acid bacterium that is commonly used in dairy product fermentation and probiotics [24]. *L. paracasei* has been isolated from the healthy gastrointestinal tract and human feces [25]. In our previous study, *L. paracasei* JS1 showed an inhibitory effect on cytokine-mediated inflammation in the large intestine, as well as has beneficial effects on the skin due to its ability to produce equol [26]. Thus, it was hypothesized that *L. paracasei* JS1 could be effective in the prevention of osteoporosis, which has a strong relationship with estrogen deficiency. Here, we evaluated the efficacy of the *L. paracasei* JS1-induced fermentation products of *P. lobata* (FPE) which contain various flavonoids, in the prevention of osteoporosis. The efficacies of DHD and EQ, which are derivatives of FPE, and 17β-estradiol (E_2) were compared with the efficacy of FPE.

2. Materials and Methods

2.1. Sample Preparation

P. lobata roots were provided by Professor Tae-Ho Park from the green-house at Daegu University (Daegu, Korea). A voucher specimen (KHU-BMRI-PL-2016) was deposited in the Biomedical Research Institute of Kyung Hee University. *P. lobata* roots were chopped into small pieces and extracted three times in 30% ethanol for 24 h at room temperature, respectively. The extract was subsequently filtered to remove any particulates and concentrated in vacuo. Then, the concentrated crude extract was lyophilized to obtain a powder and stored at −20 °C for subsequent experimentation. Consequently, the yield of PE was 13.2% and FPE was 4.4%, respectively.

2.2. Bacterial Culture and Fermentation

To discover an ideal growth condition, *L. paracasei* JS1 was incubated at various conditions. The bacteria were incubated at 37 °C for 72 h with various speeds of agitation.

The isoflavone contents were measured by high-performance liquid chromatography (HPLC). It turned out that incubation at 37 °C for 72 h with an agitation speed of 250 rpm was the most optimal growth condition, resulting in the most isoflavone content out of all other investigated agitation speeds (data not shown). For the preparation of *P. lobata* fermentation extract (FPE), *L. paracasei* JS1 was cultured in a liquid Gifu anaerobic medium (GAM) under aerobic conditions with 2% *P. lobata* extract at 37 °C for 72 h and an agitation speed of 250 rpm.

2.3. Metabolite Profiling of FPE Using Liquid Chromatography-Mass Spectrometry Analysis

Liquid chromatography-mass spectrometry (LC-MS) analysis was performed using a SCIEX Triple TOF 5600 (SCIEX, Framingham, MA, USA) operating positive ion mode. Mass spectrometry was performed in the MSE acquisition mode, which enabled alternation between high- and low-energy scans. The operating parameters were set as follows: MS scan type, full scan and information dependent acquisition scanning; ionization source, electrospray ionization (ESI); MS scan range, m/z 100 to 2000; MS/MS scan range, m/z 30 to 2000; nebulizing gas, 50 psi; heating gas, 50 psi; curtain gas, 25 psi; desolvation temperature, 500 °C; ion spray voltage floating, 5.5 kV; declustering potential, 60 V; collision energy, 10 V; cone voltage, 35 $\pm$ 15 V.

2.4. Cell Culture

Human MG63 cells, obtained from American Type Culture Collection (ATCC, Manassas, VA, USA) were cultured in alpha-MEM supplemented with 10% fetal bovine serum (FBS) and 1% penicillin and streptomycin (P/S).

Mouse macrophage RAW 264.7 cells purchased Korean Cell Line Bank (KCLB, Seoul, Korea) were cultured in DMEM supplemented with 10% FBS and 1% P/S. Cells were incubated at 37 °C and 5% CO_2 in a humidified atmosphere.

2.5. Osteoblast Proliferation Assay

MG63 cell proliferation was evaluated using the MTT assay. Cells were seeded at 2×10^4 cells per well in 96-well plates. After 24 h of incubation, the cells were washed with phosphate-buffered saline (PBS) and cultured in MEM containing the indicated concentrations of PE, FPE, DHD, or EQ (0.01, 0.1, 1, 10, or 100 μg/mL final concentrations), respectively. After 72 h of incubation, the MTT solution in PBS was added to a final concentration of 0.5 mg/mL, followed by incubation for 4 h at 37 °C. At the end of 4 h of incubation, the supernatant medium was removed. Cell suspension in 100 μL of DMSO was subsequently performed. Absorbance was measured at 540 nm using a microplate reader (Tecan, Mannedorf, Switzerland). Cell proliferation rates were calculated from the OD readings and reported as percentages of vehicle control.

2.6. Alkaline Phosphatase Assay

MG63 cells were maintained in a complete medium. Cells were seeded at 2×10^4 cells per well in 24-well plates. The medium was changed to a differentiation medium consisting of 10 mM glycerol 2-phosphate, 50 μg/mL ascorbic acid, and the indicated concentrations of PE, FPE, DHD, EQ, or E_2 (0.01, 0.1, 1, 10, or 100 μg/mL final concentrations) in culture medium after 24 h of incubation. After 96, 120 h of treatment, the cells were washed with PBS, and assay buffer was added. The solution was centrifuged at 12,000 rpm for 15 min at 4 °C. The supernatant was transferred to 96-well plates, and 5 mM pNPP solution was subsequently added after incubation for 1 h at 25 °C. The reaction was stopped by adding a stop solution, and the absorbance was measured at 405 nm. ALP activity was presented as a percentage, which was compared with ALP activity in control cells.

2.7. Real-Time PCR

Total RNA was extracted using the TRizol reagent (Thermo Fisher Scientific, Waltham, MA, USA) according to the manufacturer's protocols after 72 h of incubation. The expression of osteoporosis-related genes in MG63 cells was detected by real-time quantitative PCR using SYBR Green technology. For each RNA sample, the expression of β-actin was quantified by RT-PCR. The differential gene expression between samples was also estimated using the CT method. The primer sequences were obtained from Takara (Kyoto, Japan). All target gene-specific primer sequences are listed in Table 1.

Table 1. Target gene-specific primer sequences.

Target Genes	5′ to 3′	Sequence
RANKL	Forward	ATGGCGTCCTCTCTGCTTG
	Reverse	TGAAAGGTCAGCGTATGGCTT
Runx2	Forward	CCGGTCTCCTTCCAGGAT
	Reverse	GGGAACTGCTGTGGCTTC
Osteocalcin	Forward	CCGGTCTCCTTCCAGGAT
	Reverse	GGGAACTGCTGTGGCTTC
Osterix	Forward	ATGGCGTCCTCTCTGCTTG
	Reverse	TGAAAGGTCAGCGTATGGCTT
β-actin	Forward	CGCTGATGCATGCCTATGA
	Reverse	AGAGGTCCACAGAGCTGATTCC

2.8. Osteoclast Differentiation Assay

RAW 264.7 cells were cultured to 2×10^4 cells per well in 24-well plates. After 24 h of incubation, the cells were washed with PBS and cultured for 5 days after treatment with 50 ng/mL of RANKL and 30 ng/mL of M-CSF (Peprotech, Rocky Hill, NJ, USA). After treatment with R/M, 100 μg/mL of PE and FPE, 10 μg/mL of DHD and EQ, and 10^{-9} M of E_2 were added, and the cells were incubated for 48 h at 5% CO_2 and 37 °C, respectively. Out of the osteoclast population, TRAP-positive multinucleated cells (MNCs) with more than three nuclei were measured using the Acid Phosphatase Leukocyte (TRAP) Kit (Sigma-Aldrich, St.Louis, MA, USA). The cells were removed by washing with PBS, and then each well was photographed by a microspore camera (Tecan).

2.9. Animals

Female 6-week-old outbred ICR mice were purchased from RaonBio Co., Ltd. (Yongin, Korea) and were surgically ovariectomized under tiletamine/zolazapamanesthetic (RaonBio Co., Ltd.). Prior to experimentation, mice were kept for a week and provided with water and sterile standard mouse chow ad libitum. They were housed in an air-conditioned animal room in a 12 h light/dark cycle at a temperature of 22 ± 1 °C and a humidity of 50 ± 10%. All experimental protocols involving the use of animals were conducted in accordance with National Institutes of Health guidelines and approved by the Committee on Animal Care of Kyung Hee University (KHUASP(SE)-18-030, approved on 20 September 2017).

Mice were randomly divided into 5 groups with 6 mice in each group as follows: (1) sham-operated control mice (Sham) received daily oral gavage of 0.85% NaCl, (2) OVX mice received daily oral gavage of 0.85% NaCl (OVX), (3) OVX mice were treated daily with 100 mg PE per kg body weight (b.w), (4) OVX mice were treated daily with 100 mg FPE per kg b.w, and (5) OVX mice received intraperitoneal (i.p) injections of E_2 (0.1 mg/kg b.w/day) every other day. Each treatment administration was performed for 12 weeks. At the end of treatment, the mice were sacrificed, and adequate serum was provided for CT analysis at the same time.

2.10. Micro-CT Bone Analysis

The proximal and distal parts of the left tibia were scanned by micro-computer tomography (Micro-CT, Skyscan 1076, Billerica, MA, USA) to evaluate structural loss in cortical and trabecular bone. Exposures were carried out at 50 kVp, 200 mA, and 360 ms. The micro-CT scans were analyzed with Comprehensive Tex Archive Network (CTAN) topographic reconstruction software. The trabecular bone volume represented the total trabecular bone within the total bone volume. By dividing the trabecular bone volume by the total volume, the bone volume percentage was calculated. The assessed cortical bone parameters were BMD, bone volume fraction (BV), mean polar moment of inertia (MMI), and cross-section thickness (Cs.Th). Trabecular bone parameters were used to assess the bone volume fraction (BV/TV), specific bone surface (BS/BV), trabecular thickness (Tb.Th), trabecular separation (Tb.Sp), trabecular number (Tb.N), trabecular bone pattern factor (Tb.Pf), structure model index (SMI), and BMD.

2.11. *Serum Analysis*

At the end of the study, all animals fasted for 6 h, and blood was collected from the abdominalena cava under anesthesia with isoflurane. Blood was centrifuged at 12,000 rpm for 20 min at 4 °C to obtain serum. The serums were analyzed with enzyme-linked immunosorbent assay (ELISA) kits for RANKL, osteoprotegerin (OPG, R&D systems, Minneapolis, MN, USA), and osteocalcin (Alfa Aesar, Ward Hill, MA, USA). All ELISA procedures were performed according to the manufacturer's protocols.

2.12. *Statistical Analysis*

Each experiment was repeated three or six times, and the results of the most representative experiment are shown. The results are expressed as the mean ± S.E.M and were analyzed using one-way ANOVA followed by Tukey's method (GraphPad Prism 5.0, San Diego, CA, USA). A statistical probability of $p < 0.05$ was considered significant.

3. Results and Discussion

3.1. *Identification of Isoflavones in FPE*

P. lobata is used as herbal medicine and food [14,15]. *P. lobata* has demonstrated a variety of results related to osteoporosis [17–21]. Therefore, we evaluated the efficacy of fermented *P. lobata*, which contains various flavonoids, in the prevention of osteoporosis. In previous studies, it is confirmed that daidzein is converted to dihydrodaidzein and equol [27]. Furthermore, LC/MS qualitative analyses were performed to reveal other active components of FPE, especially isoflavones which are responsible for the strong inhibitory activities on osteoporosis. LC-MS analysis was conducted using the SCIEX Triple TOF 5600 (SCIEX) in positive ion mode. Using an HPLC system with a Kinetex F5 C18 column, the FPE were separated at a flow rate of 0.3 mL/min within 0 to 30 min. Each data set was processed with Analyst TF1.7 (SCIEX, Framingham, MA, USA) software. The total ion chromatograms (TIC) of FPE, including the name of each metabolite, retention time (RT), molecular formula, mass value, and accuracy, are shown in Figure 1. Particularly, puerarin, genistein, daidzein, calycosin, and equol are already known for their efficacy of anti-osteoporosis, respectively [28–32]. Therefore, it is believed that FPE, which includes various effective components, can be valuable candidate for anti-osteoporosis.

3.2. *Effects of FPE on Proliferation, ALP Activities, and Gene Expression in MG 63 Cells*

Osteoporosis is a metabolic disease characterized by relatively low bone density and bone mass in comparison to healthy individuals [2–4]. It is also related to a homeostatic disproportion between bone formation and resorption. Even though the major cause of the disease is unclear, genetic, endocrinological, or nutritional factors are believed to be associated with osteoporosis.

The underlying mechanisms of the cellular effects of FPE have been studied in osteoblasts and osteoclasts. Osteoblasts and osteoclasts are specialized cells that are responsible for bone formation and resorption, respectively. The bone-forming cells, osteoblasts, synthesize and modulate the deposition and mineralization of the extracellular matrix of bone. Osteoclasts, on the other hand, are responsible for the resorption of aged bone.

Continuous stress leads to the inhibition of osteoblast activity and enhances osteoclast-mediated bone resorption, thus triggering a decrease in bone mass. Therefore, it can potentially lead to osteoporosis in the long term [33].

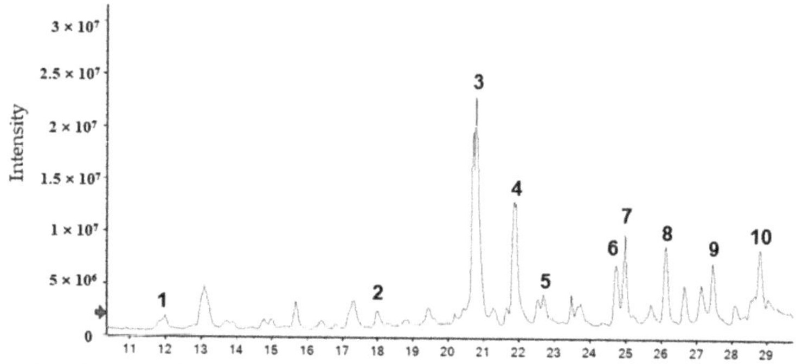

No.	RT (min)	Compound	Molecular formula	Observed Mass (Da)	Expected Mass (Da)	Mass Accuracy (ppm)
1	11.99	puerarin	$C_{21}H_{20}O_9 + H$	417.1183	417.1186	0.623
2	17.96	genistein	$C_{15}H_{10}O_5 + H$	271.0606	271.0607	0.184
3	20.78	daidzein	$C_{15}H_{10}O_4 + H$	255.0657	255.0657	0.137
4	21.91	prunetin	$C_{16}H_{12}O_5 + H$	285.0761	285.0763	0.702
5	22.71	calycosin	$C_{16}H_{12}O_5 + H$	285.0762	285.0763	0.351
6	24.72	8-hydroxydaidzein	$C_{15}H_{10}O_5 + H$	271.0599	271.0607	2.767
7	25.02	equol	$C_{15}H_{14}O_3 + H$	243.1019	243.1021	0.905
8	26.13	trimethoxyflavone	$C_{18}H_{16}O_5 + H$	313.108	313.1076	1.278
9	27.45	luteone	$C_{20}H_{18}O_6 + H$	355.1179	355.1182	0.746
10	28.80	pratol	$C_{16}H_{12}O_4 + H$	269.0814	269.0814	0.056

Figure 1. HPLC-QTOF/MS chromatograms of FPE, and mass data in total ion chromatograms (TIC) scan mode. Analysis was carried out on a Kinetex F5 C18 column (2.6 μm, 2.1 × 100 mm) with gradient elution using solvents A (0.1% formic acid in water) and B (0.1% formic acid in acetonitrile). The elution gradients were as follows: 0–6 min, B 5%; 6–8 min, B 5–15%; 8–16 min, B 15–20%; 16–27 min, B 20–30%; 27–32 min, B 30–100%; and 30–34 min, B 100%. The flow rate was 300 μL/min, and the injection volume was 10 μL for each run. Mass detector settings were as follows: cone voltage 40 V; capillary 3.0 kV; source temperature 500 °C; cone gas flow 30 L/h; and desolvation gas flow at 800 L/h. The peaks of 1–10 were identified to be 10 metabolites.

The proliferative effects of FPE were evaluated in MG63 osteoblasts. PE and FPE significantly increased the proliferation rate (%) in a concentration-dependent manner (Figure 2A,B). FPE showed the ability to enhance the proliferation of cells as much as DHD (Figure 2C). As shown in Figure 2D, EQ, on the other hand, demonstrated a weaker proliferative ability than PE, FPE, and DHD. The proliferation rate of E_2 (10^{-11} M) treated cell, the positive control, increased 114.8 ± 0.77% compared to control.

The alkaline phosphatase activity, a biological marker to indicate osteoblastic differentiation, was assessed in MG 63 osteoblasts. PE, FPE, DHD, EQ, and E_2 treatments were performed for 96 or 120 h each. ALP activities of PE and E_2 were significantly increased 96 h after treatment (Figure 3A). FPE showed a higher differentiation ability compared to PE and other compounds, including E_2, which was the positive control, after 120 h (Figure 3B).

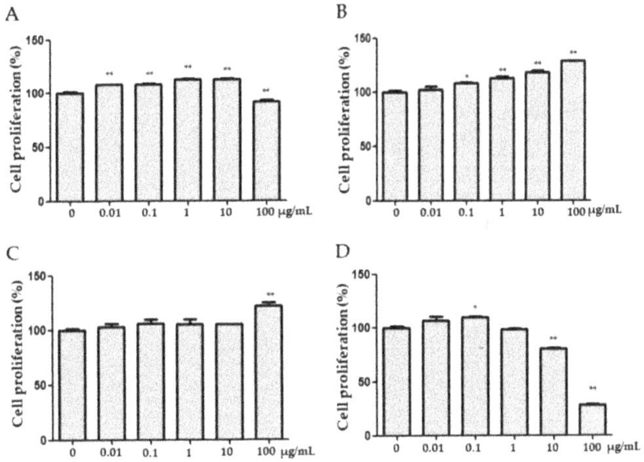

Figure 2. Proliferation activity in osteoblast MG63 cells. Effect of a variety concentration of (**A**) PE; (**B**) FPE; (**C**) DHD; (**D**) Equol on cell proliferation rate. * $p < 0.05$ compared to 0 μg/mL, ** $p < 0.01$ compared to 0 μg/mL. Data are means ± SEM of triplicates from one representative experiment.

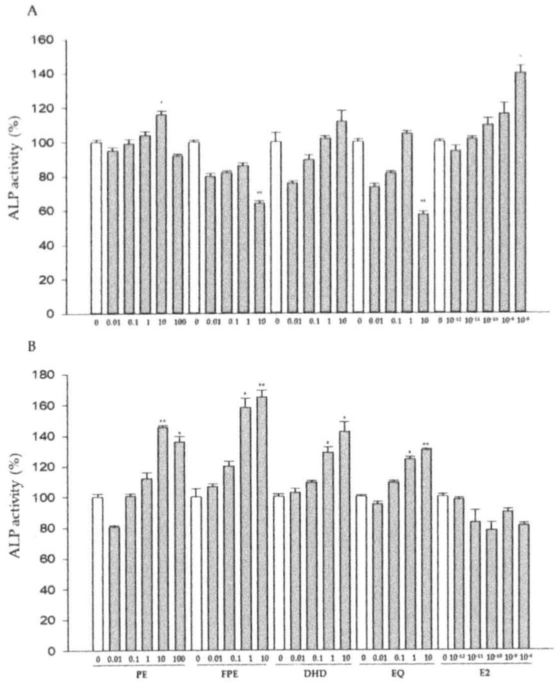

Figure 3. Effects of alkaline phosphate (ALP) activity in MG63 cells: (**A**) 96 h after treatment; (**B**) 120 h after treatment. PE: *Pueraria lobata* extract; FPE: fermented *Pueraria lobata* extract; DHD; dihydrodaidzein; EQ; equol. The results are expressed as a mean ± SEM (n = 3). * $p < 0.05$ compared to control, ** $p < 0.01$ compared to control.

The efficacy of FPE increased in a time-dependent manner. Consequently, it was demonstrated that the gene related to osteoblast differentiation was regulated by FPE.

FPE also increased calcium deposition in MG63 cells. For further exploration of the mechanisms of FPE's role in osteoblastic regulation, the markers of bone formation were evaluated in MG63 cells using RT-PCR. The levels of several bone formation biomarkers, OPG/RANKL, osteocalcin, osterix, and Runx2 mRNAs, were up-regulated by FPE treatment.

Osteoblasts synthesize and secrete OPG, which blocks the interaction between RANK and RANKL. Hence, the expression of OPG/RANKL plays a crucial role in modulating bone restoration. Additionally, OPG was able to block the interaction between RANKL and RANK, thus inhibiting osteoclastogenesis.

The efficacy of FPE in osteoblast differentiation was further elucidated by evaluating the expression of osteogenic differentiation mediator mRNA with RT-PCR. As shown in Figure 4, FPE increased the RANKL/OPG ratio in a concentration-dependent manner, but other genes (osteocalcin, osterix, and Runx2) did not have a significant effect.

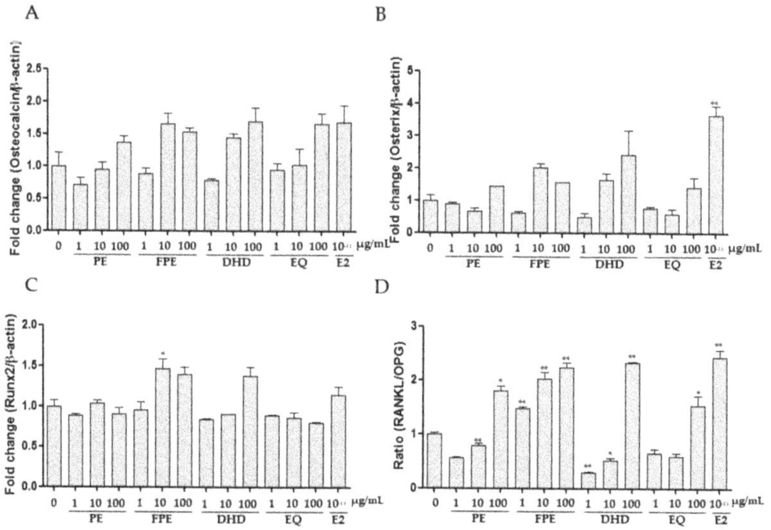

Figure 4. Expression of mRNA of osteogenic differentiation mediators, (**A**) osteocalcin/β-actin; (**B**) osterix; (**C**) Runx2; (**D**) RANKL/OPG, in MG 63 cells measured by real-time RT-PCR. The expression of β-actin was used as a loading control for RT-PCR. * $p < 0.05$ compared to normal controls and ** $p < 0.01$ compared to control. The results are expressed as a mean ± SEM ($n = 3$).

3.3. *Effect of FPE on Osteoclast Differentiation in RAW 264.7 Cells*

The effects of FPE on RANKL/M-CSF-stimulated osteoclast differentiation in osteoclast precursor RAW 264.7 cells were then investigated. After 5 days of treatment of RANKL/M-CSF alone or with PE, FPE, DHD, EQ, and E_2, TRAP-positive MNC formation was examined, and TRAP-positive MNCs that contained more than three nuclei were counted as osteoclasts. RANKL/M-CSF treatment alone induced the formation of TRAP-positive multinuclear osteoclasts. RANKL/M-CSF-induced multinuclear osteoclast formation was significantly decreased following culturing with 100 μg/mL FPE (Figure 5).

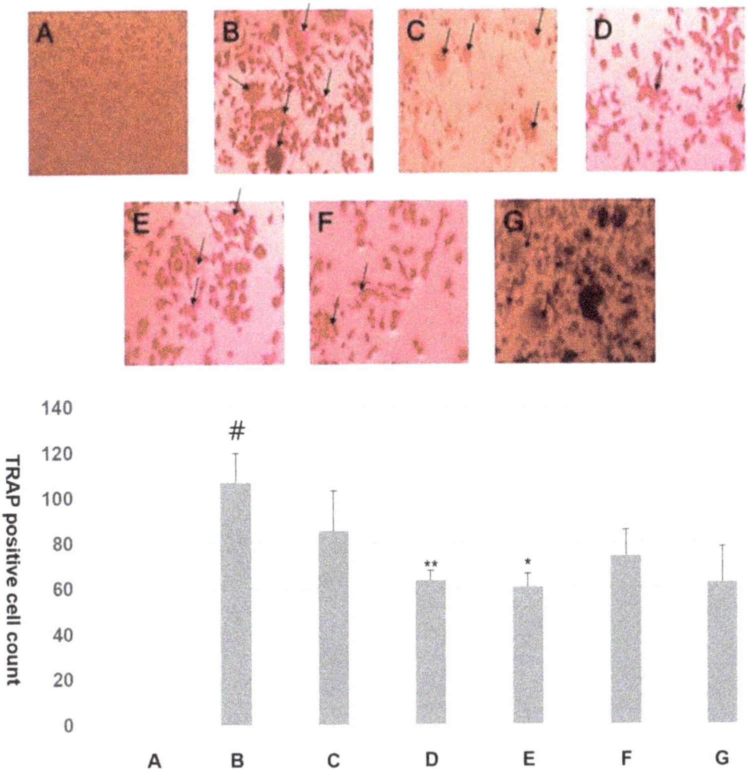

Figure 5. Osteoclast differentiation in osteoclast precursor RAW 264.7 cells: (**A**) Control; (**B**) RANKL 50 ng/mL + M-CSF 30 ng/mL; (**C**) RANKL/M-CSF + PE 100 μg/mL; (**D**) RANKL/M-CSF + FPE 100 μg/mL; (**E**) RANKL/M-CSF + DHD 10 μg/mL; (**F**) RANKL/M-CSF + EQ 10 μg/mL; (**G**) RANKL/M-CSF + E_2 10^{-9} M 48 h after treatment. # $p < 0.01$ compared to control. * $p < 0.05$ compared to RANKL 50 ng/mL + M-CSF 30 ng/mL group. ** $p < 0.01$ compared to RANKL 50 ng/mL + M-CSF 30 ng/mL group.

3.4. *Effects of FPE in OVX-Induced Mouse Model*

The OVX-induced mouse model has been widely used to investigate postmenopausal osteoporosis caused by estrogen deficiency [34]. An in vivo experiment using an OVX-induced bone loss mouse model was conducted to evaluate the anti-osteoporotic activity of FPE. As shown in Figure 6A, the body weight increase in the OVX group was higher compared with the sham operation group. On the other hand, mice given an oral administration of PE and FPE for 12 weeks after OVX showed a comparatively slower increase in body weight in comparison with the OVX group. Moreover, the decrease in uterus weight in the OVX group was lower than that in the sham operation group (Figure 6B). Mice that underwent intraperitoneal E_2 injection for 12 weeks after OVX showed a significant increase in uterus weight compared with the OVX group. However, mice given an oral administration of PE and FPE did not differ from sham-operated mice in terms of uterus weight. This result indicated that PE and FPE do not likely have the ability to stimulate or control hormone-dependent phenomena.

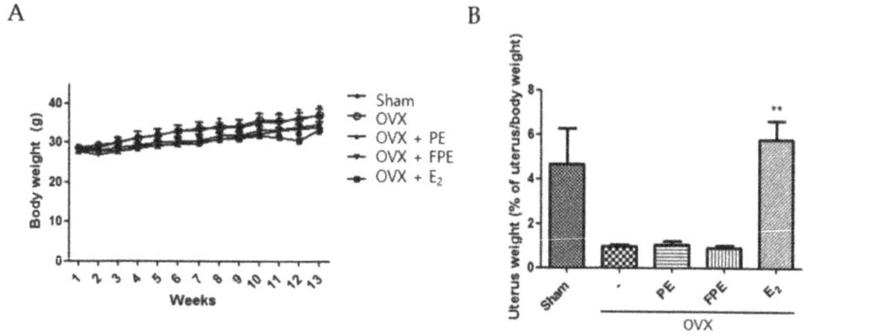

Figure 6. Effects of FPE on changes of body weight and uterus weight in OVX-induced mice. (**A**) Body weight was measured once a week; (**B**) at the end of the experiment, the uterus was removed and weighed. PE: *Pueraria lobata* extract; FPE: fermented *Pueraria lobata* extract. ** $p < 0.01$ compared to OVX induced control.

Deterioration of trabecular microarchitecture is apparent in the OVX mouse model [35]. We found that FPE prevented the deterioration of microstructural parameters in the distal femur of OVX model mice. An oral administration of FPE restored bone loss in the OVX mouse model.

To explore the structural characteristics affected by FPE, we scanned the tibia of each mouse (Figure 7A). The experimental and structural parameters for the entire cortical bone of the tibia, BV, BMD, MMI, and Cs. Th were measured and calculated in micro-CT images (Figure 7B). In the OVX group, BV, Cs. Th, MMI, and BMD did not change compared to the sham group. In addition, PE, FPE, and E_2 did not confer any significant change in the structural parameters.

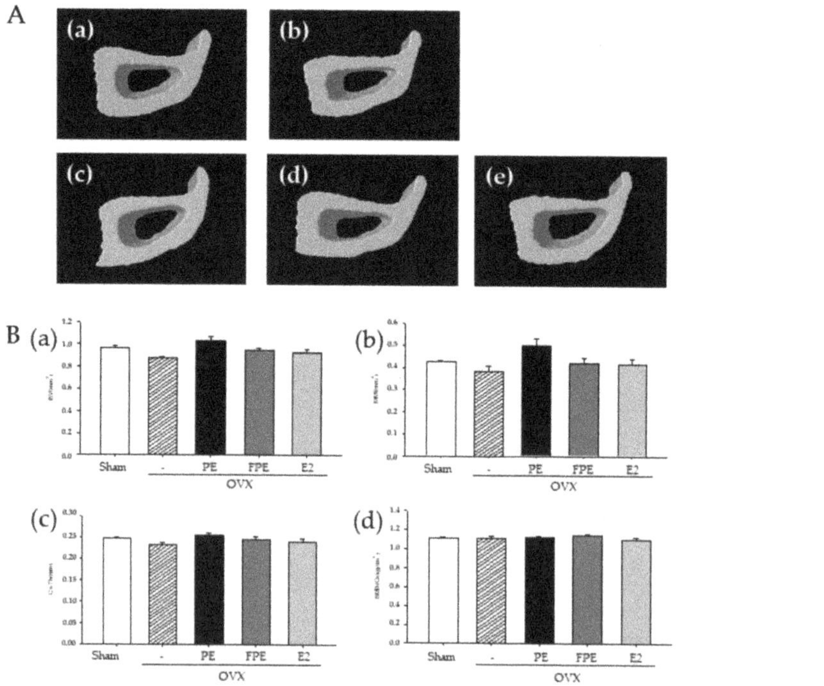

Figure 7. Micro-CT analysis of the cortical bone. (**A**) Micro-CT images of the cortical bone of the tibia;

(**a**) sham; (**b**) OVX; (**c**) OVX + PE 100 mg/kg; (**d**) OVX + FPE 100 mg/kg; (**e**) E_2; (**B**) (**a**) BV; (**b**) MMI; (**c**) Cs.Th; (**d**) BMD of OVX-induced mice treated PE, FPE, and E_2.

As illustrated in Figure 8A, OVX caused degradation of trabecular bone architecture compared to the sham group. However, treatment with PE and FPE retarded or recovered the destruction of femur trabecular bone in the OVX-induced bone loss mouse model. The protective effect on trabecular bone architecture was also clearly demonstrated by treatment with E_2, a positive control. In addition, BV/TV, Tb.Th, Tb.N, and BMD were lower, whereas trabecular separation, BS/BV, Tb.Th, Tb.Pf, and SMI were higher than the sham group (Figure 8B). In both the PE, FPE, and E_2 treatment groups, BV/TV, Tb.Th, Tb.N, and BMD were increased in sham-operated mice compared to those of the OVX group, while BS/BV, Tb.Sp, Tb.Pf, and SMI were decreased compared to those of the OVX group.

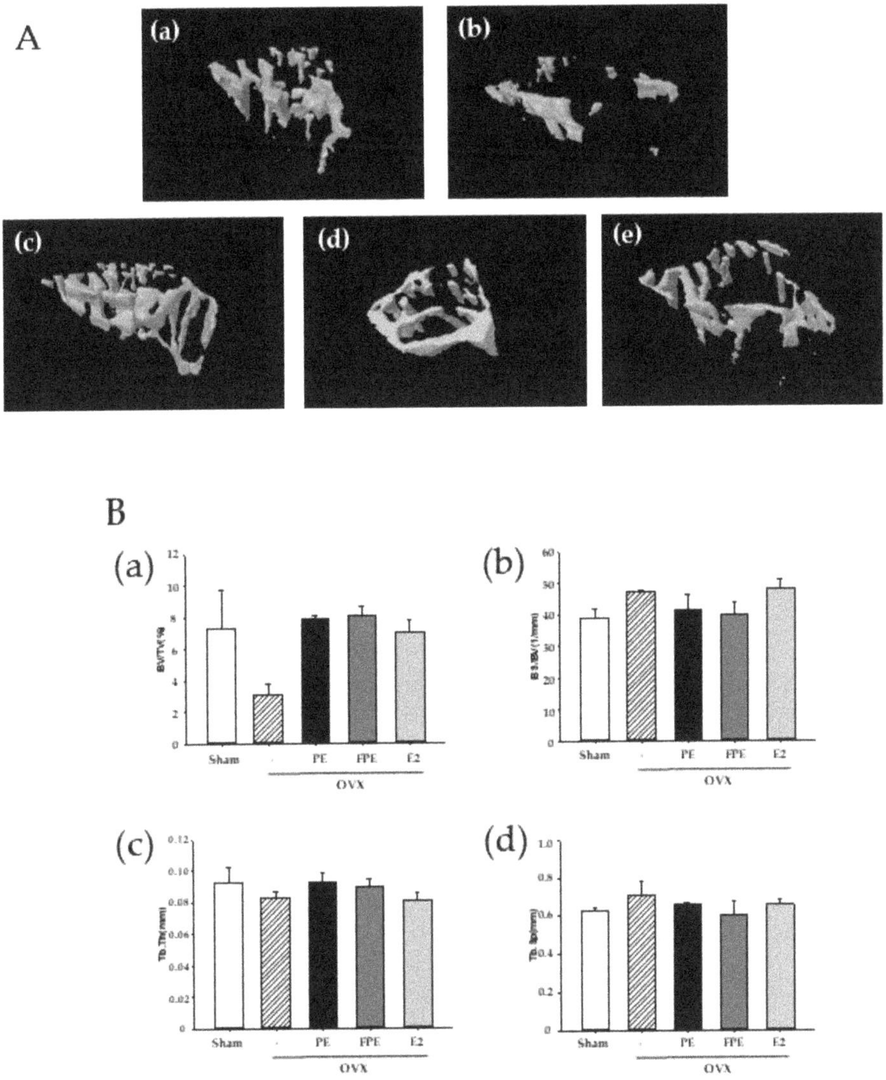

Figure 8. *Cont.*

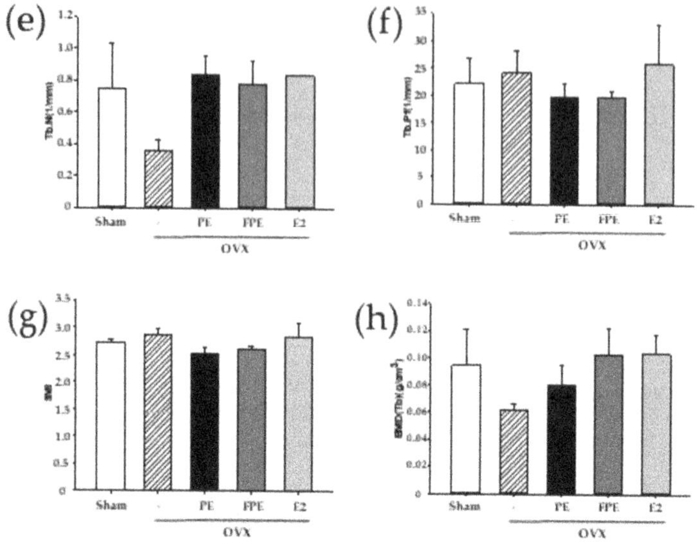

Figure 8. Micro-CT analysis of the trabecular bone. (**A**) Micro-CT images of the trabecular bone of the femur; (**a**) sham; (**b**) OVX; (**c**) OVX + PE 100 mg/kg; (**d**) OVX + FPE 100 mg/kg; (**e**) E_2; (**B**) (**a**) BV/TV; (**b**) BS/BV; (**c**) Tb.Th; (**d**) Tb.Sp; (**e**) Tb.N; (**f**) Tb.Pf; (**g**) SMI; (**h**) BMD of OVX-induced mice treated PE, FPE, and E_2.

These results suggest that FPE was effective in preserving bone mass as well as in restoring the deterioration of bone microarchitecture associated with OVX mice. Analyses of the serum levels of OPG/RANKL and osteocalcin, biomarkers of bone resorption, were shown to be significantly higher than those of the sham group. In the present study, PE and FPE showed protective efficacy against OVX-induced osteoporosis. It is hypothesized that the FPE could be effective in inhibiting the bone resorption process. Since the balance between the RANKL and OPG produced by osteoblasts is critical for osteoclast regulation, we determined serum levels of RANKL and OPG by ELISA (Figure 9). In the OVX group, serum levels of RANKL were decreased, whereas OPG was decreased compared to that of the sham group. Our results indicate that FPE treatment significantly increased the level of OPG; however, it did not affect the level of RANKL. As a consequence, the RANKL/OPG ratio was significantly decreased by FPE treatment. On the other hand, osteocalcin, which is the most abundant protein in bone matrix after its synthesis by osteoblasts, increased dramatically following FPE treatment compared to other groups. This suggests that FPE most likely prevented bone loss through decreased bone turnover.

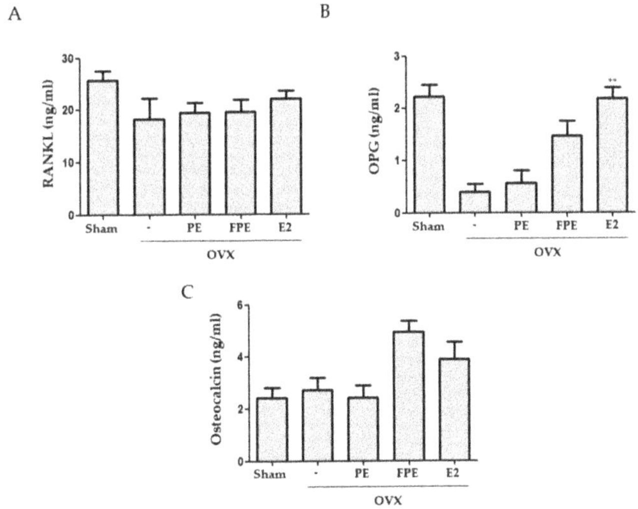

Figure 9. Serum levels of (**A**) RANKL, (**B**) OPG, and (**C**) osteocalcin measured by ELISA.

4. Conclusions

The present study suggests that FPE has anti-osteoporosis efficacy both in vitro and in vivo. In vitro studies indicated that FPE tends to have more regulatory effects on MG63 cells than PE and the positive controls. In the OVX-induced mouse model, FPE was effective in preserving bone mass and preventing the deterioration of microstructural parameters. These data provide a pharmacological basis on which FPE could be considered a therapeutic compound to prevent osteoporotic bone loss. Further research is needed to ensure the safety, specificity, and efficacy of FPE to develop its therapeutic potential. To support the present study, additional experiments such as an inhibition on osteoclast activity or mechanism analysis on RANKL-induced RAW 264.7 cells are considered. In addition, more high-quality clinical research is necessary to provide evidence for FPE as an effective anti-osteoporotic candidate.

Author Contributions: Conceptualization, J.E.K. and S.C.K.; methodology, J.E.K. and Y.-G.L.; software, Y.-G.L., T.K.; validation, S.Y.K., H.-J.L.; formal analysis, J.E.K.; investigation, J.E.K. and S.Y.K.; resources, J.E.K.; data curation, T.K.; writing—original draft preparation, J.E.K. and S.Y.K.; writing—review and editing, J.E.K., S.Y.K. and S.C.K.; visualization, H.-J.L. and T.K.; supervision, Y.-G.L. and S.C.K.; project administration, S.C.K. All authors have read and agreed to the published version of the manuscript.

Funding: This research received no external funding.

Institutional Review Board Statement: The study was conducted according to the guidelines of the Declaration of Helsinki and approved in accordance with the current ethical regulations for anima care and use at Kyung Hee University (KHUASP(SE)-18-030, approved on 20 September 2017.

Data Availability Statement: Data are contained within the article.

Conflicts of Interest: The authors declare no conflict of interest.

References

1. Magee, P.J. Is equol production beneficial to health? *Proc. Nutr. Soc.* **2011**, *70*, 10–18. [CrossRef] [PubMed]
2. Alexander, J.M.; Bab, I.; Fish, S.; Muller, R.; Uchiyama, T.; Gronowicz, G.; Nahounou, M.; Zhao, Q.; White, D.W.; Chorev, M.; et al. Human parathyroid hormone 1–34 reverses bone loss in ovariectomized mice. *J. Bone Miner. Res.* **2001**, *16*, 1665–1673. [CrossRef]
3. Chung, H.J.; Cho, L.; Shin, J.S.; Lee, J.; Ha, I.H.; Park, H.J.; Lee, S.K. Effects of JSOG-6 on protection against bone loss in ovariectomized mice through regulation of osteoblast differentiation and osteoclast formation. *BMC Complement. Altern. Med.* **2014**, *14*, 184. [CrossRef] [PubMed]

4. Raisz, L.G. Pathogenesis of osteoporosis: Concepts, conflicts, and prospects. *J. Clin. Investig.* **2005**, *115*, 3318–3325. [CrossRef] [PubMed]
5. Fu, Y.S.; Lu, C.H.; Chu, K.A.; Yeh, C.C.; Chiang, T.L.; Ko, T.L.; Chiu, M.M.; Chen, C.F. Xenograft of Human Umbilical Mesenchymal Stem Cells from Wharton's Jelly Differentiating into Osteocytes and Reducing Osteoclast Activity Reverses Osteoporosis in Ovariectomized Rats. *Cell Transplant.* **2018**, *27*, 194–208. [CrossRef]
6. Seeman, E. Invited Review: Pathogenesis of osteoporosis. *J. Appl. Physiol.* **2003**, *95*, 2142–2151. [CrossRef]
7. Kang, H.; Aryal, A.C.S.; Marini, J.C. Osteogenesis imperfecta: New genes reveal novel mechanisms in bone dysplasia. *Transl. Res.* **2017**, *181*, 27–48. [CrossRef]
8. Meunier, P.J. Hyperthyroidism and osteoporosis. *Ann. Endocrinol.* **1995**, *56*, 57–59. [PubMed]
9. Redlich, K.; Smolen, J.S. Inflammatory bone loss: Pathogenesis and therapeutic intervention. *Nat. Rev. Drug Discov.* **2012**, *11*, 234–250. [CrossRef] [PubMed]
10. Zhang, G.; Qin, L.; Hung, W.Y.; Shi, Y.Y.; Leung, P.C.; Yeung, H.Y.; Leung, K.S. Flavonoids derived from herbal Epimedium Brevicornum Maxim prevent OVX-induced osteoporosis in rats independent of its enhancement in intestinal calcium absorption. *Bone* **2006**, *38*, 818–825. [CrossRef] [PubMed]
11. Dionyssiotis, Y.; Skarantavos, G.; Papagelopoulos, P. Modern rehabilitation in osteoporosis, falls, and fractures. *Clin. Med. Insights Arthritis Musculoskelet. Disord.* **2014**, *7*, 33–40. [CrossRef] [PubMed]
12. Das, S.; Crockett, J.C. Osteoporosis—A current view of pharmacological prevention and treatment. *Drug Des. Dev. Ther.* **2013**, *7*, 435–448. [CrossRef]
13. Chow, T.H.; Lee, B.Y.; Ang, A.B.F.; Cheung, V.Y.K.; Ho, M.M.C.; Takemura, S. The effect of Chinese martial arts Tai Chi Chuan on prevention of osteoporosis: A systematic review. *J. Orthop. Transl.* **2018**, *12*, 74–84. [CrossRef] [PubMed]
14. Zheng, X.; Lee, S.K.; Chun, O.K. Soy Isoflavones and Osteoporotic Bone Loss: A Review with an Emphasis on Modulation of Bone Remodeling. *J. Med. Food* **2016**, *19*, 1–14. [CrossRef]
15. Hickman, J.E.; Wu, S.; Mickley, L.J.; Lerdau, M.T. Kudzu (*Pueraria montana*) invasion doubles emissions of nitric oxide and increases ozone pollution. *Proc. Natl. Acad. Sci. USA* **2010**, *107*, 10115–10119. [CrossRef] [PubMed]
16. Koirala, P.; Seong, S.H.; Jung, H.A.; Choi, J.S. Comparative molecular docking studies of lupeol and lupenone isolated from Pueraria lobata that inhibits BACE1: Probable remedies for Alzheimer's disease. *Asian Pac. J. Trop. Med.* **2017**, *10*, 1117–1122. [CrossRef] [PubMed]
17. Keung, W.M.; Vallee, B.L. Kudzu root: An ancient Chinese source of modern antidipsotropic agents. *Phytochemistry* **1998**, *47*, 499–506. [CrossRef]
18. Gao, Y.; Wang, X.; He, C. An isoflavonoid-enriched extract from Pueraria lobata (kudzu) root protects human umbilical vein endothelial cells against oxidative stress induced apoptosis. *J. Ethnopharmacol.* **2016**, *193*, 524–530. [CrossRef]
19. Jin, S.E.; Son, Y.K.; Min, B.S.; Jung, H.A.; Choi, J.S. Anti-inflammatory and antioxidant activities of constituents isolated from Pueraria lobata roots. *Arch. Pharm. Res.* **2012**, *35*, 823–837. [CrossRef]
20. Seong, S.H.; Roy, A.; Jung, H.A.; Jung, H.J.; Choi, J.S. Protein tyrosine phosphatase 1B and alpha-glucosidase inhibitory activities of Pueraria lobata root and its constituents. *J. Ethnopharmacol.* **2016**, *194*, 706–716. [CrossRef]
21. Cheung, D.W.; Koon, C.M.; Wat, E.; Ko, C.H.; Chan, J.Y.; Yew, D.T.; Leung, P.C.; Chan, W.Y.; Lau, C.B.; Fung, K.P. A herbal formula containing roots of Salvia miltiorrhiza (Danshen) and Pueraria lobata (Gegen) inhibits inflammatory mediators in LPS-stimulated RAW 264.7 macrophages through inhibition of nuclear factor kappaB (NFkappaB) pathway. *J. Ethnopharmacol.* **2013**, *145*, 776–783. [CrossRef] [PubMed]
22. Park, K.H.; Gu, D.R.; Jin, S.H.; Yoon, C.S.; Ko, W.; Kim, Y.C.; Lee, S.H. Pueraria lobate Inhibits RANKL-Mediated Osteoclastogenesis Via Downregulation of CREB/PGC1beta/c-Fos/NFATc1 Signaling. *Am. J. Chin. Med.* **2017**, *45*, 1725–1744. [CrossRef] [PubMed]
23. Huh, J.E.; Yang, H.R.; Park, D.S.; Choi, D.Y.; Baek, Y.H.; Cho, E.M.; Cho, Y.J.; Kang-Il, K.; Kim, D.Y.; Lee, J.D. Puerariae radix promotes differentiation and mineralization in human osteoblast-like SaOS-2 cells. *J. Ethnopharmacol.* **2006**, *104*, 345–350. [CrossRef] [PubMed]
24. Hessle, C.; Hanson, L.A.; Wold, A.E. Lactobacilli from human gastrointestinal mucosa are strong stimulators of IL-12 production. *Clin. Exp. Immunol.* **1999**, *116*, 276–282. [CrossRef]
25. Molin, G.; Jeppsson, B.; Johansson, M.L.; Ahrne, S.; Nobaek, S.; Stahl, M.; Bengmark, S. Numerical taxonomy of Lactobacillus spp. associated with healthy and diseased mucosa of the human intestines. *J. Appl. Bacteriol.* **1993**, *74*, 314–323. [CrossRef]
26. Kwon, J.E.; Lim, J.; Bang, I.; Kim, I.; Kim, D.; Kang, S.C. Fermentation product with new equol-producing Lactobacillus paracasei as a probiotic-like product candidate for prevention of skin and intestinal disorder. *J. Sci. Food Agric.* **2019**, *99*, 4200–4210. [CrossRef]
27. Matthies, A.; Loh, G.; Blaut, M.; Braune, A. Daidzein and Genistein Are Converted to Equol and 5-Hydroxy-Equol by Human Intestinal Slackia isoflavoniconvertens in Gnotobiotic Rats. *Nutr. J.* **2012**, *142*, 40–46. [CrossRef]
28. Xiao, L.; Zhong, M.; Huang, Y. Puerarin alleviates osteoporosis in the ovariectomy-induced mice by suppressing osteoclastogenesis via inhibition of TRAF6/ROS-dependent MAPK/NF-κB signaling pathways. *Aging* **2020**, *12*, 21706–21729. [CrossRef]
29. An, J.; Yang, H.; Zhang, Q. Natural products for treatment of osteoporosis: The effects and mechanisms on promoting osteoblast-mediated bone formation. *Life Sci.* **2016**, *147*, 46–58. [CrossRef]

30. Zou, S.E.; Zhang, S.F.; Zhang, R.; Zhang, J. Role of the cross-talk between estrogen receptors and peroxisome proliferator-activated receptor gamma in daidzein's prevention and treatment of osteoporosis in ovariectomized rats. *Zhonghua Yi Xue Za Zhi* **2009**, *89*, 2972–2975. [PubMed]
31. Li, N.; Tu, Y.; Shen, Y.; Qin, Y.; Lei, C.; Liu, X. Calycosin attenuates osteoporosis and regulates the expression of OPG/RANKL in ovariectomized rats via MAPK signaling. *Pharmazie* **2016**, *71*, 607–612. [CrossRef]
32. Fujioka, M.; Uehara, M.; Wu, J. Equol, a metabolite of daidzein, inhibits bone loss in ovariectomized mice. *J. Nutr.* **2004**, *134*, 2623–2627. [CrossRef] [PubMed]
33. Katagiri, T.; Takahashi, N. Regulatory mechanisms of osteoblast and osteoclast differentiation. *Oral. Dis.* **2002**, *8*, 147–159. [CrossRef] [PubMed]
34. Frolik, C.A.; Bryant, H.U.; Black, E.C.; Magee, D.E.; Chandrasekhar, S. Time-dependent changes in biochemical bone markers and serum cholesterol in ovariectomized rats: Effects of raloxifene HCl, tamoxifen, estrogen, and alendronate. *Bone* **1996**, *18*, 621–627. [CrossRef]
35. Cano, A.; Dapia, S.; Noguera, I.; Pineda, B.; Hermenegildo, C.; del Val, R.; Caeiro, J.R.; Garcia-Perez, M.A. Comparative effects of 17beta-estradiol, raloxifene and genistein on bone 3D microarchitecture and volumetric bone mineral density in the ovariectomized mice. *Osteoporos. Int.* **2008**, *19*, 793–800. [CrossRef]

fermentation

MDPI

Article

Metagenomic Analysis of Bacterial Diversity in Traditional Fermented Foods Reveals Food-Specific Dominance of Specific Bacterial Taxa

Purbajyoti Deka [1], Gajanan T. Mehetre [1], Esther Lalnunmawii [1], Kalidas Upadhyaya [2], Garima Singh [3], Abeer Hashem [4], Al-Bandari Fahad Al-Arjani [4], Elsayed Fathi Abd_Allah [5] and Bhim Pratap Singh [1,6,*]

1 Department of Biotechnology, Mizoram University, Aizawl 796004, Mizoram, India; purba562@gmail.com (P.D.); gtmehetre@gmail.com (G.T.M.); essie.zadeng@gmail.com (E.L.)
2 Department of Forestry, Mizoram University, Aizawl 796004, Mizoram, India; kumzu70@gmail.com
3 Department of Botany, Pachhunga University College (PUC), Mizoram University, Aizawl 796001, Mizoram, India; garima.singh106@gmail.com
4 Botany and Microbiology Department, College of Science, King Saud University, P.O. Box 2460, Riyadh 11451, Saudi Arabia; habeer@ksu.edu.sa (A.H.); aalarjani@ksu.edu.in (A.-B.F.A.-A.)
5 Plant Production Department, College of Food and Agricultural Sciences, King Saud University, P.O. Box 2460, Riyadh 11451, Saudi Arabia; eabdallah@ksu.edu.sa
6 Department of Agriculture and Environmental Sciences (AES), National Institute of Food Technology Entrepreneurship & Management (NIFTEM), Kundli 131028, Haryana, India
* Correspondence: bhimpratap@gmail.com; Tel.: +91-943-635-3807

Citation: Deka, P.; Mehetre, G.T.; Lalnunmawii, E.; Upadhyaya, K.; Singh, G.; Hashem, A.; Al-Arjani, A.-B.F.; Fathi Abd_Allah, E.; Singh, B.P. Metagenomic Analysis of Bacterial Diversity in Traditional Fermented Foods Reveals Food-Specific Dominance of Specific Bacterial Taxa. *Fermentation* **2021**, *7*, 167. https://doi.org/10.3390/fermentation7030167

Academic Editor: Hiroshi Kitagaki

Received: 27 July 2021
Accepted: 21 August 2021
Published: 26 August 2021

Abstract: Traditional fermented foods have been recognized by various communities to be good for health since ancient times. There is a provincial legacy of traditional fermented foods among the ethnic population of North-East India. Fermented bamboo shoots (local name: Tuaither), soybeans (Bekang), and pork fat (Sa-um) are famous in the Mizoram state and represent a primary portion of the daily diet. These foods are prepared using methods based on cultural traditions inherited from previous generations, and prepared using a relatively uncontrolled fermentation process. Analysis of the bacterial diversity in these foods can provide important information regarding the flavor and texture of the final products of fermentation. Unfortunately, studies on the microbial composition and health benefits of such traditional fermented foods have rarely been documented. Therefore, the present study aims to highlight this bacterial diversity, along with the proximate composition of different traditional fermented foods (Tuaither, Bekang and Sa-um) primarily consumed in Mizoram state, India. Samples were collected on three different days of fermentation (3rd, 5th and 7th day), and bacterial diversity analysis was performed using the V3-V4 variable region of 16S rRNA gene with Illumina sequencing. Results revealed differences in the bacterial composition of dominant group members among all of the three food types. Firmicutes (82.72–94.00%), followed by Proteobacteria (4.67–15.01%), were found to dominate to varying degrees in all three of the fermented foods. However, at genus level high variation was observed in bacterial composition among these three different types of fermented foods. *Lactobacillus* (91.64–77.16%), *Staphylococcus* (52.00–17.90%), and *Clostridium* (72.48–55.40%) exhibited the highest relative abundances in the Tuaither, Bekang and Sa-um foods, respectively, in descending order from the 3rd to 7th day of fermentation. A few of the bacterial genera such as Lactobacilli were positively correlated with fermented bamboo shoot samples, and *Staphylococcus* was positively correlated with protein, carbohydrate and crude fiber content in soybean samples. In general, Tuaither, Bekang and Sa-um exhibited distinct differences in bacterial composition. This variation may be due to differences in the raw materials and/or methods used in the preparation of the different fermented food products. This is the first study to describe the bacterial composition of these traditional fermented foods using high-throughput sequencing techniques, and could help to drive research attention to comprehensive studies on improving understanding of the role of microbial communities in the preparation of traditional foods and their health benefits.

Keywords: traditional fermented foods; metagenomics; bacterial diversity; health benefits; proximate analysis

1. Introduction

Traditional fermented foods provide many health benefits and have been prepared by various communities around the world since ancient times. Numerous beneficial microorganisms confer unique properties to finished fermented food products through their metabolic activities. Therefore, analyzing bacterial abundance and diversity in fermented foods is important to understand the role of microbial taxa in establishing flavor and taste, along with other health benefits. Fermented foods also play a role in altering the microbial composition of the digestive tract of consumers [1]. Several previous research studies which utilized microbial culturing and identification of a particular fermented food product have reported that a few dominant microbial taxa play a major role in the fermentation process [2,3]. During the fermentation process, bacterial species convert raw food constituents into products that enhance the flavor and nutraceutical value of the final fermented product [4]. A more comprehensive understanding of microbial composition in fermented food products, however, requires an approach not limited by culture-based methods. In the past decade, culture-independent approaches utilizing polymerase chain reaction (PCR)-based amplification and sequencing of 16S rRNA genes have proven to be useful for the microbiological investigation of a variety of fermented foods [5–7]. Recent advances in next generation sequencing (NGS) technologies have provided platforms to explore microbial diversity and functionality in a variety of environments [8,9]. NGS techniques have been used to characterize the microbial communities of a variety of fermented foods and beverages, such as kimchi [10], kefir grains [11], Chinese rice wine [12], etc. Considering the wide variety of fermented foods in North-East India, relatively few studies, based on NGS technologies, have been conducted on fermented foods indigenous to North-East India. Previous studies have been conducted on Bekang [13], Rawtuai rep (fermented bamboo shoot) [14], and Sa-um [15], but these have been based on microbial culture methods. Therefore, microbial analysis, utilizing NGS technologies, of traditional fermented foods provincial to this region could greatly enhance our knowledge of the bacterial diversity present in these fermented products, and contribute to preserving the ethnicity of traditional indigenous fermented foods.

Fermented foods are rich in nutrients with desirable organoleptic properties and are popular due to their unique flavors [16]. The ethnic population of North-East India is among the groups that have a long history of preparing traditional fermented food products. These foods are prepared based on ethnic knowledge that has been passed down from generation to generation for thousands of years [17,18]. The production of these fermented foods involves natural and spontaneous fermentation. A diverse range of microbes are involved in the fermentation process, some of which contribute to the unique flavors and textures of the final products [19]. Fermented bamboo shoot (FBS) (local name: Tuaither), fermented soybean (FSB) (local name: Bekang) and fermented pork fats (FPF) (local name: Sa-um) are traditional fermented foods that are widely eaten among the peoples of the Mizoram state in North-East India. All of these food products are prepared using a natural fermentation process in a relatively uncontrolled manner. Therefore, the various microbes responsible for the fermentation may be derived from the raw materials and/or other sources, including the processing equipment, air, and water used to prepare the fermented product. In this context, analyzing microbial composition using high throughput sequencing approaches could be more beneficial for understanding the health benefits of microbes from traditional foods. The bacterial diversity and composition present in Tuaither, Bekang and Sa-um fermented foods consumed in the Mizoram state have not been investigated using metagenomics approaches.

Therefore, the objective of the present study was to characterize the bacterial diversity and nutritional composition of popular traditional fermented foods from the Mizoram state of North-East India. Sampling was done in triplicates for each of Tuaither, Bekang and Sa-um fermented foods from local markets in Mizoram at three distinct days during fermentation. Bacterial diversity, based on the 16S rRNA gene (V3–V4 region) was conducted using an IlluminaMiSeq platform with further bioinformatics analysis. Proximate composition such as moisture, ash, fat, protein and carbohydrate content was also analyzed for all three types (Tuaither, Bekang, and Sa-um) of fermented food samples. Furthermore, correlation analysis of the proximate parameters with the bacterial phyla and genera was also undertaken. We assumed that we would identify differences in the nutritional compounds and bacterial community composition among the samples of the three different types of food. To the best of our knowledge, this is the first report that has investigated the microbial diversity and proximate nutritional composition of traditional fermented food from the Mizoram state of North-East India.

2. Materials and Methods

2.1. Collection of Food Samples

Three different types of fermented food products native to the Mizoram state of North-East India were sampled in triplicates. Tuaither (prepared using bamboo shoots), Bekang (prepared using soybean), and Sa-um (prepared using pork fat) are all widely-prepared and consumed by local communities (Figure 1).

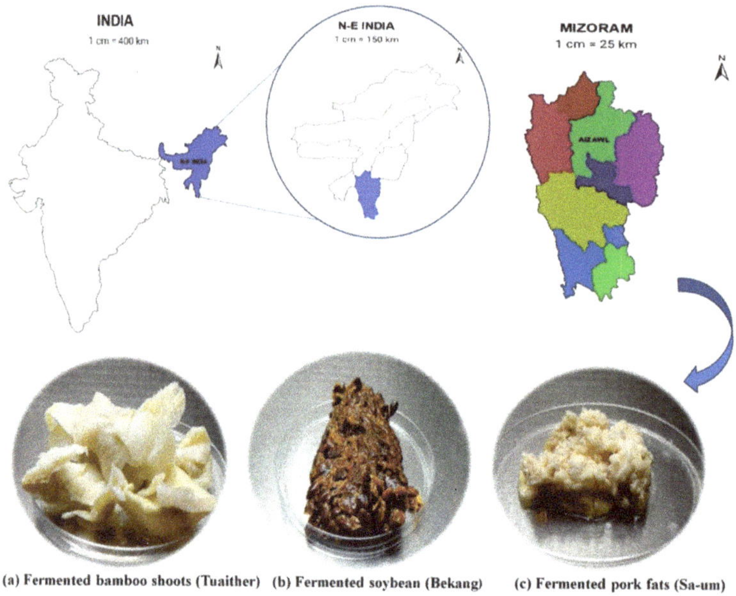

Figure 1. Maps indicating the sampling location in Aizawl in the Mizoram state of North-East India, as well as representative photographs of Tuaither (**a**), Bekang (**b**) and Sa-um (**c**) fermented foods produced and consumed by the local community.

These food products are prepared using traditional methods of processing involving spontaneous fermentation, as has been conducted since ancient times. The samples were collected from local vendors in the Aizawl city market places of Mizoram. Vendors were selected on the basis of their location and by the processing methods that were utilized. Samples were collected in triplicates on the 3rd, 5th and 7th days of fermentation, processed in the same container from the same vendor. Before processing for further analysis, triplicate sets of each sample were pooled together. Therefore, three samples from each fermented food product (Tuaither, Bekang and Sa-um) on the 3rd, 5th, and 7th days of fermentation were used for further analysis. All samples were stored at 4 °C and used immediately for estimating bacterial diversity.

2.2. Proximate Compositional Analysis

A portion of the collected samples was used to obtain the proximate composition of each product at the time of sampling. Various proximate parameters were assessed based on the methods prescribed by the Association of Official Analytical Chemists (AOAC) [20]. Briefly, protein content was determined based on nitrogen content using the micro-Kjeldahl method. Fat content was analyzed using a Soxhlet apparatus with a suitable solvent as prescribed in the AOAC methods. Moisture content and dry biomass was determined by weighing 2 g of sample, heating it to dryness in an oven at 110 °C for 2 h and then weighing the samples again. All of the proximate parameters are reported in AOAC, 2000 standard format as a percentage. Other parameters such as pH, moisture content, total ash content, crude fiber, carbohydrate, and calorific value were also determined according to AOAC methods (AOAC 2016, 20th Edition) [20].

2.3. DNA Extraction and Amplicon Sequencing

All samples (1 mL) were placed in a phosphate-buffer saline solution and centrifuged at 800× *g* for 1 min, after which the supernatants were collected. The supernatants collected from each sample were centrifuged again at 11,000× *g* for 3 min to obtain microbial cell pellets. Genomic DNA isolation was performed using a QIAGEN DNeasy Kit according to the manufacturer's instructions (QIAGEN Inc.). The quality and concentration of DNA obtained from each sample was determined spectrophotometrically using a Nanodrop Lite Spectrophotometer (Thermo Scientific, Waltham, MA, USA) and further assessed on a 1% agarose gel. The obtained high-quality DNA samples were stored at −20 °C. Extracted DNA was subjected to amplicon library preparation using a Nextera XT index kit (Illumina, San Diego, CA, USA) according to the 16S metagenomic sequencing library preparation protocol provided by Illumina. The V3-V4 region of 16S rRNA was amplified using the forward primer, 16SrRNAF (5′-GCCTACGGGNGGCWGCAG-3′), and reverse primer, 16SrRNAR (5′-ACTACHVGGGTATCTAATCC-3′). The amplicon libraries were purified using AMPure XP beads, quantified on a Qubit Fluorometer, and sequenced on an Illumina MiSeq platform (Illumina, San Diego, USA) to obtain 2 × 300 bp paired-end reads. Sequencing was conducted at a commercial sequencing facility (Eurofins, India).

2.4. Bioinformatics and Statistical Analysis

Raw sequence reads in FASTQ format were de-multiplexed into each sample based on the sample-specific barcode sequences. High-quality sequences were obtained using Trimmomatic v0.38 after removal of adapter sequences, ambiguous reads, and low quality sequences (reads with more than 10% quality threshold (QV) < 20 phred score) [21]. High-quality forward and reverse paired-end reads were merged using FLASH (v1.2.11) software [22]. Further processing of the sequence data was performed using Quantitative Insights Into Microbial Ecology (QIIME) version 1.9.1 [23]. Assignment of operational taxonomic units (OTUs) was based on sequence similarity and selection of the most representative sequence from each of the grouped reads. OTUs were clustered using a 97% similarity cutoff utilizing uclust software [24] against the Greengene database (version 13-_8) with the "pick_open_reference_otus.py" Qiime script. BLAST analysis (with default e-value) of the OTU reads against the Greengene database was used for taxonomic classification based on the RDP classifier (Version 2.2) algorithm [25]. All downstream analyses, including alpha diversity and compositional analysis were performed using Microbiome Analyst software [26]. Bacterial abundance and diversity analysis based on alpha diversity indices, such as observed species, Shannon index, and Chao1, was conducted using an OTU matrix and performed in QIIME [23]. Multivariate principal component analysis (PCA) was conducted using STAMP software [27]. The distance matrix for the PCA was generated using an unweighted UniFrac approach. OTU significance was determined using an unweighted UniFrac and Cluster Analysis of the Unweighted Pair-Group Method with Arithmetic means (UPGMA). Correspondence analysis (CCA) was conducted to determine the relationship between nutritional factors and normalized abundances of major taxonomic groups using PAST v3.14 software [28].

3. Results and Discussion

3.1. Proximate Compositional Analysis

During fermentation, microorganisms convert the chemical constituents of food materials into high value nutrients that improve the flavor and texture of the final fermented food product [29–32]. An assessment of the proximate composition analysis of the three fermented foods examined at three different stages of fermentation (3rd, 5th, and 7th day of fermentation) are presented in Table 1.

Table 1. Proximate composition of fermented bamboo shoots (Tuaither), soybean (Bekang), and pork fat (Sa-um) collected after 3, 5, and 7 days of fermentation.

Proximate Parameters (Weight in %)	Fermented Food Samples								
	Tuaither			Bekang			Sa-um		
	3rd D	5th D	7th D	3rd D	5th D	7th D	3rd D	5th D	7th D
Moisture content	86.8	86.4	87.6	58.3	57.1	56.4	3.8	3.8	2.5
Total ash content	1.1	1.1	1.1	2.7	2.5	1.1	-	-	-
Fat	1.1	0.9	1.2	4.6	5.3	5.4	90.6	94.6	95.6
Protein	2.7	3.0	2.8	19.7	19.9	18.5	1.3	1.3	1.5
Crude fiber	3.1	2.1	2.2	10.6	2.4	2.5	-	-	-
Carbohydrate	5.2	6.5	7.9	4.1	12.8	16.2	4.2	0.2	0.3
Calorific value (K cals/100 gm)	3.97	4.19	3.98	7.20	8.36	8.23	5.89	3.99	4.04

The fermented food products (Tuaither and Sa-um) exhibited an acidic pH (4.04 for Tuaither and 4.04 for Sa-um) during the fermentation process. This may be due to the lactic acid produced by the lactic-acid bacteria [33]. The Bekang sample, however, exhibited a slightly basic pH (7.93). Average moisture content was highest in the bamboo shoot samples (86.93%), followed by the soybean samples (57.26%), and was lowest in the pork fat samples (3.36%). A previous study on the proximate composition of *Bambusa balcooa* (ChingSanei-bi) also reported high moisture content (90.73–91.5%) [34]. As expected, fat content was highest in the pork fat samples (93.60%), followed by the soybean samples (5.3%), and was lowest in the bamboo shoot samples (1.06%). Interestingly, the soybean samples were found to have a higher protein (19.36%) and carbohydrate content (11.03%) than the bamboo shoot and pork fat samples. The lowest protein content was observed in the samples of the bamboo shoots (2.83%) and pork fat (1.36%). This was also true for the carbohydrate content, which was 6.53% in the bamboo shoot samples and 1.56% in the pork fat samples. A low level of crude fiber was detected in the bamboo shoot (2.46%) and soybean (1.56%) samples, and was undetectable in pork fat samples. A slight increase in protein, carbohydrate, and fat content was observed from the 3rd to 5th day of fermentation; however, it was not a significant increase as the same trend was not found during the latter stage of fermentation (Table 1).

3.2. Bacterial Diversity

A total of 93,857 (3rd day), 267,981 (5th day), and 187,190 (7th day) reads were obtained from the fermented bamboo shoot samples, 165,177 (3rd day), 252,177 (5th day), and 259,761 (7th day) for the fermented soybean samples, and, 208,302 (3rd day), 152,718 (5th day) and 126,281 (7th day) for the fermented pork fat samples. A total of 3424 OTUs were identified collectively from all the samples after clustering of all the high-quality reads. The proportion of OTUs in the samples was found to be quite similar, suggesting that the data was sufficient to compare bacterial diversity between the different fermented food products. Rarefaction was conducted on the OTU data sets of each of the different foods for each of the sampled time points (Figure 2). Results indicated that the number of reads obtained for all the samples were near saturation, suggesting that a sufficient number of reads had been obtained in the samples to accurately assess bacterial diversity.

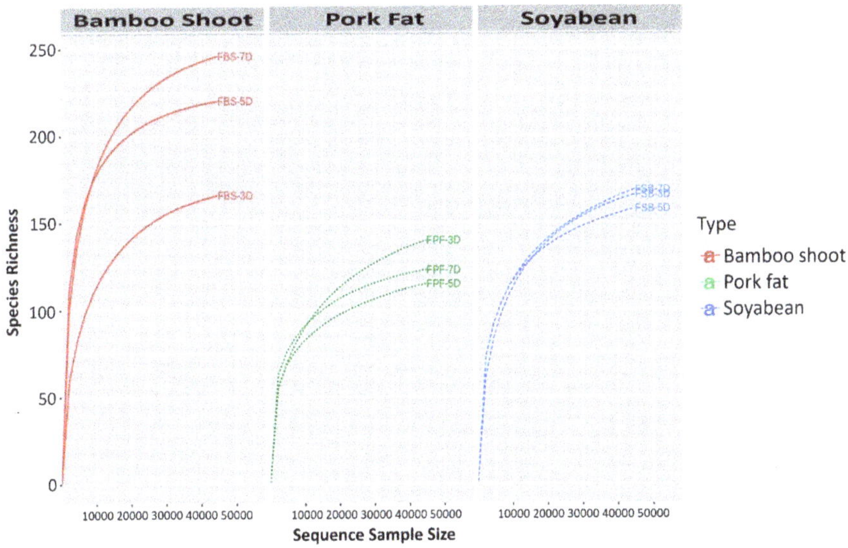

Figure 2. Rarefaction curves for the sequences obtained from fermented bamboo shoots (FBS), soybeans (FSB), and pork fat (FPF) collected at 3, 5, and 7 days of fermentation.

Simpson diversity and Chao1 indices of bacterial diversity (Figure 3) revealed that bamboo shoots (FBS) exhibited the highest level of bacterial diversity, followed by soybean (FSB) samples, while the lowest alpha-diversity was observed in pork fat (FPF) samples. Chao1 and Shannon indices are directly correlated with community diversity. The Shannon index of the FBS samples revealed that samples collected at the 7th day of fermentation had the highest species richness. However, species richness in the soybean (FSB) and pork fat (FPF) samples was found to be more evenly distributed from the 3rd day to the 7th day of the fermentation.

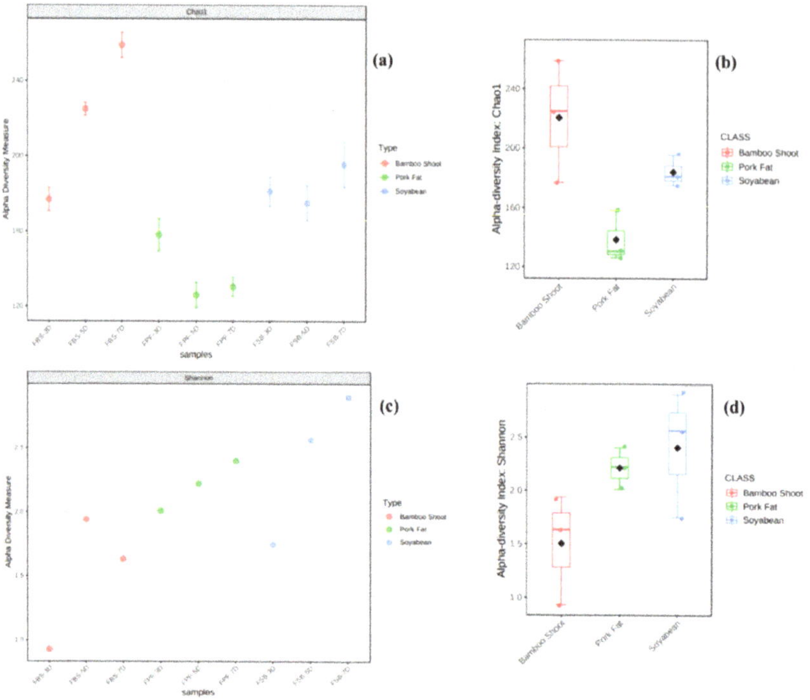

Figure 3. Alpha diversity, Chao-1, and Shannon indices of bacterial diversity in fermented food samples (Bamboo shoots, soybean, and pork fat), (**a**) and (**c**) represent the alpha diversity of the three collection dates (3rd, 5th, and 7th day) for each food-type, while b and d represent Chao-1 (**b**) and Shannon (**d**) bacterial diversity of the collective samples of each fermented food product. (The samples data was normalized at the depth of the minimum library size, i.e., sample with least reads: FBS-3D; 93,857).

3.3. Bacterial Community Composition

All of the fermented food samples possessed a diverse range of bacterial taxa. The relative abundance of the top phyla and genera is shown in Figures 4 and 5, respectively. On the 3rd day of fermentation, at the phylum level, the highest abundance was recorded for Firmicutes (94.99–82.72–92.91%), Proteobacteria (4.67–15.01–6.82%), Bacteroidetes (0.13–0.027–0.038%), Actinobacteria (0.077–2.14–0.18%), and Verrucomicrobia (0.023–0.002–0.0049%) in bamboo shoot (FBS), soybeans (FSB) and pork fat (FPF) samples, respectively. Firmicutes was the most abundant phylum in all samples, exhibiting an average relative abundance > 90.22%. Proteobacteria was the second most abundant phylum in all samples, with an average relative abundance > 8.83%. Similar results for the relative abundance of Firmicutes and Proteobacteria were reported in a recent study of 'zha-chili', a traditional Chinese fermented food [35]. Another study of a fermented vegetable, Suancai, a popular fermented food product in northern China, also reported the highest relative abundance for the bacterial phyla Firmicutes and Proteobacteria [36].

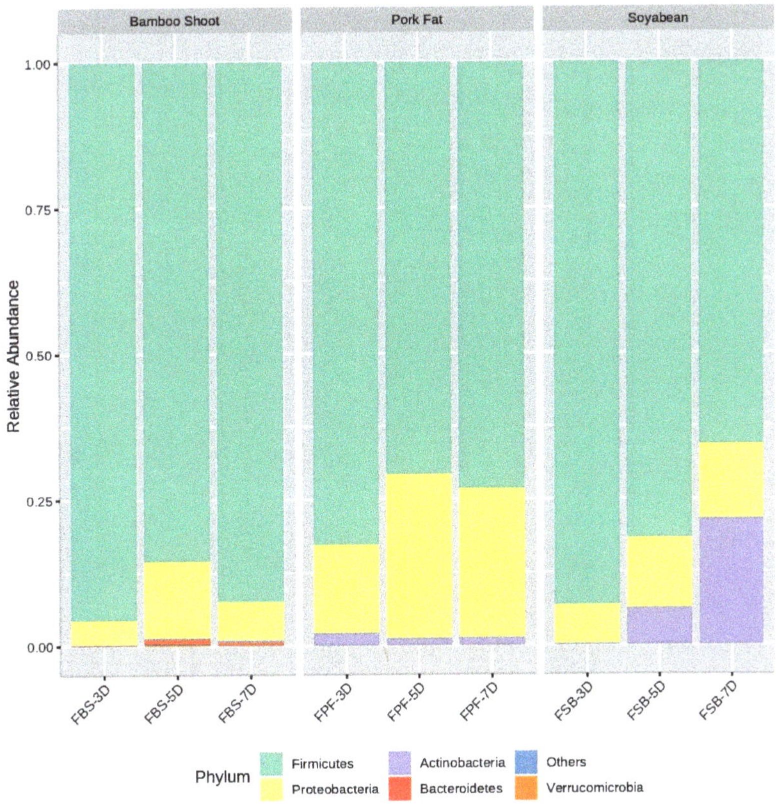

Figure 4. Relative abundance of top bacterial phyla in fermented bamboo shoots (FBS), pork fat (FPF), and soybeans (FSB) samples collected on the 3rd, 5th, and 7th day of fermentation.

Other bacterial phyla were present but at a much lower level of abundance (<1%), including Planctomycetes, Fusobacteria, Chloroflexi, Acidobacteria, Nitrospirae, Spirochaetes, Gemmatimonadetes, Synergistetes, Lentisphaerae, Elusimicrobia, and Chlorobi. On the 5th and 7th day of fermentation, the relative abundance of Firmicutes decreased slightly, while Proteobacteria increased in soybean samples (70.85%). A few other phyla were detected on the 5th and 7th day of fermentation, including Chloroflexi, Fusobacteria, Acidobacteria, Tenericutes, Fibrobacteres, and Caldiserica. An increase in bacterial diversity towards the latter stages of fermentation suggests that these bacterial taxa may play a role in determining the texture and flavor of the final fermented food product. Significant differences were observed at the genus level among all three of the fermented foods, irrespective of the days of fermentation (Figure 5).

Lactobacillus was found with the highest relative abundance (91.64%) in fermented bamboo shoots. Its relative abundance was lower (4.81%) in fermented soybeans, and lowest (0.042%) in fermented pork fat. In the case of fermented soybeans, however, Staphylococcus (52.36%), Bacillus (38.47%) and Pseudomonas (6.40%) genera were found to be abundant. Around half of the sequence reads were assigned to OTUs that belong to Staphylococcus in fermented soybeans on the 3rd day of fermentation. Notably, this genus was not detected in the other two fermented food products. In the case of fermented pork fat, Clostridium (72.48%), Sutterella (12.54%), and Lactobacillus (4.81%) were the most abundant genera observed.

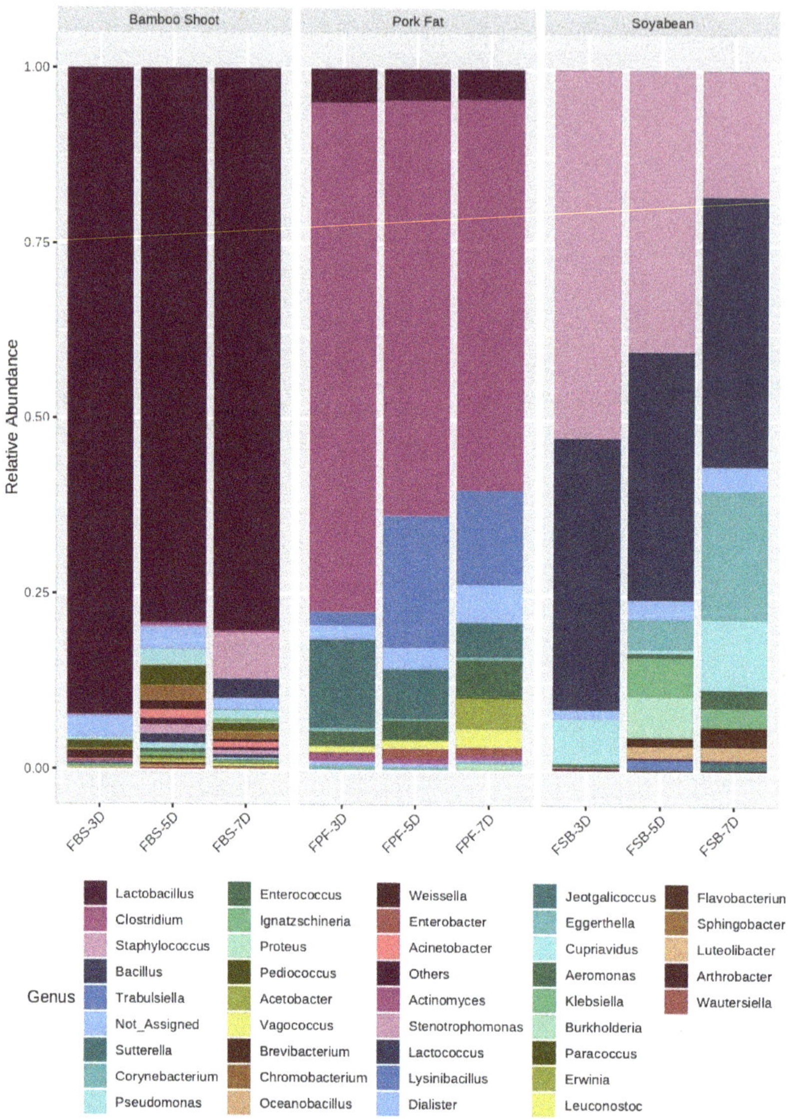

Figure 5. Relative abundance of bacteria at the genus level in fermented bamboo shoots (FBS), pork fat (FPF), and soybean (FSB) on the 3rd, 5th and 7th day of fermentation.

Lactobacillus, Staphylococcus, and Clostridium accounted for more than 60.53% of the total bacterial reads in all three fermented food products. Lactobacillus species play a significant role in bamboo and lactic acid bacteria have been reported to be the dominant bacterial taxa in several fermented bamboo products [37]. A traditional fermented bamboo shoot product, Khorisa, produced and utilized by the indigenous people of Assam, was reported to possess a variety of actobacilli with potential antimicrobial activity and, thus, high pharmacological value [38]. Studies have shown that Lactobacillus species can produce volatile compounds that enhance the aroma of fermented foods [36]. In addition to Lactobacillus, other bacterial taxa, such as Pediococcus, Enterococcus, Lactococcus,

Streptococcus and others, can also modify the flavor of original food components and improve the nutritional value of fermented foods [39,40]. In the present study, several other lactic acid bacteria and related species, including Pediococcus, Enterococcus, and Lactococcus were found to be present in in fermented bamboo shoot samples at a relative abundance > 1%, while Bacillus, Pseudomonas, and Enterococcus were found in fermented soybean samples. In fermented pork samples, Sutterella, Lactobacillus, Enterococcus, and Trabulsiella were found overall with relatively higher abundances (1.92% to 13.36%).

Staphylococcus accounted for > 50% of the relative abundance in the different samples of the fermented soybean product. The presence of Staphylococcus has been reported in a variety of fermented food products [41]. Staphylococcus and other genera of bacteria have been reported to play a role in the production of synthetic esters in fermented food products [42]. Notably, coagulase negative *Staphylococcus* species have been reported as the predominant species of bacteria in fermented foods worldwide. Species of Staphylococcus are used as microbial starters in cheese, meat, and soybean fermentation [43]. Importantly, a few species of *Staphylococcus* are also known to be food borne pathogens for humans and other animals [44]. Therefore, more detailed studies on the Staphylococcus species present in fermented foods using multiple taxonomic approaches is essential to obtain confirmed identification at the species level, especially as their proportion is relatively high in traditional fermented foods that are produced in relatively open environments [45]. Clostridium exhibited the highest relative abundances in the fermented pork fat samples. The prevalence of several enteric bacteria including Clostridiumin traditional fermented foods of North-East India also has been reported previously [46].

3.4. Comparison of Bacterial Communities during the Fermentation Process

Beta diversity represents a comparison of diversity among different samples based on the relatedness and differences in the microbial community composition of distantly related samples. The weighted UniFrac distance and un-weighted UniFrac distance were used to conduct a principal coordinate analysis (PCoA) to investigate the variation in microbial species in all samples (Figure 6).

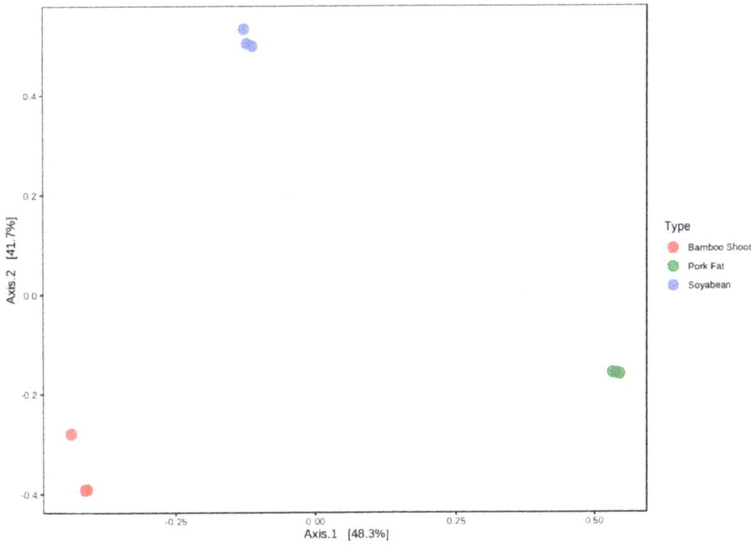

Figure 6. Principal component analysis of bacterial communities in the three different fermented foods (bamboo shoot, soybean, and pork fat).

Results indicated significant variation in the community composition of the three different food samples. PCoA analysis revealed significant difference in the microbial composition of the different fermented foods. However, each group displayed a similar composition at different stages of the fermentation process (3, 5, and 7 days of fermentation). Samples of bamboo shoots, soybean, and pork fat are all greatly separated in the graphs of the PCoA analysis. However, each of the stages of fermentation within each fermented food type clustered close to one another, indicating a similar composition (Figure 6). The first and second principal components accounted for 41.7% and 48.3% of the variance among the samples. The obtained values indicate that the three types of fermented foods examined in the study are distantly related to each other in their composition of bacterial genera. A hierarchically clustered heat map analysis based on the bacterial community profile at the genus level is shown in Figure 7.

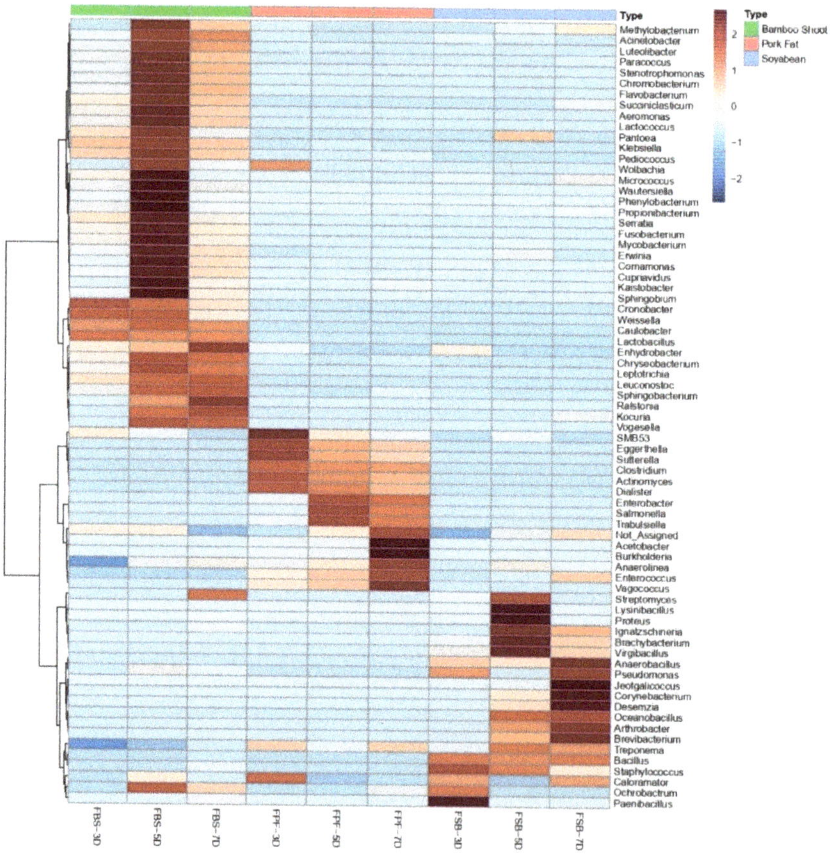

Figure 7. Heat map of bacterial community composition at the genus level of the different fermented food samples: bamboo shoots (FBS), pork fat (FPF), and soybean (FSB) collected at 3, 5, and 7 days of fermentation. Blue to red color intensity indicates a low to high relative abundance.

The dominant phylum in bamboo shoots formed a separate cluster, while the most abundant bacterial phylum in soybean and pork fat samples formed a separate cluster that was again divided into another sub-cluster. The dominant genera present in each food type formed three different distantly-related clusters. Lactobacillus, Staphylococcus, and Clostridium were the most abundant genera in the three types of fermented food samples.

The microbial composition of the fermented food changes over the course of time and different microbial groups play a functional role in adding nutritional value to the final product [47]. Relative abundance of the top most abundant bacterial genera in the samples of fermented foods collected on the 3rd, 5th, and 7th day of fermentation is given in Table 2.

Table 2. Relative abundance of the top most bacterial genera in day wise samples of the fermented foods.

	Fermented Bamboo Shoot (FBS)		
Genus	**FBS-3D**	**FBS-5D**	**FBS-7D**
Lactobacillus sp.	91.64	77.16	78.88
Weissella sp.	1.27	0.0	0.0
Pediococcus sp.	1.17	2.80	0.0
Pseudomonas sp.	0.36	2.24	1.25
Chromobacterium sp.	0.27	2.29	0.0
Acinetobacter sp.	0.0	1.35	1.14
Corynebacterium sp.	0.0	0.0	6.58
Sphingobacterium sp.	0.0	0.0	2.74
Unclassified	3.13	2.28	2.27
Others	1.86	11.58	7.11
	Fermented Pork Fats (FPF)		
Genus	**FPF-3D**	**FPF-5D**	**FPF-7D**
Clostridium sp.	72.48	59.48	55.40
Sutterella sp.	12.54	7.01	4.85
Lactobacillus sp.	4.81	4.44	4.41
Enterococcus sp.	1.92	2.67	5.28
Trabulsiella sp.	1.81	18.45	13.36
Unclassified sp.	2.07	3.26	5.24
Others	3.17	4.67	11.43
	Fermented Soybean (FSB)		
Genus	**FSB-3D**	**FSB-5D**	**FSB-7D**
Staphylococcus sp.	52.36	39.48	17.90
Bacillus sp.	38.47	35.56	37.87
Pseudomonas sp.	6.40	0.0	9.67
Enterococcus sp.	0.47	0.0	0.0
Paenibacillus sp.	0.19	0.0	0.0
Proteus sp.	0.0	5.89	0.0
Ignatzschineria sp.	0.0	5.49	0.0
Corynebacterium sp.	0.0	4.35	18.13
Brevibacterium sp.	0.0	0.0	2.77
Unclassified	1.12	2.98	3.61
Others	0.95	6.23	10.02

The bacterial composition of top genera in fermented bamboo shoots on the 3rd, 5th, and 7th day of fermentation is summarized in Table 2. The 3rd day sample of bamboo shoots (FBS_3D) exhibited a high abundance of Lactobacillus (91.64%) and a much smaller proportion of a few other genera, such as Weissella (1.27%) and Pediococcus (1.17%). The abundance of Lactobacillus in bamboo shoots, however, decreased on the 5th (FBS_5D; 77.165%) and 7th (FBS_7D; 78.88%) days of fermentation, while the abundance of Pediococcus increased to 2.8%. Differentiation of lactobacilli (LAB), such as Weissella, Pediococcus, and Lactobacillus, to species and strain levels require more definitive methods. Their rapid and accurate identification is of the utmost importance in food microbiology [48]. LAB species, such as Pediococcus, possess intriguing attributes, such as exopolysaccharides production, and the ability to enhance the aroma of fermented foods and increase the level of lactic acid present in the fermented product [49]. Interestingly, a few of the genera, such as Chromobacterium (2.29%), Pseudomonas (2.24%), Acinetobacter (1.35%), Corynebacterium (6.58%), and Sphingobacterium (2.74%) exhibited a sudden increase in their relative abundance in fermented bamboo shoots on the 7th day of fermentation.

Samples of fermented soybeans collected on the 3rd day of fermentation (FSB_3D) exhibited a high relative abundance of Staphylococcus (52.36%) and Bacillus (38.47%), as well as other genera, including Pseudomonas (6.40%), Enterococcus (0.47%), and Paenibacillus (0.19%) with a much lower level of relative abundance. Notably, the relative abundance of Staphylococcus decreased continuously (52.36% to 17.90%) from the 3rd day to the 7th day of fermentation, however, the relative abundance of Bacillus remained relatively constant (38.47–35.56–37.87%). Soybean foods are generally fermented by Bacillus spp. and impart a characteristic stickiness to the final product. The microbial composition of a variety of fermented soybean foods such as natto (from Japan) and kinema (from Nepal and North-East India) have been reported [50]. In our present study, genera other than Bacillus, such as Proteus (5.89%), Ignatzschineria (5.49%), and Clostridium (4.35%) were found to increase in relative abundance in samples collected after the 5th day of fermentation (FSB_5D). Clostridium (18.13%), Pseudomonas (9.67%), and Brevibacterium (2.77%) exhibited the greatest increase in relative abundance on the 7th day of fermentation. This was especially true for Clostridium, which increased from approximately 0.05% on the 3rd day of fermentation to 4.35% on the 5th day, and more than 18% on the 7th day of fermentation.

High populations of Clostridium have many potential consequences that may be harmful to consumers. Therefore, it is important to consider the timing of the natural fermentation of foods. Clostridium (72.48%), Sutterella (12.54%), Lactobacillus (4.81%), Enterococcus (1.92%) and Trabulsiella (1.81%) were the most abundant genera in pork fat samples on the 3rd day of fermentation (FPF_3D). The relative abundance of Clostridium, however, decreased by approximately 13% on the 5th day of fermentation (FPF_5D) and by an additional 4.08% on the 7th day of fermentation (FPF_7D). Clostridium and a few other foodborne bacterial pathogens recognized by the World Health Organization (WHO) have been previously identified in Indian fermented foods, including pork fat, using MiSeq amplicon sequencing [46]. In the present study, other genera, such as Trabulsiella (18.45%), Sutterella (7.01%), Lactobacillus (4.44%), and Enterococcus (2.67%) did not exhibit significant variation in relative abundance in any of the three food types during the fermentation process. The distribution of bacterial species in the ethnic fermented food samples native to Mizoram only exhibited variation in the relative abundance of the few dominant genera. These results suggest that the type of raw food material and the fermentation environment in these regions may impact bacterial richness but not the diversity. Similar findings have been previously reported in other types of fermented food products [51–53].

3.5. Correlation Analysis of Nutritional Parameters with Bacterial Diversity

Canonical correlation analysis (CCA) was conducted to evaluate the relationship between bacterial members (at the level of phylum and genus) and the proximate composition of the food materials (Figure 8). Results of the CCA indicated that Lactobacilli were positively correlated with fermented bamboo shoot samples, Staphylococcus was correlated with fermented soybean samples, and Clostridium was correlated with fermented pork fat samples. Staphylococcus was positively correlated in soybean samples with protein, carbohydrate and crude fiber content. Determining the correlation of microbial members with food components is important for understanding the microbial communities of related food samples and can lead to the design of synthetic microbial consortia for optimum fermentation of specific food types [54].

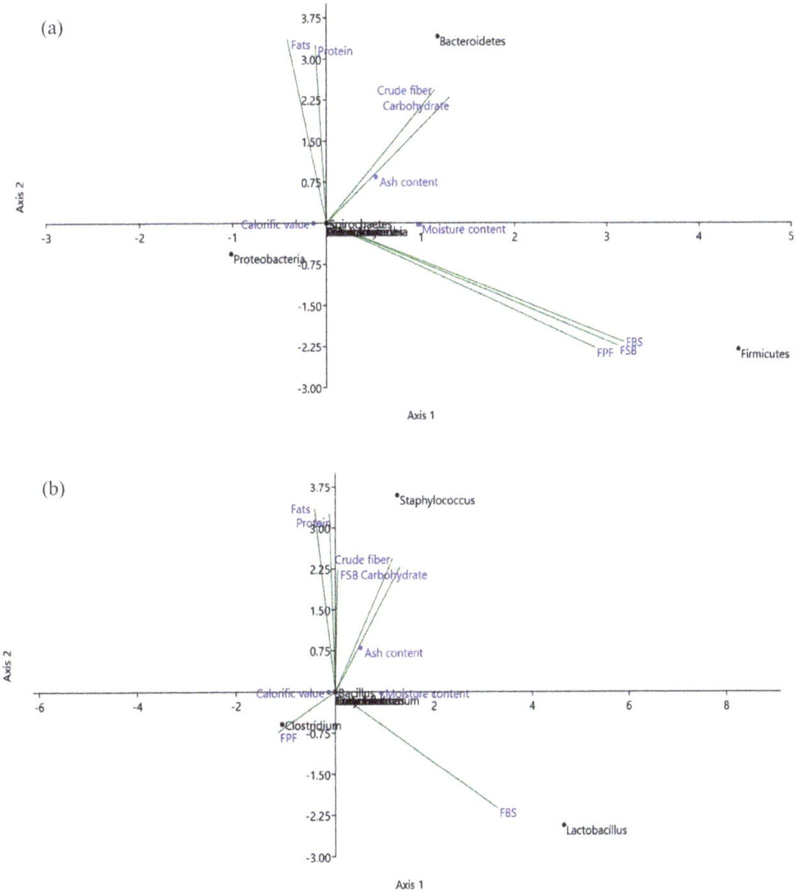

Figure 8. Triplot illustrating the results of the Canonical Correspondence Analysis (CCA) of the relative abundance of bacterial phyla (**a**), genera (**b**) and compositional attributes of the fermented food samples. Fermented bamboo shoots (FBS), fermented soybeans (FSB), and fermented pork fat (FPF).

4. Conclusions

The consumption of traditionally fermented food has a beneficial impact on consumer health due to the of presence beneficial microbial taxa. The present study examines the proximate composition and bacterial diversity of three different traditional fermented foods native to the Mizoram state of India. Samples collected at three different times (3rd, 5th, and 7th day) during the fermentation process were analyzed. Proximate composition analysis revealed the differences in the proximate content, such as fat, protein, and moisture content among the three different types of samples. Bacterial diversity analysis revealed that Firmicutes and Proteobacteria had the highest relative abundance in all the samples.

However, variation in the relative abundances was observed at the genus level. Lactobacillus, Staphylococcus, and Clostridium showed the highest dominance in fermented bamboo shoots (Tuaither), fermented soybeans (Bekang), and fermented pork fat (Sa-um) respectively. Other genera, such as Weissella, Bacillus, Lactococcus, and Pseudomonas, were also found to be present in most of the samples. On the 3rd day, the highest dominance was observed for Lactobacillus, Staphylococcus, and Clostridium respectively in fermented bamboo shoots (Tuaither), fermented soybeans (Bekang), and fermented pork fat (Sa-um) sample. However, a decrease in their abundance was seen towards the 7th day of fermentation. In addition, various other bacterial taxa were found in each of the fermented samples, showing a presence of high bacterial diversity in the fermented products. A correlative analysis exhibited a positive correlation of a few of bacterial with the proximate content of the food samples. This study is the first report on the bacterial diversity of the traditional fermented products of Mizoram using metagenomics analysis, and has revealed important information on bacterial community structure and dominance of major bacteria associated with different traditionally fermented foods. Therefore, the present study could be helpful for establishing more comprehensive investigations into the identification of indigenous bacteria with potential probiotic properties, and enable the discovery of more health benefits in traditional fermented foods, using both culture-independent and -dependent approaches.

Author Contributions: Conceptualization, B.P.S., G.T.M., K.U., E.L., P.D. wrote—original draft preparation, P.D., G.T.M.; drafting manuscript and data analysis, B.P.S. and K.U.; critically revised the final version, E.F.A., G.T.M. made intellectual contribution in drafting the manuscript, A.H., A.-B.F.A.-A., E.F.A., B.P.S., E.L., G.S., K.U. reviewed the manuscript. All authors have read and agreed to the published version of the manuscript.

Funding: The work was carried out under financial support from a joint international grant from the Department of Science and Technology (DST) of the Government of India; project number C/1756/IFD/2019-20, and the Russian Science Foundation, project number 19-46-02004. The authors would like to extend their sincere appreciation to the Researchers Supporting Project Number (RSP-2021/356), King Saud University, Riyadh, Saudi Arabia for covering APC.

Institutional Review Board Statement: Not applicable.

Informed Consent Statement: Not applicable.

Data Availability Statement: All data generated during this study are included in this article.

Acknowledgments: B.P.S. and K.U. are thankful to the Department of Science and Technology (DST), New Delhi, Government of India for financial support. The authors are thankful to the Vice-Chancellor of Mizoram University, Mizoram, India and NIFTEM, Sonepat, India for their continuous support. The authors would like to extend their sincere appreciation to the Researchers Supporting Project Number (RSP-2021/356), King Saud University, Riyadh, Saudi Arabia.

Conflicts of Interest: Authors declare no conflict of interest.

References

1. Tsafrakidou, P.; Michaelidou, A.M.; Biliaderis, C.G. Fermented Cereal-based Products: Nutritional Aspects, Possible Impact on Gut Microbiota and Health Implications. *Foods* **2020**, *9*, 734. [CrossRef]
2. Mathara, J.M.; Schillinger, U.; Kutima, P.M.; Mbugua, S.K.; Holzapfel, W.H. Isolation, identification and characterization of the dominant microorganisms of kulenaoto: The Maasai traditional fermented milk in Kenya. *Int. J. Food Microbiol.* **2004**, *94*, 269–278. [CrossRef]
3. Mokoena, M.P.; Mutanda, T.; Olaniran, A.O. Perspectives on the probiotic potential of lactic acid bacteria from African traditional fermented foods and beverages. *Food Nutr. Res.* **2016**, *60*, 29630. [CrossRef]
4. Caplice, E.; Fitzgerald, G.F. Food fermentations: Role of microorganisms in food production and preservation. *Int. J. Food Microbiol.* **1999**, *50*, 131–149. [CrossRef]
5. Silvestri, G.; Santarelli, S.; Aquilanti, L.; Beccaceci, A.; Osimani, A.; Tonucci, F.; Clementi, F. Investigation of the microbial ecology of Ciauscolo, a traditional Italian salami, by culture-dependent techniques and PCR-DGGE. *Meat Sci.* **2007**, *77*, 413–423. [CrossRef]

6. Kim, T.W.; Lee, J.H.; Kim, S.E.; Park, M.H.; Chang, H.C.; Kim, H.Y. Analysis of microbial communities in doenjang, a Korean fermented soybean paste, using nested PCR-denaturing gradient gel electrophoresis. *Int. J. Food Microbiol.* **2009**, *131*, 265–271. [CrossRef]
7. Sakamoto, N.; Tanaka, S.; Sonomoto, K.; Nakayama, J. 16S rRNApyrosequencing-based investigation of the bacterial community in nukadoko, a pickling bed of fermented rice bran. *Int. J. Food Microbiol.* **2011**, *144*, 352–359. [CrossRef] [PubMed]
8. Mehetre, G.T.; Paranjpe, A.S.; Dastager, S.G.; Dharne, M.S. Complete metagenome sequencing based bacterial diversity and functional insights from basaltic hot spring of Unkeshwar, Maharashtra, India. *Genom. Data* **2016**, *7*, 140–143. [CrossRef] [PubMed]
9. Cao, Y.; Fanning, S.; Proos, S.; Jordan, K.; Srikumar, S. A review on the applications of next generation sequencing technologies as applied to food-related microbiome studies. *Front. Microbiol.* **2017**, *8*, 1829. [CrossRef] [PubMed]
10. Jung, J.Y.; Lee, S.H.; Kim, J.M.; Park, M.S.; Bae, J.W.; Hahn, Y.; Madsen, E.L.; Jeon, C.O. Metagenomic analysis of Kimchi, a traditional Korean fermented food. *Appl. Environ. Microbiol.* **2011**, *77*, 2264–2274. [CrossRef] [PubMed]
11. Nalbantoglu, U.; Caker, A.; Dogan, H.; Abaci, N.; Ustek, D.; Sayood, K.; Can, H. Metagenomic analysis of the microbial community in kefir grains. *Food Microbiol.* **2014**, *41*, 42–51. [CrossRef]
12. Hong, X.; Chen, J.; Liu, L. Metagenomic sequencing reveals the relationship between microbiota composition and quality of Chinese Rice Wine. *Sci. Rep.* **2016**, *6*, 26621. [CrossRef]
13. Tamang, J.P.; Tamanag, B.; Schillinger, U.; Guigas, C.; Holzapfel, W.H. Functional properties of lactic acid bacteria isolated from ethnic fermented vegetables of the Himalayas. *Int. J. Food Microbiol.* **2009**, *135*, 28–33. [CrossRef] [PubMed]
14. Nongdam, P. Traditional fermented bamboo shoot foods of North-East India and their characteristics natural microbial flora. In Proceedings of the 10th World Bamboo Congress, Damyeng, Korea, 17–22 September 2015.
15. Lalthanpui, P.B.; Lalruatfela, B.; Zoramdinthara; Lalthanzara, H. Traditional food processing techniques of the Mizo people of Northeast India. *Sci. Viss.* **2015**, *15*, 39–45.
16. Marco, M.L.; Heeney, D.; Binda, S.; Cifelli, C.J.; Cotter, P.D.; Foligné, B.; Gänzle, M.; Kort, R.; Pasin, G.; Pihlanto, A.; et al. Health benefits of fermented foods: Microbiota and beyond. *Curr. Opin. Biotechnol.* **2017**, *44*, 94–102. [CrossRef] [PubMed]
17. Das, A.J.; Deka, S.C. Fermented foods and beverages of the North-East India. *Int. Food Res. J.* **2012**, *19*, 377.
18. Rawat, K.; Kumari, A.; Kumar, S.; Kumar, R.; Gehlot, R. Traditional fermented products of India. *Int. J. Curr. Microbiol. App. Sci.* **2018**, *7*, 1873–1883. [CrossRef]
19. Tamang, J.P.; Tamang, N.; Thapa, S.; Dewan, S.; Tamang, B.; Yonzan, H.; Rai, A.K.; Chettri, R.; Chakrabarty, J.; Kharel, N. Microorganisms and nutritional value of ethnic fermented foods and alcoholic beverages of North East India. *Indian J. Trad. Know.* **2012**, *11*, 7–25.
20. Association of Official Analytical Chemists (AOAC). *Official Methods of Analysis*, 20th ed.; AOAC International: Gaithersburg, MD, USA, 2016.
21. Bolger, A.M.; Lohse, M.; Usadel, B. Trimmomatic: A flexible trimmer for Illumina sequence data. *Bioinformatics* **2014**, *30*, 2114–2120. [CrossRef]
22. Magoč, T.; Salzberg, S.L. FLASH: Fast length adjustment of short reads to improve genome assemblies. *Bioinformatics* **2011**, *27*, 2957–2963. [CrossRef] [PubMed]
23. Caporaso, J.G.; Kuczynski, J.; Stombaugh, J.K.; Bittinger, K.; Bushman, F.D.; Costello, E.K.; Fierer, N.; Peña, A.G.; Goodrich, J.K.; Gordon, J.I.; et al. QIIME allows analysis of high-throughput community sequencing data. *Nat. Methods* **2010**, *7*, 335–336. [CrossRef]
24. Edgar, R.C. Search and clustering orders of magnitude faster than BLAST. *Bioinformatics* **2010**, *26*, 2460–2461. [CrossRef] [PubMed]
25. Wang, Q.; Garrity, G.M.; Tiedje, J.M.; Cole, J.R. Naive Bayesian classifier for rapid assignment of rRNA sequences into the new bacterial taxonomy. *Appl. Environ. Microbiol.* **2007**, *73*, 5261–5267. [CrossRef] [PubMed]
26. Dhariwal, A.; Chong, J.; Habib, S.; King, I.L.; Agellon, L.B.; Xia, J. Microbiome Analyst: A web-based tool for comprehensive statistical, visual and meta-analysis of microbiome data. *Nucleic. Acids. Res.* **2017**, *45*, 180–188. [CrossRef] [PubMed]
27. Parks, D.H.; Tyson, G.W.; Hugenholtz, P.; Beiko, R.G. STAMP: Statistical analysis of taxonomic and functional profiles. *Bioinformatics* **2014**, *30*, 3123–3124. [CrossRef]
28. Hammer, O.; Harper, D.A.T.; Ryan, P. PAST: Paleontological statistics software package for education and data analysis. *Palaeontol. Electron.* **2001**, *4*, 1–9.
29. McFeeters, R.F. Fermentation microorganisms and flavor changes in fermented foods. *J. Food Sci.* **2004**, *69*, 2002–2004. [CrossRef]
30. Sharma, R.; Garg, P.; Kumar, P.; Bhatia, S.K.; Kulshrestha, S. Microbial fermentation and its role in quality improvement of fermented foods. *Fermentation* **2020**, *6*, 106. [CrossRef]
31. Anal, A.K. Quality ingredients and safety concerns for traditional fermented foods and beverages from Asia: A Review. *Fermentation* **2019**, *5*, 8. [CrossRef]
32. Hu, Y.Y.; Wang, H.; Kong, B.H.; Wang, Y.; Chen, Q. The succession and correlation of the bacterial community and flavour characteristics of Harbin dry sausages during fermentation. *LWT* **2021**, *138*, 110689. [CrossRef]
33. Mugula, J.K.; Narvhus, J.A.; Sørhaug, T. Use of starter cultures of lactic acid bacteria and yeasts in the preparation of togwa, a Tanzanian fermented food. *Int. J. Food Microbiol.* **2003**, *83*, 307–318. [CrossRef]

34. Singh, H.D.; Singh, S.A.; Singh, N.R.; Singh, R. Biochemical composition of Soibum-A fermented bamboo shoot and its dynamics during fermentation in real time model. In Proceedings of the 9th International Conference on Food Engineering and Biotechnology, Bangkok, Thailand, 28–29 May 2011; Volume 9, pp. 11–18.
35. Cai, W.; Tang, F.; Wang, Y.; Zhang, Z.; Xue, Y.; Zhao, X.; Guo, Z.; Shan, C. Bacterial diversity and flavor profile of Zha-Chili, a traditional fermented food in China. *Food Res. Int.* **2021**, *141*, 110112. [CrossRef]
36. He, Z.; Chen, H.; Wang, X.; Lin, X.; Ji, C.; Li, S.; Liang, H. Effects of different temperatures on bacterial diversity and volatile flavor compounds during the fermentation of suancai, a traditional fermented vegetable food from northeastern China. *LWT* **2020**, *118*, 108773. [CrossRef]
37. Sonar, N.R.; Halami, P.M. Phenotypic identification and technological attributes of native lactic acid bacteria present in fermented bamboo shoot products from North-East India. *J. Food Sci. Technol.* **2014**, *51*, 4143–4148. [CrossRef] [PubMed]
38. Badwaik, L.S.; Borah, P.K.; Deka, S.C. Production and Purification of Anti-Bacterial Biometabolite from Wild-Type Lactobacillus, Isolated from Fermented Bamboo Shoot: Future Suggestions and a Proposed System for Secondary Metabolite Onsite Recovery during Continuous Fermentation. *Appl. Biochem. Biotechnol.* **2015**, *175*, 1915–1925. [CrossRef] [PubMed]
39. Stiles, M.E.; Holzapfel, W.H. Lactic acid bacteria of foods and their current taxonomy. *Int. J. Food Microbiol.* **1997**, *36*, 1–29. [CrossRef]
40. Salminen, S.; von Wright, A.; Ouwehand, A. *Lactic Acid Bacteria: Microbiological and Functional Aspects*, 3rd ed.; CRC: New York, NY, USA, 2004.
41. Majumdar, R.K.; Gupta, S. Isolation, identification and characterization of *Staphylococcus* sp. from Indian ethnic fermented fish product. *Lett. Appl. Microbiol.* **2020**, *71*, 359–368. [CrossRef]
42. Hasan, F.; Shah, A.A.; Javed, S.; Hameed, A. Industrial applications of microbial lipases. *Enzyme Microb. Technol.* **2006**, *39*, 235–251. [CrossRef]
43. Heo, S.; Lee, J.H.; Jeong, D.W. Food-derived coagulase-negative *Staphylococcus* as starter cultures for fermented foods. *Food Sci. Biotechnol.* **2020**, *29*, 1023–1035. [CrossRef]
44. Fetsch, A.; Johler, S. *Staphylococcus aureus* as a foodborne pathogen. *Curr. Clin. Microbiol. Rep.* **2018**, *5*, 88–96. [CrossRef]
45. Arora, M.; Baldi, A. Selective identification and characterization of potential probiotic strains: A review on comprehensive polyphasic approach. *Appl. Clin. Res. Clin. Trials Regul. Aff.* **2017**, *4*, 60–76. [CrossRef]
46. Keisam, S.; Tuikhar, N.; Ahmed, G.; Jeyaram, K. Toxigenic and pathogenic potential of enteric bacterial pathogens prevalent in the traditional fermented foods marketed in the Northeast region of India. *Int. J. Food Microbiol.* **2019**, *296*, 21–30. [CrossRef] [PubMed]
47. Zang, J.; Xu, Y.; Xia, W.; Yu, D.; Gao, P.; Jiang, Q.; Yang, F. Dynamics and diversity of microbial community succession during fermentation of Suanyu, a Chinese traditional fermented fish, determined by high throughput sequencing. *Food Res. Intern.* **2018**, *111*, 565–573. [CrossRef] [PubMed]
48. Adesulu-Dahunsi, A.T.; Sanni, A.I.; Jeyaram, K. Rapid differentiation among *Lactobacillus*, *Pediococcus* and *Weissella* species from some Nigerian indigenous fermented foods. *LWT* **2017**, *77*, 39–44. [CrossRef]
49. Adesulu-Dahunsi, A.T.; Sanni, A.I.; Jeyaram, K. Diversity and technological characterization of *Pediococcuspentosaceus* strains isolated from Nigerian traditional fermented foods. *LWT* **2021**, *140*, 110697. [CrossRef]
50. Tamang, J.P.; Watanabe, K.; Holzapfel, W.H. Diversity of microorganisms in global fermented foods and beverages. *Front. Microbiol.* **2016**, *7*, 377. [CrossRef] [PubMed]
51. Liang, R.; Huang, J.; Wu, X.; Fan, J.; Xu, Y.; Wu, C.; Jin, Y.; Zhou, R. Effect of raw material and starters on the metabolite constituents and microbial community diversity of fermented soy sauce. *J. Sci. Food Agri.* **2019**, *99*, 5687–5695. [CrossRef]
52. Leech, J.; Cabrera-Rubio, R.; Walsh, A.M.; Macori, G.; Walsh, C.J.; Barton, W.; Finnegan, L.; Crispie, F.; O'Sullivan, O.; Claesson, M.J.; et al. Fermented-Food Metagenomics Reveals Substrate-Associated Differences in Taxonomy and Health-Associated and Antibiotic Resistance Determinants. *Msystems* **2020**, *5*, e00522-20. [CrossRef]
53. Fang, G.Y.; Chai, L.J.; Zhong, X.Z.; Jiang, Y.J. Deciphering the succession patterns of bacterial community and their correlations with environmental factors and flavor compounds during the fermentation of Zhejiang rosy vinegar. *Int. J. Food Microbiol.* **2021**, *341*, 109070. [CrossRef]
54. Parente, E.; Zotta, T.; Faust, K.; De Filippis, F.; Ercolini, D. Structure of association networks in food bacterial communities. *Food Microbiol.* **2018**, *73*, 49–60. [CrossRef]

fermentation

Article

Prebiotic and Immunomodulatory Properties of the Microalga *Chlorella vulgaris* and Its Synergistic Triglyceride-Lowering Effect with Bifidobacteria

Ivana Hyrslova [1,*], Gabriela Krausova [1], Jana Smolova [1], Barbora Stankova [2], Tomas Branyik [3], Hana Malinska [4], Martina Huttl [4], Antonin Kana [5], Ivo Doskocil [6] and Ladislav Curda [7]

1 Department of Microbiology and Technology, Dairy Research Institute Ltd., 160 00 Prague, Czech Republic; krausova@milcom-as.cz (G.K.); smolova@milcom-as.cz (J.S.)
2 4th Department of Medicine-Department of Gastroenterology and Hepatology, First Faculty of Medicine, Charles University in Prague, 128 00 Prague, Czech Republic; barbora.stankova@lf1.cuni.cz
3 Department of Biotechnology, University of Chemistry and Technology, 162 00 Prague, Czech Republic; tomas.branyik@vscht.cz
4 Centre for Experimental Medicine, Institute for Clinical and Experimental Medicine, 140 21 Prague, Czech Republic; hana.malinska@ikem.cz (H.M.); martina.huttl@ikem.cz (M.H.)
5 Department of Analytical Chemistry, University of Chemistry and Technology, 162 00 Prague, Czech Republic; antonin.kana@vscht.cz
6 Department of Microbiology, Nutrition and Dietetics, Czech University of Life Sciences Prague, 165 00 Prague, Czech Republic; doskocil@af.czu.cz
7 Department of Dairy, Fat and Cosmetics, University of Chemistry and Technology, 160 00 Prague, Czech Republic; ladislav.curda@vscht.cz
* Correspondence: ivana.hyrslova@gmail.com

check for updates

Citation: Hyrslova, I.; Krausova, G.; Smolova, J.; Stankova, B.; Branyik, T.; Malinska, H.; Huttl, M.; Kana, A.; Doskocil, I.; Curda, L. Prebiotic and Immunomodulatory Properties of the Microalga *Chlorella vulgaris* and Its Synergistic Triglyceride-Lowering Effect with Bifidobacteria. *Fermentation* **2021**, *7*, 125. https://doi.org/10.3390/fermentation7030125

Academic Editor: Hiroshi Kitagaki

Received: 24 June 2021
Accepted: 20 July 2021
Published: 22 July 2021

Abstract: The microalga *Chlorella* and strains of *Bifidobacterium* have been used in human or animal food supplements for decades because of their positive health effects. The presented study assessed different properties of *C. vulgaris* and its combination with bifidobacteria with the aim to develop new functional foods. The growth of four bifidobacteria strains in milk and whey supplemented with 1.0% (*w/v*) *C. vulgaris* and the immunomodulatory effects of aqueous *Chlorella* solutions (0.5%, 1.0%, and 3.0%) on human peripheral mononuclear cells were evaluated. Furthermore, synergistic effects on lipid metabolism of rats fed a high-fat diet with *Chlorella* and *B. animalis* subsp. *lactis* BB-12® were analysed. *Chlorella* had a positive growth-promoting effect on the tested bifidobacteria ($p < 0.05$), and significantly increased the secretion of inflammatory cytokines (tumor necrosis factor-α, interleukin-10, and interleukin-6), depending on the concentration of *Chlorella* ($p < 0.05$). After 8 weeks, significant synergistic effects of *Chlorella* and bifidobacteria on triglyceride levels in rat heart, liver, and serum were observed ($p < 0.05$). These results demonstrate that various combinations of *Chlorella* and bifidobacteria have significant potential for the development of new fermented products, dependent on the algal species, probiotic strain, application form, and concentrations for acceptable sensory quality for consumers.

Keywords: *Bifidobacterium*; *Chlorella*; cytokines; functional food; health; prebiotic; triglyceride

1. Introduction

Chlorella vulgaris as a food supplement acts as a source of nutritionally-valuable substances, including proteins, carbohydrates, vitamins, pigments, antioxidants, and unsaturated fatty acids. *Chlorella* and *Arthrospira* (Spirulina), the most consumed microalgae, accumulate high-quality proteins with a balanced amino acid profile according to WHO recommendations for essential amino acids [1]. Commercial prebiotics are generally carbohydrate compounds that act as substrates for probiotic microorganisms. The most important carbohydrate in *Chlorella* cells, in terms of its usefulness for living organisms, is β-1,3-glucan, which is a branched polysaccharide of β-D-glucose units. β-1,3-glucan is also a water-soluble fiber and readily fermentable in the colon; it represents an important part of the diet because of its enhancing effect on digestion [2]. Apart from β-1,3-glucan, almost all vitamins (A, B1, B2, B6, B12, C, E, biotin, pantothenate, etc.) are provided by microalgae most commonly used in food supplements. The availability of these vitamins likely supports the observed growth increase of bacteria. Microalgae are also rich in pigments, containing on a dry mass basis mainly chlorophyll (0.5–1.0%), carotenoids (0.1–0.2%), and phycobiliproteins [3]. The positive effects of microalgae on human health, such as prebiotic, immunomodulatory, anti-oxidative, anti-cancer, and hypocholesterolemic effects, have been reported in a variety of studies; however, the mechanism imparting the positive effect is strongly dependent on the specific microalgal strain and the content of bioactive substances [4–6].

The most common bacteria used in the food and feed industry are species of *Lactobacillus* and *Bifidobacterium*, which also represent probiotics and have the status "Generally Recognized As Safe" (GRAS) [7]. Fermented dairy products with probiotics are popular and available in markets worldwide [8]. *Bifidobacterium animalis* subsp. *lactis* is one of the preferred strains used by manufacturers because of its tolerance to acid conditions (low pH) and/or low molecular oxygen arising during fermentation [7,9]. An important qualitative parameter is the viability of lactic acid bacteria and probiotic microorganisms in the final products, including until the end of their shelf-life. A value for viable cells of 10^6/g or 10^6/mL in probiotic products is considered satisfactory for fulfilling the criterion of viable counts [9]. Bifidobacteria represent one of the first colonizing bacteria in the neonatal intestine and they are known to assist in the development of adaptive immunity, as well as modulate mucosal physiology [10]. Numerous studies have assessed the use of probiotics from genera *Bifidobacterium* in the prevention or treatment of many diseases, such as gastroenteritis caused by rotavirus, enteric diseases, diarrhea, lactose intolerance, allergies, and the reduction of serum cholesterol [10–12].

Several researchers have tested the influence of additional algal biomass on yogurt, cheese, or fermented milk products [13–15]. The addition of algae, such as *C. vulgaris*, *Ch. regularis*, or *Arthrospira* has resulted in a positive prebiotic effect on the viability of lactic acid bacteria and also helped to improve the nutritional quality of fermented products. However, to develop new fermented foods enriched with probiotics and algae, it is necessary to first increase our understanding of the influence of individual algae or their components on the growth promotion of bacteria and their health effects on human cells. The prebiotic effect of *C. vulgaris* in vitro and its non-cytotoxic effects were described in our previous study [16]. In the presented study, we continued to evaluate the growth promoting potential of *C. vulgaris* on selected bifidobacteria in bovine milk and whey and also determined its immunomodulatory effect on human peripheral blood mononuclear cells (hPBMCs) isolated from healthy adult donors. The synergistic hypocholesterolemic effects of a combination of *Chlorella* and *B. animalis* subsp. *lactis* BB-12® in an animal model of Prague hereditary hypercholesterolemic rats with diet-induced hypercholesterolemia [17] were also evaluated. Our results demonstrate that Chlorella and bifidobacteria used in various combinations have significant synergistic potential for the development of new fermented products with positive health effects.

2. Materials and Methods

2.1. Microorganisms

The tested microorganism strains were selected from the Culture Collection of Dairy Microorganisms Laktoflora® (Tabor, Czech Republic) and the commercial strain *B. animalis* subsp. *lactis* BB-12® was obtained from Ch. Hansen (Hørsholm, Denmark) (Table 1). Bacterial strains were stored at −20 °C in Wilkins Chalgren anaerobic broth (Oxoid, Hampshire, UK) with 10% (*v/v*) glycerol. Before each analysis, cells were transferred twice in fresh Wilkins Chalgren anaerobic broth with *L*-cysteine hydrochloride (Merck, Darmstadt, Germany) and cultivated at 37 °C for 24 h in anaerobic jars. Heterotrophic *C. vulgaris* was obtained from the Microbiology Department of Academy of Sciences, Třeboň—ALGATECH

(Třeboň, Czech Republic). The tested *Chlorella* powder sample contained 50.3 g of protein, 8.1 g of lipid, and 15.2 g of carbohydrate per 100 g of algal biomass.

Table 1. Selected microorganisms.

Strain	Species	Origin
CCDM [1] 93	*Bifidobacterium animalis* subsp. *lactis*	Original culture
BB12®	*Bifidobacterium animalis* subsp. *lactis*	Original culture
CCDM 562	*Bifidobacterium breve*	GIT of child
CCDM 486	*Bifidobacterium breve*	Human feces

[1] CCDM, Culture Collection of Dairy Microorganisms; GIT, gastrointestinal tract.

2.2. Prebiotic Effect

To assess bacterial growth, milk and whey were supplemented with 1.0% (*w/v*) *Chlorella* powder and pasteurized at 85 °C for 10 min. Inoculated milk and whey without *Chlorella* served as negative controls. Prior to inoculation, bifidobacterial suspensions were prepared as described [18]. Inoculated samples were cultivated in anaerobic jars (Oxoid, Hampshire, UK) at 37 °C for 24 h. Counts of the tested strains were determined using 10-fold serial dilutions and plated onto MRS agar with *L*-cysteine hydrochloride (pH 6.2). The pH of the cultivated medium and the concentration of lactic and acetic acids were also measured. Concentrations of organic acids, as the main fermentation products of tested bifidobacteria, were determined by the isotachophoretic method using IONOSEP 2003 (RECMAN, Ostrava, Czech Republic). A mixture of 10 mmol/L HCl, 22 mmol/L 6-aminocaproic acid, and 0.1% hydroxyethylcellulose (Merck) was used as the leading electrolyte and 10 mmol/L caproic acid acted as the terminating electrolyte. The conditions of analysis were selected according to the manufacturer's instructions (RECMAN, Ostrava, Czech Republic). Results were obtained from three independent measurements.

2.3. Stimulation and Immunomodulation

The immunomodulatory effect of an aqueous solution of *C. vulgaris* was evaluated using Luminex multiplex assays for the simultaneous quantitative determination of multiple human cytokine concentrations in cell culture supernatants, serum, and plasma, according to a previous study with slight modifications [16]. In brief, eight samples of blood from healthy adults, for the isolation of hPBMCs via Ficoll-Hypaque gradient separation, were ordered from the Blood Transfusion Center of General Faculty Hospital (Prague, Czech Republic). Following separation and purification, hPBMCs were adjusted to a final concentration of 10^7 cells/mL. Mononuclear cells (0.1 mL) were stimulated in X-vivo medium (Cambrex, Whippany, NJ, USA) with 0.1 mL of a 0.5%, 1.0%, or 3.0% aqueous solution of *C. vulgaris* at 37 °C. The total volume was 1 mL. Unstimulated hPBMCs and X-vivo medium were used as the negative controls. Microplates with the samples were incubated for 3 d at 37 °C. Levels of cytokines produced by stimulation of hPBMCs with different concentrations of *Chlorella* were determined using the Fluorokine MAP Human Base Kit A (R&D Systems, Minneapolis, MN, USA) for interferon (IFN)-γ, interleukin (IL)-4, IL-10, IL-6, IL-17, and tumor necrosis factor (TNF)-α by multiplex analysis using a Luminex 200 Analyzer (Luminex Corp., Austin, TX, USA). The concentration of cytokines produced by hPBMCs was assessed using Luminex IS 2.3 (Luminex Corp.). Results were obtained from three independent measurements.

2.4. Experimental Animals and Diet

A total of 24 male Prague hereditary hypercholesterolemic rats with a body weight of 249 ± 16 g were obtained from Albert Weber-SEMED (Praha, Czech Republic). Animals were acclimatized to laboratory conditions for 2 weeks before the experiment by housing at room temperature (22–24 °C) and 55–60% humidity on a 12 h/12 h light-dark cycle with ad libitum access to water and food. The hypercholesterolemic diet was fortified with 2.0% (*w/w*) cholesterol. After acclimation, the rats were divided into a control group and three test groups as follows: C, rats on a hypercholesterolemic diet; GI, rats on a hypercholesterolemic diet with 1.0% *Chlorella* powder (*w/w*), GII, rats on a hypercholesterolemic diet with lyophilized *B. animalis* subsp. *lactis* BB-12® (10^6 CFU/g per one g of pellet); and GIII, rats on a hypercholesterolemic diet with lyophilized *B. animalis* subsp. *lactis* BB-12® (10^6 CFU/g per one g of pellet) and 1.0% *Chlorella* powder (*w/w*). The experimental groups received their respective diets for 8 weeks. At the end of the experiment, rats were euthanized by decapitation after light anesthetization (zoletil, 5 mg/kg body weight) in the postprandial state. Blood was collected into tubes without anticoagulant addition. Aliquots of serum and heart, liver, and aorta tissue samples were stored at −80 °C until analysis. All experiments were performed in accordance with the Animal Protection Law of the Czech Republic (311/1997) in compliance with European Community Council recommendations (86/609/ECC) for the use of laboratory animals and approved by the ethical committee of the Ministry of Education, Youth and Sports of the Czech Republic (MSMT-2309/2018-2; 2/2/2018).

2.5. Biochemical Analysis

Lipid parameters were determined as previously described [16]. Briefly, total cholesterol and lipoprotein fractions—very low-density lipoproteins (VLDL), low-density lipoproteins (LDL), intermediate-density lipoproteins (IDL), and high-density lipoproteins (HDL)—were assessed by enzymatic colorimetric methods (CHOD/PAP, direct homogeneous enzymatic-colorimetric reaction without precipitation, GPO/PAP; Lab Mark a.s., Prague, Czech Republic) using an automatic biochemical analyzer modular (Roche, Prague, Czech Republic). Subfractions of LDL were analyzed by high-performance discontinued gel electrophoresis using polyacrylamide gel tubes (Lipoprint® LDL System, Quantimetrix, Redondo Beach, CA, USA). LDL particles were separated into seven subfractions (LDL1–LDL7). The subfractions LDL1 and LDL2 represent large (buoyant) particles and LDL3–7 represent small dense LDL (sd-LDL). Concentrations of cholesterol in sd-LDL over 6 mg/dL or peak LDL particle diameter $\leq$ 26.8 nm were denoted phenotype pattern B with a predominance of sd-LDL [19]. To determine triglycerides and cholesterol in the tissues, samples were extracted in a chloroform/methanol mixture. The resulting pellet was dissolved in isopropyl alcohol, after which the triglyceride content was determined using an enzymatic assay (Erba-Lachema, Brno, Czech Republic).

2.6. Statistical Analysis

All statistical evaluations were performed using Microsoft Office Excel 2019 and Statistica 13.1CZ statistical software. Normality of data was checked using the Shapiro–Wilk test. One-way analysis of variance (ANOVA) at a significance level of $\alpha = 0.05$, followed by Tukey's test, was applied to the comparison of tested biochemical parameters. ANOVA with a post hoc least significance difference test (LSD) for multiple comparisons was used to evaluate the results of prebiotic effects, considering statistical significance at the level of $\alpha = 0.05$.

3. Results

3.1. Prebiotic Assay

Prebiotic activity was evaluated by determining the bacterial counts of four selected bifidobacterial strains, changes in pH, and production of lactic and acetic acids (Tables 2 and 3) after fermentation of whey and bovine milk with 1.0% (*w/v*) *Chlorella* powder supplementation. The production of acids was significantly increased in milk and whey enriched with *Chlorella* powder than not enriched ($p < 0.05$). The concentration of these acids was 5-fold higher in milk supplemented with *Chlorella* and 1.5-fold higher in whey supplemented with *Chlorella*. The pH values also correlated with increased concentrations of lactic and acetic acids produced by bifidobacteria (Table 3).

Table 2. Production of lactic acid and acetic acid after 24 h cultivation.

Strain	Acid Production (mg/L)	Milk	Milk + 1.0% *Chlorella*	Whey	Whey + 1.0% *Chlorella*
CCDM 93	lactic acid	388 ± 20 ^{A}	1754 ± 85 ^{B}	1661 ± 80 ^{B}	2437 ± 120 ^{C}
	acetic acid	1105 ± 50 ^{A}	2354 ± 110 ^{D}	1309 ± 65 ^{B}	2058 ± 58 ^{C}
BB-12®	lactic acid	208 ± 15 ^{A}	1876 ± 90 ^{C}	1193 ± 60 ^{B}	1937 ± 55 ^{C}
	acetic acid	512 ± 25 ^{A}	2246 ± 110 ^{C}	1812 ± 90 ^{B}	2588 ± 130 ^{D}
CCDM 486	lactic acid	259 ± 15 ^{A}	1891 ± 90 ^{C}	1524 ± 75 ^{B}	2013 ± 100 ^{C}
	acetic acid	477 ± 25 ^{A}	2384 ± 110 ^{D}	1747 ± 85 ^{B}	2124 ± 105 ^{C}
CCDM 562	lactic acid	719 ± 35 ^{B}	2200 ± 110 ^{D}	431 ± 20 ^{A}	1768 ± 90 ^{C}
	acetic acid	836 ± 39 ^{B}	2272 ± 110 ^{D}	368 ± 40 ^{A}	1734 ± 85 ^{C}

Values are the means of triplicate measurements ± standard deviation (SD). ^{A,B,C,D} Data in the lines with different superscripts differ ($p < 0.05$).

Table 3. pH and cell counts after 24 h cultivation.

Tested Parameter	Strain	Milk	Milk + 1.0% *Chlorella*	Whey	Whey + 1.0% *Chlorella*
pH	CCDM 93	5.50 ± 0.03 ^{C}	4.85 ± 0.01 ^{B}	4.79 ± 0.01 ^{B}	4.15 ± 0.02 ^{A}
	BB12	5.90 ± 0.03 ^{D}	4.89 ± 0.00 ^{C}	4.57 ± 0.00 ^{B}	4.38 ± 0.12 ^{A}
	CCDM 486	5.90 ± 0.06 ^{D}	4.87 ± 0.07 ^{C}	4.50 ± 0.16 ^{B}	4.24 ± 0.04 ^{A}
	CCDM 562	5.13 ± 0.35 ^{B}	4.78 ± 0.06 ^{AB}	4.86 ± 0.64 ^{AB}	4.21 ± 0.12 ^{A}
cell counts (CFU/mL)	CCDM 93	6.24 ± 0.76 ^{A}	8.34 ± 1.26 ^{B}	7.39 ± 0.12 ^{AB}	7.54 ± 0.55 ^{AB}
	BB12	7.90 ± 1.01 ^{A}	8.37 ± 0.19 ^{A}	8.37 ± 0.34 ^{A}	8.35 ± 0.35 ^{A}
	CCDM 486	8.59 ± 1.01 ^{A}	8.31 ± 0.08 ^{A}	8.26 ± 0.06 ^{A}	8.38 ± 0.27 ^{A}
	CCDM 562	8.16 ± 0.61 ^{A}	7.76 ± 0.33 ^{A}	7.02 ± 1.02 ^{A}	7.31 ± 0.53 ^{A}

Values are the means of triplicate measurements ± SD. ^{A,B,C,D} Data in the lines with different superscripts differ ($p < 0.05$).

3.2. Immunomodulatory Effect

The immunomodulatory effects of three different aqueous solutions of Chlorella (0.5, 1.0, and 3.0% *w/v*) on hPBMCs were compared based on the production of pro-inflammatory and regulatory cytokines. Levels of selected ILs (IL-4, IL-10, IL-17, IL-6), TNF-α, and INF-γ were determined using multiplex analysis. In contrast with other studies where the levels of tested cytokines are usually evaluated by the method of flow cytometry or ELISPOT assay [20,21]. Our results showed that the production of all tested cytokines was significantly different ($p < 0.05$) after stimulation with the lowest concentration of Chlorella (0.5% *w/v*) compared with the other Chlorella concentrations (Table 4). The production of TNF-α, IL-10, and IL-6 by hPBMCs was dependent on the concentration of Chlorella added.

Table 4. Cytokine production by mononuclear cells.

Chlorella Concentration (*w/v*)	Cytokines (pg/mL)					
	TNF-α	IL-17	IL-10	IL-6	IL-4	IFN-γ
0.5%	187.57 ± 62.02 ^{C}	1.16 ± 0.09 ^{B}	12.73 ± 7.58 ^{D}	4976.43 ± 781.21 ^{D}	2.68 ± 0.10 ^{B}	2.34 ± 0.37 ^{B}
1.0%	23.85 ± 8.98 ^{B}	0.65 ± 0.10 ^{A,B}	0.74 ± 0.11 ^{B}	202.18 ± 103.13 ^{C}	2.68 ± 0.00 ^{A}	0.82 ± 0.21 ^{A}
3.0%	1.85 ± 0.67 ^{A}	0.65 ± 0.33 ^{A}	0.11 ± 0.07 ^{A}	55.13 ± 18.91 ^{B}	3.22 ± 0.13 ^{A}	1.01 ± 0.17 ^{A}
control	0.93 ± 0.91 ^{A}	0.00 ± 0.90 ^{A}	1.66 ± 0.33 ^{C}	7.7 ± 3.1 ^{A}	5.35 ± 2.70 ^{A}	3.66 ± 1.00 ^{A}

Values are the means of triplicate measurements ± SD. ^{A,B,C,D} Data in the column with different superscripts differ ($p < 0.05$).

3.3. Serum Lipid Profile

Serum total cholesterol (TC), triacylglycerides, VLDL, LDL, IDL, and HDL levels were evaluated in a rat model after feeding a high-cholesterol diet supplemented with Chlorella powder and/or B. animalis subsp. lactis BB-12 for 8 weeks (Table 5). In the group fed the diet with B. animalis subsp. lactis BB-12® (GII), all tested serum lipid parameters, apart from the HDL concentration, significantly increased ($p < 0.05$) in comparison with all the other diet groups. Values among the rest of the diet groups and the control group did not statistically differ. Nevertheless, levels of triglycerides in liver and heart tissues, as well as serum, significantly decreased ($p < 0.05$) in rats fed a diet fortified with Chlorella powder plus B. animalis subsp. lactis BB-12® (GIII), in contrast to the control and the rest of the groups tested (Table 6). Thus, a positive synergistic effect of Chlorella and B. animalis subsp. lactis BB-12 was observed.

Table 5. Serum lipid profiles in a rat model after feeding a high-cholesterol diet supplemented with *Chlorella* powder and/or *B. animalis* subsp. *lactis* BB-12 for 8 weeks.

Tested Group	VLDL (mg/dL)	IDL-C (mg/dL)	IDL-B (mg/dL)	IDL-A (mg/dL)	HDL (mg/dL)	TC (mg/dL)	LDL (mg/dL)
C	74.8 ± 6.5 ^{A}	29.8 ± 5.2 ^{A}	16.5 ± 3.8 ^{A}	3.7 ± 0.8 ^{A}	52.2 ± 18.5 ^{A}	184.5 ± 13.5 ^{A}	55.7 ± 10.7 ^{A}
GI	76.0 ± 18.9 ^{A}	33.8 ± 6.6 ^{A}	23.7 ± 5.7 ^{A}	6.3 ± 2.0 ^{B}	71.7 ± 26.0 ^{A}	219.6 ± 13.4 ^{B}	69.5 ± 11.4 ^{A}
GII	106.0 ± 13.1 ^{B}	43.2 ± 5.9 ^{B}	26.7 ± 5.7 ^{B}	7.0 ± 2.2 ^{B}	79.8 ± 10.4 ^{A}	272.6 ± 20.3 ^{C}	85.7 ± 5.3 ^{B}
GIII	86.5 ± 7.7 ^{A}	36.8 ± 4.6 ^{A}	22.2 ± 4.0 ^{A}	5.3 ± 1.4 ^{A}	67.7 ± 23.7 ^{A}	224.7 ± 13.3 ^{B}	69.7 ± 11.8 ^{A}

VLDL, very low-density lipoprotein; LDL, low-density lipoprotein; HDL, high-density lipoprotein; TC, total cholesterol; IDL, intermediate-density lipoprotein; C, control; GI, *Chlorella*; GII, *B. animalis* subsp. *lactis* BB-12®; GIII, *Chlorella* + *B. animalis* subsp. *lactis* BB-12®. Values are the mean ± SD, $n = 6$. ^{A,B,C} Data in the column with different superscripts differ ($p < 0.05$).

Table 6. Accumulation of triglycerides in tissues.

Tested Group	Tissues			
	Liver (µmol/g)	Aorta (µmol/g)	Heart (µmol/g)	Serum (mmol/L)
C	5.4 ± 0.3 ^{A}	1.5 ± 0.4 ^{A}	3.1 ± 0.3 ^{A}	1.0 ± 0.3 ^{A}
GI	5.4 ± 0.3 ^{A}	1.4 ± 0.2 ^{A}	3.1 ± 0.3 ^{A}	1.4 ± 0.2 ^{B}
GII	5.3 ± 0.4 ^{A}	1.1 ± 0.2 ^{A}	2.9 ± 1.0 ^{A}	1.2 ± 0.2 ^{A}
GIII	4.8 ± 0.3 ^{B}	1.1 ± 0.3 ^{A}	1.9 ± 0.4 ^{B}	0.7 ± 0.1 ^{C}

C, control; GI, Chlorella; GII, *B. animalis* subsp. *lactis* BB12; GIII, *Chlorella* + *B. animalis* subsp. *lactis* BB12. Values are the mean ± SD, $n = 6$. ^{A,B,C}. Data in the column with different superscripts differ ($p < 0.05$).

4. Discussion

In the present study, we first investigated the growth-promoting effect of *C. vulgaris* on four *Bifidobacterium* strains in bovine milk and whey. Fermentation ability was evaluated based on the production of metabolic compounds, such as acetic acid and lactic acid, which are important metabolic products produced by bifidobacteria. Levels of both acids were significantly increased in milk and whey supplemented with *Chlorella* vs. without. The prebiotic or growth-promoting effects of *C. vulgaris* biomass on probiotics or intestinal bacteria has also been described in several other studies. For example, Pulz and Gross [22] reported that, compared to the control, the growth rate of lactobacilli (*Lbc. acidophilus*) increased up to 10-fold with the addition of algal biomass. The cell wall of *Chlorella* contains a wide range of oligo- and polysaccharides, which can serve as an energy source for bacteria. Their fermentability may be influenced by factors such as the chain length of individual carbohydrates, their structure (e.g., branching, glycosidic linkage, type of monosaccharide moieties, etc.), and cultivation conditions. The enzymatic reactions used by each bacterium may play a key role. Another study by Scieszka and Klewicki [23] evaluated the protective effect of *C. vulgaris* on the survival of four *L. brevis* strains under adverse environmental conditions in the human gastrointestinal tract, such as low pH and phenolic or bile salts. The addition of the alga had a positive effect on the increased survival of tested lactobacilli; nevertheless, the protective effect at low pH was strain-specific and dependent on the features of individual strains. Cantú-Bernal et al. [16] reported enhanced viability and antiviral effects of *B. longum* and *L. plantarum* against rotavirus in combination with *C. sorokiniana* in dairy products. Beheshtipour et al. [8] observed the influence of *C. vulgaris* and *Arthrospira platensis* (Spirulina) addition on the viability of selected probiotic microorganisms in yogurt. Their results showed that the addition of the alga stimulated growth and maintained the viability of *L. acidophilus* LA-5 and *B. animalis* subsp. *lactis* BB-12®, both at the end of fermentation and during cold storage for 28 d. Sciezska and Klewicka [24] tested the influence of *C. vulgaris* (0.1% and 1.5% (*w/v*)) on the growth and metabolic activity of four *L. brevis* strains isolated from vegetable silage. Selected lactobacilli showed a relatively high production of *L*-lactic acid and lower *D*-lactic acid after cultivation in medium with *Chlorella*. Changes in enzymatic activity were also observed, for instance, *L. brevis* LOCK 0980 demonstrated higher enzymatic activity of valine arylamidase, α-galactosidase, and α-glucosidase when cultivated with *Chlorella*.

Chlorella polysaccharides have also shown strong immunomodulatory activities, primarily aqueous soluble polysaccharides. This activity has been reported in mice, human blood cells, and human clinical trials [5,25–27].TNF-α is an important initiator of the inflammatory response and mediator, together with IL-1β and IL-6, which are produced by activated monocytes and macrophages that influence cell proliferation stimulation in various types of cells [21]. Anti-inflammatory IL-6, which inhibits TNF-α, is produced in response to increased TNF-α levels, helping to maintain the Th1/Th2 balance [28]. Our results showed that the three tested aqueous solutions of *C. vulgaris* (0.5, 1.0, and 3.0% *w/v*) influenced production of TNF, IL-6, and IL-10. The mononuclear cells produced the highest amounts of these cytokines after stimulation with the lowest concentration of *Chlorella* tested. This trend could have been caused by the decreased viability of the mononuclear cells after incubation with various concentrations of aqueous *Chlorella* solutions. The inhibition of inflammatory mediators and cytokines by aqueous extracts of *C. vulgaris* was first described by Sibi and Rabina [28]. They determined the in vitro anti-inflammatory activities of different solvent fractions (*n*-hexane, chloroform, ethanol, and water) from *Chlorella*; however, the effect of the aqueous extract was not significant. Notably, they reported that the concentrations of observed parameters were the highest after using the lowest concentration of all tested solvent fractions of *Chlorella*, similar to our study. Ewart et al. [21] stimulated the production of cytokines by using three different concentrations of aqueous *C. pyrenoidosa* extract (1, 10, and 100 μg/mL) for 24 h. Levels of IL-10, TNF-α, and IFN-γ were markedly increased compared to that of the control; however, in our study, the levels of INF-γ did not significantly increase.

Recently, the prevalence of obesity, diabetes, inflammatory bowel syndrome, cancer, and cardiovascular diseases (CVD) have rapidly increased worldwide [29]. Oxidative stress and hypercholesterolemia are risk factors for the development of CVD and are closely related to these diseases [30,31]. Therefore, new strategies have been developed to inhibit the growing incidence or prevention of these diseases using natural sources. Several studies have tested the effect of *Chlorella* or other microalgae on lipid metabolism in different animal models, such as rats, mice or broilers [30,32–36] Shibata et al. (2001) assessed the hypocholesterolemic effects of the indigestible fraction (5.7% *w/w*) of *Chlorella* powder (12.7% *w/w*) compared with the digestible fraction, and reported opposite conclusions. Their results proved the positive influence of the *Chlorella* fraction on decreasing serum cholesterol levels, but no effect on the levels of serum triglycerides and phospholipids. Lee et al. [30] investigated the effect of *C. vulgaris* on lipid metabolism in Wistar rats fed a high-fat diet containing 5.0% or 10.0% (*w/w*) *Chlorella* powder. In this case, serum total lipid and liver triglyceride concentrations were also significantly lower than those in the control group. Chovancikova and Simek [32] examined a mouse model fed a high-fat diet supplemented with 1.0% (*w/w*) *C. vulgaris*. After 10 weeks, the levels of TC and triglycerides in the serum and liver were significantly inhibited.

In the present study, the addition of *Chlorella* and *B. animalis subsp. lactis* BB-12 significantly decreased the concentration of triglycerides in the serum, liver, and heart of the treated rats, in contrast to the control and other tested groups (Table 6). This effect could be associated with inhibition of hepatic fatty acid synthesis and triglyceride production; thus, the output of VLDL was limited [32]. Bifidobacteria also show bile salt hydrolase activity and produce extracellular polysaccharides, which are relevant to the industrial production of human food or medicine [37,38]. Bile salt hydrolase activity, together with binding of cholesterol to the probiotic cellular surface and incorporation into the cell membrane, contributes to the cholesterol-lowering mechanisms of probiotics [39] Zanotti et al. [40] showed that daily supplementation with bifidobacteria modifies the fecal microbiota of mice toward those bacteria involved in the metabolism of cholesterol. Although the effect of the cholesterol-lowering activity of probiotics in human trials does not show consistent results, there are some successful human trials for bifidobacteria [41]. Lee et al. [42] suggested that the lipid-lowering effect of probiotics may be limited to populations with high total cholesterol and LDL-C levels.

5. Conclusions

In summary, the results demonstrated the prebiotic and immunomodulatory effect of the Chlorella powder and a synergic effect of *Chlorella* and *B. animalis* subsp. *lactis* BB-12 combination to decrease the level of triglycerides in the serum, liver, and heart of the treated rats. Consequently, the incorporation and/or combination of *Chlorella* together with bifidobacteria to functional food or fermented dairy products may have a positive effect on the viability of bifidobacteria and their properties, thus influencing human or animal health. Nevertheless, the limitations of this study include the lack of detailed analysis of the chemical composition of the tested alga relative to the content and influence of specific saccharides, lipids, or other functional compounds and their bioactive effects. Therefore, the isolation of individual fractions and determination of their functional properties will be the main aim of future studies. The mechanisms of the hypocholesterolemic effect of microalgae are also not fully understood, and other in vivo and in vitro studies are needed to elucidate this. Notably, the positive effects of algae and bacterial combinations seem to be dependent on the selected bacterial strains, the type of algae, their functional properties, and their synergistic effects.

Author Contributions: Conceptualization, I.H. and J.S.; methodology, I.H., B.S. and M.H.; formal analysis, I.H., J.S., B.S., H.M., M.H. and I.D.; investigation, A.K., G.K. and I.H.; resources, I.H.; data curation, I.H., A.K. and B.S.; writing—original draft preparation, I.H., A.K. and J.S.; writing—review and editing, T.B., L.C. and M.H.; supervision, I.H., G.K. and A.K.; project administration, I.H.; and funding acquisition, I.H. and G.K. All authors have read and agreed to the published version of the manuscript.

Funding: This research was funded by the Ministry of Agriculture of the Czech Republic Institutional support, grant numbers MZE-RO1421 and QK1910300 and the METROFOOD-CZ research infrastructure project, MEYS grant number LM2018100, which includes access to its facilities.

Institutional Review Board Statement: All experiments were performed in accordance with the Animal Protection Law of the Czech Republic (311/1997) in compliance with European Community Council recommendations (86/609/ECC) for the use of laboratory animals, and approved by the ethical committee of the Ministry of Education, Youth and Sports of the Czech Republic (MSMT-2309/2018-2; 2/2/2018).

Informed Consent Statement: Not applicable.

Data Availability Statement: All data are presented in the paper.

Conflicts of Interest: The authors declare no conflict of interest.

References

1. Caporgno, M.P.; Mathys, A. Trends in microalgae incorporation into innovative food products with potential health benefits. *Front. Nutr.* **2018**, *5*. [CrossRef]
2. Mišurcová, L.; Skrovankova, S.; Samek, D.; Ambrožová, J.; Machů, L. Health benefits of algal polysaccharides in human nutrition. *Adv. Food Nutr. Res.* **2012**, *66*, 75–145. [CrossRef]
3. Spolaore, P.; Joannis-Cassan, C.; Duran, E.; Isambert, A. Commercial applications of microalgae. *J. Biosci. Bioeng.* **2006**, *101*, 87–96. [CrossRef]
4. Cherng, J.-Y.; Shih, M.-F. Improving glycogenesis in Streptozocin (STZ) diabetic mice after administration of green algae *Chlorella*. *Life Sci.* **2006**, *78*, 1181–1186. [CrossRef]
5. Qi, J.; Kim, S.M. Characterization and immunomodulatory activities of polysaccharides extracted from green alga *Chlorella ellipsoidea*. *Int. J. Biol. Macromol.* **2017**, *95*, 106–114. [CrossRef]
6. Barboríková, J.; Šutovská, M.; Kazimierová, I.; Jošková, M.; Fraňová, S.; Kopecký, J.; Capek, P. Extracellular polysaccharide produced by *Chlorella vulgaris*—Chemical characterization and anti-asthmatic profile. *Int. J. Biol. Macromol.* **2019**, *135*. [CrossRef]
7. Camacho, F.; Macedo, A.; Malcata, F. Potential industrial applications and commercialization of microalgae in the functional food and feed industries: A short review. *Mar. Drugs* **2019**, *17*, 312. [CrossRef] [PubMed]
8. Beheshtipour, H.; Mortazavian, A.M.; Haratian, P.; Khosravi-Darani, K. Effects of *Chlorella vulgaris* and *Arthrospira platensis* addition on viability of probiotic bacteria in yogurt and its biochemical properties. *Eur. Food Res. Technol.* **2012**, *235*, 719–728. [CrossRef]
9. Beheshtipour, H.; Mortazavian, A.M.; Mohammadi, R.; Sohrabvandi, S.; Khosravi-Darani, K. Supplementation of *Spirulina platensis* and *Chlorella vulgaris* algae into probiotic fermented milks. *Compr. Rev. Food Sci. Food Saf.* **2013**, *12*, 144–154. [CrossRef]
10. Pyclik, M.; Srutkova, D.; Schwarzer, M.; Górska, S. Bifidobacteria cell wall-derived exopolysaccharides, lipoteichoic acids, peptidoglycans, polar lipids and proteins—Their chemical structure and biological attributes. *Int. J. Biol. Macromol.* **2020**, *147*, 333–349. [CrossRef] [PubMed]
11. Fox, M.J.; Ahuja, K.D.K.; Robertson, I.K.; Ball, M.J.; Eri, R.D. Can probiotic yogurt prevent diarrhoea in children on antibiotics? A double-blind, randomised, placebo-controlled study. *BMJ Open* **2015**, *5*. [CrossRef]
12. Delcaru, C.; Alexandru, I.; Podgoreanu, P.; Cristea, V.C.; Bleotu, C.; Chifiriuc, M.C.; Bezirtzoglou, E.; Lazar, V. Antagonistic activities of some Bifidobacterium sp. strains isolated from resident infant gastrointestinal microbiota on Gram-negative enteric pathogens. *Anaerobe* **2016**, *39*, 39–44. [CrossRef]
13. Heo, J.-Y.; Shin, H.-J.; Oh, D.-H.; Cho, S.-K.; Yang, C.-J.; Kong, I.-K.; Lee, S.-S.; Choi, K.-S.; Choi, S.-H.; Kim, S.-C.; et al. Quality properties of appenzeller cheese added with Chlorella. *Food Sci. Anim. Resour.* **2006**, *26*, 525–531. [CrossRef]
14. Niccolai, A.; Zittelli, G.C.; Rodolfi, L.; Biondi, N.; Tredici, M.R. Microalgae of interest as food source: Biochemical composition and digestibility. *Algal Res.* **2019**, *42*. [CrossRef]
15. Cantú-Bernal, S.; Domínguez-Gámez, M.; Medina-Peraza, I.; Aros-Uzarraga, E.; Ontiveros, N.; Flores-Mendoza, L.; Gomez-Flores, R.; Tamez-Guerra, P.; González-Ochoa, G. Enhanced viability and anti-rotavirus effect of *Bifidobacterium longum* and *Lactobacillus plantarum* in combination with *Chlorella sorokiniana* in a dairy product. *Front. Microbiol.* **2020**, *11*. [CrossRef] [PubMed]
16. Hyrslova, I.; Krausova, G.; Smolova, J.; Stankova, B.; Branyik, T.; Malinska, H.; Huttl, M.; Kana, A.; Curda, L.; Doskocil, I. Functional properties of *Chlorella vulgaris*, Colostrum, and *Bifidobacteria*, and their potential for application in functional foods. *Appl. Sci.* **2021**, *11*, 5264. [CrossRef]
17. Vlachová, M.; Heczková, M.; Jirsa, M.; Poledne, R.; Kovar, J. The response of hepatic transcriptome to dietary cholesterol in prague hereditary hypercholesterolemic (PHHC) rat. *Physiol. Res.* **2014**, *63*, 429–437. [CrossRef] [PubMed]
18. Hyrslova, I.; Krausova, G.; Michlova, T.; Kana, A.; Curda, L. Fermentation ability of bovine colostrum by different probiotic strains. *Fermentation* **2020**, *6*, 93. [CrossRef]
19. Gazi, I.; Tsimihodimos, V.; Filippatos, T.; Bairaktari, E.; Tselepis, A.D.; Elisaf, M. Concentration and relative distribution of low-density lipoprotein subfractions in patients with metabolic syndrome defined according to the National Cholesterol Education Program criteria. *Metabolism* **2006**, *55*, 885–891. [CrossRef]
20. Ewart, H.S.; Bloch, O.; Girouard, G.S.; Kralovec, J.; Barrow, C.J.; Ben-Yehudah, G.; Suarez, E.R.; Rapoport, M.J. Stimulation of cytokine production in human peripheral blood mononuclear cells by an aqueous Chlorella extract. *Planta Med.* **2007**, *73*, 762–768. [CrossRef]
21. Sibi, G.; Rabina, S. Inhibition of Pro-inflammatory mediators and cytokines by Chlorella Vulgaris extracts. *Pharmacogn. Res.* **2016**, *8*. [CrossRef]
22. Pulz, O.; Gross, W. Valuable products from biotechnology of microalgae. *Appl. Microbiol. Biotechnol.* **2004**, *65*, 635–648. [CrossRef]
23. Ścieszka, S.; Klewicka, E. Algae Chlorella vulgaris as a factor conditioning the survival of *Lactobacillus* spp. in adverse environmental conditions. *LWT* **2020**, *133*. [CrossRef]
24. Ścieszka, S.; Klewicka, E. Influence of the Microalga Chlorella vulgaris on the Growth and Metabolic Activity of Lactobacillus spp. Bacteria. *Foods* **2020**, *9*, 959. [CrossRef] [PubMed]
25. Tabarsa, M.; Shin, I.-S.; Lee, J.H.; Surayot, U.; Park, W.; You, S. An immune-enhancing water-soluble α-glucan from *Chlorella vulgaris* and structural characteristics. *Food Sci. Biotechnol.* **2015**, *24*, 1933–1941. [CrossRef]
26. Qi, J.; Kim, S.M. Effects of the molecular weight and protein and sulfate content of *Chlorella ellipsoidea* polysaccharides on their immunomodulatory activity. *Int. J. Biol. Macromol.* **2018**, *107*, 70–77. [CrossRef] [PubMed]

27. Suárez, E.R.; Kralovec, J.A.; Noseda, M.D.; Ewart, H.S.; Barrow, C.J.; Lumsden, M.D.; Grindley, T.B. Isolation, characterization and structural determination of a unique type of arabinogalactan from an immunostimulatory extract of *Chlorella pyrenoidosa*. *Carbohydr. Res.* **2005**, *340*, 1489–1498. [CrossRef]
28. An, H.J.; Rim, H.K.; Lee, J.H.; Seo, M.J.; Hong, J.W.; Kim, N.H.; Kim, H.M. Effect of Chlorella Vulgaris on Immune-Enhancement and Cytokine Production In Vivo and In Vitro. *Food Sci. Biotechnol.* **2008**, *17*, 953–958.
29. Jain, S.; Yadav, H.; Sinha, P.R. Antioxidant and cholesterol assimilation activities of selected lactobacilli and lactococci cultures. *J. Dairy Res.* **2009**, *76*. [CrossRef]
30. Lee, H.S.; Park, H.J.; Kim, M.K. Effect of Chlorella vulgaris on lipid metabolism in Wistar rats fed high fat diet. *Nutr. Res. Pr.* **2008**, *2*, 204–210. [CrossRef]
31. Ma, C.; Zhang, S.; Lu, J.; Zhang, C.; Pang, X.; Lv, J. Screening for cholesterol-lowering probiotics from lactic acid bacteria isolated from corn silage based on three hypothesized pathways. *Int. J. Mol. Sci.* **2019**, *20*, 2073. [CrossRef] [PubMed]
32. Chovančíková, M.; Šimek, V. Effects of high–fat and *Chlorella vulgaris* feeding on changes in lipid metabolism in mice. *Biol. Bratisl.* **2001**, *56*, 661–666.
33. Shibata, S.; Oda, K.; Onodera-Masuoka, N.; Matsubara, S.; Kikuchi-Hayakawa, H.; Ishikawa, F.; Iwabuchi, A.; Sansawa, H. Hypocholesterolemic effect of indigestible fraction of *Chlorella regularis* in cholesterol-fed rats. *J. Nutr. Sci. Vitaminol.* **2001**, *47*, 373–377. [CrossRef]
34. Shibata, S.; Hayakawa, K.; Egashira, Y.; Sanada, H. Hypocholesterolemic mechanism of *Chlorella*: *Chlorella* and its indigestible fraction enhance hepatic cholesterol catabolism through up-regulation of cholesterol 7α-hydroxylase in rats. *Biosci. Biotechnol. Biochem.* **2007**, *71*, 916–925. [CrossRef] [PubMed]
35. Chitsaz, M.; Mozaffari-Khosravi, H.; Salman-Roghani, H.; Zavar-Reza, J.; Lotfi, M. Effect of *chlorella vulgaris* vs. spirulina supplementation on lipid profile and liver function in patients with nonalcoholic fatty liver disease: A randomized controlled trial. *Int. J. Probiotics Prebiotics* **2016**, *11*, 127–136.
36. Kang, H.K.; Park, S.B.; Kim, C.H. Effects of dietary supplementation with a chlorella by-product on the growth performance, immune response, intestinal microflora and intestinal mucosal morphology in broiler chickens. *J. Anim. Physiol. Anim. Nutr.* **2017**, *101*, 208–214. [CrossRef] [PubMed]
37. Kim, G.-B.; Miyamoto, C.M.; Meighen, E.A.; Lee, B.H. Cloning and characterization of the bile salt hydrolase genes (*bsh*) from *Bifidobacterium bifidum* strains. *Appl. Environ. Microbiol.* **2004**, *70*, 5603–5612. [CrossRef] [PubMed]
38. Bottacini, F.; van Sinderen, D.; Ventura, M. Omics of *Bifidobacteria*: Research and insights into their health-promoting activities. *Biochem. J.* **2017**, *474*, 4137–4152. [CrossRef]
39. Ishimwe, N.; Daliri, E.B.; Lee, B.H.; Fang, F.; Du, G. The perspective on cholesterol-lowering mechanisms of probiotics. *Mol. Nutr. Food Res.* **2015**, *59*, 94–105. [CrossRef]
40. Zanotti, I.; Turroni, F.; Piemontese, A.; Mancabelli, L.; Milani, C.; Viappiani, A.; Prevedini, G.; Sanchez, B.; Margolles, A.; Elviri, L.; et al. Evidence for cholesterol-lowering activity by *Bifidobacterium bifidum* PRL2010 through gut microbiota modulation. *Appl. Microbiol. Biotechnol.* **2015**, *99*, 6813–6829. [CrossRef] [PubMed]
41. Oner, O.; Aslim, B.; Aydas, S.B. Mechanisms of cholesterol-lowering effects of *Lactobacilli* and *Bifidobacteria* strains as potential probiotics with their *bsh* gene analysis. *J. Mol. Microbiol. Biotechnol.* **2014**, *24*, 12–18. [CrossRef] [PubMed]
42. Lee, Y.; Ba, Z.; Roberts, R.F.; Rogers, C.J.; Fleming, J.A.; Meng, H.; Furumoto, E.J.; Kris-Etherton, P.M. Effects of Bifidobacterium animalis subsp. lactis BB-12® on the lipid/lipoprotein profile and short chain fatty acids in healthy young adults: A randomized controlled trial. *Nutr. J.* **2017**, *16*. [CrossRef] [PubMed]

fermentation

MDPI

Article

Fermentation Profile and Probiotic-Related Characteristics of *Bifidobacterium longum* MC-42

Anna V. Begunova [1], Irina V. Rozhkova [1], Olga A. Glazunova [2,*], Konstantin V. Moiseenko [2], Olga S. Savinova [2] and Tatyana V. Fedorova [2,*]

[1] Federal State Budgetary Scientific Institution "All-Russian Research Institute of Dairy Industry", 115093 Moscow, Russia; abegunova@yandex.ru (A.V.B.); irina.v.rozhkova@gmail.com (I.V.R.)
[2] Research Center of Biotechnology of the Russian Academy of Sciences, A.N. Bach Institute of Biochemistry, 119071 Moscow, Russia; mr.moiseenko@gmail.com (K.V.M.); savinova_os@rambler.ru (O.S.S.)
* Correspondence: olga.a.glas@gmail.com (O.A.G.); fedorova_tv@mail.ru (T.V.F.)

Abstract: This article presents new data on *Bifidobacterium longum* MC-42—a strain that has been actively used for the preparation of commercial dairy products in Russia for almost 40 years. It was demonstrated that this strain possesses high activities of β-galactosidase, α-glucosidase, and leucine arylaminidase; inhibits the growth of pathogens such as *Salmonella typhimurium*, *Staphylococcus aureus*, and *Escherichia coli*; and can efficiently remove cholesterol from the cultural medium. The resistance of *B. longum* MC-42 determined for 15 commonly used antibiotics was in agreement with those previously reported for *Bifidobacterium* spp. The absence of frequently transmittable antibiotic resistance genes in the genome and the lack of undesirable activity of β-glucuronidase proved the safe use of *B. longum* MC-42 as a probiotic and starter culture. Additionally, the impact of two growth-promoting additives—yeast extract or milk protein hydrolysate containing supplementation—on the *B. longum* MC-42 fermentation profile was assessed. The introduction of these additives increases the maximum attainable viable cell count by orders of magnitude, significantly changed the profile of aminopeptidase activities in extracellular extracts, and influenced the antioxidant and antihypertensive properties of the obtained fermented products.

Keywords: bifidobacteria; *Bifidobacterium longum*; aminopeptidase activity; antioxidant activity; angiotensin-I-inhibitory activity; cholesterol removal capacity; antibacterial activity

Citation: Begunova, A.V.; Rozhkova, I.V.; Glazunova, O.A.; Moiseenko, K.V.; Savinova, O.S.; Fedorova, T.V. Fermentation Profile and Probiotic-Related Characteristics of *Bifidobacterium longum* MC-42. *Fermentation* **2021**, *7*, 101. https://doi.org/10.3390/fermentation7030101

Academic Editor: Hiroshi Kitagaki

Received: 26 May 2021
Accepted: 21 June 2021
Published: 26 June 2021

1. Introduction

In recent years, probiotics—living microorganisms providing health benefits upon adequate consumption—have gained incredible attention both from ordinary customers and professional scientists [1]. While it was originally thought that probiotics promote physical well-being primarily through the improvement of microbial balance, later research demonstrated that the action of probiotics on organism goes far beyond direct interaction with a consumer's indigenous microflora [2]. The number of recently discovered properties of probiotics, often with strain-specificity such as improvement of intestinal barrier [3], production of beneficial enzymes (e.g., β-galactosidase and bile salt hydrolase) [4], modulation of immune system [5], synthesis of neurochemicals [6], and production of bioactive peptides [7], stimulates the search for new probiotic strains and reevaluation of old ones.

Bifidobacterium is a widely known genus of probiotic bacteria [8]. Being first introduced with breast milk, bifidobacteria constitutes 60–70% of the fecal bacteria in infancy and 30–40% in adulthood [9]. The long history of bifidobacteria consumption (both as part of fermented milk microflora and as commercially manufactured probiotic products) firmly established their safety and positive health benefits [10]. Currently, various strains of *Bifidobacterium* spp. are used in commercial products throughout the world. Although almost all of these strains were isolated many decades ago, an investigation on their beneficial properties are still actively carried out. Recently, systematical reviews regarding properties of several commercialized bifidobacterial strains, such as *Bifidobacterium animalis* subsp. *lactis* BB-12 [11], *Bifidobacterium bifidum* BGN4 [12], and *Bifidobacterium breve* M-16V [13], were published in scientific literature.

The strain *Bifidobacterium longum* MC-42 was isolated from the feces of infants by the staff of Scientific Research Institute of the Dairy Industry and patented in 1982 (USSR patent № 863639); at the time of isolation, it was identified as *Bifidobacterium adolescentis*. In the past 40 years, this strain was actively used (both alone and as part of a mixed starter culture) for the preparation of commercial dairy products and, currently, can be purchased in lyophilized form as a dietary supplement. Although advantageous technological properties of this strain such as the high viability of a lyophilized culture and the good sensory properties of the fermented products were time-tested, many aspects of its probiotic properties remain unclear. Only recently, the genome of *B. longum* MC-42 was sequenced [14], and the response of a human gut microbiome on short-term supplementation of this strain was investigated [15].

The aim of the current study was to investigate the *B. longum* MC-42 probiotic properties by evaluation of its profile of enzymatic activities, its ability to inhibit the growth of common pathogens, its resistance to common antibiotics, its possession of transmittable antibiotic resistance genes, and its capacity for cholesterol removal. Additionally, the fermentations of skim milk and skim milk supplemented with two different growth promoting additives (yeast extract and milk protein hydrolysate containing supplementation) performed by *B. longum* MC-42 were described in terms of strain's growth characteristics, acidification capability, and proteolytic activity. The resulting fermented products were assessed *in vitro* for their antioxidant and antihypertensive properties.

2. Materials and Methods

2.1. Strain Reactivation and Preparation of Starting Inoculum

The strain *B. longum* MC-42 was obtained from the Microorganism Collection of the All-Russian Research Institute of the Dairy Industry (VNIMI, Moscow, Russia), where it was stored as a lyophilized culture at −80 °C. The strain was reactivated in commercial milk hydrolysate containing corn-lactose GMK-2 medium for bifidobacteria (Biokompas-S, Uglich, Russia) at 37 °C for 24–48 h (to achieve a turbidity of 0.5–1 McFarland standard). For all experiments, the starting inoculum was prepared by the addition of 3% (*v*/*v*) of a reactivated strain into 200 mL of a GMK-2 medium with subsequent incubation at 37 °C for 24–48 h (to achieve a turbidity of 0.5–1 McFarland standard) and adjustment to approximately 10^7 $CFU \cdot mL^{-1}$.

An evaluation of the enzymatic activities of *B. longum* MC-42 was carried out using the API ZYM kit (BioMerieux, Marcy-l'Étoile, France), according to the manufacturer instructions. For analysis, *B. longum* was grown in MRS broth for 72 h at 37 °C.

2.2. Assessment of Inhibition of Pathogens

The pathogenic bacterium strains *Staphylococcus aureus* ATCC-6538 and *Escherichia coli* ATCC-25922 were purchased from American Type Culture Collection (Manassas, VA, USA), and *Salmonella typhimurium* NCTC 00074 was purchased from National Collection of Type Cultures (Salisbury, UK). The strains were cultivated on agar slants (21.0 $g \cdot L^{-1}$ peptone, 6.5 $g \cdot L^{-1}$ NaCl, 6.25 $g \cdot L^{-1}$ glucose, 3.5 $g \cdot L^{-1}$ Na_2HPO_4, 0.6 $g \cdot L^{-1}$ KH_2PO_4, and 12.5 $g \cdot L^{-1}$ agar) for 24 h at 37 ± 2 °C. For inoculation, the cell suspension was washed off the agar slant and diluted to approximately 10^7 $CFU \cdot mL^{-1}$.

The antagonistic activity was evaluated by the co-culture method [16]. For the experimental samples, 20 mL of MRS broth was simultaneously inoculated with 1 mL of *B. longum* MC-42 and 1 mL of the pathogenic test-strain starting inocula; for the control samples, only inoculation with the pathogenic test-strain was performed. The incubation was carried out at 37 ± 2 °C, and samples were collected after 24 and 48 h. Pathogen cells were counted at SPA agar medium (Mikrogen, Moscow, Russia) at 37 °C for 24–48 h.

2.3. *Antibiotic Resistance Assays*

The antibiotic resistance of *B. longum* MC-42 was assessed by the disk diffusion method. MRS agar plates were inoculated with 1 mL of the *B. longum* MC-42 starting inoculum, and antibiotic disks (DI-PLS-50-01, NICF, St. Peterburg, Russia) were placed in the center of each plate. The plates were incubated under anaerobic conditions using Oxoid AnaeroJar 2.5 L (Thermo Fisher Scientific, USA) and GasPak sachets (BD Biosciences, San Jose, CA, USA) for 18–24 h at 37 °C, followed by measurement of the inhibition zone diameters.

2.4. *Search for Genes of Interest in Genome*

To search for genes of interest, the genome of *B. longum* MC-42 was downloaded from the NCBI GenBank database [17]; the GenBank assembly accession is GCA_001516925.1. Representative sequences for the antibiotic resistance genes were downloaded from the NCBI RefSeq database. The representative sequence of bile salt hydrolase (BSH) was that reported in [18]; GenBank accession is AF148138. The representative sequences of cholesterol transporters and cholesterol reductase were those proposed in [19]; the locus tags according to the *Bifidobacterium bifidum* PRL2010 genome (GenBank assembly accession is GCA_000165905.1) are BBPR_0676, BBPR_1704, BBPR_1348, BBPR_0146, BBPR_1508, and BBPR_0519. The search was performed using the BLAST [20] program, and the locus tags of the identified genes are shown in parentheses thorough the text.

2.5. *Cholesterol-Removal Capacity Assay*

To assess cholesterol-removal capacity, *B. longum* MC-42 was cultivated in MRS broth containing 1.5 mM of cholesterol (Sigma-Aldrich, St. Louis, MO, USA) at 37 °C for 24 h. Inoculation was performed with 1% of *B. longum* MC-42 culture. The residual concentration of cholesterol was determined by HPLC using a Gilson chromatographic system (Gilson Medical Electronics, Middleton, WI, USA) equipped with Luna C18 column (4.6 × 150 mm, 5 μm, Phenomenex, Torrance, CA, USA). The mobile phase comprised acetonitrile and water in a ratio of 90:10 (*v:v*). The flow rate of the mobile phase was 2 mL·min^{-1}. Detection was carried out using a diode array detector at the 210 nm wavelength.

2.6. *Fermentations on Milk and Milk Supplemented with Growth-Promoting Additives*

Fermentations of skim milk, skim milk containing 2 g·L^{-1} of yeast extract (YE; FBUN GNC PMB, Obolensk, Russia), and skim milk containing 2 g·L^{-1} of GMK-3 growth-promoting supplement for bifidobacteria (Biokompas-S, Uglich, Russia) were performed using the DASGIP bioreactor system (Eppendorf, Hamburg, Germany). The composition of GMK-3 was as follows (per 100 g): 66 g milk protein hydrolysate, 33 g corn extract, and 1 g ascorbic acid. For fermentation, the sterile growth media were aseptically inoculated with 1% (*v*/*v*) of the *B. longum* MC-42-starting inoculum and cultivated at 37 ± 2 °C for 72 h. Sampling was performed immediately after inoculation (0 h) and at 6, 16, 24, and 72 h of cultivation.

The extracellular extracts (EEs) were prepared as follows: if the pH of a sample was above 4.6, it was adjusted to 4.6 by the addition of trichloroacetic acid; the samples were centrifuged at 10,000× *g* for 20 min at 4 °C; the supernatant was filtered through a 0.45 μm syringe filter and stored at −80 °C until further analysis.

The protein content of EEs was determined using the Pierce BSA Protein Assay Kit (ThermoFisher, Rockford, IL, USA).

2.7. *Enzymatic Activity Profile and Peptidase Assays*

The lysine and leucine aminopeptidase activities (Lys-AA and Leu-AA, respectively) in EEs were determined with lysine *p*-nitroanilide (Lys–*p*-NA, Sigma-Aldrich, Buchs, Switzerland) and leucine *p*-nitroanilide (Leu–*p*-NA, Sigma-Aldrich, St. Louis, MO, USA) chromogenic substrates, respectively. X-prolyl-dipeptidyl aminopeptidase activity was measured with glycine–proline *p*-nitroanilide (Gly–Pro–*p*-NA, Sigma-Aldrich, St. Louis,

MO, USA) as chromogenic substrate. The assay mixture contained 50 μL of the substrate (20 mM for Lys–*p*-NA and Leu–*p*-NA, and 10 mM for Gly–Pro–*p*-NA in methanol), 500 μL of 50 mM potassium phosphate buffer (pH 7.0), and 50 μL of sample (EE). After incubation at 37 °C for 4 h, the degree of hydrolysis was determined spectrophotometrically by measuring the absorbance of the colored product (*p*-nitroaniline) at 410 nm using a Lambda 35 spectrophotometer (PerkinElmer, Waltham, MA, USA) according to [21]. One unit of enzymatic activity was defined as an amount of the enzyme required to release 1 μmol of *p*-nitroaniline (ε_{410} = 8800 $M^{-1} \cdot cm^{-1}$) per minute and per gram of protein under the assay conditions. All measurements were carried out in triplicate.

2.8. Proteolytic, Antioxidant, and Angiotensin-I-Converting Enzyme Inhibitory Activities

The antioxidant activity in EEs was determined by the oxygen radical absorbance capacity fluorescence method (ORAC) with generation of the peroxyl radical as described in [22] and Trolox equivalent antioxidant capacity (TEAC) assay with generation of the ABTS radical as described in [23]. The antioxidant capacity of samples against both ABTS and peroxyl radicals was expressed as an amount of Trolox molar equivalents.

Angiotensin-I-converting enzyme inhibitory (ACE-I) activity in EEs was determined in terms of half maximal inhibitory concentration (IC_{50}), as described in [24]. ACE activity was measured using *o*-Aminobenzoyl-Phe-Arg-Lys(dinitrophenyl)-Pro (Sigma-Aldrich, St. Louis, MO, USA) as a substrate with internal fluorescence quenching. The measurements were carried out on a Synergy 2 microplate photometer–fluorometer (BioTek, Winooski, VT, USA).

The proteolytic activity was quantified by measurement of the amount of released amino groups in EEs using the 2,4,6-trinitrobenzenesulfonic acid solution (TNBS, Sigma-Aldrich, St. Louis, MO, USA) method. The optical density at 340 nm was measured using a Synergy 2 microplate photometer–fluorimeter (BioTek, Winooski, VT, USA). A calibration curve was prepared using L-leucine (L-Leu) as a standard (0.1–2.0 mM). The results were expressed as L-Leu molar equivalents (mM (Leu)).

3. Results and Discussion

3.1. Profile of Enzymatic Activities

The enzymatic profile of probiotic strains plays an important role in both their health-promoting action on organism [25] and use for preparation of fermented food [26]. Several well-known examples of useful enzymatic activities of probiotics are glycoside hydrolase activities, which increase bioavailability of plant polysaccharides [27]; protease activities, which result in production of numerous bioactive peptides [7]; and *β*-galactosidase activity, which makes dairy products well tolerated by lactose-maldigesters [28]. In addition, in order to avoid the production of potentially toxic substances, probiotics must be evaluated for the production of undesirable enzymes, such as *β*-glucuronidase, a carcinogenic enzyme that can produce reactive metabolites negatively affecting the liver and increasing probability of colon carcinogenesis [29,30].

The semi-quantitative assessment of *B. longum* MC-42 enzymatic activities is presented in Figure 1. In total, 19 enzymes from the groups of glycoside-hydrolases, phosphatases, esterases, and proteases were assessed. In general, *B. longum* MC-42 demonstrated the high activities of leucine arylamidase, *β*-galactosidase, and *α*-glucosidase; moderate activities of esterase (C4), esterase lipase (C8), acid phosphatase, naphtol-AS-BI-phosphohydrolase, and *α*-galactosidase; and the absence of activities of alkaline phosphatase, lipase (C14), trypsin and *α*-chymotrypsin, *β*-glucosidase, N-acetyl-*β*-glucosaminidase, *α*-mannosidase, *α*-fucosidase, and most importantly, *β*-glucuronidase.

№	Enzyme	Activity	№	Enzyme	Activity
1	Control	0	11	Acid phosphatase	1.5
2	Alkaline phosphatase	0	12	Naphtol-AS-BI-phosphohydrolase	1.5
3	Esterase (C4)	3.0	13	α-Galactosidase	2.0
4	Esterase lipase (C8)	2.0	14	β- Galactosidase	⩾5.0
5	Lipase (C14)	0	15	β-Glucuronidase	0
6	Leucine arylamidase	4.5	16	α-Glucosidase	4.0
7	Valine arylamidase	⩽0.5	17	β-Glucosidase	0
8	Cystine arylamidase	⩽0.5	18	N-Acetyl-β-glucosaminidase	0
9	Trypsine	0	19	α-Mannosidase	0
10	α-Chymotrypsine	0	20	α-Fucosidase	0

Figure 1. API ZYM enzymatic profile of *B. longum* MC-42.

The enzymatic profile of *B. longum* MC-42 was generally similar to that of other *B. longum* strains. Typically, these strains have high activities of β-galactosidase and α-glucosidase, moderate-to-high activity of α-galactosidase, and almost absent activity of β-glucosidase. Additionally, weak or absent activities of N-acetyl-β-glucosaminidase, α-mannosidase, and α-fucosidase are characteristic for all *Bifidobacterium* spp. [31–34]. Weak protease activities are also typical for *Bifidobacterium* spp. As for *B. longum* MC-42 in the current study, almost no activities of trypsin and α-chymotrypsin were previously shown for strains of *Bifidobacterium animalis*, *Bifidobacterium adolescentis*, *Bifidobacterium bifidum*, *Bifidobacterium breve*, *Bifidobacterium catenulatum*, *Bifidobacterium pseudocatenulatum*, *Bifidobacterium lactis*, and *B. longum* [31–36]. At the same time, all of these strains demonstrated high activity of leucine aminopeptidase while the activities of valine and cystine aminopeptidases were more strain specific.

3.2. Inhibition of Pathogens

The ability to inhibit the growth of pathogenic and opportunistic bacteria is, arguably, the most demanded characteristic of a good probiotic strain. There are two main groups of mechanisms by which probiotics can inhibit pathogenic microflora: host-dependent and host-independent [37–39]. The mechanisms in the host-dependent group are highly reliant on the host's physiology and include competition for binding sites, promotion of mucin secretion, induction of tight junction protein expression, and stimulation of immune response. The mechanisms in the host-independent group are based on the direct interaction of probiotics with pathogens and include competition for nutrients and production of organic acids, peroxide, and bacteriocins.

In this work, the host-independent inhibition of pathogenic microorganisms by *B. longum* MC-42 was assessed. The following pathogenic bacterial strains were used: *S. typhimurium* NCTC-00074, *S. aureus* ATCC-6538, and *E. coli* ATCC-25922. As seen in Figure 2, *B. longum* MC-42 almost equally inhibited the growth of *S. typhimurium* NCTC-00074 and *E. coli* ATCC-25922: the inhibition by approximately half an order of magnitude was observed after 24 h of co-cultivation, and that by approximately three orders of magnitude was observed after 48 h. The inhibition of *S. aureus* ATCC-6538 after 24 and 48 h of co-cultivation was the same and comprised approximately one and a half orders of magnitude.

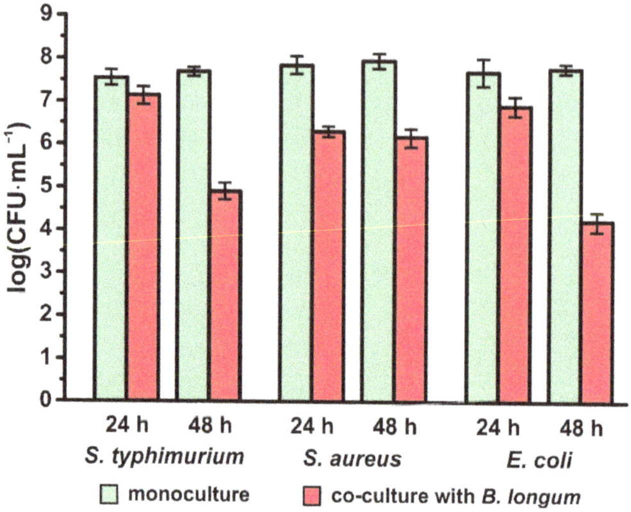

Figure 2. The growth inhibition of *S. typhimurium* NCTC-00074, *S. aureus* ATCC-6538, and *E. coli* ATCC-25922 upon co-cultivation with *B. longum* MC-42.

3.3. Resistance to Antibiotics

Due to structural or functional characteristics, many microorganisms have an innate resistance to specific classes of antibiotics [40]. This so-called intrinsic antibiotic resistance can be passed only vertically from a parent to an offspring. While intrinsic antibiotic resistance can be a danger in the case of pathogenic and opportunistic microorganisms, it does not pose a particular problem considering probiotics [41]. Moreover, the moderate innate antibiotic resistance of certain probiotic strains prevents severe dysbiosis that can occur during the antibiotic treatment of infections. On the other hand, antibiotic resistance acquired by microorganism through a horizontal gene transfer—transmissible antibiotic resistance—now presents a major concern for safe use of probiotic strains [42]. Currently, it was demonstrated that some probiotic strains harboring antibiotic resistance genes can transmit them to microorganisms causing nosocomial and common infections [43].

The antibiotic resistance (AR) of *B. longum* MC-42 was tested with 15 commonly used antibiotics by qualitative disc-diffusion assay (Table 1). It was found that *B. longum* MC-42 was resistant only to lincomycin and the antibiotics belonging to the aminoglycoside structural group. The AR profile of *B. longum* MC-42 determined was in agreement with those previously reported for *Bifidobacterium* spp. [44,45]. The analysis of the *B. longum* MC-42 genome demonstrated an absence of a commonly transmissible by gut microbiota AR genes, such as aminoglycoside AR genes encoding the aminoglycoside-modifying enzymes *aac(6′)-aph(2″)* and *aad(E)* [46]; tetracycline AR genes encoding the ribosomal protection proteins—*tet(M)*, *tet(S)*, *tet(W)*, *tet(O)*, *tet(Q)*, *tet(32)*, *tet(36)*, and *tet(T)*; the tetracycline major facilitator superfamily efflux pumps *tet(K)* and *tet(L)* [47]; amphenicol AR genes encoding the chloramphenicol acetyltransferase—*cat* [48]; macrolide–lincosamide–streptogramin (MLS) AR genes encoding the 23S rRNA methylases—*erm(A)*, *erm(B)*, *erm(C)*, *erm(F)* and *erm(T)*—and the macrolide major facilitator superfamily efflux pump *mef(A)* [49]. Hence, utilizing the previously proposed scheme for the antibiotic resistance assessment of bacteria [42], *B. longum* MC-42 can be considered an acceptable strain for use as a probiotic and starter culture.

Table 1. The antibiotic resistance profile of *B. longum* MC-42.

Antibiotic	Amount, μg	Inhibition Zone Diameter, mm	Resistance Status
	β-lactams (penams):		
Ampicillin	10	28 ± 0.5	Susceptible
Amoxicillin	20	28 ± 1	Susceptible
Oxacillin	1	12 ± 0.5	Intermediate
Penicillin G	10	28 ± 1	Susceptible
	Fosfomycins:		
Fosfomycin	200	23 ± 1	Susceptible
	Aminoglycosides:		
Gentamicin	120	10 ± 0.5	Resistant
Kanamycin A	30	11 ± 0.5	Resistant
Neomycin	30	7 ± 1	Resistant
	Tetracyclines:		
Doxycycline	30	32 ± 1	Susceptible
Tetracycline	30	28 ± 1	Susceptible
	Macrolides:		
Azithromycin	15	14 ± 1	Intermediate
	Lincosamides:		
Lincomycin	15	8 ± 1	Resistant
	Amphenicols:		
Chloramphenicol	30	26 ± 1	Susceptible
	Fluoroquinolones:		
Levofloxacin	5	14 ± 1	Intermediate
Pefloxacin	5	6 ± 1	Resistant

3.4. Cholesterol-Removal Capacity

The recently discovered hypocholesterolemic effect of probiotic consumption offers a great opportunity for drug-free management of cholesterol levels in humans [50]. Currently, two main routes by which gut microbiota can significantly influence a host's cholesterol homeostasis were proposed. The first route comprises deconjugation of bile salts by the action of bile salt hydrolases (BSHs, EC 3.5.1.24)—the microbial enzymes that catalyze hydrolysis of the amide bond in glycine/taurine-conjugated bile salts [4]. Deconjugation of bile salts decreases their reabsorption from the intestinal lumen and facilitates their excretion. To replenish the pool of bile salts, an endogenous cholesterol is used, which leads to its reduction in blood serum [51]. The second route involves decreasing the amount of cholesterol absorbed by human intestines via its microbial conversion into coprostanol [52]; however, compared to the bile salt deconjugation, less is known about this process at the molecular level.

The analysis of the *B. longum* MC-42 genome demonstrated the presence of a functional BSH encoding gene (locus_tag AS143_06365) that virtually guarantees the ability of this strain to lower a host's cholesterol level by the deconjugation of bile salts. Additionally, several genes homologous to those that were previously proposed by Zanotti et al. [19] to participate in the conversion of cholesterol into coprostanol were determined: genes of the ABC family transporter (locus_tags AS143_00305, AS143_02140, and AS143_06580), a gene of the MFS family transporter (locus_tag AS143_02910), a gene of the EIIC component of the PTS system (locus_tag AS143_05800), and a gene of cholesterol reductase (locus_tag AS143_09220). However, since the presence of these genes only suggests—but not guarantees—the ability of *B. longum* MC-42 for cholesterol conversion, this ability was experimentally confirmed. It was determined that, during the cultivation of *B. longum* MC-42 on a cholesterol-containing medium, the cholesterol concentration was decreased by 54.8% (from 1.5 mM to 0.678 ± 0.014 mM) in 24 h.

3.5. Growth Performance, Acidification Capability, and Degree of Proteolysis

As it was previously shown, in comparison with other lactic acid bacteria, *Bifidobacteria* spp. generally demonstrate slow or limited growth during milk fermentation [53]. Partially, this fact can be explained by the proposed inability of *Bifidobacteria* spp. to produce extracellular proteinases and, consequently, effectively hydrolyze proteins [54]. In agreement with this, it was demonstrated that yeast extract and hydrolysate of milk proteins are good growth promoters for various *Bifidobacteria* spp. [55].

The dynamic changes in the viable cell count, pH value, and degree of proteolysis during cultivation of *B. longum* MC-42 on skim milk, skim milk supplemented with YE, and skim milk supplemented with GMK-3 (containing hydrolysate of milk proteins and several other growth factors, see Section 2.6) are presented in Figure 3. The addition of YE and GMK-3 to the milk did not significantly alter the pH profile of fermentation. On all media, pH decreased from 6.5 to 3.7 units in a 72 h timespan, leading to the coagulation of milk proteins as a result of casein precipitation. Both YE and GMK-3 significantly stimulated the growth of *B. longum* MC-42: with these supplementations, the maximum attainable viable cell count (reached at 16 h of fermentation) comprised (2–2.3) × 10^9 CFU·mL^{-1}, while on milk, this value (reached at 48 h of fermentation) did not exceed 4 × 10^8 CFU·mL^{-1}. For both YE- and GMK-3-supplemented milk, the degree of proteolysis decreased until 16 h of fermentation, suggesting the utilization of peptides already present in these media; however, after 16 h and until the end of fermentation, the degree of proteolysis grew steadily. For skim milk, the degree of proteolysis was the same until 48 h of fermentation, after which it increased at 72 h. It should be noted that the increase in the degree of proteolysis, observed after 16 h for YE- and GMK-3-supplemented milk and after 48 h for skim milk, coincided with a decrease in the viable cell count.

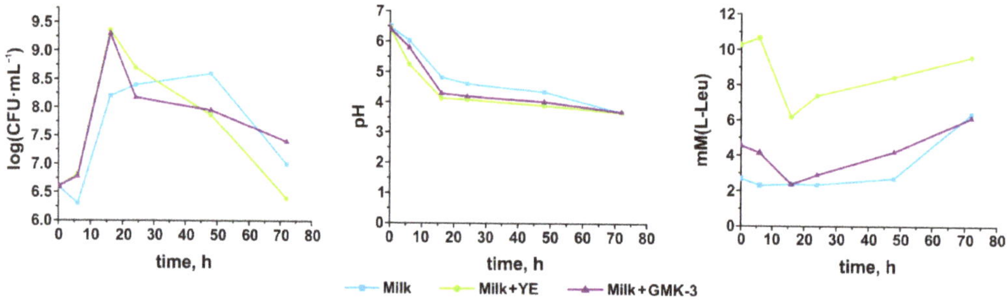

Figure 3. Growth characteristics of *B. longum* MC-42 on different media: **Milk**—skim milk; **Milk + YE**—skim milk with addition of yeast extract, and **Milk + GMK-3**—skim milk with addition of hydrolysate containing supplementation GMK-3. The initial values of the degree of proteolysis were 2.74 mM (L-Leu) for Milk, 9.88 mM (L-Leu) for Milk + YE, and 3.77 mM (L-Leu) for Milk + GMK-3.

Since no trypsin and α-chymotrypsin activities were detected in the enzymatic profile of *B. longum* MC-42 (see Section 3.1), the observed increase in the degree of proteolysis can be attributed to the acid hydrolysis of proteins and to the possible action of some intracellular aminopeptidases released as a result of cell lysis [54].

The changes in aminopeptidase activity in the EEs of *B. longum* MC-42 during fermentation were assessed with three different substrates: Gly-Pro-*p*-NA, Lys-*p*-NA, and Leu-*p*-NA. It is known that the activity toward Lys-*p*-NA and Leu-*p*-NA chromogenic substrates can be detected in the presence of generic aminopeptidases PepC and PepN, which have been proposed to play a significant role in cell growth and peptide hydrolysis [54]. The aminopeptidase activity toward Gly-Pro-*p*-NA was absent on all studied media during the entire cultivation time; and Lys-AA and Leu-AA are shown in Figure 4. On all media, both, Lys-AA and Leu-AA were detected at the stage of cell death—after 16 h for YE- and GMK-3-supplemented milk and after 48 h for skim milk. The exception was Leu-AA for GMK-3-supplemented milk, which was detected at 16 h of fermentation when the maximum viable cell count was reached. Although it is generally accepted that the majority of aminopeptidases are intracellular, their extracellular release was previously reported [56]. Additionally, the prominent leucine arylamidase (aminopeptidase) activity was detected in the enzymatic API ZYM profile of *B. longum* MC-42 (see Section 3.1).

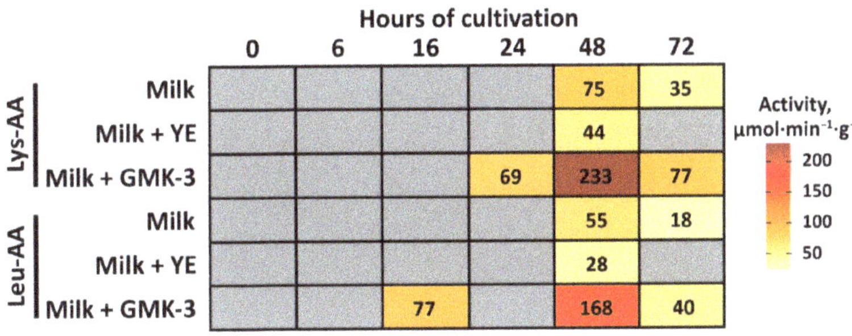

Figure 4. Lys-AA and Leu-AA of *B. longum* MC-42 during fermentation on different media: **Milk**—skim milk; **Milk + YE**—skim milk with addition of yeast extract, and **Milk + GMK-3**—skim milk with the addition of a hydrolysate containing GMK-3 supplementation. The aminopeptidase activities are expressed as µmol of *p*-NA released by aminopeptidase per minute and per gram of protein. The data are represented with a color-coded system where the highest value is shown in red and the lowest—in yellow.

Generally, Lys-AA was 1.5–2.0 times higher than Leu-AA in all studied samples collected at the same time point. Supplementation of milk with different peptide sources resulted in different patterns of the aminopeptidase activities. In EEs from fermentation of YE-supplemented milk, both Lys-AA and Leu-AA were approximately 1.8 times lower compared to milk at 48 h and were totally absent at 72 h, while in EEs from fermentation of GMK-3-supplemented milk, these activities were approximately 2 times higher at 48 h and 3 times higher at 72 h compared to milk. This could be a result of different peptide profiles of YE and GMK-3; strain-specific differential regulation of aminopeptidases by peptide sources was previously reported for *Bifidobacterium* spp. in [57].

3.6. Development of Antioxidant and Antihypertensive Properties

The antioxidant and antihypertensive properties of the fermented milk are primarily attributed to the production of bioactive peptides during the fermentation process and, hence, can be species- or even strain-specific with respect to fermenting bacteria [58]. The activities of bioactive peptides are affected by both their amino acid compositions and sequences [59]. Generally, the ACE-I peptides are more sequence-specific than antioxidant ones, since they have to perform competitive inhibition at the catalytic site of the angiotensin-I-converting enzyme. Currently, the most popularly used assays for measurements of antioxidant activity are the TEAC and ORAC assays. However, for complex samples, the correlation between the ORAC and TEAC methods is low [60].

The development of antioxidant activity measured by the TEAC and ORAC assays during cultivation of *B. longum* MC-42 on different media is shown in Figure 5. For both YE- and GMK-3-supplemented milk, the general tendency of antioxidant activity development measured by TEAC an ORAC was similar: the activity decreased at 6 h of fermentation with a subsequent increase until the end of fermentation. However, while the TEAC assay showed higher antioxidant activity of the YE-supplemented milk compared to the GMK-3-supplemented milk (at almost all time points), the ORAC assay demonstrated comparable antioxidant activity for both supplementations. Interestingly, the antioxidant activity of the skim milk measured by the TEAC assay was almost the same during the entire cultivation; on the contrary, the antioxidant activity measured by the ORAC assay steadily grew and became comparable with that for the YE- and GMK-3-supplemented milk at the end of fermentation.

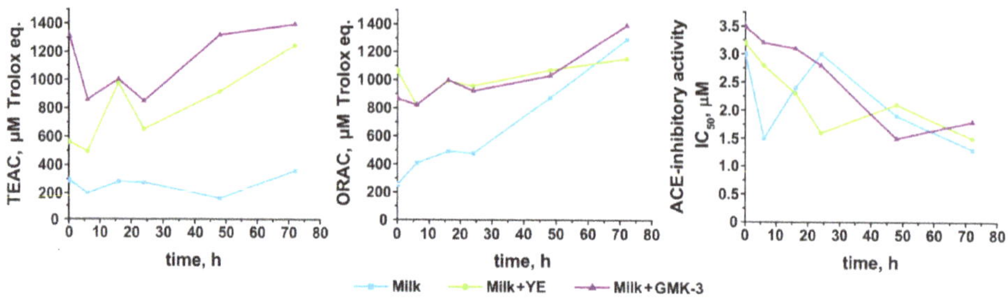

Figure 5. Antioxidant and ACE-I properties of products obtained after fermentation by *B. longum* MC-42 of **Milk**—skim milk; **Milk + YE**—skim milk with addition of yeast extract, and **Milk + GMK-3**—skim milk with addition of hydrolysate containing supplementation GMK-3.

The development of ACE-I activity is shown in Figure 5. For skim milk, ACE-I activity increased (IC_{50} decreased) at 6 h of fermentation, after which it steadily decreased until 24 h with a subsequent increase until the end of fermentation. For the YE- and GMK-3 supplemented milk, ACE-I activity steadily increased until 24 h and 48 h of fermentation, respectively, after which it stayed at almost the same value.

The antioxidant activity profile generally agreed with the changes in the degree of proteolysis, suggesting extensive utilization of peptides (including antioxidant ones) from the media at the first hours of cultivation and subsequent peptide production by *B. longum* enzymes. Moreover, the significant increase in antioxidant and ACE-I activities after 24 h can also be related to the increase in Leu-AA and other aminopeptidases contributing to protein hydrolysis.

4. Conclusions

Currently, the amount of knowledge about the action of probiotics rapidly increases. This forces a reevaluation of long-used probiotic strains for their functional properties and safe use. In this work, for the first time, the properties of the bifidobacterial strain that was commercially used for decades in Russia, *B. longum* MC-42, were systematically reassessed. It was shown that this strain does not have frequently transmittable AR genes in its genome and does not possess harmful β-glucuronidase activity and, hence, can be regarded as generally safe for use according to the modern standards. At the same time, this strain can effectively lower the amount of cholesterol in a fermentation medium and can suppress the development of harmful bacteria. It was shown that the use of different growth-promoting additives results in the development of different patterns of antioxidant and ACE-I properties in final products. This can be explained by the different patterns of aminopeptidase activities in the cultural broth, although the exact mechanism for extracellular occurrence of these activities remains unclear.

Author Contributions: Conceptualization, T.V.F. and K.V.M.; investigation, A.V.B., O.S.S., O.A.G., K.V.M., I.V.R. and T.V.F.; data curation, T.V.F. and I.V.R.; writing—original draft preparation, T.V.F. and A.V.B.; writing—review and editing, K.V.M., O.A.G., I.V.R. and T.V.F.; visualization, K.V.M. and O.A.G.; supervision, T.V.F. and I.V.R. All authors have read and agreed to the published version of the manuscript.

Funding: This research was funded by the Ministry of Science and Higher Education of Russia (project № 0578-2019-0023).

Institutional Review Board Statement: Not applicable.

Informed Consent Statement: Not applicable.

Data Availability Statement: The data are contained within the article.

Conflicts of Interest: The authors declare no conflict of interest.

References

1. Saarela, M.; Mogensen, G.; Fondén, R.; Mättö, J.; Mattila-Sandholm, T. Probiotic bacteria: Safety, functional and technological properties. *J. Biotechnol.* **2000**, *84*, 197–215. [CrossRef]
2. Markowiak, P.; Śliżewska, K. Effects of probiotics, prebiotics, and synbiotics on human health. *Nutrients* **2017**, *9*, 1021. [CrossRef] [PubMed]
3. Ohland, C.L.; MacNaughton, W.K. Probiotic bacteria and intestinal epithelial barrier function. *Am. J. Physiol. Liver Physiol.* **2010**, *298*, G807–G819. [CrossRef]
4. Begley, M.; Hill, C.; Gahan, C.G.M. Bile salt hydrolase activity in probiotics. *Appl. Environ. Microbiol.* **2006**, *72*, 1729–1738. [CrossRef]
5. Maldonado Galdeano, C.; Cazorla, S.I.; Lemme Dumit, J.M.; Vélez, E.; Perdigón, G. Beneficial effects of probiotic consumption on the immune system. *Ann. Nutr. Metab.* **2019**, *74*, 115–124. [CrossRef]
6. Wang, H.; Lee, I.-S.; Braun, C.; Enck, P. Effect of probiotics on central nervous system functions in animals and humans: A systematic review. *J. Neurogastroenterol. Motil.* **2016**, *22*, 589–605. [CrossRef]
7. Korhonen, H. Milk-derived bioactive peptides: From science to applications. *J. Funct. Foods* **2009**, *1*, 177–187. [CrossRef]
8. Lugli, G.A.; Milani, C.; Duranti, S.; Mancabelli, L.; Mangifesta, M.; Turroni, F.; Viappiani, A.; van Sinderen, D.; Ventura, M. Tracking the taxonomy of the genus *Bifidobacterium* based on a phylogenomic approach. *Appl. Environ. Microbiol.* **2017**, *84*, e02249-17. [CrossRef]
9. Arboleya, S.; Watkins, C.; Stanton, C.; Ross, R.P. Gut bifidobacteria populations in human health and aging. *Front. Microbiol.* **2016**, *7*, 1204. [CrossRef]
10. Picard, C.; Fioramonti, J.; Francois, A.; Robinson, T.; Neant, F.; Matuchansky, C. Review article: Bifidobacteria as probiotic agents—Physiological effects and clinical benefits. *Aliment. Pharmacol. Ther.* **2005**, *22*, 495–512. [CrossRef]
11. Jungersen, M.; Wind, A.; Johansen, E.; Christensen, J.; Stuer-Lauridsen, B.; Eskesen, D. The science behind the probiotic strain *Bifidobacterium animalis* subsp. lactis BB-12®. *Microorganisms* **2014**, *2*, 92–110. [CrossRef]
12. Ku, S.; Park, M.; Ji, G.; You, H. Review on *Bifidobacterium bifidum* BGN4: Functionality and nutraceutical applications as a probiotic microorganism. *Int. J. Mol. Sci.* **2016**, *17*, 1544. [CrossRef]
13. Wong, C.B.; Iwabuchi, N.; Xiao, J. Exploring the science behind *Bifidobacterium breve* M-16V in infant health. *Nutrients* **2019**, *11*, 1724. [CrossRef] [PubMed]
14. Tupikin, A.E.; Kalmykova, A.I.; Kabilov, M.R. Draft genome sequence of the probiotic *Bifidobacterium longum* subsp. *longum* strain MC-42. *Genome Announc.* **2016**, *4*, e01411-16. [CrossRef]
15. Naumova, N.; Alikina, T.; Tupikin, A.; Kalmykova, A.; Soldatova, G.; Vlassov, V.; Kabilov, M. Human gut microbiome response to short-term bifidobacterium-based probiotic treatment. *Indian J. Microbiol.* **2020**, *60*, 451–457. [CrossRef] [PubMed]
16. Coman, M.M.; Verdenelli, M.C.; Cecchini, C.; Silvi, S.; Orpianesi, C.; Boyko, N.; Cresci, A. In vitro evaluation of antimicrobial activity of *Lactobacillus rhamnosus* IMC 501®, *Lactobacillus paracasei* IMC 502® and SYNBIO® against pathogens. *J. Appl. Microbiol.* **2014**, *117*, 518–527. [CrossRef]
17. Clark, K.; Karsch-Mizrachi, I.; Lipman, D.J.; Ostell, J.; Sayers, E.W. GenBank. *Nucleic Acids Res.* **2016**, *44*, D67–D72. [CrossRef]
18. Tanaka, H.; Hashiba, H.; Kok, J.; Mierau, I. Bile salt hydrolase of *Bifidobacterium longum*—Biochemical and genetic characterization. *Appl. Environ. Microbiol.* **2000**, *66*, 2502–2512. [CrossRef] [PubMed]
19. Zanotti, I.; Turroni, F.; Piemontese, A.; Mancabelli, L.; Milani, C.; Viappiani, A.; Prevedini, G.; Sanchez, B.; Margolles, A.; Elviri, L.; et al. Evidence for cholesterol-lowering activity by *Bifidobacterium bifidum* PRL2010 through gut microbiota modulation. *Appl. Microbiol. Biotechnol.* **2015**, *99*, 6813–6829. [CrossRef] [PubMed]
20. Altschul, S.F.; Gish, W.; Miller, W.; Myers, E.W.; Lipman, D.J. Basic local alignment search tool. *J. Mol. Biol.* **1990**, *215*, 403–410. [CrossRef]
21. Exterkate, F.A. Location of peptidases outside and inside the membrane of *Streptococcus cremoris*. *Appl. Environ. Microbiol.* **1984**, *47*, 177–183. [CrossRef]
22. Agarkova, E.Y.; Kruchinin, A.G.; Glazunova, O.A.; Fedorova, T.V. Whey protein hydrolysate and pumpkin pectin as nutraceutical and prebiotic components in a functional mousse with antihypertensive and bifidogenic properties. *Nutrients* **2019**, *11*, 2930. [CrossRef] [PubMed]
23. Kruchinin, A.G.; Savinova, O.S.; Glazunova, O.A.; Moiseenko, K.V.; Agarkova, E.Y.; Fedorova, T.V. Hypotensive and hepatoprotective properties of the polysaccharide-stabilized foaming composition containing hydrolysate of whey proteins. *Nutrients* **2021**, *13*, 1031. [CrossRef] [PubMed]
24. Torkova, A.A.; Ryazantseva, K.A.; Agarkova, E.Y.; Kruchinin, A.G.; Tsentalovich, M.Y.; Fedorova, T.V. Rational design of enzyme compositions for the production of functional hydrolysates of cow milk whey proteins. *Appl. Biochem. Microbiol.* **2017**, *53*, 669–679. [CrossRef]
25. Indira, M.; Venkateswarulu, T.C.; Abraham Peele, K.; Nazneen Bobby, M.; Krupanidhi, S. Bioactive molecules of probiotic bacteria and their mechanism of action: A review. *3 Biotech* **2019**, *9*, 306. [CrossRef]

26. Marco, M.L.; Heeney, D.; Binda, S.; Cifelli, C.J.; Cotter, P.D.; Foligné, B.; Gänzle, M.; Kort, R.; Pasin, G.; Pihlanto, A.; et al. Health benefits of fermented foods: Microbiota and beyond. *Curr. Opin. Biotechnol.* **2017**, *44*, 94–102. [CrossRef]
27. Cockburn, D.W.; Koropatkin, N.M. Polysaccharide degradation by the intestinal microbiota and its influence on human health and disease. *J. Mol. Biol.* **2016**, *428*, 3230–3252. [CrossRef]
28. Oak, S.J.; Jha, R. The effects of probiotics in lactose intolerance: A systematic review. *Crit. Rev. Food Sci. Nutr.* **2019**, *59*, 1675–1683. [CrossRef]
29. Nanno, M.; Morotomi, M.; Takayama, H.; Kuroshima, T.; Tanaka, R.; Mutai, M. Mutagenic activation of biliary metabolites of benzo(a)pyrene by β-glucuronidase-positive bacteria in human faeces. *J. Med. Microbiol.* **1986**, *22*, 351–355. [CrossRef]
30. Davis, C.D.; Milner, J.A. Gastrointestinal microflora, food components and colon cancer prevention. *J. Nutr. Biochem.* **2009**, *20*, 743–752. [CrossRef]
31. Arboleya, S.; Ruas-madiedo, P.; Margolles, A.; Solís, G.; Salminen, S.; Reyes-gavilán, C.G.D.L.; Gueimonde, M. Characterization and in vitro properties of potentially probiotic *Bifidobacterium* strains isolated from breast-milk. *Int. J. Food Microbiol.* **2011**, *149*, 28–36. [CrossRef] [PubMed]
32. Delgado, S.; O'Sullivan, E.; Fitzgerald, G.; Mayo, B. In vitro evaluation of the probiotic properties of human intestinal *Bifidobacterium* species and selection of new probiotic candidates. *J. Appl. Microbiol.* **2008**, *104*, 1119–1127. [CrossRef] [PubMed]
33. Desjardins, M.L.; Roy, D.; Goulet, J. Growth of bifidobacteria and their enzyme profiles. *J. Dairy Sci.* **1990**, *73*, 299–307. [CrossRef]
34. Martínez-Villaluenga, C.; Gómez, R. Characterization of bifidobacteria as starters in fermented milk containing raffinose family of oligosaccharides from lupin as prebiotic. *Int. Dairy J.* **2007**, *17*, 116–122. [CrossRef]
35. Kim, H.; Kim, J.S.; Kim, Y.G.; Jeong, Y.; Kim, J.E.; Paek, N.S.; Kang, C.H. Antioxidant and probiotic properties of lactobacilli and bifidobacteria of human origins. *Biotechnol. Bioprocess Eng.* **2020**, *25*, 421–430. [CrossRef]
36. Kim, H.J.; Shin, S.I.; Lee, S.J.; Moon, T.W.; Lee, C.J. Screening and selection of *Bifidobacterium* strains isolated from human feces capable of utilizing resistant starch. *J. Sci. Food Agric.* **2018**, *98*, 5901–5907. [CrossRef]
37. Corr, S.C.; Hill, C.; Gahan, C.G.M. Understanding the mechanisms by which probiotics inhibit gastrointestinal pathogens. *Adv. Food Nutr. Res.* **2009**, *56*, 1–15.
38. Britton, R.A.; Versalovic, J. Probiotics and gastrointestinal infections. *Interdiscip. Perspect. Infect. Dis.* **2008**, *2008*, 1–10. [CrossRef]
39. Isolauri, E.; Kirjavainen, P.V.; Salminen, S. Probiotics: A role in the treatment of intestinal infection and inflammation? *Gut* **2002**, *50*, iii54–iii59. [CrossRef]
40. Cox, G.; Wright, G.D. Intrinsic antibiotic resistance: Mechanisms, origins, challenges and solutions. *Int. J. Med. Microbiol.* **2013**, *303*, 287–292. [CrossRef]
41. Sharma, P.; Tomar, S.K.; Goswami, P.; Sangwan, V.; Singh, R. Antibiotic resistance among commercially available probiotics. *Food Res. Int.* **2014**, *57*, 176–195. [CrossRef]
42. Álvarez-Cisneros, M.Y.; Ponce-Alquicira, E. Antibiotic resistance in lactic acid bacteria. In *Antimicrobial Resistance—A Global Threat*; IntechOpen: London, UK, 2019.
43. Kumar, S.B.; Arnipalli, S.R.; Ziouzenkova, O. Antibiotics in food chain: The consequences for antibiotic resistance. *Antibiotics* **2020**, *9*, 688. [CrossRef]
44. D'Aimmo, M.R.; Modesto, M.; Biavati, B. Antibiotic resistance of lactic acid bacteria and *Bifidobacterium* spp. isolated from dairy and pharmaceutical products. *Int. J. Food Microbiol.* **2007**, *115*, 35–42. [CrossRef]
45. Moubareck, C.; Gavini, F.; Vaugien, L.; Butel, M.J.; Doucet-Populaire, F. Antimicrobial susceptibility of bifidobacteria. *J. Antimicrob. Chemother.* **2005**, *55*, 38–44. [CrossRef]
46. Ammor, M.S.; Belén Flórez, A.; Mayo, B. Antibiotic resistance in non-enterococcal lactic acid bacteria and bifidobacteria. *Food Microbiol.* **2007**, *24*, 559–570. [CrossRef] [PubMed]
47. Roberts, M.C. Update on acquired tetracycline resistance genes. *FEMS Microbiol. Lett.* **2005**, *245*, 195–203. [CrossRef] [PubMed]
48. Schwarz, S.; Kehrenberg, C.; Doublet, B.; Cloeckaert, A. Molecular basis of bacterial resistance to chloramphenicol and florfenicol. *FEMS Microbiol. Rev.* **2004**, *28*, 519–542. [CrossRef]
49. Roberts, M.C. Update on macrolide-lincosamide-streptogramin, ketolide, and oxazolidinone resistance genes. *FEMS Microbiol. Lett.* **2008**, *282*, 147–159. [CrossRef]
50. Kumar, M.; Nagpal, R.; Kumar, R.; Hemalatha, R.; Verma, V.; Kumar, A.; Chakraborty, C.; Singh, B.; Marotta, F.; Jain, S.; et al. Cholesterol-lowering probiotics as potential biotherapeutics for metabolic diseases. *Exp. Diabetes Res.* **2012**, *2012*, 1–14. [CrossRef]
51. Chand, D.; Avinash, V.S.; Yadav, Y.; Pundle, A.V.; Suresh, C.G.; Ramasamy, S. Molecular features of bile salt hydrolases and relevance in human health. *Biochimica Biophysica Acta Gen. Subj.* **2017**, *1861*, 2981–2991. [CrossRef] [PubMed]
52. Kriaa, A.; Bourgin, M.; Mkaouar, H.; Jablaoui, A.; Akermi, N.; Soussou, S.; Maguin, E.; Rhimi, M. Microbial reduction of cholesterol to coprostanol: An old concept and new insights. *Catalysts* **2019**, *9*, 167. [CrossRef]
53. Roy, D. Technological aspects related to the use of bifidobacteria in dairy products. *Lait* **2005**, *85*, 39–56. [CrossRef]
54. Zhang, C.; Zhang, Y.; Li, H.; Liu, X. The potential of proteins, hydrolysates and peptides as growth factors for *Lactobacillus* and *Bifidobacterium*: Current research and future perspectives. *Food Funct.* **2020**, *11*, 1946–1957. [CrossRef]
55. Poch, M.; Bezkorovainy, A. Growth-enhancing supplements for various species of the genus *Bifidobacterium*. *J. Dairy Sci.* **1988**, *71*, 3214–3221. [CrossRef]
56. Shihata, A.; Shah, N.P. Proteolytic profiles of yogurt and probiotic bacteria. *Int. Dairy J.* **2000**, *10*, 401–408. [CrossRef]

57. Meli, F.; Lazzi, C.; Neviani, E.; Gatti, M. Effect of protein hydrolizates on growth kinetics and aminopeptidase activities of some *Bifidobacterium* species. *Anaerobe* **2013**, *22*, 130–133. [CrossRef]
58. Raveschot, C.; Cudennec, B.; Coutte, F.; Flahaut, C.; Fremont, M.; Drider, D.; Dhulster, P. Production of bioactive peptides by *Lactobacillus* species: From gene to application. *Front. Microbiol.* **2018**, *9*, 2354. [CrossRef]
59. Udenigwe, C.C.; Aluko, R.E. Food protein-derived bioactive peptides: Production, processing, and potential health benefits. *J. Food Sci.* **2012**, *77*, R11–R24. [CrossRef]
60. Zulueta, A.; Esteve, M.J.; Frígola, A. ORAC and TEAC assays comparison to measure the antioxidant capacity of food products. *Food Chem.* **2009**, *114*, 310–316. [CrossRef]

fermentation

MDPI

Article

Screening and Evaluation of Purine-Nucleoside-Degrading Lactic Acid Bacteria Isolated from Winemaking Byproducts In Vitro and Their Uric Acid-Lowering Effects In Vivo

Min-Wei Hsieh [1], Huey-Yueh Chen [1] and Cheng-Chih Tsai [2,*]

[1] Taiwan Tobacco & Liquor Corporation, Zhongzheng District, Taipei City 10066, Taiwan; s584131492010@gmail.com (M.-W.H.); norst.tsai@gmail.com (H.-Y.C.)

[2] Department of Food Science and Technology, Hungkuang University, Shalu District, Taichung City 43302, Taiwan

* Correspondence: tsaicc@sunrise.hk.edu.tw; Tel.: +886-4-2631-8652-5082; Fax: +886-4-2631-9176

Abstract: In Taiwan, adult hyperuricemia affects as many as 1 in 4 males and 1 in 6 females, who are predominantly young adults aged 19–45. In this study, lactic acid bacteria (LAB) with acid tolerance, bile salt tolerance and high affinity to intestinal cells were extracted from the side products of alcohol fermentation (distillers' grains). These bacteria were evaluated for their ability to lower uric acid levels. Qualitative identification and quantitative analysis were performed using high-performance liquid chromatography (HPLC) on the purine-degrading enzymes to select purine-decomposing LAB for animal testing. When the final concentration of purine compounds reached 0.1% and 1%, seven strains of LAB showed potential in degrading purine compounds. HPLC was used to analyze their purine-degrading abilities, and the three best performing LAB strains, (107) 8–16, (107) tau 1–3, and (107) 6–10 were screened for further animal testing with Wistar rats. By the third week, the results showed that strain (107) 6–10 could prevent formation and reduce the levels of blood urea nitrogen (BUN) in yeast extract/potassium oxonate-induced hyperuricemia. The multi-strain lactic acid bacteria (MLAB) performed best for uric acid reduction in the serum and down regulated BUN. Yeast extract/potassium oxonate-induced hyperuricemia had no impact on serum creatinine, while LAB did not affect the creatinine concentration. In summary, MLAB not only protects kidney function but is also effective in regulating uric acid concentration in the body. Hence, MLAB can be used as a functional food supplement that prevents or aids the treatment of hyperuricemia in a rodent model.

Keywords: hyperuricemia; lactic acid bacteria; distillers' grains; Wistar rats; blood urea nitrogen; uric acid

Citation: Hsieh, M.-W.; Chen, H.-Y.; Tsai, C.-C. Screening and Evaluation of Purine-Nucleoside-Degrading Lactic Acid Bacteria Isolated from Winemaking Byproducts In Vitro and Their Uric Acid-Lowering Effects In Vivo. *Fermentation* **2021**, *7*, 74. https://doi.org/10.3390/fermentation7020074

Academic Editor: Hiroshi Kitagaki

Received: 1 April 2021
Accepted: 7 May 2021
Published: 10 May 2021

1. Introduction

Uric acid (UA) is a metabolite of purine and is mainly excreted by the kidneys, with only a small portion removed through the intestinal tract [1]. The in vivo uric acid concentration is determined by the balance between food content, in vivo synthesis and excretion [2]. When renal function is normal, uric acid can be filtered by the glomeruli and excreted in the urine. Blood uric acid concentration rises when synthesis increases too much or excretion becomes insufficient [3]. Blood urea nitrogen (BUN) is filtered into the urine by the glomerulus. Impaired renal function raises the blood urea nitrogen concentration, which can be used as an indicator of renal function [4]. Creatinine (CRE) is a product of phosphocreatine metabolism in the muscles. The CRE level reflects the blood CRE content and glomerular filtration rate. CRE concentration increases under abnormal renal function [4].

Hyperuricemia, caused by abnormal uric acid levels in the blood, forms solid uric acid crystals in the joints, thus engendering pain [5]. Hyperuricemia is also considered a risk factor for atherosclerosis and may lead to metabolic syndrome, hypertension, chronic diabetes and kidney disease [6]. The potential causative factors for hyperuricemia include insulin resistance, hypertension, renal insufficiency, obesity, diet and alcohol consumption [7]. Hyperuricemia occurs when there is an excessively high level of uric acid in the blood. The cause of the disease may be ingesting purine-rich foods or foods that promote purine synthesis, increased in vivo synthesis and inhibition of renal metabolism [8]. In human purine metabolism, inosine and guanosine are the primary purine compounds that form uric acid. Current uric acid-lowering drugs may sometimes present gastrointestinal side effects, such as nausea and diarrhea, as well as mild skin rashes and, rarely, fatalities [9,10].

Probiotics are acid-resistant and salt-resistant, can survive in the host's digestive tract and adhere to intestinal epithelial cells to alter the gut microbiota. Probiotics can also maintain their viability under prolonged storage [11]. The intestinal epithelial cells form a natural barrier against foreign objects. Upon entering the host intestinal tract, a probiotic strain may adhere to epithelial cells to prevent expulsion by peristalsis, help maintain microbiota balance, reduce the invasion and adhesion of pathogens and promote host immune function. Therefore, adsorption can be used as a criterion for screening beneficial probiotics [12].

LAB are common probiotics in the host gut. In addition to promoting gut health, LAB also have immunoregulation and anti-allergy effects and may regulate blood pressure, blood lipids and blood glucose [13]. Recent literature has shown that LAB have the potential to lower uric acid [14]. Therefore, probiotics capable of effectively degrading purine compounds could represent a potential therapy to prevent hyperuricemia [14]. Studies have proven that a *Lactobacillus plantarum* strain, DM9218-A, reduces the serum uric acid concentration in hyperuricemic mice. This strain could thus serve as a supportive treatment for hyperuricemia patients and help to prevent hyperuricemia in healthy people [1]. In a clinical trial, the *L. gasseri* strain PA-3 was used to treat 25 hyperuricemia/gout patients. The results showed that the PA-3-containing yogurt improved serum uric acid levels [15]. Another study showed that *L. paracasei* obtained from screening 18 types of traditional pickles degraded uric acid in vitro and in vivo and demonstrated the potential of LAB to reduce renal damage in a rat model of chronic hyperuricemia [16]. The *L. brevis* strain DM9218 was shown to reduce serum uric acid levels and liver xanthine oxidative activity in rats and may have the potential to ameliorate fructose-induced hyperuricemia [17].

The present study aims to screen for strains with essential probiotic characteristics from fermented cereals, investigate the uric acid-lowering potential of those strains and compare those strains with common gut health supplements on the market. Thirty-three LAB strains were cultured in a differential medium supplemented with purines to screen for LAB strains that could degrade purines. Subsequently, a qualitative purine compound degradation assay and a quantitative HPLC analysis on the degraded purine compounds were used to identify strains capable of degrading and utilizing purines. These LAB strains were used in animal tests to evaluate the effectiveness of preventing hyperuricemia.

2. Materials and Methods

2.1. Strains, Culture Medium and Basic Growth Conditions

The 33 LAB strains used in this study were screened and isolated from winemaking byproducts obtained from the Taiwan Tobacco and Liquor Corporation (Taipei, Taiwan) and exhibited essential probiotic characteristics. The strains were suspended in an MRS broth containing 15% glycerol and stored in a −80 °C freezer. The strains were activated twice before the experiments using a Lactobacilli MRS broth (Difco Laboratories, Detroit, MI, USA) containing 0.05% L-cysteine. The culture condition was 37 °C for 24 h.

2.2. Screening for Lab Strains with Purine Degradation Abilities

The strains were activated once using Lactobacilli MRS broth (Difco Laboratories, Detroit, MI, USA) containing 0.05% *L*-cysteine at 37 °C for 16–24 h. For the second activation, 100 µL of a culture broth from the first activation was inoculated into 4 mL MRS broth without purines and cultured at 37 °C for 16–18 h. The culture was then serially diluted, and the bacterial counts were enumerated by plating. The result represented the number of normal bacteria. For the purine test, inosine and guanosine were added to 4 mL MRS broth to a final concentration of 0.1% and 1%, respectively. Next, 100 μL culture from the first activation was inoculated into the MRS broth and cultured at 37 °C for 16–18 h. The broth was then serially diluted, and bacterial counts were enumerated by plating. The solid medium used in the above experiments was Lactobacilli MRS agar (Difco Laboratories, Detroit, MI, USA) containing 0.05% L-cysteine. The culture condition was 37 °C for 24–48 h.

2.3. *HPLC Analysis of Degraded Purine by Lab Strains*

Strains obtained by screening were activated twice. After activation, the bacterial cells were collected via centrifuge at 8000× *g* for 10 min. The cells were then washed twice with PBS, and the supernatant was discarded. Next, 1 mL inosine (1.25 mM) and guanosine (1.25 mM) were added to the washed cells and mixed evenly before incubating at 37 °C/140× *g*. During the experiment, co-cultures were sampled at 0, 1, and 2 h and centrifuged at 4 °C/6000× *g*/7 min. In total, 720 μL of the supernatant was collected and mixed with 80 μL HClO4 (0.1 M) to stop the reaction. The collected supernatant was passed through a 0.22 μm filter and stored at −80 °C until HPLC analysis. The formula for calculating the degradation rate of inosine and guanosine by different probiotic strains was as follows: degradation rate (%) = (1.25 (mM) − A (mM))/1.25 (mM) × 100% (A: concentration of inosine and guanosine after reaction).

2.4. *Acid Resistance and Bile Salt Resistance Tests*

Phosphate buffer with pepsin (pH 2.0) was added to 1000 μL of LAB suspension (10^8 CFU/ mL), and the bacteria were cultured at 37 °C for 0, 1.5 and 3 h, respectively. LAB suspension mixed with phosphate buffer (pH 7.0) and cultured under the same conditions was used as the control. After culturing, the counts of viable LAB were enumerated using the pour plate method on MRS agar plates [18]. One mL of LAB suspension that underwent 3 h of acid treatment was then added to 9 mL of phosphate buffer with (or without) 0.3% oxgall bile (Sigma-Aldrich Corp., St. Louis, MO, USA) and 0.1% pancreatin [19] and cultured for 0, 1.5 and 3 h. LAB counts were enumerated afterward using the pour plate method on MRS agar plates. Strains with acid resistance of $>10^7$ CFU /mL and bile salt resistance of $>10^6$ CFU /mL were selected.

2.5. *Caco-2 Cell Adhesion Test*

One milliliter of Caco-2 cell suspension (5×10^4 cell /mL) in DMEM (Dulbecco's Modified Eagle Medium) was added per well in a 24-well plate and cultured at 37 °C for 40–45 h in a 5% CO_2 incubator. One milliliter of LAB suspension was added to an Eppendorf tube and centrifuged at 8000× *g* for 10 min. The supernatant was then discarded, and 1 mL 1 × PBS was added and mixed evenly before another centrifugation. Following that, the pellet was resuspended in 1 mL DMEM.

The old culture medium was aspirated from the 24-well plates containing Caco-2 cells, and the plates were washed twice with 1 × PBS. Nine-hundred microliters of fresh culture medium and 100 μL PBS washed LAB suspension were then added to the plates. The plates were co-cultured at 37 °C in an incubator with 5% CO_2 for 2 h. To fix the bacteria and cells, the culture medium was aspirated, and the plates were washed twice with 1 × PBS, followed by adding 200 μL 6–10% formalin to each well and letting the plates stand for 30 min. The formalin was then aspirated, and the plates were washed twice with 1 × PBS. Following that, 200 μL crystal violet solution was added and allowed to react for 5 min. The number of bacteria that adhered to the cells was counted using an inverted microscope [20]. The bacteria were considered to have an adhesion ability if 15 or more bacteria adhered to each cell [21].

2.6. Animal Tests on the Uric Acid-Lowering Effects of Lab Strains

The animal experiment protocol approval number is HK-10805 (date: 5 July 2019) from HungKuang University, Taichung, Taiwan. The experimental animals included 56 30-day-old (about four weeks) male Wistar rats, which were purchased from BioLASCO Taiwan Co. Ltd. (Taipei, Taiwan). The temperature in the animal room was controlled at 23 ± 1 °C and the relative humidity at 40–60%. An automatic controller was used to create a 12/12 h light-dark cycle (light period: 8:00 to 20:00; dark period: 20:00 to 8:00). The animals were kept in individually ventilated cages (IVC) and given ad libitum access to feed and sterile distilled water. After one week of acclimatization, S-shaped cages were used for grouping, with eight rats per group. A total of seven groups were established to test the uric acid-lowering effects of LAB strains from different sources. These groups included a blank control (C), a multi-strain LAB blank control (CM), a hyperuricemia induction group (HUA), a hyperuricemia induction + strain (107) 8–16 treatment group, a hyperuricemia induction + strain (107) 1–3 treatment group, a hyperuricemia induction + strain (107) 6–10 treatment group and a hyperuricemia induction + multi-strain LAB treatment group (MLAB). Intraperitoneal injection of potassium oxonate (0.35 mg/100 g BW/day) and gavage with yeast extract (15 g/kg BW/day) were used to induce hyperuricemia in the rats (Figure 1).

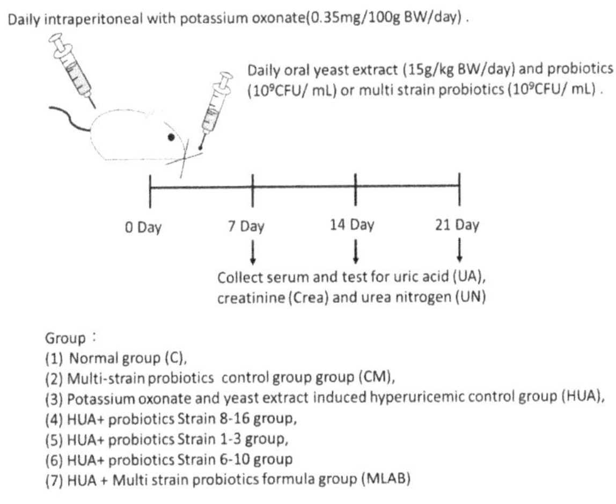

Figure 1. Animal model experimental structure and schedule.

Strains obtained from screening were activated twice. After activation, the culture medium was centrifuged at 9000× *g* for 10 min to collect the bacteria. The bacterial cells were washed twice with PBS, and the supernatant was discarded. The washed bacterial cells were then resuspended in PBS, the volume of which was the same as that of the MRS used in the second culture. The suspension was stored at 4 °C for up to 3 days. The dose of MLAB used in each group was 10^9 CFU/mL, and the ratio of MLAB used was 1:1:1. One milliliter of LAB was fed to the rats after the daily induction procedure. Weight and food intake were recorded every week, and fasting venous blood was collected. The isolated serum was measured for uric acid, creatinine and urea nitrogen contents (Figure 1).

2.7. Statistical Analysis

The results of this study are based on statistical analyses performed using SPSS 23.00 for Windows (IBM Corporation, Armonk, NY, USA). The experimental results are expressed as the mean ± standard deviation (SD) based on a single factor one-way analysis of variance (one-way ANOVA) test. Duncan's Multiple Range Test was used to test the difference between the averages of the experimental groups, and $p < 0.05$ was considered as a significant difference with a two-tailed test.

3. Results

3.1. Screening for LAB Strains with Potential Purine Degradation Abilities

When the final concentration of purines was 0.1% and 1%, the bacterial counts of most strains were not severely affected (Table 1). However, the bacterial counts for (107) 8–16, (107) tau 3–5, (107) tau 2–1, (107) 6–10 and (107) tau 2–8 were slightly increased (from 0.99 to 0.74 Log CFU/mL) compared to the original bacterial count. Moreover, the bacterial counts for (107) tau 6–2 and (107) tau 1–3 were increased by 1 log CFU/mL (Table 1). If the addition of purines causes the bacterial count to remain constant or increase, the strain may be able to utilize purines for growth. Therefore, we hypothesized that the above seven probiotic strains might be able to degrade purines. Subsequently, HPLC was used to determine the ratio of degraded purines in the above seven strains.

Table 1. Tolerance of lactic acid bacteria to the purine compound medium.

Strain	Number of LAB [a] (Log CFU [b]/mL)		
	0% Purine Compound	0.1% Purine Compound	1% Purine Compound
(106) 6-1	10.24	10.19	10.21
(106) 11-6	10.14	10.22	10.13
(106) 3-9	10.17	10.05	10.00
(106) 6-4	10.16	10.20	10.17
(106) 8-5	10.12	10.13	9.98
(106) 3-21-2	10.11	10.17	10.21
(106) 3-11	9.96	10.04	9.95
(106) 1-17	9.94	10.09	9.86
(106) 3-14	10.23	10.22	10.16
(106) 7-2	9.94	10.03	9.85
(107) tau 6-2	9.21	10.25	10.24
(107) jia 6-7	9.84	9.84	9.94
(107) 8-16	9.42	10.32	10.39
(107) jia 6-5	9.82	9.83	9.98
(107) tau 1-3	9.27	10.26	10.30
(107) tau 3-5	9.27	10.21	10.01
(107) tau 2-1	9.62	10.48	10.49
(107) 6-10	9.21	10.18	10.02
(107) tau 2-8	9.57	10.41	10.40
(107) tau 6-7	9.97	9.97	10.03
(106) 6-9	9.95	9.97	10.01
(106) 6-18	10.16	10.21	10.29
(106) 1-5	10.03	10.09	10.10
(106) 1-4	10.04	10.08	10.12
(106) 3-21-1	9.89	10.01	9.97
(106) 1-1	10.03	10.11	10.22
(107) jia 6-6	9.94	9.85	10.03
(107) jia 5-7	10.04	9.97	10.01
(107) jia 6-1	9.99	9.87	10.03
(107) tau 1-1	10.40	10.32	10.47
(107) tau 4-2	9.99	9.95	9.94
(107) tau 3-7	10.24	10.21	10.30
(107) tau 2-4	10.33	10.21	10.15

[a] LAB: lactic acid bacteria. [b] CFU: colony forming unit.

The excessive intake of purines in the diet will increase serum uric acid levels, which is a risk factor for hyperuricemia and gout. Lifestyle guides recommend nutritional therapy, such as caloric restriction, to reduce obesity and limit purine intake, as well as to restrict alcoholic beverage intake and engage in adequate exercise to avoid raising serum uric acid levels [22]. The ability of probiotics to degrade purine compounds in food to prevent hyperuricemia was previously reported [16].

3.2. HPLC Analysis of the Ratio of Inosine and Guanosine Degraded by Lab Strains

Microorganisms degrade nucleosides mainly through the biosynthesis of nucleoside hydrolases. The existence of nucleoside hydrolase in the strain can be proven by the degradation of nucleoside substances, such as guanosine and inosine [16,23]. The addition of various ratios of purines was used to screen for LAB strains capable of degrading purines. Seven strains were isolated: (107) tau 6–2, (107) tau 2–1, (107) 6–10, (107) tau 2–8, (107) 8–16, (107) tau 1–3 and (107) tau 3–5.

Figure 2 showed the inosine degradation experiment results for the LAB strains. At all time points, the inosine degradation rates of all strains were 95% or higher. The preferred strains were (107) 8–16 and (107) 6–10. Both strains maintained stable degradation rates of 99% across all time points. The next best strains were (107) tau 1–3 and (107) tau 3–5. Although (107) tau 1–3 had a degradation rate lower than that of the preferred strains, it reached 95% and above and showed an increasing trend. On the other hand, (107) tau 3–5showed a reduction to 97% at 1 h but then rose to 99% at 2 h. Therefore, both strains were retained as candidates. The analysis of guanosine degradation by LAB found that the guanosine degradation rates of all tested strains reached 95% or higher (Figure 3). The preferred strain was (107) tau 1–3, with a degradation rate reaching 99.5% and a stable increase. The second was (107) tau 2–8, with an overall stable degradation rate of 99.5%. In third place were (107) 8–16 and (107) tau 2–1. Although the rate of (107) 8–16 dropped to 99.36% at 2 h, the overall degradation rate was high. There was a stable degradation rate in (107) tau 2–1 at 0 and 1 h, which increased at 2 h (Figure 3). The guanosine degradation rates of the (107) 6–10 strains were 99.2, 98.8 and 98.9% at different time point (Figure 3). In summary, the HPLC analysis of inosine degradation tested strains (107) 8–16, (107) 6–10, (107) tau 1–3 and (107) tau 3–5, while the analysis of guanosine degradation tested strains (107) tau 1–3, (107) tau 2–8, (107) 8–16 and (107) tau 2–1.

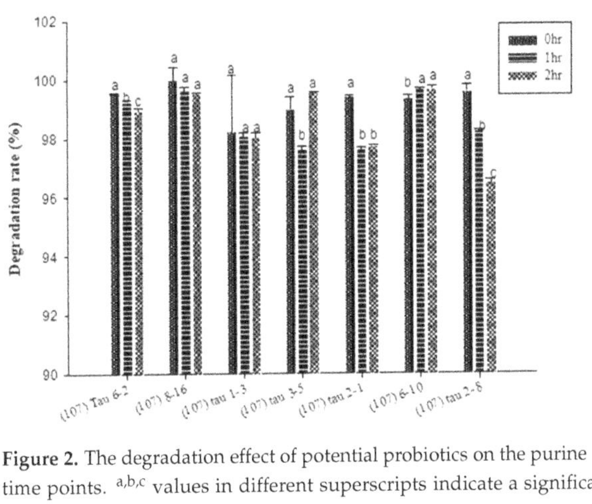

Figure 2. The degradation effect of potential probiotics on the purine compound inosine at different time points. ^{a,b,c} values in different superscripts indicate a significant difference ($p < 0.05$) using Duncan's multiple range test.

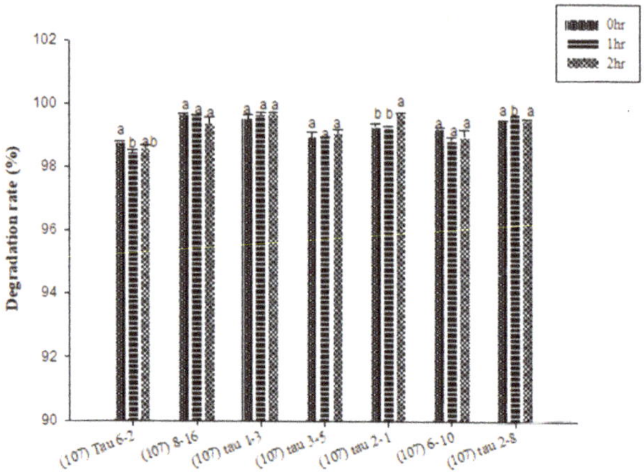

Figure 3. The degradation effect of potential probiotics on the purine compound guanosine at different time points. ^{a,b} values in different superscripts indicate a significant difference ($p < 0.05$) using Duncan's multiple range test.

3.3. *Bacteria Identification*

The API kit was used for strain identification during the HPLC analysis. As shown in the appendix, (107) tau 2–1, (107) tau 2–8, and 6–10 were identified as *Lactobacillus paracasei*, (107) 8–16 as *L. plantarum*, (107) tau 3–5 as *L. curvatus* and (107) tau 1–3 as *Pediococcus acidilactici*. However, the API results showed other possible bacterial species and streaking was used to isolate single strains for further identification. The current literature proves that the LAB able to reduce hyperuricemia includes *L. plantarum*, *L. brevis*, *L. gasseri* and *L. paracasei* [1,15–17]. In this study, we found that *L. curvatus* and *P. acidilactici* isolated from distillers' grains might also have the ability to degrade purine and reduce uric acid.

3.4. *Acid and Bile Salt Resistance and Intestinal Cell Adhesion Tests*

The bacterial counts of LAB strains (107) 8–16 and (107) 6–10 at 3 h were at least 10^8 CFU/mL, indicating acid resistance (Table 2). The bacterial counts of LAB strains (107) 8–16 and (107) 6–10 at 3 h were at least 10^7 CFU/mL, indicating bile salt resistance (Table 2). After 2 h of co-culture with Caco-2 cells, the adhesion capacity of LAB strains (107) 8–16 and (107) 6–10 was 26.8 ± 3.77 CFU/cell and 27.8 ± 9.94 CFU/cell, respectively, which was higher than the 15 CFU/cell threshold, indicating that both strains could adhere to intestinal epithelial cells (Figure 4).

Table 2. Acid and bile salt tolerance of LAB strains.

	Acid Tolerance			Bile Salt Tolerance		
Strain	**0 h**	**1.5 h**	**3 h**	**0 h**	**1.5 h**	**3 h**
(107) 8–16	9.1 ± 0.03	9.12 ± 0.04	8.98 ± 0.02	8.76 ± 0.12	8.49 ± 0.02	8.62 ± 0.08
(107) 6–10	9.11 ± 0.02	9.03 ± 0.01	9.11 ± 0.02	7.84 ± 0.04	7.88 ± 0.11	7.73 ± 0.01

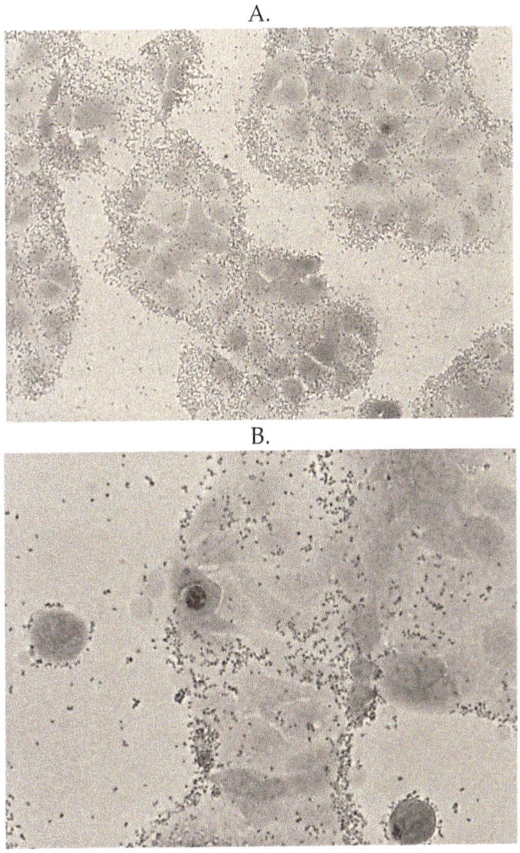

Figure 4. Adhesion of LAB (**A**) strain (107) 6–10 and (**B**) strain (107) 8–16 to the Caco-2 intestinal epithelium cells.

The biological characteristics of tolerance to acid and bile salt and cell adhesion abilities were important to determine whether the tested LAB strains could survive in and successfully colonize the gastrointestinal tract [1]. Current studies demonstrate that the factors facilitating Lactobacillus adhesion on epithelial cells include hydrophobic effects, lipoteichoic acids (LTA), lectins, proteins and exopolysaccharides (EPA) [24–27]. EPA often attaches to cell surfaces or is secreted into an extracellular medium. One study noted that the removal of EPA from the cell surface of *L. plantarum* significantly decreased adhesion [28]. The same study also showed that adhesion was affected by trypsin treatment [28]. Moreover, probiotics mainly exert their effects in the posterior segment of the gut. Therefore, probiotics must be able to resist gastric acid and bile acid before reaching the site of effect. Studies show that gastric acid pH changes with the time of gastric juice secretion and type of food, usually fluctuating between pH 1.5 and 4.58 over 4 h. The survival of LAB is easily affected by an extremely low pH. The main constituents of bile salts are mixtures of glycine or taurine and bile acids, as well as chenodeoxycholic acid, deoxycholic acid, lithocholic acids and their derivatives. The presence of bile salts changes the permeability of the outer membrane of the intestinal bacteria, and bile acids in bile salts also inhibit the growth of intestinal bacteria [29,30].

3.5. Animal Tests to Determine the Uric-Acid-Lowering Effects of the Lab Strains

The HPLC analysis of inosine degradation tested strains (107) 8–16, (107) 6–10, (107) tau 1–3, and (107) tau 3–5, while the analysis of guanosine degradation tested strains (107) tau 1–3, (107) tau 2–8, (107) 8–16 and (107) tau 2–1. The three probiotic strains used for animal experiments were (107) 8–16, (107) tau 1–3 and (107) 6–10. The criterium for selecting these three strains was their high inosine and guanosine degradation rates in the in vitro experiments. Therefore, these three probiotic strains were expected to show positive effects in the animal tests.

The uric acid results in the experimental animals are shown in Figure 5. There were no significant differences among the different groups at week 0 ($p > 0.05$). At week 2, the uric acid concentration in the HUA group was 17.86% higher than that in group C ($p < 0.05$). After induction and feeding with LAB strains from different sources, the serum uric acid concentrations in the treatment groups, including in strain (107) 8–16, (107) tau 1–3, (107) 6–10, and MLAB, were significantly lower than those in the HUA group ($p < 0.05$) by 16.43%, 20% and 32.14%, respectively. Among these groups, uric acid reduction was the greatest in the MLAB group, while there was no significant difference between the strain (107) tau 1–3 treatment group and the HUA group ($p > 0.05$). There was no significant difference between the CM and C groups ($p > 0.05$). At week 3, the uric acid concentration in the HUA group was 23% higher than that in the C group ($p < 0.05$). After induction and feeding with LAB strains from different sources, the serum uric acid concentrations in the treatment groups, including strains (107) 8–16, (107) tau 1–3, (107) 6–10, and MLAB, were significantly lower than those in the HUA group ($p < 0.05$) by 14.5%, 12.25%, and 28%, respectively. Among these groups, the uric acid reduction was the greatest in the MLAB group, while there was no significant difference between the CM and C groups ($p > 0.05$). The results showed that mixed probiotics could prevent and alleviate the uric acid issue in hyperuricemia induced by potassium oxonate and yeast extract.

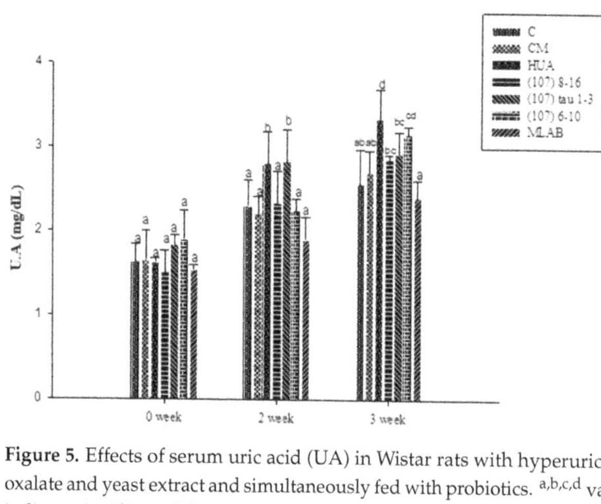

Figure 5. Effects of serum uric acid (UA) in Wistar rats with hyperuricemia induced by potassium oxalate and yeast extract and simultaneously fed with probiotics. ^{a,b,c,d} values in different superscripts indicate significant differences ($p < 0.05$) using a Duncan's multiple range test.

The BUN results in the experimental animals are shown in Figure 6. There were no significant differences among the different groups at week 0 ($p > 0.05$). At week 2, the BUN concentration in the HUA group was 13.64% higher than that in the C group ($p < 0.05$). After induction and feeding with LAB strains from different sources, the serum BUN concentrations in the treatment groups, including strains (107) 8–16, (107) tau 1–3, (107) 6–10, and MLAB, were significantly lower than those in the HUA group ($p < 0.05$) by 22.42%, 19.07%, 27.27% and 23.12%, respectively. Among these groups, BUN reduction was the greatest in the 6–10 group, while there was no significant difference observed between the CM and C groups ($p > 0.05$). At week 3, the BUN concentration in the HUA group was 13.19% higher than that in the C group ($p < 0.05$). After induction and feeding with LAB strains from different sources, the serum BUN concentrations in the treatment groups including strains (107) 8–16, (107) tau 1–3, (107) 6–10, and MLAB, were significantly lower than those in the HUA group ($p < 0.05$) by 11.1%, 17.02%, 19.63%, and 16.11%, respectively. Among these groups, BUN reduction was the greatest in the 6–10 group, while there was no significant difference observed between the CM and C groups ($p > 0.05$). The results showed that strains 6–10 could prevent and decrease BUN in hyperuricemia induced by potassium oxonate and yeast extract.

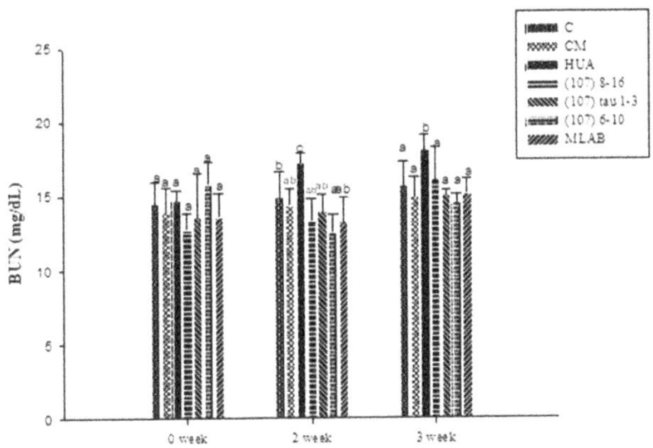

Figure 6. Effects of blood urea nitrogen (BUN) in Wistar rats with hyperuricemia induced by potassium oxalate and yeast extract and simultaneously fed with probiotics. $^{a, b, c}$ values in different superscripts indicate significant differences ($p < 0.05$) using a Duncan's multiple range test.

Figure 7 shows the blood creatinine results for the experimental animals. There were no significant differences observed between the groups at week 0 ($p > 0.05$). At weeks 2 and 3, there were no significant differences found between groups after induction and feeding with LAB strains from different sources ($p > 0.05$). The results showed that hyperuricemia induced by potassium oxonate and yeast extract did not affect serum creatinine and that probiotics from different sources did not affect the creatinine concentration.

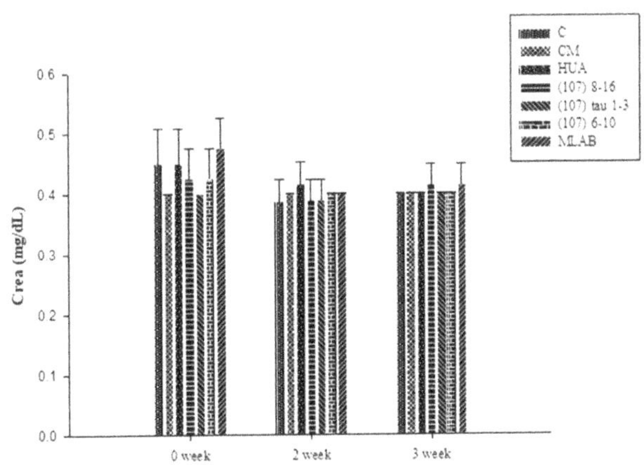

Figure 7. Effects of creatinine (Crea) in Wistar rats with hyperuricemia induced by potassium oxalate and yeast extract and simultaneously fed with probiotics. Values showed no significant differences ($p > 0.05$) based on a Duncan's multiple range test.

The literature review found that at least two to three weeks is required for a uric acid-lowering test in animals with probiotics alone [1]. Rodents such as rats and mice are used in most animal experiments. These animals possess uric acid (oxidation) enzymes, such as uricase (urate oxidase), that convert uric acid into more water-soluble allantoin for excretion. Therefore, rodents are less susceptible to hyperuricemia [31,32]. There are three main methods for inducing hyperuricemia in experimental animals. The first is to increase the uric acid level in vivo (i.e., inducing the body to generate a large amount of uric acid, leading to hyperuricemia) such as by using yeast extract, a high-purine diet or uric acid precursors [33]. The second method is to inhibit the uric acid excretion pathways, thus impairing renal uric acid metabolism and causing hyperuricemia through treatment with adenine, niacin, or ethambutol. The final method is uricase inhibition, which involves using uric acid analogs, such as potassium oxonate, to competitively inhibit uricase and induce hyperuricemia [9]. However, inducing hyperuricemia by inhibiting uric acid excretion or uricase activity may lead to renal impairment in the lab animals and does not match the mechanism of hyperuricemia in humans. On the other hand, it is difficult to maintain long term hyperuricemia using yeast extracts or high-purine diets alone, which may prolong the research period. Therefore, in the present study, we combined potassium oxonate and yeast extract to induce hyperuricemia in rats. This method not only inhibited uricase activity but also increased in vivo purine levels, thereby causing hyperuricemia [34].

4. Conclusions

We screened LAB from distillers' grains and indicated their ability to degrade purine in vitro and reduce UA in vivo. Based on the above results, when probiotics from different sources were fed to Wistar rats with hyperuricemia induced by potassium oxonate and yeast extract, the MLAB group showed the best serum uric acid-lowering effect. MLAB also reduced BUN without affecting creatinine. Although strain (107) 6–10 showed the best BUN-lowering effects, it did not control uric acid to the same extent as MLAB did. Therefore, employing a fixed ratio of mixed probiotics not only presented some nephroprotective effects but also well-regulated in vivo uric acid. In the future, we will study the effectiveness of probiotics through clinical research.

Author Contributions: H.-Y.C. and C.-C.T. designed the study and acquired funding; M.-W.H. performed the experiments. All authors have read and agreed to the published version of the manuscript.

Funding: This research was funded by Taiwan Tobacco and Liquor Corporation, Taipei, Taiwan under grant contract case number 107-0620-1-001.

Institutional Review Board Statement: The animal experiment protocol approval number is HK-10805 (date: 2019/07/05) from HungKuang University, Taichung, Taiwan.

Data Availability Statement: The data presented in this study are available in the article.

Acknowledgments: This study was funded by the contract case number 107-0620-1-001 project from Taiwan Tobacco and Liquor Corporation, Taipei, Taiwan.

Conflicts of Interest: The authors declare no conflict of interest.

References

1. Cicero, A.F.G.; Fogacci, F.; Kuwabara, M.; Borghi, C. Therapeutic strategies for the treatment of chronic hyperuricemia: An evidence-based update. *Medicina* **2021**, *57*, 58–76. [CrossRef] [PubMed]
2. Lima, W.G.; Martins-Santos, M.E.; Chaves, V.E. Uric acid as a modulator of glucose and lipid metabolism. *Biochimie* **2015**, *116*, 17–23. [CrossRef]
3. Giordano, C.; Karasik, O.; King-Morris, K.; Asmar, A. Uric acid as a marker of kidney disease: Review of the current literature. *Dis. Markers* **2015**, *2015*, 382918. [CrossRef]
4. Akkasilpa, S.; Avihingsanon, Y.; Hanvivadhanakul, P.; Wonchinsri, J. Clinical manifestations of patients with hyperuricemia. *J. Med. Assoc. Thai.* **2004**, *87*, 41–44.

5. Li, M.; Yang, D.; Mei, L.; Yuan, L.; Xie, A.; Yuan, J. Screening and characterization of purine nucleoside degrading lactic acid bacteria isolated from Chinese sauerkraut and evaluation of the serum uric acid lowering effect in hyperuricemic rats. *PLoS ONE* **2014**, *9*, e105577. [CrossRef]
6. Zhang, S.; Wang, Y.; Cheng, J.; Huangfu, N.; Zhao, R.; Xu, Z.; Zhang, F.; Zheng, W.; Zhang, D. Hyperuricemia and cardiovascular disease. *Curr. Pharm. Des.* **2019**, *25*, 700–709. [CrossRef]
7. Li, C.; Hsieh, M.C.; Chang, S.J. Metabolic syndrome, diabetes, and hyperuricemia. *Curr. Opin. Rheumatol.* **2013**, *25*, 210–216. [CrossRef]
8. Chalès, G. How should we manage asymptomatic hyperuricemia? *Jt. Bone Spine* **2019**, *86*, 437–443. [CrossRef]
9. García-Arroyo, F.E.; Gonzaga, G.; Muñoz-Jiménez, I.; Blas-Marron, M.G.; Silverio, O.; Tapia, E.; Soto, V.; Ranganathan, N.; Ranganathan, P.; Vyas, U.; et al. Probiotic supplements prevented oxonic acid-induced hyperuricemia and renal damage. *PLoS ONE* **2018**, *13*, e0202901. [CrossRef]
10. Strilchuk, L.; Fogacci, F.; Cicero, A.F. Safety and tolerability of available urate-lowering drugs: A critical review. *Expert. Opin. Drug Saf.* **2019**, *18*, 261–271. [CrossRef]
11. Vinderola, G.; Binetti, A.; Burns, P.; Reinheimer, J. Cell viability and functionality of probiotic bacteria in dairy products. *Front. Microbiol.* **2011**, *2*, 70. [CrossRef] [PubMed]
12. Papadimitriou, K.; Zoumpopoulou, G.; Foligné, B.; Alexandraki, V.; Kazou, M.; Pot, B.; Tsakalidou, E. Discovering probiotic microorganisms: In vitro, in vivo, genetic and omics approaches. *Front. Microbiol.* **2015**, *6*, 58. [CrossRef]
13. Markowiak, P.; Śliżewska, K. Effects of Probiotics, Prebiotics, and Synbiotics on human health. *Nutrients* **2017**, *9*, 1021. [CrossRef]
14. Sedaghat, A.; Karimi Torshizi, M.A. Immune responses, intestinal microbiota, performance and blood characteristics of Japanese quail fed on diets containing camphor. *Animal* **2017**, *11*, 2139–2146. [CrossRef] [PubMed]
15. Yamanaka, H.; Taniguchi, A.; Tsuboi, H.; Kano, H.; Asami, Y. Hypouricaemic effects of yoghurt containing *Lactobacillus gasseri* PA-3 in patients with hyperuricaemia and/or gout: A randomised, double-blind, placebo-controlled study. *Mod. Rheumatol.* **2019**, *29*, 146–150. [CrossRef]
16. Xiao, Y.; Zhang, C.; Zeng, X.; Yuan, Z. Microecological treatment of hyperuricemia using *Lactobacillus* from pickles. *BMC Microbiol.* **2020**, *20*, 195–204. [CrossRef]
17. Wang, H.; Mei, L.; Deng, Y.; Liu, Y.; Wei, X.; Liu, M.; Zhou, J.; Ma, H.; Zheng, P.; Yuan, J.; et al. *Lactobacillus brevis* DM9218 ameliorates fructose-induced hyperuricemia through inosine degradation and manipulation of intestinal dysbiosis. *Nutrition* **2019**, *62*, 63–73. [CrossRef]
18. Gómez Zavaglia, A.; Kociubinski, G.; Pérez, P.; De Antoni, G. Isolation and characterization of Bifidobacterium strains for probiotic formulation. *J. Food Prot.* **1998**, *61*, 865–873. [CrossRef]
19. Gilliland, S.E.; Walker, D.K. Factors to consider when selecting a culture of *Lactobacillus acidophilus* as a dietary adjunct to produce a hypocholesterolemic effect in humans. *J. Dairy Sci.* **1990**, *73*, 905–911. [CrossRef]
20. Banerjee, P.; Merkel, G.J.; Bhunia, A.K. Lactobacillus delbrueckii ssp. bulgaricus B-30892 can inhibit cytotoxic effects and adhesion of pathogenic Clostridium difficile to Caco-2 cells. *Gut Pathog.* **2009**, *1*, 8. [CrossRef]
21. Pedersen, K.; Tannock, G.W. Colonization of the porcine gastrointestinal tract by lactobacilli. *Appl. Environ. Microbiol.* **1989**, *55*, 279–283. [CrossRef]
22. Kaneko, K.; Takayanagi, F.; Fukuuchi, T.; Yamaoka, N.; Yasuda, M.; Mawatari, K.I.; Fujimori, S. Determination of total purine and purine base content of 80 food products to aid nutritional therapy for gout and hyperuricemia. *Nucleosides Nucleotides Nucleic Acids* **2020**, *39*, 1449–1457. [CrossRef]
23. Guimaraes, A.P.; Oliveir, A.A.; Ramalho, T.C. Analysis of bacillus anthracis nucleoside hydrolase via in silico docking with inhibitors and molecular dynamics simulation. *J. Mol. Model.* **2011**, *17*, 2939–2951. [CrossRef] [PubMed]
24. Babot, J.D.; Argañaraz-Martínez, E.; Saavedra, L.; Apella, M.C.; Chaia, A.P. Compatibility and safety of five lectin-binding putative probiotic strains for the development of a multi-strain protective culture for poultry. *Benef. Microbes* **2018**, *9*, 927–935. [CrossRef]
25. Wang, G.; Zhang, M.; Zhao, J.; Xia, Y.; Lai, P.F.; Ai, L. A surface protein from *Lactobacillus plantarum* increases the adhesion of *Lactobacillus* strains to human epithelial cells. *Front. Microbiol.* **2018**, *9*, 2858. [CrossRef]
26. Castro-Bravo, N.; Wells, J.M.; Margolles, A.; Ruas-Madiedo, P. Interactions of surface exopolysaccharides from *Bifidobacterium* and *Lactobacillus* within the intestinal environment. *Front. Microbiol.* **2018**, *9*, 2426. [CrossRef] [PubMed]
27. Deepika, G.; Charalampopoulos, D. Surface and adhesion properties of lactobacilli. *Adv. Appl. Microbiol.* **2010**, *70*, 127–152.
28. Sun, J.; Le, G.W.; Shi, Y.H.; Su, G.W. Factors involved in binding of *Lactobacillus plantarum* Lp6 to rat small intestinal mucus. *Lett. Appl. Microbiol.* **2007**, *44*, 79–85. [CrossRef]
29. Gilliland, S.E. Acidophilus milk products: A review of potential benefits to consumers. *J. Dairy Sci.* **1989**, *72*, 2483–2494. [CrossRef]
30. Ridlon, J.M.; Harris, S.C.; Bhowmik, S.; Kang, D.J.; Hylemon, P.B. Consequences of bile salt biotransformations by intestinal bacteria. *Gut Microbes* **2016**, *7*, 22–39. [CrossRef]
31. Sánchez-Lozada, L.G.; Tapia, E.; Soto, V.; Avila-Casado, C.; Franco, M.; Zhao, L.; Johnson, R.J. Treatment with the xanthine oxidase inhibitor febuxostat lowers uric acid and alleviates systemic and glomerular hypertension in experimental hyperuricaemia. *Nephrol. Dial. Transplant.* **2008**, *23*, 1179–1185. [CrossRef]

32. Sánchez-Lozada, L.G.; Soto, V.; Tapia, E.; Avila-Casado, C.; Sautin, Y.Y.; Nakagawa, T.; Franco, M.; Rodríguez-Iturbe, B.; Johnson, R.J. Role of oxidative stress in the renal abnormalities induced by experimental hyperuricemia. *Am. J. Physiol. Renal. Physiol.* **2008**, *295*, 1134–1141. [CrossRef]
33. Haryono, A.; Nugrahaningsih, D.A.A.; Sari, D.C.R.; Romi, M.M.; Arfian, N. Reduction of serum uric acid associated with attenuation of renal injury, inflammation and macrophages M1/M2 ratio in hyperuricemic mice model. *Kobe J. Med. Sci.* **2018**, *64*, 107–114.
34. Wu, P.; Li, J.; Zhang, X.; Zeng, F.; Liu, Y.; Sun, W. Study of the treatment effects of compound tufuling granules in hyperuricemic rats using serum metabolomics. *Evid. Based Complement Alternat. Med.* **2018**, *2018*, 3458185. [CrossRef] [PubMed]

fermentation

Article

Bacteriocinogenic *Bacillus* spp. Isolated from Korean Fermented Cabbage (*Kimchi*)—Beneficial or Hazardous?

Joanna Ivy Irorita Fugaban , Jorge Enrique Vazquez Bucheli , Wilhelm Heinrich Holzapfel and Svetoslav Dimitrov Todorov *

ProBacLab, Department of Advanced Convergence, Handong Global University, Pohang, Gyeongbuk 37554, Korea; jifugaban@up.edu.ph (J.I.I.F.); jorge_jbv@hotmail.com (J.E.V.B.); wilhelm@woodapple.net (W.H.H.)
* Correspondence: slavi310570@abv.bg; Tel.: +82-10-3490-3152

Abstract: *Bacillus velezensis* ST03 and ST32, *Bacillus amyloliquefaciens* ST06 and ST109, and *Bacillus subtilis* ST08 were isolated from artisanal-produced *kimchi* and were identified based on 16S rRNA partial sequencing. DNA obtained from the investigated bacilli generated positive results for lichenicidin, iturin, subtilosin, and surfactin on a strain-specific basis. The strains were found to produce antimicrobial metabolites with activity levels ranging between 800 and 1600 AU/mL on a strain-specific basis, as determined against *Listeria monocytogenes* ATCC15313. Moreover, all tested strains in this study were still active after treatment with proteolytic enzymes, even with reduced inhibition zones compared to the controls, pointing to additional antimicrobial activity possibly related to a non-proteinaceous molecular structure. Most probably these strains may express surfactin as an additional factor in their complex antimicrobial activity. *B. amyloliquefaciens* ST09 and *B. velezensis* ST03 and ST32 were characterized as positive for β-hemolysis. *B. subtilis* ST08 was shown to be positive for *hbl*C and *nhe*C and *B. amyloliquefaciens* ST109 for *nhe*B. *B. amyloliquefaciens* ST109 generated positive results for gelatinase activity. The ability of the studied *Bacillus* strains to metabolize different carbohydrate sources was done based on the API50CHB test, while the enzyme production profile was recorded by the APIZym kit. All studied strains were positive producers for biogenic amines production. Studied *Bacillus* spp. strains were resistant to some of the evaluated antibiotics, tested according to recommendations of CLSI and EFSA.

Keywords: *Bacillus velezensis*; *Bacillus amyloliquefaciens*; *Bacillus subtilis*; *Listeria monocytogenes*; bacteriocins; safety; antibiotic resistance

Citation: Irorita Fugaban, J.I.; Vazquez Bucheli, J.E.; Holzapfel, W.H.; Todorov, S.D. Bacteriocinogenic *Bacillus* spp. Isolated from Korean Fermented Cabbage (*Kimchi*)—Beneficial or Hazardous? *Fermentation* **2021**, *7*, 56. https://doi.org/10.3390/fermentation7020056

Academic Editor: Hiroshi Kitagaki

Received: 19 March 2021
Accepted: 4 April 2021
Published: 7 April 2021

1. Introduction

Evaluation of microbial diversity in different traditional fermented food products may open ways for isolating beneficial microbial strains. Probiotic cultures, bacteriocinogenic strains, producers of enzymes, exopolysaccharides, and other bioactive metabolites were obtained from fermented food products of all continents and were explored by biotechnical, pharmaceutical, and food industries [1,2]. Asian countries are considered as a promising source of different beneficial microbial strains. This huge potential is suggested by several remote regions on this continent and the fact that local groups have been producing typical fermented food products according to conventional artisanal practices and without commercial standardized starter cultures [3]. Moreover, the extent of biological diversity can be considered as an indicator for the health of the planet.

Kimchi is perhaps the most well-known among several traditional fermented Korean food products, and it forms part of the heritage of Asian culture, especially of the Korean peninsula. According to Encyclopedia Britannica there are hundreds of varieties of *kimchi* based on the methods of preparation and the specific raw materials used including vegetables such as cabbage, Korean radish, leak; and supplemented with a varying selection of seasonings including salts, chili powder, spring onions, garlic, fermented seafood (*jeotkal*), and ginger [4,5]. However, due to lack of standardization, preparation of traditional fermented food products can be hiding unexpected problems related to their safety and quality. In general, good manufacturing practices are an assurance for the safety of the products. Food fermentation processes were developed and established over centuries. Fermentation processes of food preparation normally involve participation of microorganisms that can produce different arsenals of antimicrobial metabolites, including organic acids, hydrogen peroxide, carbon dioxide, antibiotics, antimicrobial proteins, bioactive peptides, and low molecular antimicrobial organic molecules [6,7]. Changes in water activity, redox potential, and pH are some of the key factors involved in food safety [8] and stability, and can change in a positive way during fermentation of *kimchi*. Food fermentation can be considered as a dynamic process in which implicit factors related to microbial interactions (competition, symbiosis, and synergism) play a key role. However, food safety assurance not only implies the absence of pathogenic or food spoilage microorganisms, but also of toxic or detrimental metabolites or residues. Some microbial cultures can be producers of biogenic amines by amino acid decarboxylase activity [9]. Potentially pathogenic strains can be characterized by either hemolytic [10] and/or gelatinase or mucus degrading activities [11]. These characteristics can serve as indicator(s) of the potential virulence of a specific strain. Some virulent characteristics of bacterial cultures maybe expressed only under specific conditions, e.g., at a typical body temperature of 37 °C, and even escape detection during fermentation processes normally occurring at room temperature [12].

Representative strains of *Bacillus* spp. are found to be biotechnological application as producers of a variety of bioactive metabolites, including antimicrobial proteins, antibiotics, enzymes, exopolysaccharides, vitamins, and surfactants [13–17]. Industrial exploring of particular *Bacillus* strains is widely practiced [18,19] and has been part of the preparation of different fermented food products including *kimchi* [20]. Even when the beneficial potential of different *Bacillus* strains has been widely proven, some representatives of this genus have also been clearly identified as human and animal pathogens, including *Bacillus anthracis* [21] and opportunistic pathogens such as some strains of *Bacillus cereus* [22]. On the beneficial side, different strains of the species *B. clausii*, *B. coagulans*, *B. licheniformis*, *B. polyfermenticus*, *B. pumilus*, and *B. subtilis* have been widely commercialized as probiotics and their efficacy proven [23–25]. In Italy, a *B. clausii* spore suspension has been available since 1958 as part of endospore-based probiotics for human consumption [26].

In our study, we attempted to isolate producers lactic acid bacteria (LAB) and *Bacillus* spp.) of antimicrobial metabolites from *kimchi* of locally produced fermented cabbage (Pohang region, Republic of Korea). However, based on the predominate yield of *Bacillus* spp. isolated in preliminary screening, the present work was focused on obtained bacilli. In addition to the antibacterial properties of the expressed antimicrobial(s) and their specific potential application in the control of food borne pathogen organisms, represented by *Listeria monocytogenes*, a well-known, highly contagious pathogen able to survive at low pH and refrigeration temperatures, an additional goal has been to collect more information about the safety of *kimchi*.

2. Materials and Methods

2.1. Isolation of Antagonistic Strains and Evaluation of Antimicrobial Activity

Kimchi, a traditional Korean fermented cabbage, was evaluated as a source for isolation of bacteriocinogenic strains. *Kimchi* samples were obtained from artisanal sources prepared by local producers (Pohang, Republic of Korea). Each sample of around 50 g was collected aseptically and according to good microbiological practices, and investigated for the presence of bacterial strains with the potential ability to produce antimicrobial metabolites. Twenty-five grams of each *kimchi* sample were homogenized with 9 volumes of sterile saline solution (0.85% NaCl, *m*/*v*) in sterile sample bags (Whirl-Pak stomacher sampling bag, Neogen, Lansing, MI, USA) and homogenized in a stomacher (Stomacher 400 Circulator, Worthing, West Sussex, UK) for 3–5 min. After 10x serial dilution, 100 μL were surface-plated in triplicates on de Man, Rogosa and Sharpe (MRS) (Difco, Detroit, MI, USA) supplemented with 2% agar (Difco). An extra layer of 2% agar was added to each plate and then incubated for 48 h at 37 °C. Plates were covered with an additional layer of Brain Heart Infusion (BHI) (Difco) supplemented with 1% agar and *L. monocytogenes* ATCC15313 at a final concentration of around 10^5 CFU/mL and incubated for an additional 24 h at 37 °C. Colonies surrounded by inhibition zones larger than 5 mm in diameter were selected, isolated, and cultured in MRS broth for 24 h at 37 °C. Bacterial cultures were evaluated based on recommendations for good microbiological practices for purity, macroscopic observation of colonies, microscopic observation, and Gram-staining. The production of oxidase and catalase enzymes was determined in addition to tests recommended by Bergey's Manual of Systematic Bacteriology of Archaea and Bacteria [27]. All pure bacterial cultures were stored at −80 °C in presence of 20% glycerol. All additional bacterial cultures applied in this study and microorganisms part of the spectrum of activity evaluation experiments (Table 1) were grown under their respective recommended growth conditions (media and temperatures) and stored at −80 °C in the presence of 20% glycerol as the cryoprotector. Before each experiment, all cultures were propagated at least 2 times.

Table 1. Spectrum of activity of supernatants from *B. velezensis* ST03, *B. amyloliquefaciens* ST06, *B. subtilis* ST08, *B. velezensis* ST32, and *B. amyloliquefaciens* ST109 against selected bacteria from different culture collections.

	Activity of Bacteriocins Produced by				
	ST03	**ST06**	**ST08**	**ST32**	**ST109**
***Bacillus* spp.**	**0/12 ***	**2/12**	**1/12**	**1/12**	**2/12**
Bacillus amyloliquefaciens	0/2	0/2	0/2	0/2	0/2
Bacillus velezensis	0/1	0/1	0/1	0/1	0/1
Bacillus subtilis	0/4	2/4	1/4	0/4	2/4
Bacillus cereus	0/1	0/1	0/1	0/1	0/1
Bacillus pumilus	0/1	0/1	0/1	0/1	0/1
Bacillus sonorensis	0/1	0/1	0/1	0/1	0/1
Bacillus megaterium	0/1	0/1	0/1	0/1	0/1
Bacillus firmus	0/1	0/1	0/1	1/1	0/1
***Enterococcus* spp.**	**0/51**	**1/51**	**8/51**	**3/51**	**5/51**
Enterococcus faecium	0/12	1/12	2/12	1/12	1/12
Enterococcus avium	0/3	0/3	0/3	0/3	1/3
Enterococcus faecalis	0/2	0/2	0/2	0/2	0/2
Enterococcus thailandicus	0/1	0/1	1/1	0/1	0/1
Enterococcus durans	0/1	0/1	0/1	0/1	0/1
Enterococcus gallinarum	0/1	0/1	1/1	1/1	0/1
VRE (vancomycin resistant enterococci)	0/31	0/31	4/31	1/31	3/31
***Lactobacillus* spp.**	**0/50**	**0/50**	**0/50**	**6/50**	**0/50**
Lactobacillus coryniformis	0/1	0/1	0/1	0/1	0/1
Lactobacillus plantarum	0/9	0/9	0/9	0/9	0/9
Lactobacillus brevis	0/3	0/3	0/3	2/3	0/3
Lactobacillus curvatus	0/2	0/2	0/2	2/2	0/2
Lactobacillus fermentum	0/3	0/3	0/3	0/3	0/3
Lactobacillus paracasei	0/5	0/5	0/5	1/5	0/5
Lactobacillus plantarum	0/9	0/9	0/9	1/9	0/9
Lactobacillus reuteri	0/2	0/2	0/2	0/2	0/2
Lactobacillus rhamnosus	0/5	0/5	0/5	0/5	0/5
Lactobacillus sakei	0/4	0/4	0/4	0/4	0/4
Lactobacillus salivarius	0/3	0/3	0/3	0/3	0/3
Lactobacillus sanikiri	0/4	0/4	0/4	0/4	0/4
***Lactococcus* spp.**	**0/2**	**0/2**	**0/2**	**0/2**	**0/2**
Lactococcus lactis	0/2	0/2	0/2	0/2	0/2
***Leuconostoc* spp.**	**0/8**	**0/8**	**0/8**	**2/8**	**0/8**
Leuconostoc citreum	0/1	0/1	0/1	0/1	0/1
Leuconostoc mesenteroides	0/7	0/7	0/7	2/7	0/7
***Listeria* spp.**	**5/5**	**5/5**	**5/5**	**5/5**	**5/5**
Listeria innocua	2/2	2/2	2/2	2/2	2/2
Listeria monocytogenes	3/3	3/3	3/3	3/3	3/3

Table 1. *Cont.*

		Activity of Bacteriocins Produced by				
		ST03	ST06	ST08	ST32	ST109
***Pediococcus* spp.**		**0/9**	**0/9**	**2/9**	**0/9**	**0/9**
	Pediococcus acidilactici	0/3	0/3	0/3	0/3	0/3
	Pediococcus lolii	0/1	0/1	0/1	0/1	0/1
	Pediococcus pentosaceus	0/5	0/5	2/5	0/5	0/5
***Weissella* spp.**		**0/1**	**0/1**	**0/1**	**0/1**	**0/1**
	Weissella cibaria	0/1	0/1	0/1	0/1	0/1
Total tested strains		**5/138**	**11/138**	**16/138**	**16/138**	**12/138**

* The number of strains sensitive to the tested antimicrobial(s) is shown against the total number of tested strains. Data on each genus are summarized in bold; representatives (species) are shown below each genus.

2.2. Differentiation and Identification of Selected Isolates

Bacterial isolates of interest with confirmed production of antimicrobial metabolites, were subjected to DNA fingerprinting through Random Amplification of Polymorphic DNA (RAPD)-PCR and subsequently identified on basis of tests recommended by Bergey's manual [27] and by 16S rRNA sequencing. Bacterial cultures were grown in 50 mL of MRS broth for 24 h at 37 °C, and for future isolation of DNA, cells were collected by centrifugation (8000× *g*, 10 min, 4 °C). Total DNA was obtained using the ZR Fungal/Bacterial DNA kit (Zymo Research, Irvine, CA, USA) in accordance with the manufacturer's protocol and quantified by SPECTROstar Nano nanodrop (BMG LABTECH, Ortenberg, Germany). Differentiation between the evaluated cultures was based on RAPD-PCR performed with primers OPL-01 (5′-GGC ATG ACC T-3′), OPL-09 (5′-TGC GAG AGT C-3′), and OPL-11 (5′-ACG ATG AGC C-3′) according to de Moraes et al. [28] on a Veriti 96 well Thermal Cycler, Applied Biosystems (Thermo Scientific, Waltham, MA, USA). Specificity of the obtained amplicons was evaluated via gel electrophoresis on 1.0% (*w*/*v*) agarose gels for 1 h in the presence of 0.02 μL/mL of SYBR®Safe (Thermo Fisher) in 1x TAE buffer at 100 V (GH-200 Genera Biosystems, Victoria, Australia; Elite 300 Plus Power Supply, Wealtec Bioscience Co., Ltd., Taiwan) and visualized using Omega Lum™ G gel documenter (Aplegen, Inc., Pleasanton, CA, USA). The RAPD profiles generated from the isolates were analyzed and specific profiles and their representative strains were selected, and subjected to identification based on 16S rRNA [28] on Veriti 96 well Thermal Cycler, Applied Biosystems, and further sequenced at the SolGent Analysis Services (Solgent Co. Ltd., Daejeon, Republic of Korea). The generated sequences were analyzed using the Basic Local Alignment Search Tool (BLAST, GenBank, National Center for Biotechnology Information, Bethesda, MD, USA) for identification.

2.3. Sugar Fermentation Profile (API50CHB) and Production of Enzymes (APIZym)

Sugar fermentation abilities of the evaluated strains were studied by culturing in API50CHB strips in accordance with the manufacturer's recommendations (bioMérieux, Marcy l'Etoile, France) at 37 °C for 24, 48, and 72 h. In a separate experiment, the enzymatic activity profiles of studied strains, previously grown in MRS for 24 h at 37 °C were determined using APIZYM strips (bioMérieux), in accordance with the manufacturer's instructions and incubated at 37 °C for 4 h. The evaluation of the enzymatic activity was based on changes in the intensity of coloration in accordance with the manufacturer's instructions (bioMérieux).

2.4. Detection of Bacteriocin Genes and Other Antimicrobial Genes

DNA from the selected strain, isolated as described before, was scanned for presence of different genes related to previously described antimicrobials, including Class IIa bacteriocins (Oli48/105: 5′- TAY GGI AAY GGI GTI TAY TG -3′ and 5′- CYT CDA TNG CRT TRT C -3′); lichenicidin (bli: 5′- GGA AAT GAT TCT TTC ATG G -3′ and 5′- TTA GTT ACA GCT TGG CAT G-3′), thuricin (thu: 5′- GTA GGT CAA ATG GAA ACA C-3′ and 5′- TTA ACT TGC AGT ACT AGC TC-3′), iturin (ituc: 5′- GGC TGC TGC AGA TGC TTT AT-3′

and 5′- TCG CAG ATA ATC GCA GTG AG-3′), surfactin (srfa: 5′- TCG GGA CAG GAA GAC ATC AT-3′ and 5′- CCT CTC AAA CGG ATA ATC CTG A-3′), coagulin (coa: 5′- GGT GGT AAA TAC TAC GGT AAT GGG GT-3′ and 5′- GTG TCT AAA TTA CTG GTT GAT TCG T-3′), subtilosin (sbo: 5′- GGT TGT GCA ACA TGC TCG AT-3′ and 5′- CTC AGG AAG CTG GTG AAC TC-3′), and pediocin PA-1(ped: 5′- CAA GAT CGT TAA CCA GTT T-3′ and 5′- CCGTTG TTC CCA TAG TCT AA-3′) [29,30] on Veriti 96-well Thermal Cycler, Applied Biosystems. The generated amplicons were visualized on gel electrophoresis on 1.0–2.0% (*w*/*v*) agarose gels in 1x TAE buffer at 100 V for 1 h in the presence of 0.02 μL/mL of SYBR®Safe (Thermo Fisher) (GH-200 Genera Biosystems/Elite 300 Plus Power Supply) and visualized using OmegaLum™ G gel documenter (Aplegen).

2.5. *Determination of Antimicrobial Activity and Nature of Antimicrobial Agent(s)*

The selected and previously identified producers of antimicrobial metabolites were grown in MRS at 37 °C for 24 h. The pH of the cell-free supernatant (CFS), obtained by centrifugation (8000× *g*, 10 min, 20 °C), was adjusted to pH 5.5–6.0 with sterile 1 M of NaOH, subjected to 80 °C for 10 min, and filtered via 0.20-μm syringe filters (Sartorius Ministart Syringe Filter, Göttingen, Germany). Antimicrobial properties of the obtained CFS were tested for activity against *L. monocytogenes* ATCC15313 formerly used in the initial screening process for selection of producers of antimicrobials and, in addition, against selected test strains from the culture collections of HEM Inc. (Holzapfel Effective Microbes Inc., Pohang, Republic of Korea), KCTC (Korean Culture Type Collection, Jeongeup, Republic of Korea), KACC (Korean Agricultural Culture Collection, Jeollabuk-do, Republic of Korea), and ATCC (American Type Culture Collection, Manassas, VA, USA), listed in Table 1. These strains were incorporated in the recommended growth media supplemented with 1% agar at a final concentration of 10^5 CFU/mL, and 10 μL of previously prepared CFS were deposited on the surface. Plates were kept for 1 h at room temperature in order to facilitate diffusion of the tested CFS and then incubated at temperatures recommended for the test microorganisms according to dos Santos et al. [31]. Diameters of the inhibition zones greater than 2 mm were considered as positive evidence for potential production of antimicrobial metabolites, including bacteriocins. For evaluating the nature of the inhibitory agent, possible inhibition due to organic acids and H_2O_2 was excluded based on the specificity of the pretreated CFS. All experiments were performed in triplicate on at least two independent occasions.

For evaluation of the titer of produced extracellular antimicrobial metabolites, the approach recommended by Todorov et al. [32] was applied. CFS obtained as described before was serially diluted two-fold in sterile 100-mM phosphate buffer (K_2HPO_4/KH_2PO_4) with a pH of 6.5, and 10 μL of each dilution were tested against *L. monocytogenes* ATCC15313 as previously described, with inhibition zones of at least 2 mm in diameter considered as positive. Levels of antimicrobial activity were expressed as arbitrary units per milliliter (AU/mL), taking into consideration a value corresponding to the reciprocal of the highest dilution that presented inhibitory halos from each experiment [32].

The proteinaceous nature of the produced extracellular antimicrobial compounds in the CFS was analyzed according to recommendations of dos Santos et al. [31]. CFS samples were subjected to treatment with different proteolytic enzymes (proteinase K, pronase, pepsin, trypsin, and α-chymotrypsin for confirmation of proteinaceous nature), α-amylase (for evaluation of the potential presence of carbohydrate moieties, which are part of the active molecular complex), and catalase (for excluding possible H_2O_2 production) (all enzymes from Sigma–Aldrich, St. Louis, MO, USA) at a final concentration of 0.1 mg/mL at 37 °C for 2 h. To stop the enzymatic reactions, set-ups were heat treated (95 °C for 3 min) and then evaluated for evidence of active antimicrobial proteins, as described above [31]. Antimicrobial activity was determined against *L. monocytogenes* ATCC15313 as the test microorganism. In addition, CFS of the examined strain not treated with enzymes, and pure enzymes were evaluated as controls.

2.6. Stability of the Antimicrobial Substance(s)—pH, Temperature, and Chemicals

Antimicrobial metabolites produced by the studied strain were evaluated for its activity in the presence of some chemicals, and at different temperatures and pH values [31]. Previously prepared CFS was mixed with 10 mg/mL of Tween 20, Tween 80, NaCl, SDS, EDTA (all provided from Sigma–Aldrich), milk (Difco), and glycerol (Junsei Chemical Co., Ltd., Nihonbashi-honcho, Chuo-ku, Tokyo, Japan) to a final volume of 2 mL and incubated for 60 and 120 min at 37 °C. Before testing for antimicrobial activity, the pH of all experimental set-ups was adjusted to 5.5–6.5 if needed. In the experiment for evaluation of the effect of pH, CFS of the selected strains was prepared as described previously, and adjusted to pH values of 2.0, 4.0, 6.0, 8.0, and 10.0, respectively (with 1 M of HCl or 1 M of NaOH), and incubated for 60 and 120 min at 37 °C. Effect of temperature on antimicrobial metabolites stability was evaluated by incubating previously prepared CFS of the evaluated strain (with adjusted pH 5.5–6.5) at 30, 37, 45, 60, 80, and 100 °C, respectively, for 60 and 120 min, and by autoclaving at 121 °C for 20 min. Activity of all set-ups were tested against *L. monocytogenes* ATCC15313, a sensitive indicator strain. Untreated CFSs of the examined strains at the examined temperatures, pH, or chemicals and solutions of applied chemicals were evaluated as controls.

2.7. Screening of Virulence Genes

In these experiments, genes, parts of the expression of hemolysin BL, and non-hemolytic enterotoxin were evaluated by PCR reactions [33], performed on the Veriti 96-well Thermal Cycler, Applied Biosystems. The selected strains were tested for the presence of virulence genes encoding hemolysin BL (hblA: 5′- AAG CAA TGG AAT ACA ATG GG -3′and 5′- AGA ATC TAA ATC ATG CCA CTG C -3′; hblB: 5′- AAG CAA TGG AAT ACA ATG GG -3′ and 5′- AAT ATG TCC CAG TAC ACC CG -3′; hblC: 5′- GAT ACY AAT GTG GCA ACT GC -3′ and 5′- TTG AGA CTG CTC GYT AGT TG -3′) and nonhemolytic enterotoxin (nheA: 5′- GTG AGG ATC ACA ATC ACC GC -3′ and 5′- ACG AAT GTA ATT TGA GTC GTC GC -3′; nheB: 5′- TTT AGT GGA TCT GTA CGC -3′ and 5′- TTA ATG TTC GTT AAT CCT GC -3′; nhe C5′- TGG ATT CCA AGA TGT AAC G -3′and 5′- ATT ACG ACT TCT GCT TGT GC -3′), in accordance with the PCR protocols of Guinebretière et al. [33]. The generated amplicons were separated via gel electrophoresis on 1.0–2.0% (*w*/*v*) agarose gels in 1x TAE buffer at 100 V for 1 h in presence of 0.02 µL/mL of SYBR®Safe (Thermo Fisher) as described before.

2.8. Virulence Activity—Physiological Approach

The evaluated strains were investigated by selected phenotypic tests to identify their hemolytic activity and gelatinase production according to Colombo et al. [11]. All tests were performed in at least three independent replicates.

The representatives of the evaluated culture were streaked on the surface of trypticase soy agar (TSA) supplemented with 5% (*v*/*v*) defibrinated sheep blood (Synergy Innovation, Seongnam-si, Republic of Korea). Plates were incubated for 24 h at 37 °C and the hemolytic activity of each isolate was evaluated as total or β-hemolysis (clear halos around the colonies), partial or α-hemolysis (greenish halos around the colonies), or γ-hemolysis (absence of hemolysis). *B. cereus* ATCC27348, *Escherichia coli* ATCC25922, and *Lactobacillus plantarum* ATCC14197 were applied as positive and negative controls, respectively.

For the evaluation for gelatinase activity of the investigated strains, the approach proposed by Colombo et al. [11] was followed. For the purpose of the experiment, 1-µL aliquots of the evaluated cultures were spotted on the surface of Luria–Bertani (LB) agar (BD, Franklin Lakes, NJ, USA) supplemented with 3% (*w*/*v*) gelatin (BD). Plates were incubated at 37 °C for 48 h, followed by maintaining the plates at 4 °C for 4 h. Evidence for gelatin hydrolysis was recorded based on the formation of opaque halos around the growing colonies. *E. coli* ATCC25922 served as a positive control, while *Lb. plantarum* ATCC14197 and *E. faecalis* HEM200 served as negative controls.

The potential to produce biogenic amines from the evaluated strains was performed according to Bover–Cid and Holzapfel [9]. Evaluated cultures were grown at least five times consecutively to MRS broth supplemented with 0.1% (*w*/*v*) amino acid precursors for the production of biogenic amines. The biogenic amine precursors were added individually and comprised lysine, tyrosine, ornithine, and histidine (Sigma–Aldrich) and 0.005% (*w*/*v*) pyridoxal-5-phosphate (Sigma–Aldrich). The incubation was performed for 24 h at 37 °C, and the last transfer performed was streaked in duplicate on a modified MRS agar according to Bover–Cid and Holzapfel [9] which was supplemented with one of each biogenic amine precursor described as above (1% *w*/*v*). All plates were incubated for 4 days at 37 °C and the change of the color from yellow to purple was considered as a positive result for the production of a biogenic amine. *E. coli* ATCC25922 and *Lb. plantarum* ATCC14197 served as positive and negative controls, respectively.

2.9. Antimicrobial Susceptibility Profiling

The antimicrobial susceptibility testing for the *Bacillus* strains was performed according to recommendations by the Clinical and Laboratory Standards Institute (CLSI) on the Performance Standards for Antimicrobial Susceptibility Testing M45 for *Bacillus* spp. Using the broth micro-dilution method, minimum inhibitory concentrations (MIC) were determined for the antibiotics ampicillin, gentamycin, kanamycin, streptomycin, erythromycin, clindamycin, tetracycline, chloramphenicol (all from Sigma–Aldrich), vancomycin (Cheil Jedang Pharma Co., Republic of Korea), and ciprofloxacin (Sigma–Aldrich), which is commonly applied as an antibiotic agent in treatment of *Bacillus* infections (https://clsi.org/, accessed on 25 March 2020).

Cation-adjusted Mueller–Hinton broth, additionally supplemented with 5.0 g/L (*v*/*v*) MRS was distributed in a sterile 96-well micro-dilution plate (SPL Life Sciences, Pocheon-si, Gyeongi-do, Republic of Korea). Fifty microliters in different concentrations of each antibiotic, previously prepared to cover expected ranges of MIC values, were added to 50 µL of a bacterial suspension, adjusted to approximately 10^5 CFU/mL based on the McFarland scale. The plates were incubated for 24 h at 35 °C ± 1, according to recommendations of EUCAST [34]. The MIC values, defined as the lowest concentration that completely inhibits bacterial growth, were read in a spectrophotometer and compared to the cut-off set of EUCAST. Each antibiotic was assessed in at least two individual experiments.

2.10. Detection of Vancomycin Resistant Genes

The DNA previously isolated from the evaluated strains was subjected to PCR analysis for investigating the presence of vancomycin resistance genes *van*A (vanAB/F: 5′ –GTA GGC TGC GAT ATT CAA AGC - 3′; vanA/R: 5′ – CGA TTC AAT TGC GTA GTC CAA - 3′), *van*B (vanAB/F: 5′ – GTA GGC TGC GAT ATT CAA AGC - 3′; vanB/R: 5′ – GCC GAC AAT CAA ATC CTC - 3′), *van*C (vanC/F: 5′ – ATC CAA GCT ATT GAC CCG CT - 3′; vanC/R: 5′ –TGT GGC AGG ATC GTT TTC AT - 3′), *van*D (vanD/F: 5′ – TGT GGG ATG CGA TAT TCA A - 3′; vanD/R: 5′ – TGC AGC CAA GTA TCC GGT AA - 3′), *van*E (vanE/F: 5′ – TGT GGT ATC GGA GCT GCA G - 3′; vanE/R: 5′ – GTC GAT TCT CGC TAA TCC - 3′) and *van*G (vanG/F: 5′ – GAA GAT GGT ACT TTG CAG GGC A - 3′; vanG/R: 5′ – AGC CGC TTC TTG TAT CCG TTT T - 3′) previously reported by Valledor et al. [35]. PCR products were separated on 2.0% (*w*/*v*) agarose gels in 1x TBE and visualized as described before.

3. Results

Based on the application of the triple layer approach, applying MRS supplemented with 2% agar as principal growth medium resulted in predominant growth of colonies with morphology typical for representatives from genus *Bacillus*. Preliminary screening of bacterial strains with potential antimicrobial properties against *L. monocytogenes* ATCC15314 resulted in selection of 21 isolates. Based on preliminary observations, all isolates were catalase positive, showed specific *Bacillus* colony morphology on solid media, and Gram-positive, and thereby pre-identified as *Bacillus* spp. Additional antimicrobial activity tests confirmed that 12 isolates were producers of antimicrobial compounds based on activity against *L. monocytogenes* ATCC153131.

3.1. Differentiation and Identification

Twelve isolates with confirmed production of antimicrobial metabolites were differentiated based on fingerprinting generated by RAPD-PCR and 5 distinct clusters were identified (Figure 1). Biochemical, morphological, and physiological results were taken into account in the selection of 5 representatives out of 12 isolates for 16S rRNA sequencing and specific identification of these strains. Representatives from each cluster, ST03, ST06, ST08, ST32, and ST109, were identified to belong to *Bacillus velezensis*, *Bacillus amyloliquefaciens*, and *Bacillus subtilis*; and named *Bacillus velezensis* ST03, *Bacillus velezensis* ST32, *Bacillus amyloliquefaciens* ST06, *Bacillus amyloliquefaciens* ST109, and *Bacillus subtilis* ST08.

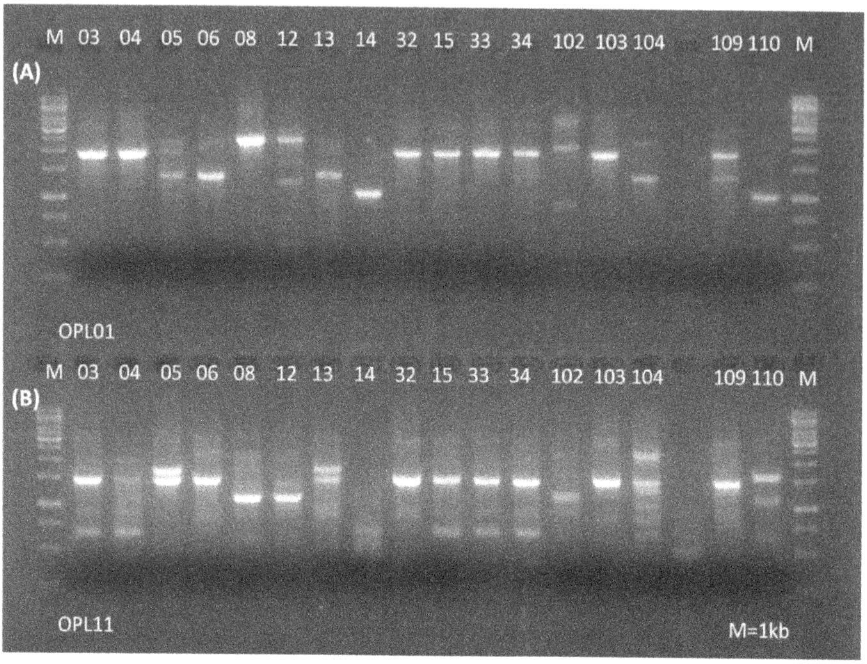

Figure 1. RAPD-PCR products generated with (**A**) primer OPL01 and (**B**) primer OPL11 for various isolates differentiated in this study; lane M: O'GeneRuler 1 kb DNA Ladder (Fermentas).

3.2. Sugar Fermentation Profile (API50CHB) and Enzymes Production (APIZym)

The ability of the studied *Bacillus* strains (ST03, ST06, ST08, ST32, and ST109) to metabolize different carbohydrates sources was evaluated based on the API50CHB test, and all tested strains were able to ferment glycerol, L-arabinose, ribose, D-xylose, D-glucose, D-fructose, D-mannose, inositol, mannitol, sorbitol, α-methyl-D-glucosamine, amygdalin, arbutin, esculin, salicin, cellobiose, maltose, saccharose, trehalose, D-raffinose, amidon, and glycogen (Table 2). In addition, only *B. amyloliquefaciens* ST109 fermented L-xylose and α-methyl-D-mannoside; *B. amyloliquefaciens* ST06 and ST109 galactose and β-gentiobiose, *B. velezensis* ST03 and ST32 and *B. amyloliquefaciens* ST06 and ST109 lactose, *B. amyloliquefaciens* ST06 and ST109 and *B. subtilis* ST08 melibiose and D-turanose, and *B. subtilis* ST08 inulin (Table 2).

Table 2. Carbohydrate fermentation profile recorded for *B. velezensis* ST03, *B. amyloliquefaciens* ST06, *B. subtilis* ST08, *B. velezensis* ST32, and *B. amyloliquefaciens* ST109 based on API50CHB (BioMerieux).

Carbohydrates	ST03	ST06	ST08	ST32	ST109
control (no carbohydrate)	−	−	−	−	−
glycerol, L-arabinose, ribose, D-xylose, D-glucose, D-fructose, D mannose, inositol, mannitol, sorbitol, α-methyl-D-glucosamine, amygdalin, arbutin, esculin, salicin, cellobiose, maltose, saccharose, trehalose, D-raffinose, amidon, glycogen	+	+	+	+	+
L-xylose, α-methyl-D-mannoside	−	−	−	−	+
galactose, β-gentiobiose	−	+	−	−	+
lactose	+	+	−	+	+
melibiose, D-turanose	−	+	+	−	+
inulin	−	−	+	−	−
erythritol, D-arabinose, adonitol, β-methyl-xyloside, L-sorbose, rhamnose, dulcitol, N-acetylglucosamine, melezitose, xylitol, D-lyxose, D-tagatose, D-fucose, L-fucose, D-arabitol, L-arabitol, gluconate, 2-keto-gluconate, 5-keto-gluconate	−	−	−	−	−

+ Bacterial growth, evidence for fermenting the evaluated carbohydrate; − no bacterial growth.

Expression of different enzymes (alkaline phosphatase, esterase (C4), lipase esterase (C8), acidphosphatase, naphthol phosphohydrolase, α-galactosidase, α-glucosidase, and β-glucosidase) were recorded for the studied strains (ST03, ST06, ST08, ST32, and ST109) (Table 3).

Table 3. Enzyme production profile recorded for *B. velezensis* ST03, *B. amyloliquefaciens* ST06, *B. subtilis* ST08, *B. velezensis* ST32, and *B. amyloliquefaciens* ST109 based on APIZym (BioMerieux).

Target Enzymes	ST03	ST06	ST08	ST32	ST109
Control	0 *	0	0	0	0
alkaline phosphatase	5	3	3	3	3
esterase (C4)	4	4	4	5	3
lipase esterase (C8)	5	4	4	5	3
acid phosphatase	0	0	1	0	3
naphthol phosphohydrolase	3	3	3	3	3
α-galactosidase	0	0	1	0	0
α-glucosidase	3	3	5	3	3
β-glucosidase	1	3	4	4	4
lipase (C14), leucinearylamidase, valinearylamidase, cystinarylamidase, trypsin, α-chymotrypsin, β-galactosidase, β-glucuronidase, N-acetyl-β-glucosaminidase, α-mannosidase, α-fucosidase	0	0	0	0	0

* 3–5: Strong enzymatic activity, 1–2: Weak enzymatic activity. 0: No evidence for enzymatic activity (interpretation was performed in accordance with the manual provided by BioMerieux).

3.3. Detection of Bacteriocin and Other Antimicrobial Genes

DNA samples obtained from the studied strains (ST03, ST06, ST08, ST32, and ST109) generated positive results for the presence of lichenicidin in *B. subtilis* ST08, iturin in *B. velezensis* ST03, *B. amyloliquefaciens* ST06 and *B. amyloliquefaciens* ST109, subtilosin in *B. velezensis* ST03, *B subtilis* ST08 and *B. amyloliquefaciens* ST109, and surfactin in all studied bacillus strains. No evidence based on the PCR reactions wasrecorded for the presence of Class IIa bacteriocins thuricin, coagulin, orpediocin PA-1 (Table 4).

Preliminary results evaluating bacterial growth and production of bacteriocins suggested that Luria–Bertani (LB) and nutrition broth can support bacterial growth; however, bacteriocin production was recorded only at 200 AU/mL or lower for the studied strains. Higher levels of expressed bacteriocins, between 800 and 1600 AU/mL, were recorded when strains were cultured in MRS broth. Based on these observations, MRS was selected as the growth medium for evaluation of bacteriocinogenic properties of *B. velezensis* ST03, ST32, *B. amyloliquefaciens* ST06 and ST109, and *B. subtilis* ST08. The selected strains were found to produce antimicrobial metabolites at an activity level of 800 AU/mL for *B. velezensis* ST03, *B. amyloliquefaciens* ST06, and *B. velezensis* ST32, and 1600 AU/mL for *B. subtilis* ST08 and *B. amyloliquefaciens* ST109, as determined against *L. monocytogenes* ATCC15313. Moreover, the spectrum of activity of these antimicrobial metabolites produced by strains ST03, ST06, ST08, ST32, and ST109 was determined against selected test organisms obtained from different collections (Table 1).

Table 4. Spread of antimicrobials, virulence- and vancomycin-resistant related genes in DNA obtained from *B. velezensis* ST03, *B. amyloliquefaciens* ST06, *B. subtilis* ST08, *B. velezensis* ST32, and *B. amyloliquefaciens* ST109.

	***Bacillus velezensis* ST03**	***Bacillus amyloliquefaciens* ST06**	***Bacillus subtilis* ST08**	***Bacillus velezensis* ST32**	***Bacillus amyloliquefaciens* ST109**
Antimicrobials related genes					
Class IIa bacteriocins	−	−	−	−	−
Lichenicidin	−	−	+	−	−
Thuricin	−	−	−	−	−
Surfactin	+	+	+	+	+
Iturin	+	+	−	−	+
Coagulin	−	−	−	−	−
Subtilosin	+	−	+	−	+
Pediocin PA-1	−	−	−	−	−
Virulence genes					
*hbl*A	−	−	−	−	−
*hbl*B	−	−	−	−	−
*hbl*C	−	−	+	−	−
*nhe*A	−	−	−	−	−
*nhe*B	−	−	−	−	+
*nhe*C	−	−	+	−	−
Vancomycin resistance genes					
*van*A	−	−	−	−	−
*van*B	−	−	−	−	−
*van*C	−	−	−	−	−
*van*D	−	−	−	−	−
*van*E	−	−	−	−	−
*van*G	−	−	−	−	−

+ Detection of targeted gen; − absence of targeted gene.

All tested strains in this study (ST03, ST06, ST08, ST32, and ST109) remained partially active after proteolytic enzyme treatment, albeit with reduced inhibition zones compared to the controls (non-treated supernatants) suggesting that antimicrobial activity may be related to a combination of antimicrobial protein (bacteriocin) in a combination with a non-proteinaceous antimicrobial molecular structure, or partial resistance of bacteriocins (antimicrobial peptide) to the effect of applied specific proteolytic enzymes. PCR analysis of the DNA from strains ST03, ST06, ST08, ST32, and ST109 showed evidence for the presence of a surfactin-related gene in all tested *Bacillus* spp. strains in this study.

3.4. Stability of the Antimicrobial Substances—Effect of pH, Temperature, and Chemicals

Antimicrobials produced by the strains ST03, ST06, ST08, ST32, and ST109 generally presented stability of the activity after being exposed to the effect of different pH levels (pH of 2.0, 4.0, 6.0, 8.0, and 10.0) and temperatures (30, 37, 45, 60, 80, and 100 °C, respectively, up to 120 min, and at 121 °C for 20 min). In addition, their stability was not affected notably in the presence of 10 mg/mL of Tween 20, Tween 80, NaCl, SDS, EDTA, milk, or glycerol, respectively.

3.5. In Vitro Hemolytic Activity and Presence of Genes Related to Hemolysin and Non-Hemolytic Enterotoxin Virulence Factors; Gelatinase and Biogenic Amine Production

Strains ST03, ST06, ST08, ST32, and ST109 generated different results and were evaluated for their hemolysis behavior on blood agar plates. ST03, ST09, and ST32 were tested positive for β-hemolysis. On the other side, results from PCR analysis targeting *hbl*A, *hbl*B, *hbl*C, *nhe*A, *nhe*B, and *nhe*C showed that strains *B. subtilis* ST08 could carry *hbl*C and *nhe*C and *B. amyloliquefaciens* ST109 was positive for *nhe*B (Table 2).

From the tested strains ST03, ST06, ST08, ST32, and ST109, only ST109 generated positive results for gelatinase activity. However, for strains ST03, ST06, ST08, ST32, and ST109 we recorded production of all 4 tested biogenic amines, histamine, tyramine, putrescine, and cadaverine, derived from histidine, tyrosine, ornithine, and lysine, respectively.

3.6. Antibiotic Resistance

Based on PCR analysis, no evidence for the investigated vancomycin resistance genes (*vanA*, *vanB*, *vanC*, *vanD*, *vanE*, and *vanG*) could be recorded in the studied *Bacillus* strains

(Table 2). Based on the *in vitro* test with vancomycin, all tested *Bacillus* strain were susceptible with cut-off values of 0.5 μg/mL, suggesting their susceptibility according to the recommendations of CLSI (≤4 μg/mL) and EFSA (4 μg/mL) (Table 5).

Table 5. Antibiotic susceptibility for *B. velezensis* ST03, *B. amyloliquefaciens* ST06, *B. subtilis* ST08, *B. velezensis* ST32, and *B. amyloliquefaciens* ST109.

Antibiotics	Experimental MIC (μg/mL) Values for Strains					CLSI Cut-Offs (S, I, R) for *Bacillus* spp. (μg/mL)	EFSA Cut-Offs for *Bacillus* spp. (μg/mL)
	ST03	ST06	ST08	ST32	ST109		
ampicillin	128	16	≤0.25	128	64	≤0.25, -, ≥0.5	n.r.
vancomycin	0.5	0.5	0.5	0.5	0.5	≤4, *, *	4
gentamycin	0.5	0.5	2	0.5	0.5	≤4, 8, ≥16	4
kanamycin	2	2	8	4	16	≤1, 2, ≥4	8
streptomycin	8	8	64	8	2	n.r.	8
erythromycin	2	2	2	2	2	≤0.5, 1-2, ≥4	4
clindamycin	0.5	≤0.25	1	0.5	0.5	≤0.5, 1-2, ≥4	4
tetracycline	4	4	4	8	≤0.25	<4, 8, >16	8
chloramphenicol	4	2	4	1	2	≤8, 16, ≥32	8
ciprofloxacin	≤0.25	≤0.25	≤0.25	≤0.25	≤0.25	≤1, 2, ≥4	n.r.

S: Sensitive; I: Intermediate; R: Resistant; * no set values; n.r.: Antibiotic not listed in the recommendations of cited agency.

Results of the performed *in vitro* tests for determination of interaction of the studied *Bacillus* strains and selected antibiotics are summarized in Table 5. All tested *Bacillus* strains (ST03, ST06, ST08, ST32, and ST109) were susceptible to vancomycin, gentamycin, chloramphenicol, and ciprofloxacin according to breakpoint values recommendations of CLSI and EFSA (Table 5). The other tested antibiotics showed different interference with activity of the studied *Bacillus* strains. Only strain ST08 showed susceptibility to ampicillin with MIC value of ≤0.25 μg/mL, while the other *Bacillus* strains were resistant to ampicillin. All strains were resistant to kanamycin according to EFSA (with a cut-off value of 8 μg/mL), however, intermediate for ST03 and ST06 according to CLSI (with cut-off of 2 μg/mL). Based on the EFSA recommendations, only ST109 was susceptible to streptomycin and the other strains were resistant. In addition, according to the recommendations of EFSA, the studied strains were susceptible to clindamycin and tetracycline, but according to CLSI, ST08 was intermediate to clindamycin (with cut-off of 1–2 μg/mL), and ST109 was intermediate to tetracycline (with cut-off of 8 μg/mL). All tested strains were classified as intermediate to the effect of erythromycin (with cut-off of 1–2 μg/mL) (Table 5).

4. Discussion

Preliminary screening of bacterial strains with anti-listerial activity resulted in selection of 21 isolates. Selection of *L. monocytogenes* as principal test organism in screening for producer(s) of antimicrobial metabolites strains was based on relevance of that food-borne pathogen for human and other animals' health. *L. monocytogenes* is highly contagious microorganisms, able to survive in low pH and under refrigeration, and was listed as having "zero tolerance" for different food products and by several countries [6,32,35]. All selected potential bacteriocin producers were pre-identified as *Bacillus* spp. Different *Bacillus* strains are well known as producers of an arsenal of antimicrobial metabolites, including bacteriocins, antibiotics, and enzymes [13–19,36–39]. These features are frequently explored for industrial microbiological applications, and different *Bacillus* strains are used for large scale production of antibiotics and enzymes [13,15,16]. Despite the fact that bacteriocin production by *Bacillus* spp. is well documented [18,19,36–39], industrial production of such bacteriocins has not been explored yet. On the other side, *Bacillus* starter cultures are well known and applied in the production of different fermented food products in different regions of Asia and Africa, including soybean paste [40], *bikalga*, an alkaline fermented food [40], *lanhouin*, a traditional fish-based condiment from West Africa [41], *kimchi* and *chongkukjang* from Korean Peninsula [42], and *natto*, soybean products from Japan [43,44]. Some of these *Bacillus* starter cultures are described as producers of bacteriocin [18,19] and antibiotics [15]. On the other hand, application of antimicrobial metabolites produced by *Bacillus* spp. in the control of spoilage or pathogenic organisms is well explored in the food industry. However, scientific questions are related to the identification and characterization of the produced antimicrobials and their role in the effective control of spoilage organisms and pathogens.

From 21 initially selected potential bacteriocin producers, 12 isolates with confirmed production of antimicrobial metabolites were differentiated based on fingerprinting generated by RAPD-PCR, grouped in 5 distinct clusters (Figure 1) and were identified as *Bacillus velezensis* ST03, *Bacillus velezensis* ST32, *Bacillus amyloliquefaciens* ST06, *Bacillus amyloliquefaciens* ST109, and *Bacillus subtilis* ST08. Strains belonging to *B. velezensis* and *B. subtilis* subsp. *subtilis* were previously isolated from *kimchi* [25,45], even if their role in the production of this fermented cabbage product is still not very well established [25]. *Bacillus* spp. are generally described as soil-related organisms, and their presence in *kimchi* may most probably be the consequence of low hygienic practices in the preparation process. On other side, *Bacillus* spp. were described as important microbial cultures, taking part in the fermentation processes of traditional fermented food products of plant origin [25,40,45,46]. Still, the putative role of *Bacillus* strains in the fermentation process of *kimchi* merits more precise scientific evaluation.

The ability of the studied *Bacillus* strains to metabolize different carbohydrate sources was evaluated based on the API50CHB test (Table 2). These results represent additional relevant information for confirmation differences between the evaluated strains. Furthermore, the carbohydrate fermentation profile can be applied as important information in the design of effective and low-cost growth-production media for the production of bacteriocins in future potential biotechnological applications.

Expression of different enzymes were recorded for the studied strains by APIZym test (Table 3). *Bacillus* spp. are known for their biologically active molecules, including enzymatic production abilities, and this is well explored by biotechnological industry [13–17]. Specific enzymatic activity can be discussed both as a positive and negative feature. Montel et al. [47] pointed out that aldehydes, alcohols, and acids derived from the breakdown of some amino acids (leucine, valine, phenylalanine, and methionine) have a minimal effect on the sensory quality (i.e., taste and smell) of the fermented products, but can be more relevant in the reduction of allergenic characteristics of some proteins, and, related to this, can be considered as beneficial. El Mecherfi et al. [48] suggested the production of proteolytic and lipolytic enzymes is to be beneficial in the reduction of allergenicity in dairy, meat, and even gluten in cereal-based products. Park et al. [49] evaluated the production of lipolytic enzymes of *Bacillus* strains with regard to their potential application in the development of anti-obesity probiotics. George [50] pointed to the fact that β-glucuronidase is a carcinogenic bacterial enzyme that exerts negative effects on the liver and that such enzymatic activity needs to be considered with special attention to the safety evaluation of bacterial cultures.

Based on the bio-molecular approach, studied strains (ST03, ST06, ST08, ST32, and ST109) generated positive results for the potential production of lichenicidin in *B. subtilis* ST08, iturin in *B. velezensis* ST03, *B. amyloliquefaciens* ST06, and *B. amyloliquefaciens* ST109, subtilosin in *B. velezensis* ST03, *B subtilis* ST08, and *B. amyloliquefaciens* ST109, and surfactin in all studied *Bacillus* strains (Table 2). An overview of the literature showed that *Bacillus* spp. can be multi-antimicrobial producers. Yang and Chang [51] reported on two bacteriocins of 2.4 and 4.5 kDa produced by *B. subtilis* MJP1, isolated from *meju*. Wu et al. [52] investigated two bacteriocins produced by *B. subtilis* JM4 isolated from soil in Korea, and found a molecular mass of 1422.71 and 1422.65 Da, respectively, and differences only in the seventh amino acid, as determined after purification, mass spectrometry, and amino-acid sequencing. Salazar et al. [53] reported on two bacteriocins produced by *B. amyloliquefaciens* ELI149 isolated from soil in Mexico.

The most widely studied bacteriocins produced by different representatives of the genus *Bacillus* are subtilin [54,55] and subtilosin A [56,57]. Some reports focused on the application of bacteriocins in the control of food spoilage and pathogenic organisms, with producing strains belonging to *B. subtilis*, *B. cereus*, *B. thuringiensis*, and other *Bacillus* spp. [58,59]. Most bacteriocins produced by *Bacillus* spp. strains were classified as class I (lantibiotics), with some exceptions of a class II pediocin-like bacteriocin, such as coagulin [60].

The ability to produce and express antimicrobial metabolites, including antimicrobial peptides, needs to be evaluated as an adaptive process, giving selective benefits for better survival and colonization opportunity to the producing strain. This adaptive ecological specificity also applies to representative strains of *Bacillus* spp. Generally, as part of the soil–plant ecosystem, *Bacillus* spp. have to compete with several other groups of organisms, and an arsenal of different antimicrobial metabolites can provide competitive benefits to a producing strain, resulting in domination in a mixed culture. However, from a metabolic point, additional research needs to be performed in order to clarify which specific antimicrobials are expressed and/or in what proportions, based on purification and mass spectrometry identification and/or following the specific RNA expression related to appropriate genes involved in the production of reported antimicrobials. Thereby, it should also be considered that environmental conditions may play a regulatory role in the expression of these genes, as well as investigating how these processes are regulated. As suggested by Chikindas et al. [6] investigations should also be directed to information on activation of these genes either by inhibitory metabolites or whether these bacteriocins play an additional role in the life cycle of the producer organism(s).

Studied strains were able to express bacteriocins between 800 and 1600 AU/mL when cultured in MRS broth at 30 °C for 24 h, as determined against *L. monocytogenes* ATCC15313 and against some other test organisms obtained from different collections (Table 1). Most of the studied antimicrobials showed a narrow spectrum of activity, generally inhibiting only a few of the test strains. The specificity in the spectrum of activity is one of the principal differences between these two classes of antimicrobial agents [61,62]. Therefore, these results suggest that the evaluated strains are expressing bacteriocins rather than antibiotics.

The protein nature is a primary characteristic of bacteriocins, and by definition they are considered as polypeptides [6,7,32]. However, while some classes of antibiotics are proteins by nature as well, their final formation is by additional posttranslational modification [61]. In addition, they are characterized as peptides with significantly higher molecular mass compared to other bacteriocins. Taking into consideration that preparation of the cell-free supernatant was also included, heat treatment for 10 min at 80 °C by larger proteins was supposed to be inactivated. Thus, the antimicrobials produced by strains ST03, ST06, ST08, ST32, and ST109 may be considered as typical bacteriocins. However, all tested strains in this study (ST03, ST06, ST08, ST32, and ST109) remained active after proteolytic enzyme treatment, albeit with reduced inhibition zones compared to the controls (non-treated supernatants) pointing that antimicrobial activity may be related to a non-proteinaceous molecular structure, or partial resistance of bacteriocins (antimicrobial peptide) to the effect of applied specific proteolytic enzymes. PCR analysis of the DNA from strains ST03, ST06, ST08, ST32, and ST109 showed evidence for the presence of a surfactin related gene in all tested *Bacillus* spp. strains in this study. Most probably the set strains may express surfactin, a substance that has been previously shown to have antimicrobial activity [29]. Moreover, this is not surprising, since *Bacillus* spp. are known for their ability to produce a variety of bioactive metabolites [13–17].

For scientific and technological information, the stability of antimicrobial compounds at different temperatures, levels of pH, and in the presence of chemicals is a relevant property in the characterization of newly isolated metabolites. Interactions with surfactants and chelating agents can provide information related to the structures of proteins. Moreover, interaction with common chemical additives applied in the food industry can provide information about the relevance of application of specific antimicrobial metabolites in food systems. Thereby, pH and temperature stability are important parameters for the practical application of such antimicrobials as food additives and preservatives, and are important characteristics in the development of antimicrobials for human and veterinary applications. Antimicrobials produced by the strains ST03, ST06, ST08, ST32, and ST109 generally presented stability of the activity after being exposed to the effect of different pH levels (pH of 2.0–10.0) and temperatures (30–100 °C, respectively, up to 120 min, and at 121 °C for 20 min). In addition, their stability was not affected notably in the presence of 10 mg/mL of Tween 20, Tween 80, NaCl, SDS, EDTA, milk, or glycerol, respectively. Stability in the presence of different chemicals, exposure to temperatures, and various levels of pH has been reported for numerous bacteriocins produced by LAB and *Bacillus* spp. [54–57].

Only ST03, ST09, and ST32 tested positive for β-hemolysis. Moreover, results from PCR analysis targeting *hbl*A, *hbl*B, *hbl*C, *nhe*A, *nhe*B, and *nhe*C showed that strains *B. subtilis* ST08 can carry *hbl*C and *nhe*C and *B. amyloliquefaciens* ST109 was positive for *nhe*B (Table 4). However, in order to be considered as positive for expression of hemolysin, a strain needs to be positive for *hbl*A, *hbl*B, and *hbl*C. For expression of non-hemolytic enterotoxin(s), a strain needs to carry *nhe*A, *nhe*B, and *nhe*C genes. Safety evaluation is a priority in the evaluation of beneficial organisms. As part of these tests, hemolytic activity is considered as an important virulence factor and a key in the safety evaluation of newly isolated bacterial cultures towards their characterization as beneficial organisms. Hemolytic activity can play a role in pathogenesis of harmful bacteria and was associated with anemia or edema in the host based on facilitating acquisition of iron to the pathogen [63].

Only ST109 from all five tested strains generated positive results for gelatinase activity. Production of the gelatinase enzyme is considered as a virulence factor in the evaluation of safety of beneficial organisms, being related to the ability to hydrolyze collagens and thereby maybe directly involved in the initiation of inflammatory responses [64]. Barbosa et al. [64] reported in a study evaluating 76 enterococcal cultures that high numbers of them (26%) were gelatinase producers. Franz et al. [65] previously reported similar observations in the evaluation of safety for different beneficial cultures.

Biogenic amines are typically formed as decarboxylation products of amino acids. For strains ST03, ST06, ST08, ST32, and ST109, we have recorded production of all 4 tested biogenic amines, histamine, tyramine, putrescine, and cadaverine, derived from histidine, tyrosine, ornithine, and lysine, respectively. Production of biogenic amines is an important safety concern, and is frequently associated with the presence of food spoilage organisms [66,67], but are also commonly produced during fermentation [68,69]. Uptake of high levels of biogenic amines, and especially of histamine and tyramine, can result in health issues and be associated with severe symptoms such as abdominal pain, diarrhea, headache, hypertension, migraine, and possibly other neurological complications [70]. Characterization of the strains ST03, ST06, ST08, ST32, and ST109 as producers of biogenic amines may compromise their application as beneficial cultures. On the other hand, appropriate detection and quantification of the levels of biogenic amines need to be clearly assessed in the products (*kimchi*) from where these strains were isolated in order to compare levels of the produced biogenic amines in the fermented food products with the international safety standards.

Biogenic amines are considered as toxic compounds and their levels in food products should not exceed specific levels (histamine, 100 mg/kg; tyramine, 100–800 mg/kg; β-phenylethylamine, 30 mg/kg; total biogenic amines, 1000 mg/kg) [71,72]. Some polyamines such as putrescine and cadaverine were reported to potentially have toxic consequences when present in food [71], especially for sensitive persons. The ability of the specific strain to produce biogenic amines is genetically determined; however, it is dependent on the specific growth conditions and presence of precursors for this production. Levels of expression of specific genes by studied strains and levels of accumulation of biogenic amines in the food products need to be evaluated in the future to confirm if this will generate a concern issues related to safety of studied strains.

Protein metabolism and formation of different free amino acids and/or small peptides were intensively studied in different groups of microorganisms and their role as spoilage or beneficial processes were evaluated [73–75]. Atanasova et al. [73] reported on the role of starter cultures in the formation and free amino acids and different bioactive peptides in cheeses fermented with *Lactobacillus delbrueckii* subsp. *bulgaricus* and *Streptococcus thermophilus*; Stribny et al. [74] detailed the evaluated role of *Saccharomyces* spp. in the production of aroma-related higher alcohols and acetate esters in metabolism of amino acid precursors; Perea–Sanz et al. [75] investigated the involvement of *Debaryomyces hansenii* in the production of volatile sulfur compounds from sulfur amino acids and the related metabolic pathway in different strains from food origins. Different bacterial cultures can play an essential role as starters in the formation of organoleptic characteristics of the final products; however, they can be stated as spoilage, when they were leading to formation of unpleasant or toxic metabolites, including different sulfur compounds or biogenic amines.

Some of types of *kimchi* can contain additional ingredients, and high levels of proteins and free amino acids can be recorded, and as consequence of fermentation processes, accumulation of biogenic amines may occur during *kimchi* preparation [76]. *Kimchi* is a highly popular side dish in the Korean Peninsula and, besides different beneficial properties, its safety evaluation has been the subject of different studies, including the presence of biogenic amines and their quantification in the final, ready-to-eat products [77–80].

No evidence for the presence of *vanA*, *vanB*, *vanC*, *vanD*, *vanE*,or *vanG* was recorded in the studied *Bacillus* strains (Table 4). Based on the *in vitro* test with vancomycin, all tested *Bacillus* strains were susceptible with cut-off values of 0.5 μg/mL, suggesting their susceptibility according to the recommendations of CLSI ($\leq$4 μg/mL) and EFSA (4 μg/mL) (Table 5). Moreover, additional results regarding the interaction of the studied *Bacillus* strains and selected antibiotics are summarized in Table 5.

Kang et al. [81] reported on the isolation of LAB from *kimchi* and the identification of high levels of tetracycline resistance. Ten out of 50 different batches of *kimchi* were sources of isolation of antibiotic-resistant LAB with MICs of tetracycline ranging between 25 and up to higher than 100 μg/mL, in addition to the identification of specific genetic determinants as well [81]. Kang et al. [81] even alarmed that "kimchi can be considered as a potential vehicle for the spread of antibiotic-resistant lactic acid bacteria along the food chain to the consumers". Antibiotic resistance is a highly relevant topic in the safety evaluation of beneficial organisms. The possible spread of antibiotic-resistant genes between beneficial organisms and pathogens or other inhabitants of the gastro intestinal tract (GIT) is a justified argument in discussing the safety features of probiotics and starter cultures. However, Suvorov [82] pointed out that scenarios of horizontal transfer of antibiotics resistance genes in *Enterococcus* spp. are most probably somewhat exaggerated, and deeper studies on this topic need to be conducted. On the other side, present knowledge is pointing to the fact that strains carrying antibiotic resistance factors need to be considered as a potential hazard, and according EFSA and CLSI recommendations, such strains should not be recommended for application as starter cultures or as live probiotics.

5. Conclusions

Different bacteriocinogenic *Bacillus* strains were isolated from *kimchi*. Bacteriocin production is generally considered as a beneficial property by which starter cultures may have selective advantage in a mixed population for conquering an ecological niche. This may also include inhibition and elimination of sensitive pathogenic bacterial strains. However, when bacteriocinogenic strains are characterized to carry virulence or antibiotic resistance determinants, or found to possess virulence characteristics, then production of antimicrobial peptides may be considered as enhancing factors of virulence potential. The presence of strains with hemolytic activity in *kimchi* needs to be regarded as undesired and a potential health hazard to the consumers. On other side, antimicrobial properties of the strains may serve as basis for the development of appropriate biotechnological processes, and for exploring the application of the expressed antimicrobials.

Author Contributions: Concept: S.D.T.; experimental work: J.I.I.F., J.E.V.B., S.D.T.; data analysis: S.D.T.; funds: W.H.H.; writing of the manuscript: S.D.T.; corrections and editing: W.H.H., S.D.T. All authors have read and agreed to the published version of the manuscript.

Funding: This research was funded by grants from the National Research Foundation (NRF) funded by the Ministry of Science & ICT (NRF-2016M3A9A5923160 and NRF-2018M3A9F3021964), Seoul, Republic of Korea.

Data Availability Statement: All data generated or analyzed during this study are included in this published article and comply with research standards.

Acknowledgments: Grants from the National Research Foundation (NRF) were funded by the Ministry of Science & ICT (NRF-2016M3A9A5923160 and NRF-2018M3A9F3021964), Seoul, Republic of Korea. The authors acknowledge the support from Handong Global University, Pohang, Republic of Korea. Special thanks to Sungmin Jung, Chaerin Woo, and Samantha Joy D. Valledor, students of Handong Global University, Pohang, Republic of Korea, for technical assistance.

Conflicts of Interest: The authors declare no conflict of interest.

References

1. Ouoba, L.I.; Diawara, B.; Amoa-Awua, W.K.; Traoré, A.S.; Møller, P.L. Genotyping of starter cultures of *Bacillus subtilis* and *Bacillus pumilus* for fermentation of African locust bean (*Parkia biglobosa*) to produce Soumbala. *Int. J. Food Microbiol.* **2004**, *90*, 197–205. [CrossRef]
2. Aliakbarpour, H.R.; Chamani, M.; Rahimi, G.; Sadeghi, A.A.; Qujeq, D. The *Bacillus subtilis* and lactic acid bacteria probiotics influences intestinal mucin gene expression, histomorphology and growth performance in broilers. *Asian Australas. J. Anim. Sci.* **2012**, *25*, 1285–1293. [CrossRef]
3. Tamang, J.P.; Watanabe, K.; Holzapfel, W.H. Diversity of microorganisms in global fermented foods and beverages. A review. *Front. Microbiol.* **2016**, *7*, 377. [CrossRef] [PubMed]
4. Encyclopedia Britannica. 2008. Available online: https://www.britannica.com/ (accessed on 25 March 2020).
5. Jung, J.Y.; Lee, S.H.; Kim, J.M.; Park, M.S.; Bae, J.-W.; Hahn, Y.; Madsen, E.L.; Jeon, C.O. Metagenomic analysis of kimchi, a traditional Korean fermented food. *Appl. Environ. Microbiol.* **2011**, *77*, 2264–2274. [CrossRef]
6. Chikindas, M.L.; Weeks, R.; Drider, D.; Chistyakov, V.A.; Dicks, L.M. Functions and emerging applications of bacteriocins. *Curr. Opin. Biotechnol.* **2018**, *49*, 23–28. [CrossRef]
7. Todorov, S.D.; Franco, B.D.G.M.; Tagg, J.R. Bacteriocins of Gram positive bacteria having activity spectra extending beyond closely-related species. *Benef. Microbes* **2019**, *10*, 315–328. [CrossRef] [PubMed]
8. Syamaladevi, R.M.; Tang, J.; Villa-Rojas, R.; Sablani, S.; Carter, B.; Campbell, G. Influence of water activity on thermal resistance of microorganisms in low-moistre foods: A review. *Compr. Rev. Food Sci. Food Saf.* **2016**, *15*, 353–370. [CrossRef]
9. Bover-Cid, S.; Holzapfel, W.H. Improved screening procedure for biogenic amine production by lactic acid bacteria. *Int. J. Food Microbiol.* **1999**, *53*, 33–41. [CrossRef]
10. Dos Anjos, T.R.; Cavicchioli, V.Q.; Lima, J.A.S.; Vasconcellos, A.N.; Vaz, A.C.N.; Rossi, G.A.M.; Campos-Galvão, M.E.M.; Todorov, S.D.; Mathias, L.A.; Schocken-Iturrino, R.P.; et al. Unsatisfactory microbiological aspects of UHT goat milk, soy milk and dairy beverage of goat milk and soy protein: A public health issue. *Food Sci. Technol.* **2019**, *40*, 349–354. [CrossRef]
11. Colombo, M.; Nero, L.A.; Todorov, S.D. Safety profile of beneficial lactic acid bacteria isolated from dairy systems. *Braz. J. Microbiol.* **2020**, *51*, 787–795. [CrossRef]
12. Perin, L.M.; Miranda, R.O.; Todorov, S.D.; Franco, B.D.G.M.; Nero, L.A. Virulence, antibiotic resistance and biogenic amines of bacteriocinogenic lactococci and enterococci isolated from goat milk. *Int. J. Food Microbiol.* **2014**, *185*, 121–126. [CrossRef]
13. Priest, F.G. Extracellular enzyme synthesis in the genus *Bacillus*. *Bacteriol. Rev.* **1977**, *41*, 711–753. [CrossRef] [PubMed]
14. Barg, H.; Malten, M.; Jahn, M.; Jahn, D. Protein and vitamin production in Bacillus megaterium. In *Microbial Processes and Products. Methods in Biotechnology*; Barredo, J.L., Ed.; Humana Press: Totowa, NJ, USA, 2005; Volume 18. [CrossRef]
15. Stein, T. Bacillus subtilis antibiotics: Structures, syntheses and specific functions. *Mol. Microbiol.* **2005**, *56*, 845–857. [CrossRef] [PubMed]
16. Latorre, J.D.; Hernandez-Velasco, X.; Wolfenden, R.E.; Vicente, J.L.; Wolfender, A.D.; Menconi, A.; Bielke, L.R.; Hargis, B.M.; Tellez, G. Evaluation and selection of *Bacillus* species based on enzyme production, antimicrobial activity, and biofilm synthesis as direct-fed microbial candidates for poultry. *Front. Vet. Sci.* **2016**, *3*, 95. [CrossRef] [PubMed]
17. Daliri, E.B.; Oh, D.H.; Lee, B.H. Bioactive Peptides. *Foods* **2017**, *6*, 32. [CrossRef] [PubMed]
18. Field, D.; Cotter, P.D.; Hill, C.; Ross, R.P. Bioengineering Lantibiotics for therapeutic success. *Front. Microbiol.* **2015**, *6*, 1363. [CrossRef] [PubMed]
19. Khatri, I.; Sharma, G.; Subramanian, S. Composite genome sequence of *Bacillus clausii*, a probiotic commercially available as Enterogermina®, and insights into its probiotic properties. *BMC Microbiol.* **2019**, *19*, 307. [CrossRef]
20. Hongu, N.; Kim, A.S.; Suzuki, A.; Wilson, H.; Tsui, K.C.; Park, S. Korean kimchi: Promoting healthy meals through cultural tradition. *J. Ethn. Foods* **2016**, *4*, 172–180. [CrossRef]
21. Spencer, R.C. Bacillus anthracis. *J. Clin. Pathol.* **2003**, *56*, 182–187. [CrossRef] [PubMed]
22. Messelhäußer, U.; Ehling-Schulz, M. *Bacillus cereus*—A multifaceted opportunistic pathogen. *Curr. Clin. Microbiol. Rep.* **2018**, *5*, 120–125. [CrossRef]

23. Baccigalupi, L.; Ricca, E.; Ghelardi, E. Non-LAB probiotics: Spore formers. In *Probiotics and Prebiotics: Current Research and Future Trends*; Venema, K., do Carmo, A.P., Eds.; Caister Academic Press: Poole, UK, 2015; pp. 93–104.
24. Elshaghabee, F.M.F.; Rokana, N.; Gulhane, R.D.; Sharma, C.; Panwar, H. *Bacillus* as potential probiotics: Status, concerns, and future perspectives. *Front. Microbiol.* **2017**, *8*, 1490. [CrossRef]
25. Lee, H.-W.; Yoon, S.-R.; Kim, S.-J.; Lee, H.M.; Lee, J.Y.; Lee, J.-H.; Kim, S.H.; Ha, J.-H. Identification of microbial communities, with a focus on foodborne pathogens, during kimchi manufacturing process using culture-independent and -dependent analyses. *LWT Food Sci. Technol.* **2016**, *81*, 153–159. [CrossRef]
26. Senesi, S.; Celandroni, F.; Tavanti, A.; Ghelardi, E. Molecular characterization and identification of *Bacillus clausii* strains marketed for use in oral bacteriotherapy. *Appl. Environ. Microbiol.* **2001**, *67*, 834–839. [CrossRef]
27. De Vos, P.; Garrity, G.M.; Jones, D.; Kreig, N.R.; Ludwig, W.; Rainey, F.A.; Schleifel, K.-H.; Whitman, W.B. *Bergey's Manual of Systematic Bacteriology*; The Firmicutes; Wiley Publishing Group: Hoboken, NJ, USA, 2009; Volume 3. [CrossRef]
28. De Moraes, G.M.D.; de Abreu, L.R.; do Egito, A.S.; Salles, H.O.; da Silva, L.M.F.; Nero, L.A.; Todorov, S.D.; dos Santos, K.M.O. Functional properties of *Lactobacillus mucosae* strains isolated from Brazilian goat milk. *Probiotics Antimicrob. Proteins* **2016**, *9*, 235–245. [CrossRef] [PubMed]
29. Chopra, L.; Singh, G.; Choudhary, V.; Sahoo, D.K. Sonorensin: An antimicrobial peptide, belonginig to the heterocycloantharicin subfamily of bacteriocins, from a new marine isolate, *Bacillus sonorensis* MT93. *Appl. Environ. Microbiol.* **2014**, *80*, 2981–2990. [CrossRef] [PubMed]
30. Todorov, S.D.; Stojanovski, S.; Iliev, I.; Moncheva, P.; Nero, L.A.; Ivanova, I.V. Technology and safety assessment for lactic acid bacteria isolated from traditional Bulgarian fermented meat product "Lukanka". *Braz. J. Microbiol.* **2017**, *48*, 576–586. [CrossRef] [PubMed]
31. Dos Santos, K.M.O.; de Matos, C.R.; Salles, H.O.; Franco, B.D.G.M.; Arellano, K.; Holzapfel, W.H.; Todorov, S.D. Exploring beneficial/virulence properties of two dairy related strains of *Streptococcus infantarius* subsp. *infantarius*. *Probiotics Antimicrob. Proteins* **2020**, *12*, 1524–1541. [CrossRef]
32. Todorov, S.D.; Cavicchioli, V.Q.; Ananieva, M.; Bivolarski, V.P.; Vasileva, T.A.; Hinkov, A.V.; Todorov, D.G.; Shishkov, S.; Haertlé, T.; Iliev, I.N.; et al. Expression of coagulin A with low cytotoxic activity by *Pediococcus pentosaceus* ST65ACC isolated from raw milk cheese. *J. Appl. Microbiol.* **2019**, *128*, 458–472. [CrossRef]
33. Guinebretiere, M.-H.; Broussolle, V.; Nguyen-The, C. Enterotoxigenic profile of food-poisoning and food-borne *Bacillus cereus* strains. *J. Clin. Microbiol.* **2002**, *40*, 3053–3056. [CrossRef]
34. EUCAST. Available online: https://www.eucast.org/ast_of_bacteria/ (accessed on 25 March 2020).
35. Valledor, S.J.D.; Bucheli, J.E.V.; Holzapfel, W.H.; Todorov, S.D. Exploring beneficial properties of the bacteriocinogenic *Enterococcus faecium* ST10Bz strain isolated from *boza*, a Bulgarian cereal-based beverage. *Microorganisms* **2020**, *8*, 1474. [CrossRef]
36. Ciffo, F. Determination of the spectrum of antibiotic resistance of the *Bacillus subtilis* strains of Enterogermina. *Chemioterapia* **1984**, *3*, 45–52.
37. McAuliffe, O.; Ryan, M.P.; Ross, R.P.; Hill, C.; Breeuwer, P.; Abee, T. Lacticin 3147, a broad-spectrum bacteriocin which selectively dissipates the membrane potential. *Appl. Environ. Microbiol.* **1998**, *64*, 439–445. [CrossRef] [PubMed]
38. Shelburne, C.E.; An, F.Y.; Dholpe, V.; Ramamoorthy, A.; Lopatin, D.E.; Lantz, M.S. The spectrum of antimicrobial activity of the bacteriocin subtilosin A. *J. Antimicrob. Chemother.* **2006**, *59*, 297–300. [CrossRef] [PubMed]
39. Masuda, Y.; Ono, H.; Kitagawa, H.; Ito, H.; Mu, F.; Sawa, N.; Zendo, T.; Sonomoto, K. Identification and characterization of leucocyclicin Q, a novel cyclic bacteriocin produced by *Leuconostoc mesenteroides* TK41401. *Appl. Environ. Microbiol.* **2011**, *77*, 8164–8170. [CrossRef] [PubMed]
40. Compaore, C.S.; Nielsen, D.S.; Sawadogo-Lingani, H.; Berner, T.S.; Nielsen, K.F.; Adimpong, D.B.; Diawara, B.; Ouedraogo, G.A.; Jakobsen, M.; Thorsen, L. *Bacillus amyloliquefaciens* subsp. *plantarum* strains as potential protective starter cultures for the production of Bikalga, an alkaline fermented food. *J. Appl. Microbiol.* **2013**, *115*, 133–146. [CrossRef]
41. Anihouvi, V.B.; Kpoclou, E.Y.; Hounhouigan, J.D. Use of starter cultures of *Bacillus* and staphylococcus in the controlled fermentation of Lanhouin, a traditional fish-based condiment from West Africa. *Afr. J. Microbiol. Res.* **2012**, *6*, 4767–4774. [CrossRef]
42. Chang, H.C. Healthy and safe Korean traditional fermented foods: Kimchi and chongkukjang. *J. Ethn. Foods.* **2018**, *5*, 161–166. [CrossRef]
43. Katz, E.; Demain, A.L. The peptide antibiotics of *Bacillus*: Chemistry, biogenesis, and possible functions. *Biotechnol. Rev.* **1977**, *41*, 449–474. [CrossRef]
44. Hosoi, T.; Kiuchi, K. Natto—A food made by fermenting cooked soybeanswith *Bacillus subtilis* (natto). In *Fermented Functional Foods*; Farnworth, E.R., Ed.; CRC Press: Boca Raton, FL, USA, 2003; pp. 227–251.
45. Cho, M.S.; Jin, Y.J.; Kang, B.K.; Park, Y.K.; Kim, C.K.; Park, D.S. Understanding the ontogeny and succession of *Bacillus velezensis* and *B. subtilis* subsp. *subtilis* by focusing on kimchi fermentation. *Sci. Rep.* **2018**, *8*, 7045. [CrossRef]
46. Jeon, H.H.; Jung, J.Y.; Chun, B.H.; Kim, M.D.; Baek, S.Y.; Moon, J.Y.; Yeo, S.H.; Jeon, C.O. Screening and characterization of potential *Bacillus* starter cultures for fermenting low-salt soybean paste (Doenjang). *J. Microbiol. Biotechnol.* **2016**, *26*, 666–674. [CrossRef]
47. Montel, M.C.; Masson, F.; Talon, R. Bacterial role in flavor development. *Meat Sci.* **1998**, *49*, S111–S123. [CrossRef]
48. El Mecherfi, K.-E.; Todorov, S.D.; de Albuquerque, M.C.; Denery-Papini, S.; Lupi, R.; Haertlé, T.; Franco, B.D.G.M.; Larré, C. Allergenicity of fermented foods: Emphasis on seeds protein based product. *Foods* **2020**, *9*, 792. [CrossRef]

49. Park, H.; Lee, M.; Ji, Y.; Todorov, S.D.; Holzapfel, W.H. Safety evaluation and in vivostrain-specific functionality of *Bacillus* strains isolated from Korean traditional fermented foods. *Probiotics Antimicrob. Proteins* **2020**, *13*, 60–71. [CrossRef]
50. George, J. Elevated serum β-glucuronidase reflects hepatic lysosomal fragility following toxic liver injury in rats. *Biochem. Cell Biol.* **2008**, *86*, 235–243. [CrossRef] [PubMed]
51. Yang, E.J.; Chang, H.C. Characterization of bacteriocin-like substances produced by *Bacillus subtilis* MJP1. *Korean J. Microbiol. Biotechnol.* **2007**, *35*, 339–346.
52. Wu, S.; Jia, S.; Sun, D.; Chen, M.; Chen, X.; Zhong, J.; Huan, L. Purification and characterization of two novel antimicrobial peptides subpeptinJM4-A and subpeptinJM4-B Produced by *Bacillus subtilis* JM4. *Curr. Microbiol.* **2005**, *51*, 292–296. [CrossRef]
53. Salazar, F.; Ortiz, A.; Sansinenea, E. Characterisation of two novel bacteriocin-like substances produced by *Bacillus amyloliquefaciens* ELI149 with broad-spectrum antimicrobial activity. *J. Glob. Antimicrob. Resist.* **2017**, *11*, 177–182. [CrossRef]
54. Entian, K.-D.; de Vos, W.M. Genetics of subtilin and nisin biosyntheses. *Antonie Van Leeuwenhoek* **1996**, *69*, 109–117. [CrossRef]
55. Parisot, J.; Carey, S.; Breukink, E.; Chan, W.C.; Narbad, A.; Bonev, B. Molecular mechanism of target recognition by subtilin, a class I lanthionine antibiotic. *Antimicrob. Agents Chemother.* **2008**, *52*, 612–618. [CrossRef]
56. Zheng, G.; Yan, L.Z.; Vederas, J.C.; Zuber, P. Genes of the sbo-alb locus of *Bacillus subtilis* are required for production of the antilisterial bacteriocin subtilosin. *J. Bacteriol.* **1999**, *181*, 7346–7355. [CrossRef]
57. Nakano, M.M.; Zheng, G.; Zuber, P. Dual control of sbo-alb operon expression by the Spo0 and ResDE systems of signal transduction under anaerobic conditions in *Bacillus subtilis*. *J. Bacteriol.* **2000**, *182*, 3274–3277. [CrossRef] [PubMed]
58. Lee, N.-K.; Kim, W.-S.; Paik, H.-D. *Bacillus* strains as human probiotic: Characterization, safety, microbiome, and probiotic carrier. *Food Sci. Biotechnol.* **2019**, *28*, 1297–1305. [CrossRef]
59. Lee, H.; Kim, H.-Y. Lantibiotics, Class I Bacteriocins from the Genus *Bacillus*. *J. Microbiol. Biotechnol.* **2011**, *21*, 229–235. [CrossRef] [PubMed]
60. Le Marrec, C.; Hyronimus, B.; Bressollier, P.; Verneuil, B.; Urdaci, M.C. Biochemical and genetic characterization of coagulin, a new antilisterial bacteriocin in the pediocin family of bacteriocins, produced by *Bacillus coagulans* I(4). *Appl. Environ. Microbiol.* **2000**, *66*, 5213–5220. [CrossRef] [PubMed]
61. Cotter, P.D.; Ross, R.P.; Hill, C. Bacteriocins—A viable alternative to antibiotics? *Nat. Rev. Microbiol.* **2013**, *11*, 95–105. [CrossRef] [PubMed]
62. Joerger, R.D. Alternatives to antibiotics: Bacteriocins, antimicrobial peptides and bacteriophages. *Poult. Sci.* **2003**, *82*, 640–647. [CrossRef] [PubMed]
63. Vesterlund, S.; Vankerckhoven, V.; Saxelin, M.; Goossens, H.; Salminen, S.; Ouwehand, A.C. Safety assessment of *Lactobacillus* strains: Presence of putative risk factors in faecal, blood and probiotic isolates. *Int. J. Food Microbiol.* **2007**, *116*, 325–331. [CrossRef]
64. Barbosa, J.; Gibbs, P.A.; Teixeira, P. Virulence factors among enterococci isolated from traditional fermented meat products produced in the north of Portugal. *Food Control* **2010**, *21*, 651–656. [CrossRef]
65. Franz, C.M.A.P.; Muscholl-Silberhorn, A.B.; Yousif, N.M.K.; Vancanneyt, M.; Swings, J.; Holzapfel, W.H. Incidence of virulence factors and antibiotic resistance among enterococci isolated from food. *Appl. Environ. Microbiol.* **2001**, *67*, 4385–4389. [CrossRef]
66. Simon-Sarkadi, L.; Holzapfel, W.H.; Halász, A. Biogenic amine content and microbial contamination of leafy vegetables during storage at 5 °C. *J. Food Biochem.* **1994**, *17*, 407–418. [CrossRef]
67. Özogul, Y.; Özogul, F. Chapter 1: Biogenic amines formation, toxicity, regulations in food. In *Biogenic Amines in Food: Analysis, Occurrence and Toxicity*; Royal Society of Chemistry: London, UK, 2019; pp. 1–17. [CrossRef]
68. Halász, A.; Barath, A.; Simon-Sarkadi, L.; Holzapfel, W.H. Biogenic amines and their production by microorganisms in food. *Trends Food Sci. Technol.* **1994**, *5*, 42–49. [CrossRef]
69. Halász, A.; Baráth, Ă.; Holzapfel, W.H. The influence of starter culture selection on sauerkraut fermentation. *Z. Lebensm. Forsch. A* **1999**, *208*, 434–438. [CrossRef]
70. Barbieri, F.; Montanari, C.; Gardini, F.; Tabanelli, G. Biogenic amine production by lactic acid bacteria: A Review. *Foods* **2019**, *8*, 17. [CrossRef] [PubMed]
71. Santos, M.H.S. Biogenic amines: Their importance in foods. *Int. J. Food Microbiol.* **1996**, *29*, 213–231. [CrossRef]
72. Brink, B.T.; Damink, C.; Joosten, H.M.L.J.; Veld, J.H.I. Occurrence and formation of biologically active amines in foods. *Int. J. Food Microbiol.* **1990**, *11*, 73–84. [CrossRef]
73. Atanasova, J.; Dalgalarrondo, M.; Iliev, I.; Moncheva, P.; Todorov, S.D.; Ivanova, I.V. Formation of free amino acids and bioactive peptides during the ripening of Bulgarian white brined cheeses. *Probiotics Antimicrob. Proteins* **2020**, *13*, 261–272. [CrossRef] [PubMed]
74. Stribny, J.; Gamero, A.; Pérez-Torrado, R.; Querol, A. *Saccharomyces kudriavzevii* and *Saccharomyces uvarum* differ from *Saccharomyces cerevisiae* during the production of aroma-active higher alcohols and acetate esters using their amino acidic precursors. *Int. J. Food Microbiol.* **2015**, *205*, 41–46. [CrossRef]
75. Perea-Sanz, L.; Peris, D.; Belloch, C.; Flores, M. *Debaryomyces hansenii* metabolism of sulfur amino acids as precursors of volatile sulfur compounds of interest in meat products. *J. Agric. Food Chem.* **2019**, *67*, 9335–9343. [CrossRef]
76. Mah, J.-H.; Kim, Y.J.; No, H.-K.; Hwang, H.-J. Determination of biogenic amines in kimchi, Korean traditional fermented vegetable products. *Food Sci. Biotechnol.* **2004**, *13*, 826–829.
77. Tsai, Y.-H.; Kung, H.-F.; Lin, Q.-L.; Hwang, J.-H.; Cheng, S.-H.; Wei, C.-I.; Hwang, D.-F. Occurrence of histamine and histamine-forming bacteria in kimchi products in Taiwan. *Food Chem.* **2005**, *90*, 635–641. [CrossRef]

78. Jeong, D.-W.; Lee, J.-H. Antibiotic resistance, hemolysis and biogenic amine production assessments of *Leuconostoc* and *Weissella* isolates for kimchi starter development. *LWT Food Sci. Technol.* **2015**, *64*, 1078–1084. [CrossRef]
79. Jin, Y.H.; Lee, J.H.; Park, Y.K.; Lee, J.-H.; Mah, J.-H. The occurrence of biogenic amines and determination of biogenic amine-producing lactic acid bacteria in kkakdugi and chonggak kimchi. *Foods* **2019**, *8*, 73. [CrossRef] [PubMed]
80. Kim, M.-J.; Kim, K.-S. Tyramine production among lactic acid bacteria and other species isolated from kimchi. *LWT Food Sci. Technol.* **2014**, *56*, 406–413. [CrossRef]
81. Kang, H.-J.; Kim, B.-C.; Park, W. Isolation of tetracycline-resistnt lactic acid bacteria from kimchi. *Korean J. Microbiol.* **2004**, *40*, 1–6. (In Korean)
82. Suvorov, A. What is wrong with enterococcal probiotics? *Probiotics Antimicrob. Proteins* **2020**, *12*, 1–4. [CrossRef] [PubMed]

fermentation

Article

Anti-Inflammatory Effect on Colitis and Modulation of Microbiota by Fermented Plant Extract Supplementation

Manabu Sugimoto [1,*], Toshiro Watanabe [2], Motoko Takaoka [3], Kyoko Suzuki [3], Tadatoshi Murakami [4], Nobutada Murakami [4] and Shoichi Sumikawa [4]

1 Institute of Plant Science and Resources, Okayama University, 2-20-1 Chuo, Kurashiki, Okayama 710-0046, Japan
2 Department of Food and Nutrition, Sonoda Women's University, 7-29-1 Minamitsukaguchi, Amagasaki, Hyogo 661-8520, Japan; watnb-ts@sonoda-u.ac.jp
3 Department of Biosphere Sciences, Kobe College, 4-1 Okadayama, Nishinomiya, Hyogo 662-8505, Japan; takaoka@mail.kobe-c.ac.jp (M.T.); suzuki@mail.kobe-c.ac.jp (K.S.)
4 Functional Food Creation Research Institute Co., Ltd., 4422-1 Yoshikawa, Kibichuo, Kaga, Okayama 716-1241, Japan; murakami@kinousei.co.jp (T.M.); n-murakami@kinousei.co.jp (N.M.); sumikawa@kinousei.co.jp (S.S.)
* Correspondence: manabus@okayama-u.ac.jp; Tel.: +81-86-424-1661

Abstract: Although results of recent studies suggest that fermented foods strongly affect the gut microbiota composition and that they relieve inflammatory bowel disease symptoms, some reports have described that fermented foods increase some inflammation markers based on differences in fermented food materials. This study evaluated the effects of fermented plant extract (FPE) on dextran sulfate sodium (DSS)-induced colitis in mice and the effects on fecal microbiota composition in humans. Mice fed 5% FPE with 3% DSS (FPE group) showed no body weight loss, atrophy of colonic length, or bloody stool, similar to mice fed a basal diet (negative group), whereas mice fed 3% DSS (positive group) exhibited those effects. Concentrations of inflammation markers IL-6 and TNF-α were not significantly different between FPE and negative groups; however, those concentrations became higher in the positive group. 16S ribosomal RNA gene sequencing was used to characterize fecal microbiota in healthy women before and after 3-month FPE supplementation. The FPE supplementation induced increases in *Firmicutes* phyla and in *Clostridiales* order, which play a central role in inflammation suppression. These results suggest that FPE enhances *Clostridiales* growth in the gut and that it has an anti-inflammatory effect.

Keywords: fermented plant extract; microbiota; dextran sulfate sodium; inflammatory; *Clostridiales*

Citation: Sugimoto, M.; Watanabe, T.; Takaoka, M.; Suzuki, K.; Murakami, T.; Murakami, N.; Sumikawa, S. Anti-Inflammatory Effect on Colitis and Modulation of Microbiota by Fermented Plant Extract Supplementation. *Fermentation* **2021**, 7, 55. https://doi.org/10.3390/fermentation7020055

Academic Editor: Hiroshi Kitagaki

Received: 12 March 2021
Accepted: 3 April 2021
Published: 6 April 2021

1. Introduction

Inflammatory bowel disease, as typified by diseases such as ulcerative colitis and Crohn's disease, affects the gastrointestinal tract. No satisfactory treatment has been found because its causes remain unidentified. Moreover, the Japanese Ministry of Health, Labor, and Welfare has specified these diseases as intractable diseases. The number of patients with these diseases has been increasing in Japan [1,2]. Lack of dietary fiber and the consumption of western foods are implicated as reasons. Actually, dietary changes alter gut microbiota composition. In fact, dysbiosis and disruption of immunological homeostasis have been associated with inflammatory diseases [3–13] because the gut microbiota affects immunologic, nutritional, and metabolic processes of the human body [14], indicating that diet strongly affects inflammatory bowel disease.

Functional foods are part of the human diet. They have been demonstrated to provide health benefits and to decrease the risk of chronic diseases beyond those provided by adequate nutrition [15]. Functional foods include naturally occurring bioactive substances, supplemented bioactive substances, and derived food ingredients. Especially, specialized ingredients in functional foods promoting the growth or activity of specific bacteria, such as *Lactobacillus* and *Bifidobacterium* genera, have been specifically examined as prebiotics. Recent scientific advances show that modifying the bacterial composition of the intestinal ecosystem induces functional changes such as the host physiology [16–19]. A diet with greater amounts of dietary fiber increases the composition of *Clostridiales* in mice. This is a dominant class of commensal microbes that produces butyric acid to induce colonic regulatory T (Treg) cells, which play a central role in suppressing inflammatory and allergic responses [20–22]. Therefore, it is crucially important to evaluate foods to ascertain whether they increase the composition of bacteria that produce bioactive molecules.

Fermented foods have been a fundamentally important part of the human diet for centuries in most parts of the world. Fermented foods of many types are produced from animal and plant materials, some of which have the potential to provide additional health benefits through fermentation as functional foods [23,24]. Some fermented foods have beneficial immune, glycemic, and anti-inflammatory activities [25–28], whereas other fermented foods increase some inflammation markers [29]. These results indicate the necessity for assessment of each fermented food to assess its health-promoting activities because of differences in the materials and microbes used for fermentation.

For this study, we used a fermented plant extract (FPE) produced from whole plants to examine the effects on dextran sulfate sodium (DSS)-induced colitis in a mouse model. Bloody stools and atrophy of colonic length were not observed. The TNF-α and IL-6 concentrations were suppressed by FPE supplementation. Subsequently, 16S rRNA amplicon sequencing revealed that the composition of *Clostridiales* order had increased significantly in the fecal samples of young women after 3-month FPE supplementation with daily diet.

2. Materials and Methods

2.1. Preparation of Fermented Plant Extract

Whole fruits and vegetables with peels and seeds were minced and extracted using muscovado sugar in separate barrels (Table S1). For animal experimentation, the extracts were mixed with mushroom, pulse, cereals, seaweed, and decoction of loquat leaves, to make a starting material of 75 kinds, consisting of 50.0% of sugar, 23.9% of fruits, 15.6% of vegetables and wild herbs, 2.0% of mushrooms, 1.8% of seaweed, and 6.7% of pulse and cereals. Lactic acid bacteria, *Lactiplantibacillus pentosus, L. plantaru, Pediococcus pentosaceus, P. acidilactici, Lacticaseibacillus paracasei, L. casei, Lactococcus lactis, Latilactobacillus curvatus, Leuconostoc mesenteroides,* and *Levilactobacillus brevis* were added to the starting material and fermented at room temperature. After three year-fermentation and maturation, the fermented extract was strained and heated at 80 °C for 10 min to obtain FPE. pH, water content, and brix of were 3.8–4.8, 33–39%, and 59–65%, respectively. For microbiota analyses, the materials, which were acceptable for exporting to international countries, were mixed to make a starting material of 40 kinds, consisting of 42.0% of sugar, 24.6% of fruits, 19.4% of vegetables and wild herbs, 5.0% of mushrooms, 3.0% of seaweed, and 6.0% of pulse and cereals. The mixture was fermented using the same lactic acid bacteria and method as that described above. pH, water content, and brix of the fermented extract were 3.8–4.5, 34–42%, and 57–62%, respectively. As an FPE package, 5 g of the mixture, consisting of 40% of FPE, 59% of apple extract, and 1% of plum juice, was packed in an aluminum bag. The mixture was prepared to consist of 43.8% of apple juice, 26.3% of sugar, 29.0% of dextrin, and 0.9% of caramel color. Then, 5 g of the mixture was packed in an aluminum bag as a placebo package.

2.2. Animals

Male BALB/c mice were purchased from CLEA Japan, Inc. (Tokyo, Japan). They were kept under standard conditions in a room at 21–24 °C with a constant 12 h light/dark cycle. The ethics committee of Sonoda Women's University approved the experiments, which were performed in accordance with relevant guidelines and regulations.

2.3. *Induction of Colitis*

Five-week-old mice were divided into three groups: a negative group ($n = 8$), a positive group ($n = 8$), and an FPE group ($n = 8$). The positive group was fed diet with 3% dextran sulfated sodium 5000 (DSS) solution. The FPE group was fed diet containing 5% FPE with 3% DSS solution. The negative group was fed diet with water for 7 days. The diet composition provided to each group is presented in Table S2.

2.4. *IL-6 and TNF-α Assays*

IL-6 and TNF-α concentrations in mouse sera were found using Mouse IL-6 Assay Kit (Immuno-Biological Laboratories Co., Ltd., Gunma, Japan) and Quantikine ELISA Mouse TNF-α Kit (R&D Systems, Minneapolis, MN, USA), respectively, according to the instructions of the manufacturers.

2.5. *Dietary Supplementation and Fecal Collection*

We recruited 20 healthy women (age of 20–24 years, 21.4 ± 1.2 years) from the School of Human Sciences, Kobe College. All participants gave written, informed consent before participating in the study. This study was approved by the review board of Kobe College. Participants were assigned randomly into two groups. Participants in the FPE group were administered one FPE package three times after meals every day for 3 months in addition to their habitual diet. Participants in the placebo group were administered one placebo package in the same way as that of FPE group. No dietary restriction was instructed on participants. Participants collected fecal materials before and after 3-month administration. The materials were suspended in a solution (DNA/RNA Shield™; Zymo Research Corp., Irvine, CA, USA).

2.6. *Microbial DNA Extraction*

Total microbial DNA was extracted from each fecal suspension using a kit (Quick-DNA™ Fecal/Soil Microbe Miniprep; Zymo Research Corp., Irvine, CA, USA) according to the manufacturer's instruction. The DNA samples were examined using a spectrophotometer (Nanodrop ND-1000; Thermo Scientific, Waltham, MA, USA) by measuring absorbance values at 260 and 280 nm.

2.7. *16S rRNA Gene Sequencing*

The V3–V4 region of the microbial 16S rRNA gene in each DNA sample was amplified by PCR using the universal primer set 341F/806R according to a protocol described elsewhere [30]. A dual-indexing amplification and sequencing approach was used. The resulting amplicons were purified and subjected to sequencing with 2 × 301 bp paired-end reads on the MiSeq systems (Illumina, Inc., San Diego, CA, USA) using MiSeq v3 reagent kit (Illumina) according the protocols described by Illumina (https://jp.support.illumina.com/content/dam/illumina-support/documents/documentation/chemistry_documentation/16s/16s-metagenomic-library-prep-guide-15044223-b.pdf, accessed on 5 April 2021). The raw sequence data quality was checked using the FastQC quality-control tool (Babraham Bioinformatics, Cambridge, United Kingdom; http://www.bioinformatics.babraham.ac.uk/projects/fastqc/, accessed on 5 April 2021). The datasets were analyzed with QIIME 1.9.1 pipeline (Quantitative Insights Into Microbial Ecology; http://qiime.org, accessed on 5 April 2021) using GreenGenes 13.5 [31].

2.8. *Statistical Analysis*

Data analysis was performed using the analysis of Bonferroni's multiple comparison test and Student's *t*-test, with a statistical significance set at $p < 0.05$.

3. Results and Discussion

3.1. Anti-Colitis Effect of FPE in Mice

Plant materials of fermented foods, such as asparagus, garlic, chicory, onion, wheat, barley, rye, soybean, peas, beans, banana, tomato, seaweeds, and microalgae, contain lactulose, galactooligosaccharides, fructooligosaccharides, inulin, maltoorogosaccharides, and resistant starch, known as "prebiotics", adversely affect the growth and activity of specific bacteria producing immunomodulatory products, particularly a short-chain fatty acid such as butyric acid. In addition, fermentation of these materials produces novel bioactive compounds working as prebiotics [25–28,32]. Therefore, we specifically examine the potential of plant materials and the produced FPE. FPE produced from extract of 75 kinds was rich in phytochemicals such as dietary fiber and polyphenol, and physiological function such as anti-oxidative, anti-inflammatory, and anti-allergy activities [33].

To assess the anti-colitis effects of FPE, acute colitis was induced by DSS feeding for 7 days to evaluate the effects of FPE on gut inflammation. The daily dietary intakes of negative group, positive group, and FPE group, which were, respectively, 2.1 ± 0.2, 1.8 ± 0.2, and 2.0 ± 0.3 g/day, were not found to be significantly different. The positive group was found to have significantly reduced body weight after 7-day DSS administration (21.9 ± 0.31 g) compared with results found for the negative group (24.6 ± 0.48 g), although the FPE group was found to have no significant difference in body weight (23.5 ± 0.41 g) compared with the negative group (Figure S1). The bloody stool appeared at 5 days after DSS administration in the positive group. Its color was brown. It progressed at 6 and 7 days. The color was red. In contrast, the bloody stool found in the FPE group was not observed as it was in the negative group, whereas the fecal color of the FPE group was black (Table 1). The colon length in the positive group (5.9 ± 0.17 cm) was significantly shorter than that in the negative group (8.3 ± 0.24 cm), but the FPE group exhibited no difference in the colon length, which was 7.7 ± 0.21 cm compared with the negative group (Figure 1 and Figure S2). The concentrations of IL-6 and TNF-α, inflammatory cytokines, and markers of inflammation were increased significantly, five-fold and four-fold, respectively, compared with the negative group after 5-day DSS administration. However, those in the FPE group were not significantly different from those in the negative group (Figure 2). These results demonstrate that our FPE has an anti-inflammatory effect on DSS-induced mouse colitis.

Table 1. Blood stool score of feces.

Day after DSS Administration	Negative Group	Positive Group	FPE Group
4	-	-	- (BK)
5	-	+	- (BK)
6	-	++	- (BK)
7	-	++	- (BK)

-, normal; +, brown bloody stool; ++, red bloody stool; BK, black color.

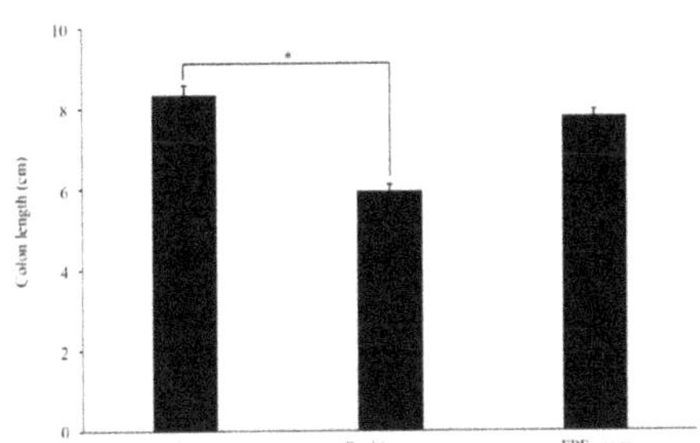

Figure 1. Colon length of mice after 7 day-administration. The values are expressed as means ± SEM (n = 8). * $p < 0.05$, accessed using Bonferroni's multiple comparison test.

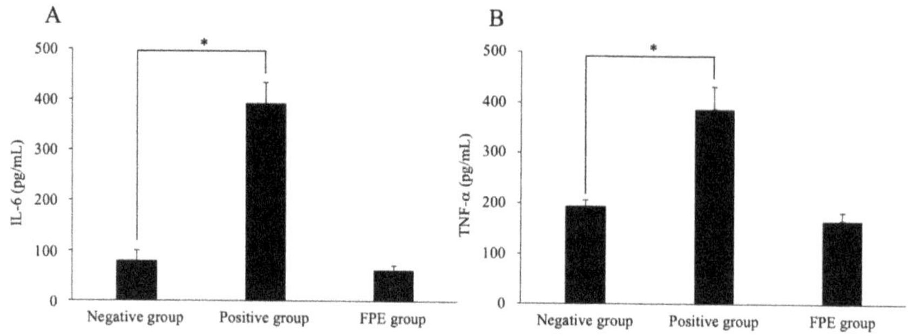

Figure 2. IL-6 (**A**) and TNF-α (**B**) concentrations of mice serum after 7-day administration. The values are expressed as means ± SEM ($n = 8$). * $p < 0.05$, assessed using Bonferroni's multiple comparison test.

3.2. Change of Gut Microbiota Composition in Human

To investigate changes of gut microbiota composition attributable to dietary supplementation of FPE, FPE was produced from extracts of 40 kinds, of which plant materials were included in those of 75 kinds and total content of fruits, vegetables, and wild herbs, which are sources of prebiotic [34], was 44% in extract of 40 kinds and dietary fiber was 4.6 g/100 g extract in FPE from extracts of 40 kinds, while total content of them in extract of 75 kinds was 39.5% and FPE from extract of 75 kinds contained 1.9 g dietary fiber/100 g extract. FPE from extract of 40 kinds contained same kinds of nutritional compounds such as amino acid and organic acid as FPE from extract of 75 kinds (Table S3). Each of 20 participants supplemented one FPE package or placebo package was administered three times after meals every day for 3 months. The changes of body weight, body fat mass, muscle mass, and body mass index (BMI) of the FPE group after 3-month administration were not significantly different from those of the placebo group (Figure S3). Based on data collected through a questionnaire, the scores of health conditions of both the FPE group and the placebo group after 3-month administration were not significantly different compared with those before administration. These results indicate that FPE supplementation did not affect health or physical condition.

Microbial DNA was isolated from fecal samples collected from FPE and placebo groups before and after 3-month administration. The DNA sequencing of the V3–V4 amplicons from 40 samples revealed 5,295,873 paired-end sequence reads with an average of 132,397 reads per sample. After filtering to eliminate poor quality sequences, finally 4,890,551 sequences with an average of 122,264 reads per sample were generated. The composition of bacterial phyla in feces of the FPE group before supplementation was not significantly different from that of the placebo group, of which *Firmicutes* was the most abundant, followed by *Bacteroidetes* and *Actinobacteria*. After 3 months of supplementation, the composition of *Firmicutes* in the FPE group increased along with reduction in *Bacteroidetes*. It was found to be significantly different from that in the placebo group ($p = 0.001$) (Table 2, Figure S4). Reportedly, dietary fiber can be designed to favor beneficial bacterial groups; certain *Firmicutes* access insoluble matrix fibers and resistant starch [35,36]. Plant materials in FPE, which contain fiber, starch, and oligosaccharides, can induce *Firmicutes* phyla in the gut. At the bacterial order level, the composition of *Clostridiales*, 39.0 ± 1.8%, increased significantly to 43.6 ± 2.5% after 3-month administration ($p = 0.03$) in the FPE group, whereas those in the placebo group before and after 3 months were 37.8 ± 1.6% and 38.0 ± 2.1%, respectively (Figure 3, Table S4). *Clostridiales* order in FPE group after 3-month administration consisted of 49% of *Lachnospiraceae* family and 33% of *Ruminococcaceae* family, which produce butyric acid [37,38]. *Clostridiales* induces colonic Treg cells and generates short-chain fatty acid, butyric acid, by fermentation of dietary fiber [39,40]. Butyric acid promotes to increase the expression of IL-10 in Treg cells and inhibits the activation of the transcription factor, NF-κβ, in gut cells. Increase of IL-10 in Treg cells and decrease of NF-κβ lead the intestinal immune homeostasis, suppressing systemic and mucosal immune activation to control intestinal inflammation and contributing to maintaining tolerance towards gut microbiota [20–22,41]. Taken together, results show that FPE enhances *Clostridiales* growth in the gut and show that it has an anti-inflammatory effect on colitis.

Table 2. Microbiota composition at the phyla level in feces of placebo and fermented plant extract (FPE) groups before and after 3 month-administration (%).

Phyla	Placebo Group		FPE Group	
	0 Month	3 Months	0 Month	3 Months
Actinobacteria	16.3 ± 1.7	15.8 ± 1.7	16.8 ± 1.8	15.5 ±1.1
Bacteroidetes	36.1 ± 1.3	35.7 ± 1.5	32.2 ± 2.6	31.3 ± 1.2
Cyanobacteria	0	0	0.1 ± 0.1	0
Firmicutes	43.1 ± 1.5	42.6 ± 2.1	46.7 ± 1.5	50.2 ± 1.5 **
Fusobacteria	0.2 ± 0.1	0.1 ± 0.1	0.2 ± 0.1	0.1 ± 0.1
Lentisphaerae	0	0.1 ± 0.1	0.3 ± 0.2	0
Nitrospirae	0	0	0	0
Proteobacteria	3.7 ± 1.6	2.2 ± 0.4	3.7 ± 0.9	2.7 ± 0.5
Spirochaetes	0	0	0	0
Synergistetes	0	0	0	0
TM7	0	0	0	0
Tenericutes	0	0.3 ± 0.3	0	0
Verrucomicrobia	1.5 ± 1.4	3.2 ± 2.1	0	0

** $p < 0.01$, assessed using Student's t-test.

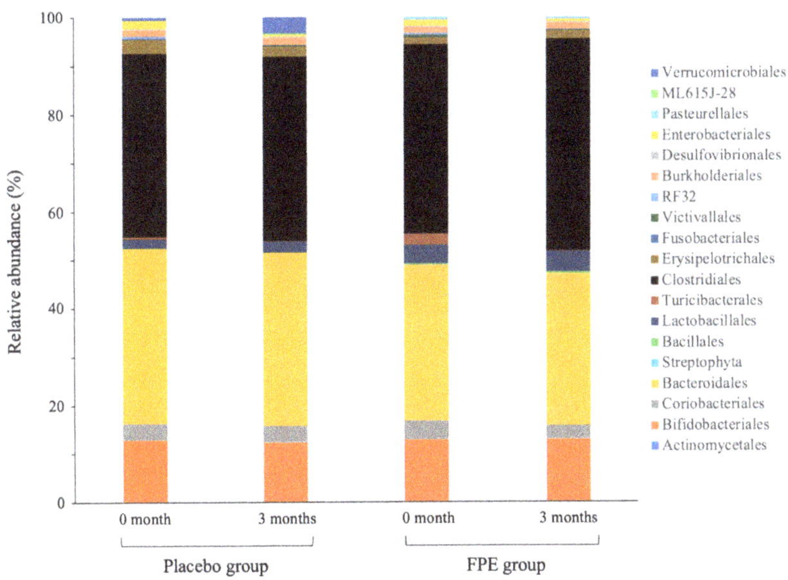

Figure 3. Microbiota composition at the order level in feces of placebo and FPE groups before and after 3-month administration.

Inflammatory bowel disease is generally treated with medicine. Lifestyle habits, particularly diet, have been implicated in the pathogenesis of inflammatory bowel disease [42]. Dietary therapy is assumed to lack life-threatening and severe side effects. However, a low-fat diet or a rich-fiber diet takes time and effort to prepare. Our study demonstrated that the daily diet supplemented with FPE directly prevented inflammation and changed the microbial composition in the gut, especially the increase of *Clostridiales* order, which plays a central role in suppressing inflammation, suggesting that FPE could improve the quality of human life against inflammatory bowel disease.

4. Conclusions

The daily diet was supplemented with FPE. Mice fed FPE with DSS for 7 days showed no body weight loss, atrophy of colonic length, or bloody stool. The concentrations of inflammation markers IL-6 and TNF-α did not increase in the mice, demonstrating that FPE supplementation suppressed inflammation. The microbiota composition in feces from humans eating a diet supplemented with FPE for 3 months showed increased *Clostridiales* order, which produces short-chain fatty acids to suppress inflammation. These results indicate that FPE supplementation could enhance *Clostridiales* growth in the gut and indicate that it can have anti-inflammatory effects on colitis.

Supplementary Materials: The following are available online at https://www.mdpi.com/article/10.3390/fermentation7020055/s1. Table S1: Materials of FPE for animal and human. Table S2: Composition of experimental diet (%). Table S3: Nutritional compound and amino acid content in FPE produced from 40 kinds of extracts. Table S4: Microbiota composition at the order level in feces of placebo and FPE groups before and after 3 month-administration (%). Figure S1: Body weight of mice after 7-day administration. The values are expressed as means ± SEM (n = 8). * $p < 0.05$, assessed using Bonferroni's multiple comparison test. Figure S2: Colon length of mice after 7-day administration. Figure S3: Change of body weight (A), fat mass (B), muscle mass (C), BMI (D), and health condition (E) of FPE group or placebo group. Values are expressed as means ± SD (n = 8) assessed using Student's t-test. Figure S3: Microbiota composition at the phyla level in feces of placebo and FPE groups before and after 3-month administration.

Author Contributions: M.S.: Conceptualization, Formal analysis, Investigation, Methodology, Writing—original draft, review and editing. T.W.: Conceptualization, Formal analysis, Investigation, Methodology, Writing—review and editing. M.T.: Conceptualization, Formal analysis, Investigation, Methodology, Writing—review and editing. K.S.: Formal analysis, Investigation, Methodology. T.M.: Resources, Writing—review and editing. N.M.: Resources, Writing—review and editing. S.S.: Resources, Methodology. All authors have read and agreed to the published version of the manuscript.

Funding: This research received no external funding.

Institutional Review Board Statement: Ethical review and approval were waived for this study. It was conducted according to legally established internal rules based on research ethics recommendations and with the informed consent of all study participants.

Informed Consent Statement: Informed consent was obtained from all study participants.

Data Availability Statement: The data described as a result of this study are available in the article and Supplementary Materials.

Conflicts of Interest: The authors declare that they have no conflict of interest.

References

1. Okabayashi, S.; Kobayashi, T.; Hibi, T. Inflammatory bowel disease in Japan—Is it similar to or different from Westerns? *J. Anus Rectum Colon* **2020**, *4*, 1–13. [CrossRef]
2. Japan Intractable Diseases Information Center. [Ulcerative Colitis (Designated Intractable Disease No. 97)]. Available online: https://www.nanbyou.or.jp/entry/62 (accessed on 5 April 2021). (In Japanese).
3. Bäckhed, F.; Ding, H.; Wang, T.; Hooper, L.V.; Koh, G.Y.; Nagy, A.; Semenkovich, C.F.; Gordon, J.I. The gut microbiota as an environmental factor that regulates fat storage. *Proc. Natl. Acad. Sci. USA* **2004**, *101*, 15718–15723. [CrossRef]
4. Wang, J.; Linnernbrink, M.; Künzel, S.; Fernandes, R.; Nadeau, M.J.; Rosenstiel, P.; Baines, J.F. Dietary history contributes to enterotype-like clustering and functional metagenomic content in the intestinal microbiome of wild mice. *Proc. Natl. Acad. Sci. USA* **2014**, *111*, 2703–2710. [CrossRef]
5. Willing, B.; Vörös, A.; Roos, S.; Jones, C.; Jansson, A.; Lindberg, J.E. Changes in faecal bacteria associated with concentrate and forage-only diets fed to horses in training. *Equine Vet. J.* **2009**, *41*, 908–914. [CrossRef]
6. Cani, P.D.; Bibiloni, R.; Knauf, C.; Waget, A.; Neyrinck, A.M.; Delzenne, N.M.; Burcelin, R. Changes in gut microbiota control metabolic endotoxemia-induced infammation in high-fat diet-induced obesity and diabetes in mice. *Diabetes* **2008**, *57*, 1470–1481. [CrossRef]

7. Delzenne, N.M.; Cani, P.D. Gut microbiota and the pathogenesis of insulin resistance. *Curr. Diab. Rep.* **2011**, *11*, 154–159. [CrossRef]
8. Turnbaugh, P.J.; Hamady, M.; Yatsunenko, T.; Cantarel, B.L.; Duncan, C.A.; Ley, R.E.; Sogin, M.L.; Jones, W.J.; Roe, B.A.; Affourtit, J.P.; et al. A core gut microbiome in obese and lean twins. *Nature* **2009**, *457*, 480–484. [CrossRef]
9. Belkaid, Y.; Hand, T.W. Role of the microbiota in immunity and infammation. *Cell* **2014**, *157*, 121–141. [CrossRef]
10. Cani, P.D.; Osto, M.; Geurts, L.; Everard, A. Involvement of gut microbiota in the development of low-grade inflammation and type 2 diabetes associated with obesity. *Gut Microbes* **2012**, *3*, 279–288. [CrossRef]
11. Cani, P.D.; Amar, J.; Iglesias, M.A.; Poggi, M.; Knauf, C.; Bastelica, D.; Neyrinck, A.M.; Fava, F.; Tuohy, K.M.; Chabo, C.; et al. Metabolic endotoxemia initiates obesity and insulin resistance. *Diabetes* **2007**, *56*, 1761–1772. [CrossRef]
12. Fecteau, M.E.; Pitta, D.; Vecchiarelli, B.; Indugu, N.; Kumar, S.; Gallagher, S.C.; Fyock, T.L.; Sweeney, R.W. Dysbiosis of the Fecal Microbiota in Cattle Infected with *Mycobacterium avium* subsp. paratuberculosis. *PLoS ONE* **2016**, *11*, e0160353.
13. Ley, R.E.; Bäckhed, R.; Turnbaugh, P.; Lozupone, C.A.; Knight, R.D.; Gordon, J.I. Obesity alters gut microbial ecology. *Proc. Natl. Acad. Sci. USA* **2005**, *102*, 11070–11075. [CrossRef] [PubMed]
14. Sommer, F.; Backhed, F. The gut microbiota—Masters of host development and physiology. *Nat. Rev. Microbiol.* **2013**, *11*, 227–238. [CrossRef] [PubMed]
15. Grajek, W.; Olejnik, A.; Sip, A. Probiotics, prebiotics and antioxidants as functional foods: A review. *Acta Biochim. Pol.* **2005**, *52*, 665–671. [CrossRef]
16. Backhed, F.; Ley, R.E.; Sonnenburg, J.L.; Peterson, D.A.; Gordon, J.I. Host-bacterial mutualism in the human intestine. *Science* **2005**, *307*, 1915–1920. [CrossRef]
17. Manson, J.M.; Rauch, M.; Gilmore, M.S. The commensal microbiology of the gastrointestinal tract. *Adv. Exp. Med. Biol.* **2008**, *635*, 15–28.
18. McCracken, V.J.; Lorenz, R.G. The gastrointestinal ecosystem: A precarious alliance among epithelium, immunity and microbiota. *Cell Microbiol.* **2001**, *3*, 1–11. [CrossRef]
19. Lievin-Le Moal, V.; Servin, A.L. The front line of enteric host defense against unwelcome intrusion of harmful microorganisms: Mucins, antimicrobial peptides, and microbiota. *Clin. Microbiol. Rev.* **2006**, *19*, 315–337. [CrossRef]
20. Furusawa, F.; Obata, Y.; Fukuda, S.; Endo, T.A.; Nakato, G.; Takahashi, D.; Nakanishi, Y.; Uetake, C.; Kato, K.; Kato, T.; et al. Commensal microbe-derived butyrate induces the differentiation of colonic regulatory T cells. *Nature* **2013**, *504*, 446–450. [CrossRef]
21. Atarashi, K.; Tanoue, T.; Oshima, K.; Suda, W.; Nagano, Y.; Nishikawa, H.; Fukuda, S.; Saito, T.; Narushima, S.; Hase, K.; et al. Treg induction by a rationally selected mixture of Clostridia strains from the human microbiota. *Nature* **2013**, *500*, 232–236. [CrossRef]
22. Atarashi, K.; Tanoue, T.; Shima, T.; Imaoka, A.; Kuwahara, T.; Momose, Y.; Cheng, G.; Yamasaki, S.; Saito, T.; Ohba, Y.; et al. Induction of colonic regulatory T cells by indigenous Clostridium species. *Science* **2011**, *331*, 337–341. [CrossRef]
23. Marco, M.L.; Heeney, D.; Binda, S.; Cifelli, C.J.; Cotter, P.D.; Foligné, B.; Ganzle, M.; Kort, R.; Pasin, G.; Pihlanto, A.; et al. Health benefits of fermented foods: Microbiota and beyond. *Curr. Opin. Biotechnol.* **2017**, *44*, 94–102. [CrossRef] [PubMed]
24. Bell, V.; Ferrao, J.; Fernandes, T. Nutritional guidelines and fermented food frameworks. *Foods* **2017**, *6*, 65. [CrossRef]
25. Moussa, L.; Bézirard, V.; Salvador-Cartier, C.; Bacquié, V.; Lencina, C.; Lévêque, M.; Braniste, V.; Ménard, S.; Théodorou, V.; Houdeau, E. A low dose of fermented soy germ alleviates gut barrier injury, hyperalgesia and faecal protease activity in a rat model of inflammatory bowel disease. *PLoS ONE* **2012**, *7*, e49547. [CrossRef]
26. Bondia-Pons, I.; Nordlund, E.; Mattila, I.; Katina, K.; Aura, A.M.; Kolehmainen, M.; Orešič, M.; Mykkänen, H.; Poutanen, K. Postprandial differences in the plasma metabolome of healthy Finnish subjects after intake of a sourdough fermented endosperm rye bread versus white wheat bread. *Nutr. J.* **2011**, *10*, 116. [CrossRef]
27. Mueller, T.; Voigt, W. Fermented wheat germ extract—Nutritional supplement or anticancer drug? *Nutr. J.* **2011**, *10*, 89. [CrossRef] [PubMed]
28. Ryan, E.P.; Heuberger, A.L.; Weir, T.L.; Barnett, B.; Broeckling, C.D.; Prenni, J.E. Rice bran fermented with *Saccharomyces boulardii* generates novel metabolite profiles with bioactivity. *J. Agric. Food. Chem.* **2011**, *59*, 1862–1870. [CrossRef] [PubMed]
29. Ibrahim, F.; Ruvio, S.; Granlund, L.; Salminen, S.; Viitanen, M.; Ouwehand, A.C. Probiotics and immunosenescence: Cheese as a carrier. *FEMS Immunol. Med. Microbiol.* **2010**, *59*, 53–59. [CrossRef] [PubMed]
30. Caporaso, J.G.; Lauber, C.L.; Walters, W.A.; Berg-Lyons, D.; Lozupone, C.A.; Turnbaugh, P.J.; Fierer, N.; Knight, R. Global patterns of 16S rRNA diversity at a depth of millions of sequences per sample. *Proc. Natl. Acad. Sci. USA* **2011**, *108*, 4516–4522. [CrossRef] [PubMed]
31. Ariefdjohan, M.W.; Savaiano, D.A.; Nakatsu, C.H. Comparison of DNA extraction kits for PCR-DGGE analysis of human intestinal microbial communities from fecal specimens. *Nutr. J.* **2010**, *9*, 23. [CrossRef]
32. Davani-Davari, D.; Negahdaripour, M.; Karimzadeh, I.; Seifan, M.; Mohkam, M.; Masoumi, S.J.; Berenjian, A.; Ghasemi, Y. Prebiotics: Definition, types, sources, mechanisms, and clinical applications. *Foods* **2019**, *8*, 92. [CrossRef] [PubMed]
33. Kuwaki, S.; Nakajima, N.; Tanaka, H. Plant-based paste fermented by lactic acid bacteria and yeast: Functional analysis and possibility of application to functional foods. *Biochem. Insights* **2012**, *5*, 21–29. [CrossRef]
34. Al-Sheraji, S.H.; Ismail, A.; Manap, M.Y.; Mustafa, S.; Yusof, R.M.; Hassan, F.A. Prebiotics as functional foods: A review. *J. Funct. Foods* **2013**, *5*, 1542–1553. [CrossRef]

35. Ze, X.; David, Y.B.; Laverde-Gomez, J.A.; Dassa, B.; Sheridan, P.O.; Duncan, S.H.; Louis, P.; Henrissat, B.; Juge, N.; Koropatkin, N.M.; et al. Unique organization of extracellular amylases into amylosomes in the resistant starch-utilizing human colonic *Firmicutes* bacterium *Ruinococcus Bromii*. *mBio* **2015**, *6*, e01058-15. [CrossRef]
36. Kaur, A.; Chen, T.; Green, S.J.; Mutlu, E.; Martin, B.R.; Rumpagaporn, P.; Patterson, J.A.; Keshavarzian, A.; Hamaker, B.R. Physical inaccessibility of a resistant starch shifts mouse gut microbiota to butyrogenic Firmicutes. *Mol. Nutr. Food Res.* **2019**, *63*, e1801012. [CrossRef]
37. Flint, H.J.; Duncan, S.H.; Scott, K.P.; Louis, P. Links between diet, gut microbiota composition and gut metabolism. *Proc. Nutl. Soc.* **2015**, *74*, 13–22. [CrossRef]
38. Vital, M.; Karch, A.; Pieper, D.H. Colonic butyrate-producing communities in humans: An overview using omics data. *mSystems* **2017**, *2*, e00130-17. [CrossRef]
39. Kanauchi, O.; Andoh, A.; Iwanaga, T.; Fujiyama, T.; Mitsuyama, K.; Toyonaga, A.; Bamba, T. Germinated barley foodstuffs attenuate colonic mucosal damage and mucosal nuclear factor kappa B activity in a spontaneous colitis model. *J. Gastroenterol.* **1999**, *14*, 1173–1179. [CrossRef]
40. Andoh, A.; Bamba, T.; Sasaki, M. Physiological and anti-inflammatory roles of dietary fiber and butyrate in intestinal functions. *J. Parenter. Enteral. Nutr.* **1999**, *23*, S70–S73. [CrossRef]
41. Lopetuso, L.R.; Scaldaferri, F.; Petito, V.; Gasbarrini, A. Commensal Clostridia: Leading players in the maintenance of gut homeostasis. *Gut Phathogens* **2013**, *5*, 23. [CrossRef]
42. Levi, A.J. Diet in the management of Crohn's disease. *Gut* **1985**, *26*, 985–988. [CrossRef] [PubMed]

fermentation

MDPI

Article

Chinese Traditional Fermented Soy Sauce Exerts Protective Effects against High-Fat and High-Salt Diet-Induced Hypertension in Sprague-Dawley Rats by Improving Adipogenesis and Renin-Angiotensin-Aldosterone System Activity

Bao Zhong [1,2], Eun-Gyung Mun [1], Jin-Xi Wang [1] and Youn-Soo Cha [1,*]

1 Department of Food Science and Human Nutrition, Jeonbuk National University and Obesity Research Center, Jeonju-si 54896, Korea; zhongbao19870626@163.com (B.Z.); egmun1982@gmail.com (E.-G.M.); jinxiwang3@gmail.com (J.-X.W.)

2 School of Food Engineering, Jilin Agriculture Science and Technology University, Hanlin Road, Changyi District, Jilin 132000, China

* Correspondence: cha8@jbnu.ac.kr; Tel.: +82-63-270-3822

Citation: Zhong, B.; Mun, E.-G.; Wang, J.-X.; Cha, Y.-S. Chinese Traditional Fermented Soy Sauce Exerts Protective Effects against High-Fat and High-Salt Diet-Induced Hypertension in Sprague-Dawley Rats by Improving Adipogenesis and Renin-Angiotensin-Aldosterone System Activity. *Fermentation* **2021**, *7*, 52. https://doi.org/10.3390/fermentation7020052

Academic Editor: Hiroshi Kitagaki

Received: 16 March 2021
Accepted: 3 April 2021
Published: 5 April 2021

Abstract: Although high-fat and high-salt diets are considered risk factors for hypertension, the intake of salty soybean-based fermented foods has beneficial effects. This study explored the potential of Chinese traditional fermented soy sauce (CTFSS) in preventing hypertension by analyzing its effects on adipogenesis and the renin-angiotensin-aldosterone system (RAAS). Male Sprague-Dawley (SD) rats were divided into four groups ($n = 6$): normal diet (ND), high-fat diet (HD), high-fat diet with saline (HDS, NaCl-8%), and high-fat diet with Chinese traditional soy sauce (HDCTS, NaCl-8%). Each group is administrated 12 weeks by oral gavage as 10 mL/kg dose, respectively. CTFSS supplementation resulted in significantly lower body weight, epididymal fat weight, and systolic blood pressure. Additionally, it decreased the serum total cholesterol (TC), triglyceride (TG), alanine aminotransferase (ALT), aspartate aminotransferase (AST), renin, angiotensin II (Ang II), angiotensin-converting enzyme (ACE), and aldosterone levels. It also increased the urinary volume and improved sodium and potassium ion balance. The gene levels showed significant enhancements in the mRNA levels of renin-angiotensin-aldosterone system-related and adipogenesis-related genes. In addition, CTFSS may prevent hypertension-associated kidney injury. Therefore, this study demonstrates that CTFSS has no harmful effects on hypertension. In contrast, the beneficial effects of CTFSS intake in ameliorating hypertension were shown.

Keywords: soy sauce; salt; hypertension; adipogenesis; renin-angiotensin-aldosterone system

1. Introduction

Hypertension is a growing public health challenge worldwide. It affects approximately 40% of the world population over the age of 25, and by 2025, the global prevalence of hypertension will increase by 60%, and cardiovascular disease (CVD) is becoming an increasingly common health problem worldwide [1]. Hypertension can induce more than half of the coronary heart disease burden, leading to premature death [2]. A high-salt and high-fat diet is an important risk factor for hypertension, atherosclerosis, and coronary heart disease [3]. The World Health Organization (WHO) recommends less than 5 g/day salt intake, whereas the Chinese Nutrition Society (CNS) recommends a salt intake of less than 6 g/day. However, the daily salt intake in Chinese population is approximately 10.5 g/day. Long-term intake of large amounts of sodium can easily lead to the retention of sodium and water, leading to increased blood pressure [4]. Obesity is a metabolic syndrome, which accounts for hypertension in approximately 72% of men and women [5].

High-fat diet may increase lipid levels and promote angiotensinogen (AGT) production by adipocytes, leading to renin-angiotensin-aldosterone system (RAAS) activation [6,7]. High-fat diet-fed mice showed increased expression of angiotensin II (Ang II), angiotensin-converting enzyme (ACE), and aldosterone in the kidney, indicating increased RAAS activity [8]. High sodium intake leads to RAAS dysfunction, and hypertension is associated with activated RAAS [9]. High salt sensitivity has been shown to lead to obesity and vascular dysfunction via RAAS activation [10]. Salt increases the metabolites involved in the RAAS, such as aldosterone synthase (Cyp11b2) and mineralocorticoid receptor (MR), and promotes rising serum aldosterone levels, sodium retention and volume expansion, renal disease, and elevated blood pressure [11,12]. High-salt diets accelerate hypertension and renal inflammation/injury in male rats such as low-grade renal histological injury, glomerular hyaline casts, interstitial fibrosis (peritubular), glomerular sclerosis, tubular atrophy [13].

Soy sauce originated in China over 2500 years ago. Presently, soybean-fermented foods are popular worldwide, especially in most East Asian countries [14]. Soybean fermentation can increase the content of isoflavones in aglycones, thereby increasing the bioavailability of isoflavones [15]. Koji is made from wheat and soybeans fermented with *Aspergillus oryzae* and is essential for the production of Chinese traditional fermentation soy sauce (CTFSS) [16]. A placebo-controlled double-blind trial showed that supplementation with *koji* decreased systolic and diastolic pressure significantly in patients with hypertension [17]. Other than just a seasoning, soy sauce is also a potential functional food with anti-obesity and anti-hypertensive effects. In previous studies, soy sauce intake was found to alter the fat metabolism of *Caenorhabditis elegans* [18]. Traditional Korean soy sauce can ameliorate hypertension in rats [19].

The RAAS is a hormonal cascade that functions in the homeostatic control of arterial pressure, and the imbalance of the RAAS is crucial in the pathogenesis of cardiovascular and renal disorders [20]. Salty fermented soybean foods, such as Japanese soy sauce, can effectively inhibit ACE activity, thereby regulating blood pressure [21]. Korean soy sauce is manufactured using meju, unlike its Chinese counterpart. Although there are differences in the manufacturing methods of Chinese soy sauce and Korean soy sauce, the fermentation bacteria are similar, and all the fermented metabolites contain lactic acid bacteria and *Bacillus* [22,23]. Chinese soy sauce and Japanese soy sauce were prepared using the same method [24]. However, consumption of Japanese soy sauce does not increase the blood pressure [25]. Most studies on Chinese soy sauce have focused on analysis of aroma and composition content. However, research on the functionality of Chinese fermented soy sauce is limited. We hypothesized that the intake of Chinese traditional soybean-fermented foods has a regulatory effect on high-fat and high-salt diet-induced hypertension. Therefore, in this study, the antihypertensive effects of CTFSS were investigated in Sprague-Dawley (SD) rats.

2. Results

2.1. Metabolic Characterization, Na^+ and K^+ Ion Concentration, and Serum Chemistry

The results are shown in Table 1. The initial body weight was insignificantly different between all the groups. The final body weight significantly decreased in the HDCTS group compared to that in the HD and HDS groups. The epididymal fat weight of the HDCTS group was significantly lower than that of the HD and HDS groups. Diet intake in ND, HD, and HDCTS groups compared to the HDS group were significantly different, while same was insignificantly different between the ND, HD, and HDCTS groups. Results of the 24-h metabolic cage experiment showed that urine metabolism significantly increased in the HDCTS group compared to the other groups. No significant difference in fecal excretion in all groups. Water intake was significantly increased in the HDS and HDCTS groups than in the ND and HD groups.

Table 1. Metabolic characterization, Na^+ and K^+ ion concentration, and serum chemistry.

Group		ND	HD	HDS	HDCTS
Metabolic characterization					
Initial body weight (g)		116.23 ± 3.18	116.98 ± 8.33	117.22 ± 7.06	116.12 ± 5.63
Final body weight (g)		454.52 ± 6.94 [c]	556.69 ± 28.14 [a]	547.65 ± 33.82 [ab]	534.98 ± 35.78 [b]
Epididymal fat weight (g)/BW (g)%		1.84 ± 0.25 [c]	3.64 ± 0.48 [a]	3.46 ± 0.26 [a]	2.82 ± 0.38 [b]
Food intake (g/day)		17.22 ± 0.19 [a]	17.05 ± 0.10 [a]	16.51 ± 0.39 [b]	17.14 ± 0.07 [a]
Water intake (mL/day)		16.33 ± 1.18 [c]	17.71 ± 1.48 [c]	25.08 ± 2.16 [a]	21.65 ± 2.50 [b]
Urinary volume (mL/day)		9.58 ± 1.88 [c]	9.63 ± 0.77 [c]	13.50 ± 1.22 [b]	15.08 ± 0.66 [a]
Fecal excretion (g/day)		3.31 ± 0.14	3.24 ± 0.15	3.16 ± 0.19	3.32 ± 0.14
Na^+ and K^+ ion concentration in urine and feces (ppm)					
Urine	Na^+	2130.33 ± 698.16 [b]	1844.92 ± 620.88 [b]	9314.08 ± 978.00 [a]	8778.68 ± 354.74 [a]
	K^+	9292.65 ± 2594.60 [ab]	10,084.34 ± 1359.43 [a]	6837.20 ± 459.40 [b]	7046.87 ± 1079.26 [ab]
Feces	Na^+	722.82 ± 317.74	796.94 ± 339.65	1116.71 ± 200.52	603.91 ± 283.51
	K^+	6387.93 ± 1357.67	6232.49 ± 756.63	4734.04 ± 246.97	4795.09 ± 708.70
Serum chemistry					
TC (mg/dL)		66.44 ± 6.09 [c]	95.33 ± 4.35 [a]	91.16 ± 8.97 [a]	78.28 ± 5.83 [b]
TG (mg/dL)		51.52 ± 4.19 [d]	116.91 ± 4.66 [a]	97.83 ± 5.90 [b]	60.44 ± 7.03 [c]
AST (IU/L)		39.41 ± 2.43 [d]	56.42 ± 2.93 [a]	52.59 ± 1.70 [b]	42.68 ± 1.29 [c]
ALT(IU/L)		11.10 ± 1.28 [b]	18.66 ± 1.65 [a]	17.70 ± 0.74 [a]	11.34 ± 1.67 [b]

Values are shown as the means ± standard deviation and statistical difference from each other at $p < 0.05$ by Duncan's multiple range test (a > b > c > d). ND: normal diet; HD: high-fat diet; HDS: high-fat diet + saline; HDCTS: high-fat diet + Chinese traditional soy sauce.

Compared to the ND and HD groups, urine concentration of Na^+ was significantly increased in the HDS group, whereas the HDCTS group showed a downward trend in the concentration of urine Na^+ compared to the HDS group, although this difference was statistically insignificant. The concentration of urine K^+ was significantly higher in the HDCTS group than in the HDS group. In feces, the HDCTS group when compared with the HDS group shows the Na^+ and K^+ to be statistically insignificant.

In HD and HDS groups showed significantly higher levels of serum TC, TG, AST, and ALT compared to the HDCTS group.

2.2. Systolic Blood Pressure

The alterations in systolic blood pressure (SBP) are shown in Figure 1. The initial SBP did not differ significantly between the groups. At six weeks of experiments, in the HDCTS group, blood pressure was significantly attenuated compared to the HD and HDS groups. SBP was significantly higher in the HD and HDS groups than in the other groups at 12 weeks. However, the final SBP was significantly attenuated in the HDCTS group. (Supplementary Figure S1).

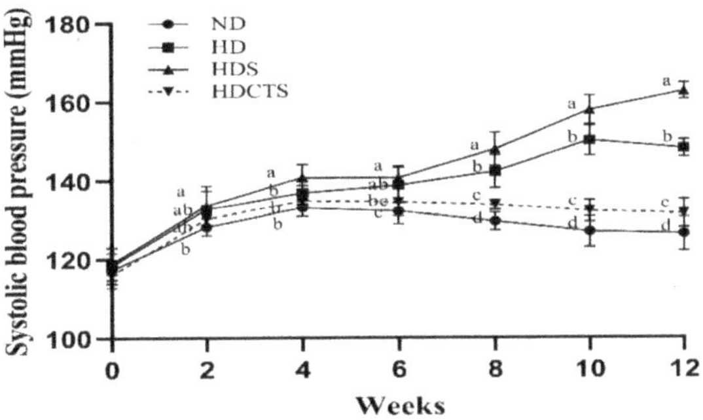

Figure 1. Changes in SBP during the experiment. Values are shown as the means ± standard deviation, and statistical differences are expressed in letters (a > b > c > d). ND: normal diet; HD: high-fat diet; HDS: high-fat diet + saline; HDCTS: high-fat diet + Chinese traditional soy sauce.

2.3. Renin, Ang II, ACE, and Aldosterone Levels in Serum

The test results are shown in Figure 2. The serum levels of renin, Ang II, ACE, and aldosterone were significantly higher in the HD and HDS groups than in the ND group, and a significant increase was observed in the HDCTS group.

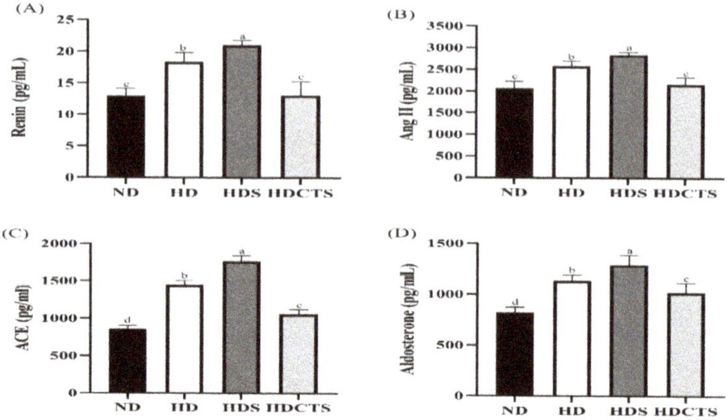

Figure 2. Renin-angiotensin-aldosterone relative levels in serum. (**A**) Renin level of serum; (**B**) Ang II level of serum; (**C**) ACE level of serum; (**D**) Aldosterone level of serum. Values are shown as the means ± standard deviation. Values with different superscripts letters (a > b > c > d) are significantly different among groups. ND: normal diet; HD: high-fat diet; HDS: high-fat diet + saline; HDCTS: high-fat diet + Chinese traditional soy sauce.

2.4. Histopathology of Kidney

Histological examination confirmed significant glomerular hypertrophy and slight hyaline degeneration in the kidney sections from rats in the HD and HDS groups compared to the ND group. The glomerular area and hyaline degeneration in the HDCTS group were significantly lower than the HD and HDS groups (Figure 3).

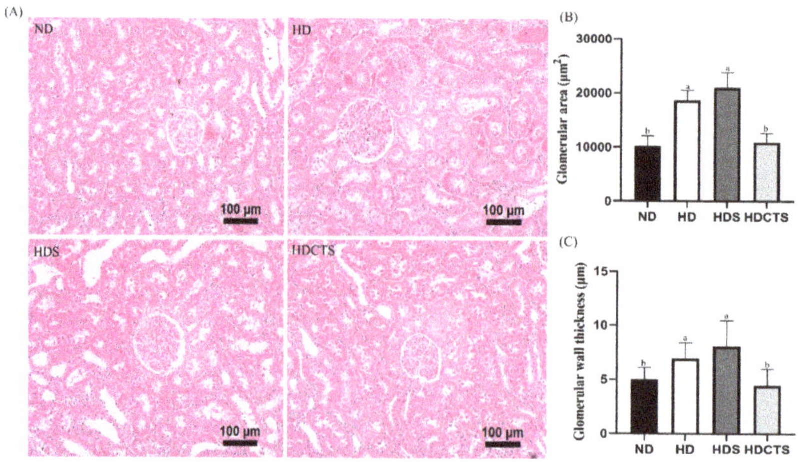

Figure 3. Histological analysis of kidney tissue. (**A**) Histological examination (20×) in kidney; (**B**) Glomerular area; and (**C**) Glomerular wall thickness. Values are shown as the means ± standard deviation. Values with different superscripts letters (a > b) are significantly different among groups. ND: normal diet, HD: high-fat diet, HDS: high-fat diet + saline, and HDCTS: high-fat diet + Chinese traditional soy sauce.

2.5. Expression of Adipogenesis-Regulating and RAAS-Related Genes in Liver Tissues

The mRNA expression of adipogenesis-related genes, leptin and PPARγ, significantly increased in the HD and HDS groups compared to the ND group. However, the levels of these genes in the HDCTS group were lower than those in the HD and HDS groups. The mRNA level of adiponectin was significantly higher in the HDCTS group than in the HD and HDS groups. The expression of the RAAS-related gene AGT was significantly increased in the HD and HDS groups compared to the ND group, and was downregulated in the HDCTS group (Figure 4).

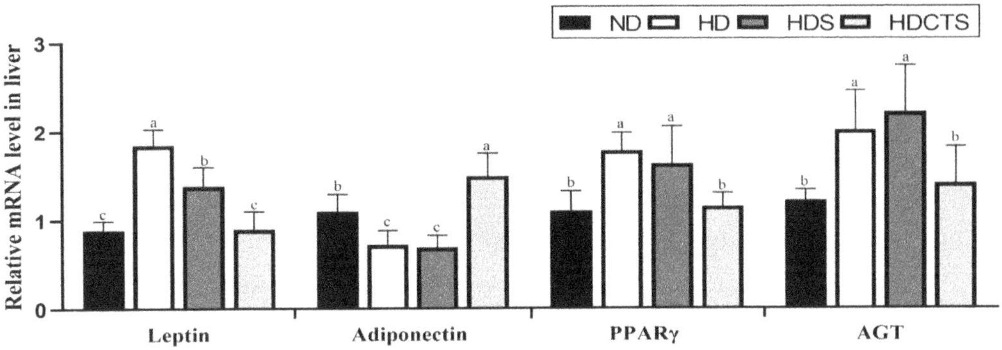

Figure 4. The mRNA expression of adipogenesis- and RAAS-related genes in the liver tissue. Values are shown as the means ± standard deviation. Values with different superscripts letters (a > b > c) are significantly different among groups. ND: normal diet; HD: high-fat diet; HDS: high-fat diet + saline; HDCTS: high-fat diet + Chinese traditional soy sauce; PPARγ: Peroxisome proliferator-activated receptor gamma; AGT: Angiotensinogen.

2.6. Expression of RAAS-Related Genes in Kidney Tissues

The mRNA expression of RAAS-related genes, renin, ACE, and AT1 was significantly overexpressed in the HD and HDS groups compared to the ND group, whereas the HDCTS group significantly downregulated the expression levels of these genes. In the HD and HDS groups, Cyp11a1, Cyp11b2, Hsd3b1, Star, and MR levels were significantly increased compared to the ND group, and in the HDCTS group, these genes were significantly decreased. Notably, the expression of Rnls was significantly reduced in the HD and HDS groups compared to the ND group, and increased in the HDCTS group (Figure 5).

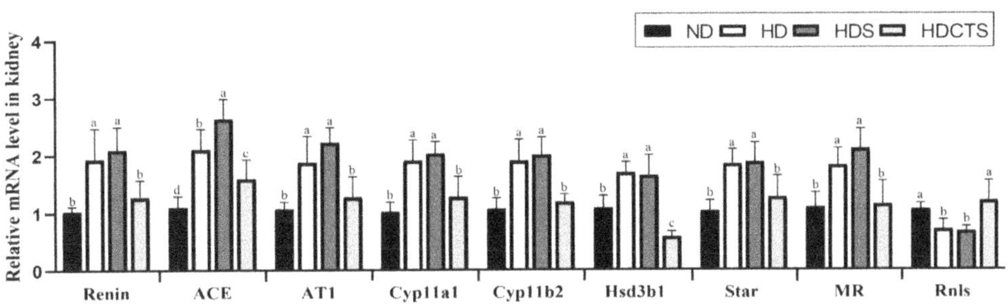

Figure 5. The RAAS-related mRNA expression levels in the kidney tissue. Values are shown as the means ± standard deviation. Values with different superscripts letters (a > b > c > d) are significantly different among groups. ND: normal diet; HD: high-fat diet; HDS: high-fat diet + saline; HDCTS: high-fat diet + Chinese traditional soy sauce; ACE: Angiotensin-converting enzyme; AT1: Angiotensin type 1 receptor; Cyp11a1: Cholesterol side-chain cleavage enzyme; Cyp11b2: Aldosterone synthase; Hsd3b1: 3β-hydroxysteroid dehydrogenase type 1; Star: Steroidogenic acute regulatory protein; MR: Mineralocorticoid receptor; Rnls: Renalase.

3. Discussion

The occurrence of high blood pressure is related to changes in lifestyle, such as increase sodium intake, decreased potassium intake, and decreased vegetarian diet [26]. High-fat diet consumption, which causes increases the risk of hypertension [27]. The high sodium intake and the increase in SBP levels are related to water and sodium retention and modification of sympathetic activity [28]. Some evidence suggests that soybean-fermented foods, such as Korean paste, soy sauce, and Japanese miso, can regulate obesity and blood pressure [19,29,30]. In this study, we demonstrated that the intake of CTFSS for 12 weeks prevented body weight gain, epididymal fat weight gain, and increased blood pressure.

Salty soybean fermentation food supplementation is related to decreases in serum lipid levels, which may be partially due to the higher content of soybean isoflavones in soybean fermentation food [31,32]. In this study, the serum levels of TC, TG, ALT, and AST levels significantly decreased in the HDCTS group than in the HD and HDS groups. The amount of renin in the serum is the key rate-limiting step determining the level of ACE [33]. Increased aldosterone levels in serum are a risk of hypertension [34]. In this study, CTFSS supplementation significantly lowered the expression of serum renin, ACE, Ang II, and aldosterone.

Early manifestations of renal injury caused by hypertension include structural changes, such as glomerular hypertrophy, followed by thickening of the glomerular walls and hyaline degeneration [13]. Numerous studies have demonstrated that Ang II induces glomerular hypertrophy and glomerulosclerosis by activating its specific receptor [35]. In the present study, significant hypertrophy and slight hyaline degeneration of glomeruli were observed in the HD and HDS groups. However, significant prevention of glomerular hypertrophy and hyaline degeneration was observed in the HDCTS group. Therefore, in this study, we demonstrated that CTFSS may prevent hypertension-associated renal injury.

High leptin and low adiponectin levels are characteristic of obesity. PPARγ plays a vital role in the early stage of adipose differentiation [36]. In this study, CTFSS suppressed leptin and PPARγ gene expression, and increased adiponectin gene levels. Therefore, the results of our study showed that CTFSS exerts lipid-lowering effects. Liver-derived AGT is the precursor to all angiotensin and is converted via renin to form angiotensin I (Ang I) and induces RAAS activation, thereby increasing blood pressure [37]. Previous studies have shown that the expression level of AGT in obese mice increased significantly, and AGT knockout showed lower blood pressure regulation [38]. Increased leptin and PPARγ can modulate the local production of AGT, thereby increasing blood pressure by activating the RAAS [39]. Decreased adiponectin expression causes an increase in angiotensinogen, thereby increasing blood pressure [40]. Salt intake has an inappropriate augmentation of AGT, which may contribute to hypertension [41]. In this study, AGT levels were significantly increased in the HD and HDS groups compared to the ND group, and a significant downregulation of AGT levels was observed in the HDCTS group.

The renin activity may be a marker of CVD risk in hypertensive patients [42]. ACE can generate the vasoactive peptide Ang II by cleaving 2 amino acids from the C-terminus of the inactive precursor Ang I [43]. Obese subjects, significant increases in ACE activity, and Ang II, thereby induced elevated SBP [44,45]. High-salt diet-induced hypertension in rats and renin and ACE levels increased significantly [46]. However, miso sauce intake has potent angiotensin-converting enzyme inhibitory effects and can reduce nighttime blood pressure [47]. In this study, CTFSS supplementation decreased the gene expression of renin and ACE.

The high-salt diet stimulates glomerular oxidative stress, which leads to AT1 receptor upregulation, subsequently causing sodium retention and hypertension [48]. Ang II may activate Cyp11b2 and Star expression and promote aldosterone secretion [49,50]. Cyp11a1 catalyzes the side-chain cleavage of cholesterol, and Cyp11b2 catalyzes the final steps in the biosynthesis of MR [51]. Increased levels of Hsd3b1 and Star genes may upregulate blood pressure via changes in aldosterone [52,53]. Furthermore, MR-mediated changes in the cardiovascular system are potentiated by activation of AT1 receptors [54]. A previous study reported that Korean soy sauce might decrease AT1, aldosterone, and MR levels, and downregulate the expression of Na^+/K^+ ATPaseα1, and improve Na^+ reabsorption, thereby decreasing blood pressure [19]. The present study demonstrated that CTFSS supplementation significantly decreased the mRNA levels of AT1, Cyp11a1, Cyp11b2, Hsd3b1, Star, and MR. In the HDCTS group, the balance of sodium and potassium ions also showed an adjustment, and urine volume increased significantly.

Previous studies have shown that sympathetic nervous system activity is heightened by the activation of brain regions controlling autonomic function due to a high-fat diet, salt, and Ang II [55]. Inappropriate activation of intrarenal RAAS may not contribute directly to the decreased expression of renalase (Rnls) in the rats that received a high-salt diet. However, activated RAAS promotes sympathetic nervous activation, which may decrease the renal Rnls expression [56]. Our research found that the expression level of Rnls was significantly reduced in the HD and HDS groups and increased significantly in the HDCTS group. In the present study, the results suggest that the consumption of CTFSS may have a lower risk of hypertension (Figure 6).

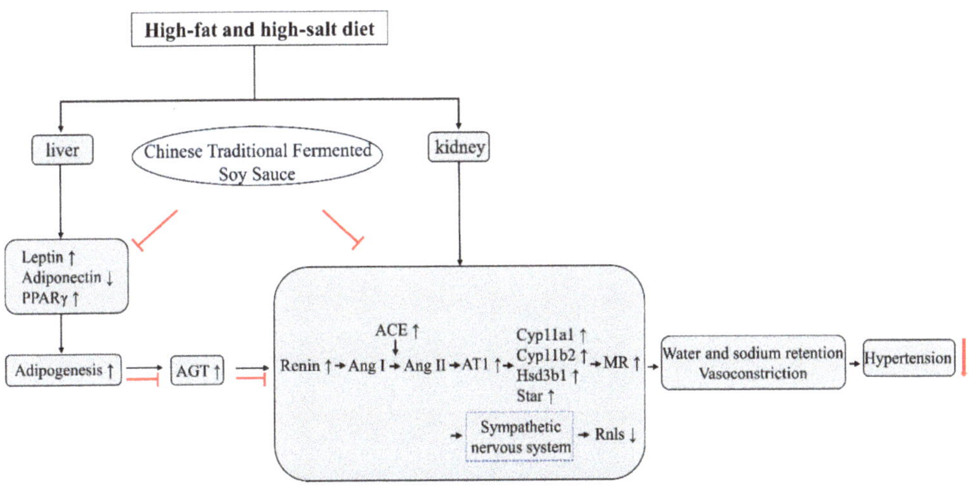

Figure 6. Mechanism underlying the hypertension-regulating action of CTFSS. When compared to high-fat and high-salt intake, adipogenesis and RAAS-relative levels were regulated with CTFSS. PPARγ: Peroxisome proliferator-activated receptor gamma; AGT: Angiotensinogen; Ang I: Angiotensin I; Ang II: Angiotensin II; ACE: Angiotensin-converting enzyme; AT1: Angiotensin type 1 receptor; Cyp11a1: Cholesterol side-chain cleavage enzyme; Cyp11b2: Aldosterone synthase; Hsd3b1: 3β-hydroxysteroid dehydrogenase type 1; Star: Steroidogenic acute regulatory protein; MR: Mineralocorticoid receptor; Rnls: Renalase. "T" bars: The inhibitory effect of Chinese traditional fermented soy sauce.

Chinese soybean fermented foods, such as soybean paste and soy sauce, are important sources of salt intake [57,58]. The intake of a high-fat and high-salt diet is considered a risk factor for hypertension [28,59]. However, the present study showed that CTFSS has no harmful effects, such as elevated blood pressure. In contrast, the intake of CTFSS in rats fed with a high-fat and high-salt diet might alleviate hypertension.

4. Materials and Methods

4.1. Preparation of Soy Sauce

The CTFSS was supplied by JiangCheng Brewing Group Co., Ltd. (Jilin, China). Soybeans were washed and soaked overnight, then steamed at 121 °C for 10 min, cooled to room temperature, mixed with bran-free raw wheat flour, inoculated with *Aspergillus oryzae*, and kept at 30 °C for 48 h. Koji was mixed with 2.5 times volume of brine solution, fermented and separated into liquid phases, and soy sauce was harvested (Figure 7). The salinity of the soy sauce was adjusted to 8% using distilled water by analysis using Mohr's method [60].

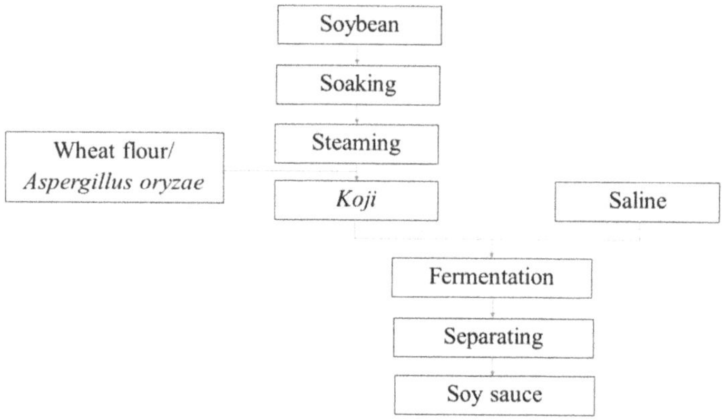

Figure 7. Procedure for manufacturing Chinese traditional fermented soy sauce.

4.2. Animal Experiment

Twenty-four male, five-week-old SD rats were purchased from DooYeol Biotech (Seoul, Korea). After adaptation for 1 week, SD rats were divided into four groups (n = 6) by non-significantly different blood pressure and body weight: normal diet (ND), high-fat diet (HD), high-fat diet and saline (HDS, NaCl-8%), and high-fat diet and Chinese traditional soy sauce (HDCTS, NaCl-8%). Wooden chips (JRS, Rosenberg, Germany) were used as bedding in the cages. The animal room temperature was maintained at 23 ± 1 °C, and the relative humidity was maintained at 65 ± 5%. The lighting was adjusted at 12 h light/dark cycle. The rats had free access to food and tap water. Diet intake was recorded every 2 d. All the rats were orally gavage administered by gastric intubation for 12 weeks, with the dose 10 mL/kg body weight. (Supplementary Figure S2).

4.3. Measurement of Body Weight and Blood Pressure

During feeding, body weight was measured once a week, and blood pressure was measured every 2 weeks using the tail-cuff method (BP-2000 series II blood pressure analysis system, Apex, NC, USA) after 3 h of oral gavage.

4.4. Metabolic Cages Experiment and Urine and Feces Analysis

During the last 4 weeks of the experiment, rats were kept in metabolic cages for 24 h a week, all feces and urine were collected, and water intake was recorded. Concentrations of sodium and potassium ions in urine and feces were analyzed at the Center for University-wide Research Facilities (CURF) at Jeonbuk National University by inductively coupled plasma-mass spectrometry (ICP-MS; 7500A, Agilent Technologies, Germantown, MD, USA).

4.5. Sacrifice and Administration

After 12 weeks of the experimental period, the SD rats were sacrificed by anesthetization after 12 h of overnight fasting. Blood was drawn from the abdominal artery, and the serum was separated by centrifugation at 3000 rpm for 15 min at 4 °C. Collected epididymal fat and examined the weight without testis. The tissue samples were snap-frozen in liquid nitrogen and stored at −80 °C until analyzed.

4.6. Analysis of Serum Biochemical Parameters

Serum total cholesterol (TC), triglyceride (TG), alanine aminotransferase (ALT), and aspartate aminotransferase (AST) levels were analyzed by using commercial kits (Asan Pharmaceutical Co., Seoul, Korea).

The serum concentrations of renin, Ang II, ACE, and aldosterone were analyzed by using enzyme-linked immunosorbent assay (ELISA) kits (MyBioSource, San Diego, CA, USA; Enzo Life Sciences, Farmingdale, NY, USA).

4.7. Kidney Histology

Kidney tissue samples were fixed with 10% formalin solution overnight and embedded in paraffin. The tissue samples were then cut into 5-μm-thick sections and stained with hematoxylin and eosin (H&E). Stained areas were viewed using a Leica Microsystems CMS GmbH (Wetzlar, Germany) at 20× magnification, and images were analyzed using SIS 3.2 software (Soft-Imaging System), and glomerular area and glomerular wall thickness were measured using Image J software (National Institutes of Health, Bethesda, MD, USA).

4.8. Gene Expression Analysis by Real-Time Polymerase Chain Reaction (RT-PCR)

Total RNA was extracted from kidney and liver tissues using the RNAiso Plus reagent (Takara, Japan). Total RNA was reverse-transcribed into cDNA using a cDNA synthesis kit (Takara, Kusatsu, Japan). PCR was performed in a 7500 Real-Time PCR system (Applied Biosystems, Foster City, CA, USA) using SYBR Green PCR Master Mix (TOYOBO, Japan). Relative gene expression was calculated using β-actin as an internal control. The primer sequences used in this were obtained from PrimerBank. 4.9. Statistical analyses

The results are expressed as the mean ± standard deviation. Data significant values ($p < 0.05$) were analyzed using one-way ANOVA with SPSS 23 (IBM, New York, NY, USA). Differences among the groups were determined using Duncan's multiple range test (a > b > c > d).

5. Conclusions

CTFSS supplementation decreases body weight, epididymal fat weight, and blood pressure in rats, as well as decreased serum renin, Ang II, ACE, aldosterone, and lipid levels, and promotes sodium and potassium ion balance. In addition, the expression levels of adipogenesis- and RAAS-related genes were improved in the tissues. The results of this study suggest that improvement of RAAS-related mRNA levels and adipogenesis mRNA levels might be an underlying mechanism involved in the amelioration of the hypertension by CTFSS. In addition, CTFSS ameliorated hypertension-associated kidney injury. Therefore, further experiments are needed to explore the beneficial effects of soy sauce to beneficial value on hypertension.

Supplementary Materials: The following are available online at https://www.mdpi.com/article/10.3390/fermentation7020052/s1, Figure S1: Changes in rat's heart rate during the experiment. Figure S2: Experimental design schedule.

Author Contributions: B.Z. completed the experiment, analyzed the results, and wrote the original draft of manuscript; E.-G.M. reviewed and revised the manuscript; J.-X.W. provided assistance; Y.-S.C. was associated with project administration, funding acquisition, and supervision. All authors have read and agreed to the published version of the manuscript.

Funding: This work was supported by the National Research Foundation of Korea (NRF) grant funded by the Korean government (MSIT). (No. 2018R1A2B6006477).

Institutional Review Board Statement: Institutional Animal Care and Use Committee of Chonbuk National University approved the animal experimental protocol. (CBNU 2016-0030).

Informed Consent Statement: Not applicable.

Data Availability Statement: The data presented in this study are available on request from the corresponding author.

Conflicts of Interest: The authors declare no conflict of interest.

References

1. Garfinkle, M.A. Salt and essential hypertension: Pathophysiology and implications for treatment. *J. Am. Soc. Hypertens.* **2017**, *11*, 385–391. [CrossRef] [PubMed]
2. Shi, H.; Yang, D.; Qiao, J.; Sun, R.; Li, R.; Zhu, C.; Jing, R.; Liu, L.; Huang, L.; Li, L. Effects of sang-qi granules on blood pressure and endothelial dysfunction in stage I or II hypertension: Study protocol for a randomized double-blind double-simulation controlled trial. *Trials* **2020**, *21*, 12. [CrossRef] [PubMed]
3. Ketonen, J.; Mervaala, E. Effects of dietary sodium on reactive oxygen species formation and endothelial dysfunction in low-density lipoprotein receptor-deficient mice on high-fat diet. *Heart Vessel.* **2008**, *23*, 420–429. [CrossRef] [PubMed]
4. Whitworth, J.A.; World Health Organization, International Society of Hypertension Writing Group. 2003 World Health Organization (WHO)/International Society of Hypertension (ISH) statement on management of hypertension. *J. Hypertens.* **2003**, *21*, 1983–1992. [CrossRef] [PubMed]
5. Chung, S.; Park, C.W.; Shin, S.J.; Lim, J.H.; Chung, H.W.; Youn, D.Y.; Kim, H.W.; Kim, B.S.; Lee, J.H.; Kim, G.H.; et al. Tempol or candesartan prevents high-fat diet-induced hypertension and renal damage in spontaneously hypertensive rats. *Nephrol. Dial. Transplant.* **2010**, *25*, 389–399. [CrossRef] [PubMed]
6. Sapouckey, S.A.; Deng, G.; Sigmund, C.D.; Grobe, J.L. Potential mechanisms of hypothalamic renin-angiotensin system activation by leptin and DOCA-salt for the control of resting metabolism. *Physiol. Genom.* **2017**, *49*, 722–732. [CrossRef] [PubMed]
7. Engeli, S.; Schling, P.; Gorzelniak, K.; Boschmann, M.; Janke, J.; Ailhaud, G.; Teboul, M.; Massiera, F.; Sharma, A.M. The adipose-tissue renin-angiotensin-aldosterone system: Role in the metabolic syndrome? *Int. J. Biochem. Cell Biol.* **2003**, *35*, 807–825. [CrossRef]
8. Li, C.; Culver, S.A.; Quadri, S.; Ledford, K.L.; Al-Share, Q.Y.; Ghadieh, H.E.; Najjar, S.M.; Siragy, H.M. High-fat diet amplifies renal renin angiotensin system expression, blood pressure elevation, and renal dysfunction caused by Ceacam1 null deletion. *Am. J. Physiol. Endocrinol. Metab.* **2015**, *309*, E802–E810. [CrossRef]
9. Spradling-Reeves, K.D.; Shade, R.E.; Haywood, J.R.; Cox, L.A. Primate response to angiotensin infusion and high sodium intake differ by sodium lithium countertransport phenotype. *J. Am. Soc. Hypertens.* **2017**, *11*, 178–184. [CrossRef]
10. Shibata, S.; Mu, S.; Kawarazaki, H.; Muraoka, K.; Ishizawa, K.; Yoshida, S.; Kawarazaki, W.; Takeuchi, M.; Ayuzawa, N.; Miyoshi, J.; et al. Rac1 GTPase in rodent kidneys is essential for salt-sensitive hypertension via a mineralocorticoid receptor-dependent pathway. *J. Clin. Investig.* **2011**, *121*, 3233–3243. [CrossRef]
11. Lee, M.; Sorn, S.R.; Lee, Y.; Kang, I. Salt Induces Adipogenesis/Lipogenesis and Inflammatory Adipocytokines Secretion in Adipocytes. *Int. J. Mol. Sci.* **2019**, *20*, 160. [CrossRef]
12. El-Atat, F.A.; Stas, S.N.; McFarlane, S.I.; Sowers, J.R. The relationship between hyperinsulinemia, hypertension and progressive renal disease. *J. Am. Soc. Nephrol.* **2004**, *15*, 2816–2827. [CrossRef]
13. Fernandes, R.; Garver, H.; Harkema, J.R.; Galligan, J.J.; Fink, G.D.; Xu, H. Sex Differences in Renal Inflammation and Injury in High-Fat Diet-Fed Dahl Salt-Sensitive Rats. *Hypertension* **2018**, *72*, e43–e52. [CrossRef]
14. Li, J.; Zhou, L.; Feng, W.; Cheng, H.; Muhammad, A.I.; Ye, X.; Zhi, Z. Comparison of Biogenic Amines in Chinese Commercial Soy Sauces. *Molecules* **2019**, *24*, 1522. [CrossRef]
15. Donkor, O.N.; Shah, N.P. Production of beta-glucosidase and hydrolysis of isoflavone phytoestrogens by Lactobacillus acidophilus, Bifidobacterium lactis, and Lactobacillus casei in soymilk. *J. Food Sci.* **2008**, *73*, M15–M20. [CrossRef]
16. Chen, Z.Y.; Feng, Y.Z.; Cui, C.; Zhao, H.F.; Zhao, M.M. Effects of koji-making with mixed strains on physicochemical and sensory properties of Chinese-type soy sauce. *J. Sci. Food Agric.* **2015**, *95*, 2145–2154. [CrossRef] [PubMed]
17. Kushiro, T.; Hashida, J.; Kawamura, H.; Mitsubayashi, H.; Saito, T.; Suzuki, Y.; Takahashi, N.; Ishii, T.; Kimura, T.; Tsuji, K.; et al. Clinical effects of beni-koji in mild essential hypertension–a multi-center double-blind comparison with placebo. *Nihon Jinzo Gakkai Shi* **1996**, *38*, 625–633. [PubMed]
18. Sugawara, T.; Saraprug, D.; Sakamoto, K. Soy sauce increased the oxidative stress tolerance of nematode via p38 MAPK pathway. *Biosci. Biotechnol. Biochem.* **2019**, *83*, 709–716. [CrossRef]
19. Mun, E.G.; Sohn, H.S.; Kim, M.S.; Cha, Y.S. Antihypertensive effect of Ganjang (traditional Korean soy sauce) on Sprague-Dawley Rats. *Nutr. Res. Pract.* **2017**, *11*, 388–395. [CrossRef] [PubMed]
20. Atlas, S.A. The renin-angiotensin aldosterone system: Pathophysiological role and pharmacologic inhibition. *J Manag. Care Pharm.* **2007**, *13*, 9–20. [CrossRef] [PubMed]
21. Nakahara, T.; Sugimoto, K.; Sano, A.; Yamaguchi, H.; Katayama, H.; Uchida, R. Antihypertensive mechanism of a peptide-enriched soy sauce-like seasoning: The active constituents and its suppressive effect on renin-angiotensin-aldosterone system. *J. Food Sci.* **2011**, *76*, H201–H206. [CrossRef]
22. Sulaiman, J.; Gan, H.M.; Yin, W.F.; Chan, K.G. Microbial succession and the functional potential during the fermentation of Chinese soy sauce brine. *Front. Microbiol.* **2014**, *5*, 556. [CrossRef]
23. Kim, Y.S.; Kim, M.C.; Kwon, S.W.; Kim, S.J.; Park, I.C.; Ka, J.O.; Weon, H.Y. Analyses of bacterial communities in meju, a Korean traditional fermented soybean bricks, by cultivation-based and pyrosequencing methods. *J. Microbiol.* **2011**, *49*, 340–348. [CrossRef] [PubMed]

24. Cao, Z.H.; Green-Johnson, J.M.; Buckley, N.D.; Lin, Q.Y. Bioactivity of soy-based fermented foods: A review. *Biotechnol. Adv.* **2019**, *37*, 223–238. [CrossRef]
25. Okada, E.; Saito, A.; Takimoto, H. Association between the Portion Sizes of Traditional Japanese Seasonings-Soy Sauce and Miso-and Blood Pressure: Cross-Sectional Study Using National Health and Nutrition Survey, 2012–2016 Data. *Nutrients* **2018**, *10*, 1865. [CrossRef] [PubMed]
26. Appel, L.J. The Effects of Dietary Factors on Blood Pressure. *Cardiol. Clin.* **2017**, *35*, 197–212. [CrossRef] [PubMed]
27. Oishi, J.C.; Castro, C.A.; Silva, K.A.; Fabricio, V.; Carnio, E.C.; Phillips, S.A.; Duarte, A.; Rodrigues, G.J. Endothelial Dysfunction and Inflammation Precedes Elevations in Blood Pressure Induced by a High-Fat Diet. *Arq. Bras. Cardiol.* **2018**, *110*, 558–567. [CrossRef]
28. Emamian, M.H. Salt intake and blood pressure. *J. Hypertens.* **2019**, *37*, 1530–1531. [CrossRef] [PubMed]
29. Mun, E.G.; Park, J.E.; Cha, Y.S. Effects of Doenjang, a Traditional Korean Soybean Paste, with High-Salt Diet on Blood Pressure in Sprague-Dawley Rats. *Nutrients* **2019**, *11*, 2745. [CrossRef] [PubMed]
30. Ito, K.; Miyata, K.; Mohri, M.; Origuchi, H.; Yamamoto, H. The Effects of the Habitual Consumption of Miso Soup on the Blood Pressure and Heart Rate of Japanese Adults: A Cross-sectional Study of a Health Examination. *Intern. Med.* **2017**, *56*, 23–29. [CrossRef]
31. Kwak, C.S.; Park, S.C.; Song, K.Y. Doenjang, a fermented soybean paste, decreased visceral fat accumulation and adipocyte size in rats fed with high fat diet more effectively than nonfermented soybeans. *J. Med. Food* **2012**, *15*, 1–9. [CrossRef]
32. Cha, Y.S.; Kim, S.R.; Yang, J.A.; Back, H.I.; Kim, M.G.; Jung, S.J.; Song, W.O.; Chae, S.W. Kochujang, fermented soybean-based red pepper paste, decreases visceral fat and improves blood lipid profiles in overweight adults. *Nutr. Metab.* **2013**, *10*, 24. [CrossRef]
33. Peach, M.J. Renin-angiotensin system: Biochemistry and mechanisms of action. *Physiol. Rev.* **1977**, *57*, 313–370. [CrossRef] [PubMed]
34. Vasan, R.S.; Evans, J.C.; Larson, M.G.; Wilson, P.W.; Meigs, J.B.; Rifai, N.; Benjamin, E.J.; Levy, D. Serum aldosterone and the incidence of hypertension in nonhypertensive persons. *N. Engl. J. Med.* **2004**, *351*, 33–41. [CrossRef] [PubMed]
35. Almeida, L.F.; Tofteng, S.S.; Madsen, K.; Jensen, B.L. Role of the renin-angiotensin system in kidney development and programming of adult blood pressure. *Clin. Sci.* **2020**, *134*, 641–656. [CrossRef] [PubMed]
36. Mota de Sa, P.; Richard, A.J.; Hang, H.; Stephens, J.M. Transcriptional Regulation of Adipogenesis. *Compr. Physiol.* **2017**, *7*, 635–674. [CrossRef] [PubMed]
37. Mirabito Colafella, K.M.; Bovee, D.M.; Danser, A.H.J. The renin-angiotensin-aldosterone system and its therapeutic targets. *Exp. Eye Res.* **2019**, *186*, 107680. [CrossRef]
38. Kim, S.; Urs, S.; Massiera, F.; Wortmann, P.; Joshi, R.; Heo, Y.R.; Andersen, B.; Kobayashi, H.; Teboul, M.; Ailhaud, G.; et al. Effects of high-fat diet, angiotensinogen (agt) gene inactivation, and targeted expression to adipose tissue on lipid metabolism and renal gene expression. *Horm. Metab. Res.* **2002**, *34*, 721–725. [CrossRef] [PubMed]
39. Nakagami, H.; Morishita, R. Obesity and gastrointestinal hormones-dual effect of angiotensin II receptor blockade and a partial agonist of PPAR-gamma. *Curr. Vasc. Pharmacol.* **2011**, *9*, 162–166. [CrossRef] [PubMed]
40. DeClercq, V.; Taylor, C.G.; Zahradka, P. Isomer-specific effects of conjugated linoleic acid on blood pressure, adipocyte size and function. *Br. J. Nutr.* **2012**, *107*, 1413–1421. [CrossRef]
41. Kobori, H.; Nishiyama, A.; Abe, Y.; Navar, L.G. Enhancement of intrarenal angiotensinogen in Dahl salt-sensitive rats on high salt diet. *Hypertension* **2003**, *41*, 592–597. [CrossRef]
42. Volpe, M.; Unger, T. Plasma renin and cardiovascular risk: What is the evidence for an association? *Cardiology* **2013**, *125*, 50–59. [CrossRef]
43. Corvol, P.; Williams, T.A.; Soubrier, F. Peptidyl dipeptidase A: Angiotensin I-converting enzyme. *Methods Enzymol.* **1995**, *248*, 283–305. [CrossRef]
44. Engeli, S.; Sharma, A.M. The renin-angiotensin system and natriuretic peptides in obesity-associated hypertension. *J. Mol. Med.* **2001**, *79*, 21–29. [CrossRef] [PubMed]
45. Quadri, S.S.; Culver, S.; Ramkumar, N.; Kohan, D.E.; Siragy, H.M. (Pro)Renin receptor mediates obesity-induced antinatriuresis and elevated blood pressure via upregulation of the renal epithelial sodium channel. *PLoS ONE* **2018**, *13*, e0202419. [CrossRef]
46. Shimoura, C.G.; Lincevicius, G.S.; Nishi, E.E.; Girardi, A.C.; Simon, K.A.; Bergamaschi, C.T.; Campos, R.R. Increased Dietary Salt Changes Baroreceptor Sensitivity and Intrarenal Renin-Angiotensin System in Goldblatt Hypertension. *Am. J. Hypertens.* **2017**, *30*, 28–36. [CrossRef] [PubMed]
47. Kondo, H.; Sakuyama Tomari, H.; Yamakawa, S.; Kitagawa, M.; Yamada, M.; Itou, S.; Yamamoto, T.; Uehara, Y. Long-term intake of miso soup decreases nighttime blood pressure in subjects with high-normal blood pressure or stage I hypertension. *Hypertens. Res.* **2019**, *42*, 1757–1767. [CrossRef] [PubMed]
48. Banday, A.A.; Lokhandwala, M.F. Oxidative stress-induced renal angiotensin AT1 receptor upregulation causes increased stimulation of sodium transporters and hypertension. *Am. J. Physiol. Renal. Physiol.* **2008**, *295*, F698–F706. [CrossRef]
49. Takeda, Y.; Demura, M.; Wang, F.; Karashima, S.; Yoneda, T.; Kometani, M.; Hashimoto, A.; Aono, D.; Horike, S.I.; Meguro-Horike, M.; et al. Epigenetic Regulation of Aldosterone Synthase Gene by Sodium and Angiotensin II. *J. Am. Heart Assoc.* **2018**, *7*. [CrossRef] [PubMed]

50. Zhang, W.W.; Zheng, R.H.; Bai, F.; Sturdivant, K.; Wang, N.P.; James, E.A.; Bose, H.S.; Zhao, Z.Q. Steroidogenic acute regulatory protein/aldosterone synthase mediates angiotensin II-induced cardiac fibrosis and hypertrophy. *Mol. Biol. Rep.* **2020**, *47*, 1207–1222. [CrossRef]
51. Schiffer, L.; Brixius-Anderko, S.; Hannemann, F.; Zapp, J.; Neunzig, J.; Thevis, M.; Bernhardt, R. Metabolism of Oral Turinabol by Human Steroid Hormone-Synthesizing Cytochrome P450 Enzymes. *Drug Metab. Dispos.* **2016**, *44*, 227–237. [CrossRef] [PubMed]
52. Tripodi, G.; Citterio, L.; Kouznetsova, T.; Lanzani, C.; Florio, M.; Modica, R.; Messaggio, E.; Hamlyn, J.M.; Zagato, L.; Bianchi, G.; et al. Steroid biosynthesis and renal excretion in human essential hypertension: Association with blood pressure and endogenous ouabain. *Am. J. Hypertens.* **2009**, *22*, 357–363. [CrossRef]
53. Pezzi, V.; Mathis, J.M.; Rainey, W.E.; Carr, B.R. Profiling transcript levels for steroidogenic enzymes in fetal tissues. *J. Steroid. Biochem. Mol. Biol.* **2003**, *87*, 181–189. [CrossRef]
54. Sztechman, D.; Czarzasta, K.; Cudnoch-Jedrzejewska, A.; Szczepanska-Sadowska, E.; Zera, T. Aldosterone and mineralocorticoid receptors in regulation of the cardiovascular system and pathological remodelling of the heart and arteries. *J. Physiol. Pharmacol.* **2018**, *69*. [CrossRef]
55. Yang, T.; Richards, E.M.; Pepine, C.J.; Raizada, M.K. The gut microbiota and the brain-gut-kidney axis in hypertension and chronic kidney disease. *Nat. Rev. Nephrol.* **2018**, *14*, 442–456. [CrossRef]
56. Wang, Y.; Xie, B.Q.; Gao, W.H.; Yan, D.Y.; Zheng, W.L.; Lv, Y.B.; Cao, Y.M.; Hu, J.W.; Yuan, Z.Y.; Mu, J.J. Effects of Renin-Angiotensin System Inhibitors on Renal Expression of Renalase in Sprague-Dawley Rats Fed with High Salt Diet. *Kidney Blood Press. Res.* **2015**, *40*, 605–613. [CrossRef]
57. Zhang, J.G.; Wang, Z.H.; Du, W.W.; Su, C.; Jiang, H.R.; Huang, F.F.; Jia, X.F.; Ouyang, Y.F.; Li, L.; Wang, Y.; et al. Dietary sodium intake of adult residents in 15 provinces of China in 2015. *Zhonghua Yu Fang Yi Xue Za Zhi [Chin. J. Prev. Med.]* **2019**, *53*, 455–458. [CrossRef] [PubMed]
58. Yu, L.; Li, S.; Zhao, J.; Zhang, J.; Wang, L.; Wang, K. Secular trends in salt and soy sauce intake among Chinese adults, 1997–2011. *Int. J. Food Sci. Nutr.* **2018**, *69*, 215–222. [CrossRef] [PubMed]
59. Phang, S.C.W.; Palanisamy, U.D.; Kadir, K.A. Effects of geraniin (rambutan rind extract) on blood pressure and metabolic parameters in rats fed high-fat diet. *J. Integr. Med.* **2019**, *17*, 100–106. [CrossRef]
60. Meija, J.; Michalowska-Kaczmarczyk, A.M.; Michalowski, T. Mohr's method challenge. *Anal. Bioanal. Chem.* **2016**, *408*, 1721–1722. [CrossRef]

fermentation

Article

Cordyceps cicadae NTTU 868 Mycelium with The Addition of Bioavailable Forms of Magnesium from Deep Ocean Water Prevents the Aβ40 and Streptozotocin-Induced Memory Deficit via Suppressing Alzheimer's Disease Risk Factors and Increasing Magnesium Uptake of Brain

Yan-Zhong Wu and Chun-Lin Lee *

Department of Life Science, National Taitung University, 369, Section 2, University Rd., Taitung 95092, Taiwan; pocky4913@gmail.com
* Correspondence: cllee@nttu.edu.tw; Tel./Fax: +886-89-517759

Abstract: Alzheimer's disease (AD) is a common neurodegenerative disease characterized by continuous accumulation of β-amyloid (Aβ) in the brain. Deep ocean water (DOW) with rich inorganic salts and minerals was proven to promote fungi growth and metabolism. *Cordyceps cicada*, a functional food fungus, can produce higher anti-oxidant and anti-inflammatory compounds including adenosine, polysaccharide, and N(6)-(2-Hydroxyethyl) adenosine (HEA). This study used DOW as the culture water of *C. cicadae* NTTU 868 for producing DOW-cultured *C. cicadae* (DCC), and further investigated the effects and mechanisms on improving the memory deficit and repressing risk factors expressions in Aβ40 and streptozotocin (STZ)-induced Alzheimer's disease rats model. In the results, DCC including mycelium and filtrate had adenosine, HEA, polysaccharide, and intracellular Mg^{2+} after fermentation with DOW. DCC had more effect on the improvement of memory deficit because it suppressed Aβ40 and streptozotocin (STZ) infusion caused BACE, pro-inflammatory factors expressions, and Aβ40 accumulation by increasing sRAGE expression in the brain. Furthermore, DCC enhanced the MAGT1 expression due to high organic magnesium, which can reverse Aβ40-induced cortex magnesium deficiency and further repress Aβ40 accumulation.

Keywords: deep ocean water; *Cordyceps cicadae*; Alzheimer's disease; β-amyloid; polysaccharide; N(6)-(2-Hydroxyethyl) adenosine)

Citation: Wu, Y.-Z.; Lee, C.-L. *Cordyceps cicadae* NTTU 868 Mycelium with The Addition of Bioavailable Forms of Magnesium from Deep Ocean Water Prevents the Aβ40 and Streptozotocin-Induced Memory Deficit via Suppressing Alzheimer's Disease Risk Factors and Increasing Magnesium Uptake of Brain. *Fermentation* **2021**, *7*, 39. https://doi.org/10.3390/fermentation7010039

Academic Editor: Hiroshi Kitagaki

Received: 5 February 2021
Accepted: 12 March 2021
Published: 14 March 2021

1. Introduction

Cordyceps cicadae, a Chinese traditional food fungus, has antioxidation, anti-inflammation [1], blood sugar regulation (Li et al., 2018), renal function improvement [2,3], liver protection [4], and neuroprotection [5] properties. The functional components of *C. cicadae* are adenosine, N(6)-(2-Hydroxyethyl) adenosine (HEA), and polysaccharides [6,7], which are proven to have antioxidation, anti-inflammatory, and anti-aging properties [8].

Researchers have verified that deep ocean water (DOW) can treat obesity [9] and cardiovascular diseases [10] and can fight atherosclerosis [11]. Furthermore, when DOW is applied to microbial fermentation, the nutrient content can promote the production of *Antrodia camphorata* mycelium and increase the quantity and anti-oxidation effects of its ingredients (triterpenoids, polysaccharides, and total flavonoids) [12], drive production of the functional components in *Monascus* (monascin and ankaflavin), lower the content of the toxin citrinin in kidneys and the liver [13], and increase the content of adenosine and cordycepin in *Cordyceps* fermentation products [14]. It can also increase the therapeutic effects of microbial fermentation products on diseases. Furthermore, the main ions in DOW are absorbed by microorganisms during the microbial fermentation process and are then converted to be better absorbed by the human body [15].

C. cicadae is known to have antioxidative, anti-inflammatory, and neuroprotective effects, and DOW nutrients not only treat diseases but also effectively increase the content of the functional ingredients in microbial fermentation products and thereby improve disease treatment effectiveness. During the microbial fermentation process, DOW nutrients are absorbed and utilized effectively by microorganisms; this effect allows DOW nutrients to have higher bioavailability, which endows microbial fermentation products with greater synergistic health benefits. This study is an investigation of DOW-cultivated *C. cicadae* products and their therapeutic effects on Alzheimer's disease (AD) as well as an investigation into whether $MgCl_2$ in deep ocean water is the main functional mineral increasing DOW fermentation of *C. cicadae* products. The study simultaneously explored the main mechanism through which DOW, HEA, and polysaccharides have effects on overall AD improvements.

2. Materials and Methods

2.1. Chemicals

N(6)-(2-Hydroxyethyl) adenosine (HEA) (>95% purity) and streptozotocin were purchased from Sigma Chemical Co. (St. Louis, MO, USA). Potato dextrose agar (PDA) and potato dextrose broth (PDB) were purchased from Difco Co. (Detroit, MI, USA). Ethanol (95%) was purchased from Taiwan Tobacco and Liquor Co. (Taipei, Taiwan). Aβ40 was purchased from Tocris Bioscience Co. (Ellisville, MO, USA). Mouse TNF-α protein (50349-MNAE), mouse IL-6 protein (50349-MNAE), mouse IL-1β protein (50349-MNAE), and rat iNOS protein (Q06518) were purchased from SinoBiological Inc. (North Wales, PA, USA). Rabbit iNOS antibody (FNab04325) was purchased from Cloud-Clone Corp. (Wuhan, China). Rabbit anti-mouse TNF-α polyclonal antibody (AB2148P) were purchased from EMD Millipore Corporation (Temecula, CA, USA). Mouse anti-rat IL-6 monoclonal antibody (sc-57315) and mouse anti-human IL-1β monoclonal antibody (sc-32294) were purchased from Santa Cruz Biotechnology, Inc. (Dallas, TX, USA). Goat anti-rabbit IgG, (H+L), peroxidase conjugated anti-body (31460) were purchased from Pierce Biotechnology (Rockford, IL, USA)

2.2. The Source of DOW

The concentrated DOW provided from the Eastern Taiwan Deep Sea Water Innovation and Research Center (Taitung, Taiwan) was pumped from a depth of 670 m in the Pacific Ocean near the Eastern Taiwan and processed though the electrodeionization and vacuum concentration. According to our previous study, DOW including 20.65 mg/L Mg^{2+} was defined as onefold DOW. In this study, 15-fold DOW (including 309.75 mg/L Mg^{2+}) was prepared by the dilution of concentrated DOW (including 82,640 mg/L Mg^{2+}) with UPW. The concentrations of the trace elements and minerals in 15-fold DOW included 309.75 mg/L Mg, 24.08 mg/L Na, 36.56 mg/L K, 104.95 mg/L Ca, 1.237 μg/L Zn, 2.436 μg/L Cu, 0.441 mg/L Sr, 0.375 μg/L Mo, 8.4 mg/L H_2SiO_3, 1.65 mg/L phosphate, and 0.949 g/L chloride.

2.3. Sample Preparation

C. cicadae NTTU 868 fermentation product was obtained from 1.5-L of submerged fermentation. *C. cicadae* NTTU 868 was cultured with potato dextrose broth at 23 °C for 3 days for the seed culture, and then inoculated (10%) into the fermentation broth (containing 2.4% potato dextrose broth powder and 0.2% yeast extract in 1.5 L ultra-pure water, DOW, or $MgCl_2$, pH 6.8) and cultured at 23 °C for 10 days. Whole submerged fermentation product including mycelium and filtrate were homogenized for the preparation of animal test sample.

The mycelium was separated from whole submerged fermentation product through centrifugation. Next, the mycelium was dried by using the freeze dryer. The dried *C. cicadae* NTTU 868 mycelium was extracted in ultra-pure water solution at 95 °C for an hour and then centrifuged for 10 min at 8000× *g* to get the supernate. The filtrate was collected for the analysis of extracellular polysaccharides. The supernate and filtrate were treated with 3 times the volume of 95% ethanol and extracted at 60 °C for 60 min in order to deposit intracellular and extracellular polysaccharides. The After centrifugation (8000× *g*, 10 min), the polysaccharides depositions were collected for the use of animal test sample.

2.4. Extract and Quantitative Analysis of Adenosine, HEA, and Polysaccharides

Dried *C. cicadae* NTTU 868 mycelium were extracted using 10 volumes of 20% methanol at 60 °C for 30 min and vortexed at every 10 min intervals. The supernatant was collected by centrifugation at 8000× *g* for 10 min. The extract was stored at −20 °C overnight and centrifuged at 8000× *g* for 10 min to remove polysaccharides [16]. Adenosine, HEA were determined by high performance liquid chromatography (HPLC) with a reverse-phase column (Mightysil RP-18 GP 5 μm C18, 250 × 4.6 mm, Kanto Chemical Co., Inc., Tokyo, Japan) and diode array detector (DAD, L-2000 series, Hitachi, Japan). The mobile phase (A solvent: methanol; B solvent: water) was eluted with 0.8 mL/min of flow rate and gradient condition (A solvent: methanol; B solvent: water; 0–3 min, 10% A; 4–8 min, 0% A; 8–18 min, 0% A to 100% A; 18–25 min, 10% A, 28 to 30 min). A 20 μL sample was injected each time. Absorption spectra of eluted compounds were recorded at 262 nm [1]. The polysaccharide was resolved to a suitable concentration by distilled water. The polysaccharide concentration of filtrate was analyzed according to the previous study [17]. The magnesium ion of mycelium extract was determined by ICP-OES (Perkin Elmer Optima 2100 DV, Waltham, MA, USA)

2.5. Animals Grouping and Experiment Schedule

Male Sprague Dawley (SD) rats at 6-8 weeks of age were purchased from the BioLasco Co. (Taipei, Taiwan). The animals were housed individually and allowed free access to a standard laboratory chow (Ralston Purina, St Louis, MO, USA) and water. They were kept in a temperature controlled room (23 °C) under a 12L:12D cycle (light on at 6:00) and were given free access to food and water. In the experiment, the 48 rats were randomly divided to 8 groups and fed standard chow (control group, NOR; 4.5% fat, 3.34 kcal/g).

Rats were infused with vehicle solution (Vh group) or Aβ40 and streptozotocin (STZ) solution (Aβ group) by intracerebroventricular (i.c.v.) injection without administration of test materials. Other rats with i.c.v Aβ40 and STZ infusion were administered DOW (0.0976 mL/kg/day, DOW group), Ultra pure water-cultured *C. cicadae* (UCC) (200 mg/kg/day, UCC group), DOW-cultured *C. cicadae* (DCC) (220 mg/kg/day, DCC group), $MgCl_2$ solution-cultured *C. cicadae* (MgCC) (137 mg/kg/day, MgCC group), N6-(2-hydroxyethyl)-adenosine) (1.12 mg/kg/day, HEA group), or polysaccharides (1.5 mg/kg/day, PS group). The recommendation dosage of homogenate of *C. cicadae* NTTU 868 fermented product is suggested as 250 g/day for 60 kg adult. Wet UCC, DCC, and MgCC homogenate (250 g) were freeze died, and obtained 1.93 g, 2.12 g, and 1.31 g dry products. The doses of the test substances used in this study were calculated in accordance with Boyd's Formula of Body Surface Area as recommended by the FDA (Food and Drug Administration) [18,19]. Therefore, UCC of 200 mg/kg B.W. /day, DCC of 220 mg/kg B.W. /day, MgCC of 137 mg/kg B.W./day for animal dosages were corresponding to the human dosage that a 60 kg adult daily intake 250 g homogenate of *C. cicadae* NTTU 868 fermented product. The contents of DOW, HEA, and polysaccharide of DOW group, HEA group, and PS group were equal to that in DCC group, respectively.

The experiment schedule of AD animal model was shown as follow. Aβ40 infusion on 0th day was continued for 28 days and the test substance in suspension was orally administrated to the rat from the 1st day to the 28th day. The behavioral test was started on the 19th day. The passive avoidance task was carried out from 19th day to the 21st day. Subsequent reference memory task, probe test, and working memory task were started on the 22nd day, the 24th day, and the 25th day. On the 28th day, rats were sacrificed, and the brain tissues were collected for the measurement of TNFα, IL-6, IL1β, sRAGE, *p*-tau, BACE, sAPPα, Aβ40, and MAGT1 protein expressions.

2.6. Surgery for i.c.v. Aβ40 Infusion

Rats were anesthetized with sodium pentobarbital (50 mg/kg BW i.p.). The left skull was exposed and drilled (relative to the bregma; 0.8 mm posterior, 1.4 mm lateral) according to the atlas of Paxinos and Watson [20] using a stereotaxic frame (Narishige, Tokyo, Japan). Aβ40 was prepared in the vehicle solvent of 35% (*v*/*v*) acetonitrile plus 0.1% (*v*/*v*) trifluoroacetic acid (pH 2.0). The osmotic mini-pump (2004, Durect Co., Cupertino, CA, USA) used to result in an animal model of AD with impaired memory was filled with AD solution (24.299 μg Aβ40 and 0.9 mg STZ in 180 μL) or the vehicle solution. The outlet of infusion cannula was inserted 4.0 mm into the left ventricle and attached to the skull with dental cement, and then the mini-pump was quickly implanted into the backs of the rats. AD solution of 180 μL contained in the osmotic pump was continuously infused into left ventricle by 0.28 μL/h for 28 days [21].

2.7. Apparatus of Water Maze

The Morris water maze task was used to evaluate the memory and learning ability from the 22nd day to the 27th day [21]. A black circular tank (diameter: 140 cm, height: 45 cm) was used as the apparatus of water maze in which a movable escape platform (diameter: 10 cm, height: 25 cm) was located inside the tank. The tank was filled to a height of 27.5 cm with water of temperature approximately 23 °C; thus, the surface of the platform was 2.5 cm below the surface of the water. The circular tank was divided into four quadrants (I, II, III, and IV), and a position with equal distance from center and edge in the middle of each quadrant was marked for the location of platform. The water tank was located in a test room with many cues external to maze. The room had adjustable indirect light, and camera was set at ceiling above the center of water tank. The position of the cues remained unchanged throughout the water-maze task.

2.8. Morris Water Maze Task

According to the procedure of our previous study [21], reference memory test was carried out from the 22nd day to the 24th day and included continuous 4 trials per day. Probe test was immediately carried out after the 12th training trial of reference memory task on the 24th day. Working memory test was performed from the 25th day to the 27th day and consisted of five trails per day.

2.9. Preparation of Brains

After completing the behavioral studies, the rats were anesthetized with sodium pentobarbital (65 mg/kg BW, i.p.), and the blood was collected; the cerebral cortex and hippocampus were separated from the whole brain on ice, blotted gently with filter paper to remove blood and extraneous tissue fragments, then flash-frozen with liquid nitrogen and stored at −80 °C until use. Hippocampus and cortex tissues (100 mg) were crushed with an amalgam mixer (UT-1600, Sharp, Osaka, Japan) and suspended in 1.0 mL of ice-cold Tris saline (50 mM Tris-HCl, pH, 7.6, 0.15 M NaCl) buffer containing 1% (*v*/*v*) Triton X-100 and protease inhibitor cocktail, and then sonicated for 30 sec. The homogenate was centrifuged at 100,000× *g* for 30 min and the supernatant was used for magnesium analysis using the commercial kit (Fortress Diagnostics Ltd., Antrim, UK). Regarding the protein extraction for immunoblotting, the tissue (100 mg) was homogenated in 1.0 mL of lysis buffer (1% Triton X-100, 20 mM Tris, pH 7.5, 100 mM NaCl, 40 mM NaF, 0.2% SDS, 0.5% deoxycholate, 1 mM EDTA, 1 mM EGTA, and 1 mM Na_3VO_4) and brief sonication (10 s). The homogenate was centrifuged at 100,000× *g* for 30 min and the supernatant was used for immunoblotting assay.

2.10. Enzyme-Linked Immunosorbent Assay

TNF-α, IL-1β, and IL-6 were determined by self-coating ELISA kit (DY008, R&D Systems, Inc., Minneapolis, MN, USA). TNF-α, IL-1β, and IL-6 recombinant proteins were used as the standards. The standards or tissue homogenates were added to a 96-

well plate at 37 °C for 90 min (using a microplate spectrophotometer) (Thermo Fisher Scientific Inc., Waltham, MA, USA). After removing the standard or liver homogenates, primary antibodies were added and incubated at 37 °C for 60 min and then aspirated and washed 3 times by filling each well with 350 μL of fresh wash buffer. The secondary antibodies were added and incubated at 37 °C for 30 min. After washing 6 times, color subtract (30% H_2O_2 and TMB reagent 1:1) was added in each well and incubated at 37 °C for 15 min. Subsequently, 2N H_2SO_4 as stop solution was added and the plate read at 450 nm immediately.

The other proteins including Aβ40 (ER0754, BACE (ER0756), sRAGE (MBS9338787), were determined by commercial ELISA kits (Fine Biotech Ltd., Wuhan, China). sAPPα (27419) were determined by commercial ELISA kits (Immuno Biological Laboratories Inc, Minneapolis, MN, USA). MAGT1 (MBS9338787) were determined by commercial ELISA kits (Mybiosource Inc., San Diego, CA, USA).

2.11. Statistical Analysis

Data are expressed as means ± standard deviation. Analysis of variance by Duncan's test and Pearson's product-moment correlation coefficient test were determined using SPSS version 10.0 software (SPSS Institute, Inc., Chicago, IL, USA). Differences with $p < 0.05$ were considered statistically significant.

3. Results

3.1. Effects of DOW and $MgCl_2$ on *C. cicadae* and the Production of Functional Components

DOW is abundant in nutrients and minerals. Studies have verified that cultivating *Monascus*, *Cordyceps militaris*, and *Antrodia cinnamomea* with DOW can effectively increase their mycelium growth and functional component content [13–15,22]. Table 1 compares the effects of UPW, DOW, and $MgCl_2$ solutions on *C. cicadae* and the production of their functional components on a 1.5 L cultivation scale. Compared with the other two groups, UPW-cultivated *C. cicadae* has lower HEA but exhibits no significant differences in total polysaccharides. However, DOW can significantly increase *C. cicadae* mycelium content and HEA product even though the highest $MgCl_2$ solutions content lowered biomass and adenosine content while significantly increasing HEA content ($p < 0.05$). In addition, Intracellular Mg^{2+} concentration of mycelium cultured by DOW or $MgCl_2$ was more than that by UPW, suggesting that the mycelium bio-absorbs the magnesium ions in DOW or $MgCl_2$ solution. Furthermore, Mg^{2+} may be converted into organic ions after being absorbed in cells.

Table 1. Effect of DOW and $MgCl_2$ solution on the production of dry mycelium and functional compounds of *C. cicadae* NTTU 868 (fermented capacity: 1.5 L).

Culture Water	Dry Mycelium (g/L)	Adenosine (mg/L)	HEA (mg/L)	Intracellular Polysaccharide (mg/L)	Extracellular Polysaccharide (mg/L)	Intracellular Magnesium (mg/L)
UPW	7.27 ± 0.47^{b}	7.15 ± 0.94^{b}	0.88 ± 0.08^{a}	99.3 ± 8.7^{b}	548 ± 104^{a}	10.1 ± 0.9^{a}
DOW	8.07 ± 0.73^{c}	6.67 ± 0.48^{b}	1.04 ± 0.15^{a}	88.0 ± 12.0^{ab}	518 ± 141^{a}	161.4 ± 12.0^{c}
$MgCl_2$	5.27 ± 0.59^{a}	1.58 ± 0.05^{a}	1.32 ± 0.15^{b}	74.7 ± 14.7^{a}	554 ± 131^{a}	132.0 ± 14.7^{b}

The data are presented as mean ± SD ($n = 3$). ^{a,b,c} Different letters indicate significantly different values according to a one-way ANOVA using Duncan's multiple test ($p < 0.05$). UPW: ultrapure water, DOW: deep ocean water, $MgCl_2$: magnesium chloride. HEA: N(6)-(2-Hydroxyethyl) adenosine.

3.2. Effects of DOW-Cultivated *C. cicadae* Fermentation Products on the Memory Tests and Spatial Learning of Rats with Brain Infusions of Aβ40 and STZ

This experiment involved infusions of Aβ40 and STZ to induce memory loss in rats and exploration of the effects of *C. cicadae* fermentation products cultivated with DOW on brains infused with Aβ40 and STZ, to treat AD. The main effects of DOW and

C. cicadae on AD improvements were explored through the comparison of water cultures and ingredients.

The reference memory tests were conducted by placing a hidden platform in the IV quadrant and recording the amount of time the rats spent searching the platform each time. This was one of the approaches to evaluating the learning and memory of the rats. The results are presented in Figure 1. On the first day, no significant differences were present in the amount of time spent searching the platform by each group of animals ($p > 0.05$), whereas the training results from the second and third day demonstrated that compared with the Vh group, which had normal memory, the Aβ group spent more time and exhibited significant differences ($p < 0.05$). Except the D group, which displayed a downward trend, the groups with other test substances exhibited reduced time spent on the reference memory test. The third day's results indicated that the DCC group exhibited the most satisfactory results, with significantly reduced time spent searching the platform ($p < 0.05$), and therefore demonstrated the most satisfactory improvement of long-term memory.

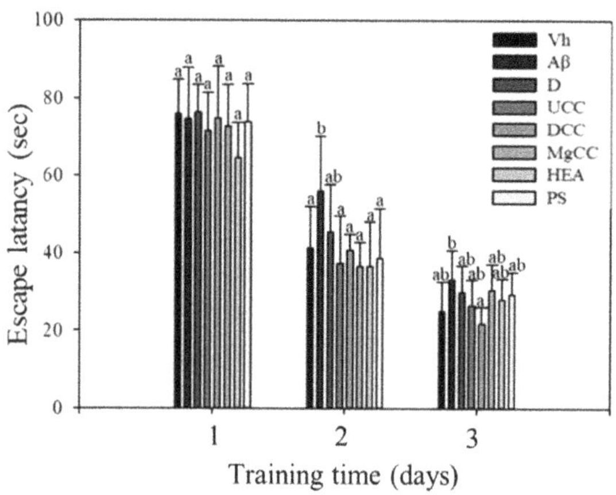

Figure 1. Effect of various *C. cicadae* NTTU 868 fermented products on the searching escape latency of rats with Alzheimer's disease (AD) induced by Aβ40 and streptozotocin (STZ) i.c.v infusion in a reference memory task. Rats were infused with vehicle solution (Vh group) or Aβ40 solution (Aβ group) by i.c.v. injection without administration of test materials. Other Aβ40- and STZ-infused rats were administered deep ocean water (0.0976 mL/kg/day, deep ocean water (DOW) group), ultra-pure water cultured *C. cicadae* (UCC) (200 mg/kg/day, UCC group), DOW-cultured *C. cicadae* (DCC) (220 mg/kg/day, DCC group), $MgCl_2$ solution-cultured *C. cicadae* (MgCC) (137 mg/kg/day, MgCC group), N6-(2-hydroxyethyl)-adenosine) (1.12 mg/kg/day, N(6)-(2-Hydroxyethyl) adenosine (HEA) group), or polysaccharides (1.5 mg/kg/day, PS group). The Mg^{2+} concentration of DCC and MgCC was 309.75 mg/L. Data are presented as mean ± SD; [a,b] Different letters indicate significantly different values according to a one-way ANOVA using Duncan's multiple test ($p < 0.05$).

After the reference memory test was completed, the platform was removed to conduct the probe test. The swimming pathway is beneficial in identifying the authenticity of the learning and memory of lab animals during spatial exploration tests. After removing the hidden platform, the time and pathway distance of the animals roaming where the platform was originally placed and the average swimming speed of the rats within the space were viewed as the memory and learning indicators of the spatial exploration tests. The results are presented in Figure 2a. The Aβ group searched for the target quadrant without direction and without aim, and their swimming pathway covered the entire pool. Conversely, the Vh, D, UCC, DCC, MgCC, HEA, and PS groups swam directly to the target quadrant and spent considerable time moving back and forth in the target quadrant. Furthermore, as Figure 2b demonstrates, the Vh, DCC, MgCC, and PS groups spent more time going back and forth in the target quadrant ($p < 0.05$) than did the Aβ group, by 64%, 54.7%, 29.9%, and 28.9%, respectively; the D, UCC, and HEA groups spent 8.17%, 20.85%, and 11.92% more time.

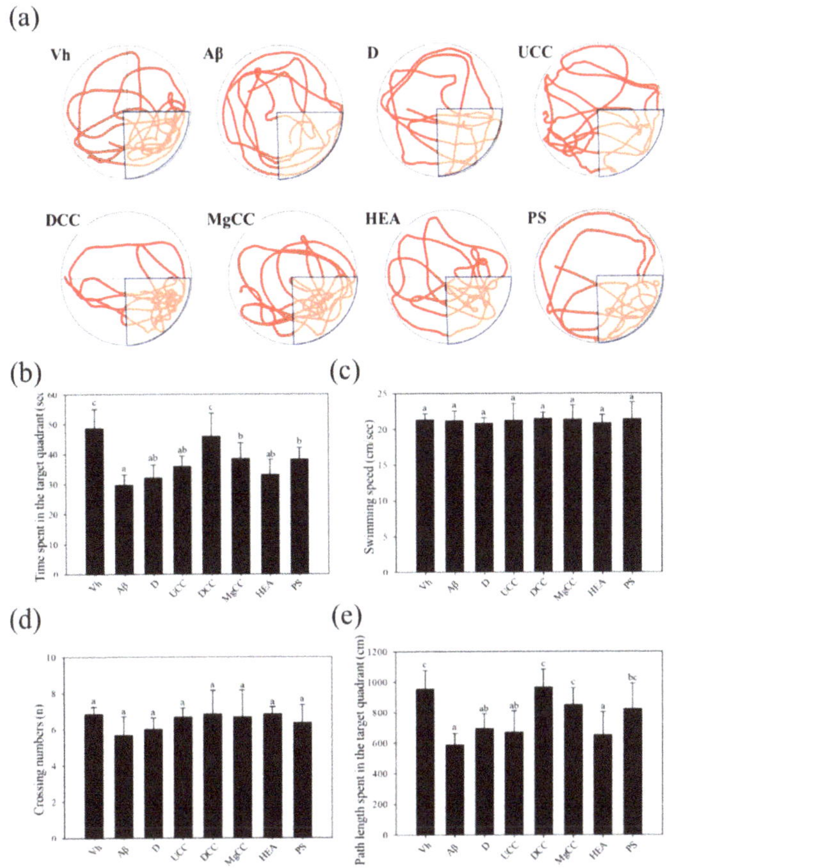

Figure 2. Effect of various *C. cicadae* NTTU 868 fermented products on the memory and learning of rats with AD induced by Aβ40 and STZ i.c.v infusion in spatial probe traces. (**a**) Swimming track (**b**) Time spent in the target quadrant (**c**) Swimming speed (**d**) Path length in the target quadrant (**e**) Crossing numbers. Groups of rats were infused with vehicle solution (Vh group) or Aβ40 solution (Aβ group) by i.c.v. injection without administration of test materials. Other Aβ40- and STZ-infused rats were administered deep ocean water (0.0976 mL/kg/day, deep ocean water (DOW) group), ultra-pure water cultured *C. cicadae* (UCC) (200 mg/kg/day, UCC group), DOW-cultured *C. cicadae* (DCC) (220 mg/kg/day, DCC group), $MgCl_2$ solution-cultured *C. cicadae* (MgCC) (137 mg/kg/day, MgCC group), N6-(2-hydroxyethyl)-adenosine) (1.12 mg/kg/day, N(6)-(2-Hydroxyethyl) adenosine (HEA) group), or polysaccharides (1.5 mg/kg/day, PS group). The Mg^{2+} concentration of DCC and MgCC was 309.75 mg/L. Data are presented as mean ± SD; [a,b,c] Different letters indicate significantly different values according to a one-way ANOVA using Duncan's multiple test ($p < 0.05$).

In addition to swimming time, the speed, number of times entering the quadrant, and swimming distances were analyzed. As can be observed in Figure 2c, no significant differences were present in the swimming speed of the groups; accordingly, swimming speed did not affect the length of the pathways lingering in the target quadrant. Figure 2d presents the number of times the rats entered or passed through the target quadrant.

Although the differences between groups were nonsignificant ($p > 0.05$), the Aβ group was observed to have fewer entries into the target quadrant, whereas the DCC group exhibited more entries. Compared with the Aβ group, the D, UCC, DCC, MgCC, HEA, and PS groups had 20.5%, 5.88%, 17.6%, 20.5%, 17.6%, 20.5%, and 11.7% more entries.

The swimming pathways of the rats entering or passing through the target quadrant are presented in Figure 2e. The Vh, DCC, MgCC, and PS groups had longer pathways moving back and forth within the target quadrant, which were respectively 62.2%, 64.2%, 44.8%, and 40% longer than the Aβ group's pathway ($p < 0.05$); the D, UCC, and HEA groups had 17.9%, 13.9%, and 10.8% longer pathways. These results are similar to the trend in Figure 2b, with the Vh, DCC, MgCC, and PS groups exhibiting significantly higher results than the Aβ group ($p < 0.05$). Therefore, given the lack of significant differences in swimming speed ($p > 0.05$), the time spent going back and forth in the target quadrant and the length of the pathway were similar, verifying that being fed DCC, MgCC, and PS can effectively improve the spatial learning and memory of rats with AD.

3.3. C. cicadae Fermentation Products and the Expression of Aβ40 and BACE Proteins in the Hippocampus of Rats with Brain Infusions of Aβ40 and STZ

After being cleaved by BACE and γ-secretase, amyloid precursor protein (APP) transforms into Aβ [23]. Increased BACE expression or enzyme activity in the brain drives Aβ production in the brain and aggravates abnormal phosphorylation of tau proteins to form neural tangles [24]. Therefore, in this study, the expression Aβ40 and BACE proteins in the hippocampus of rats with AD were assessed.

Figure 3a,b present the amount of expressed AD factors Aβ40 and BACE. Brain infusions of Aβ40 and STZ significantly increased the expression of Aβ40 and BACE proteins ($p < 0.05$). Although DOW (D group) cannot reduce Aβ40 and BACE protein expressions in the hippocampus ($p > 0.05$), the DCC group had the best results with substantially suppressed Aβ40 and BACE content. This demonstrates that DOW-cultivated DCC has the ability to reduce the expression of AD risk factors and has better results than UCC, MgCC, HEA, or polysaccharides. This effect may derive from the additive effects formed by various ingredients.

After α-secretase cleaving, APP produces sAPPα and C83. Particularly, sAPPα elevates the formation of myelin, increasing the neurotransmission of the nervous system [25]. As Figure 3c demonstrates, infusions of Aβ40 and STZ slightly lower the expression of sAPPα; of all the test groups, only the DCC group exhibited a slight upward trend ($p < 0.05$).

The tau protein is a microtubule-associated protein of mature neurons. Excessive phosphorylation produces neural tangles, resulting in impaired transmission of neural signals and affecting learning and memory [26,27]. As Figure 3d indicates, *p*-tau expression was significantly increased as a result of Aβ40 and STZ infusions ($p < 0.05$), but the D, UCC, DCC, and HEA groups did not exhibit lower *p*-tau ($p < 0.05$). The MgCC group exhibited significantly reduced *p*-tau expression ($p < 0.05$).

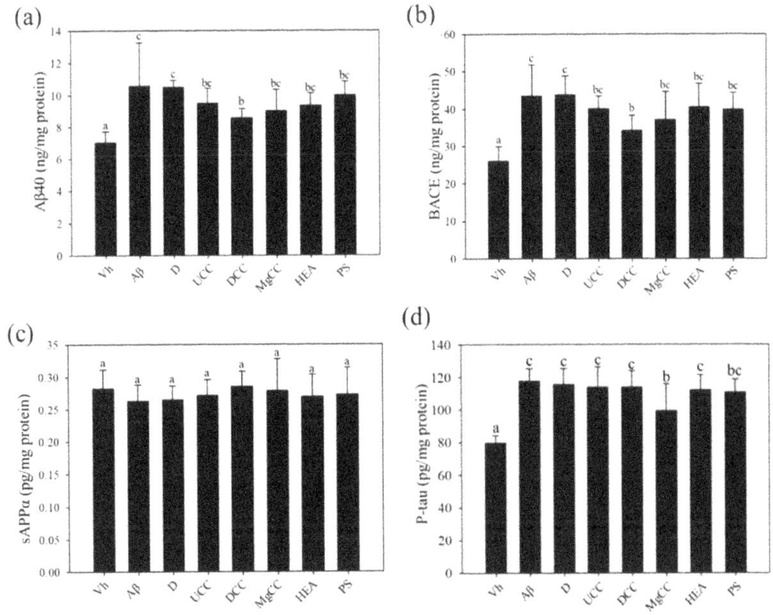

Figure 3. Effect of various *C. cicadae* NTTU 868 fermented products on Aβ40 (**a**), BACE (**b**), and sAPPα (**c**) and p-tau (**d**) protein expression of rats with AD induced by Aβ40 and STZ i.c.v infusion in the hippocampus. Groups of rats were infused with vehicle solution (Vh group) or Aβ40 solution (Aβ group) by i.c.v. injection without administration of test materials. Other Aβ40- and STZ-infused rats were administered deep ocean water (0.0976 mL/kg/day, deep ocean water (DOW) group), ultra-pure water cultured *C. cicadae* (UCC) (200 mg/kg/day, UCC group), DOW-cultured *C. cicadae* (DCC) (220 mg/kg/day, DCC group), $MgCl_2$ solution-cultured *C. cicadae* (MgCC) (137 mg/kg/day, MgCC group), N6-(2-hydroxyethyl)-adenosine) (1.12 mg/kg/day, N(6)-(2-Hydroxyethyl) adenosine (HEA) group), or polysaccharides (1.5 mg/kg/day, PS group). The Mg^{2+} concentration of DCC and MgCC was 309.75 mg/L. Data are presented as mean ± SD; [a,b,c] Different letters indicate significantly different values according to a one-way ANOVA using Duncan's multiple test ($p < 0.05$).

3.4. Effects of C. cicadae Fermentation Products on the Expression of TNF-α, IL-6, and IL-1β in the Hippocampus of Rats with Brain Infusions of Aβ40 and STZ

Excessive accumulation of Aβ in the brain activates stellate cells and macrophages to produce large amounts of proinflammatory factors (TNF-α, IL-6, and IL-1β) and ROS. This reaction results in damage to and inflammation of neuronal cells and even the apoptosis of neurons [28,29]. Figure 4 presents the amount of TNF-α, IL-6, and IL-1β expressed in the cortex and hippocampus. Expressions of TNF-α, IL-6, and IL-1β in the cortex and hippocampus demonstrate similar trends—that Aβ and STZ can induce significantly higher expressions of TNF-α, IL-6, and IL-1β ($p < 0.05$). DCC performs more favorably than UCC and MgCC in reducing the expression of TNF-α, IL-6, and IL-1β. However, by itself, DOW did not result in significant improvements ($p < 0.05$). This demonstrates that by itself, DOW does not have a significant effect in resisting inflammation of brain tissue induced by Aβ and STZ, but it does assist in increasing the anti-inflammatory effects of *C. cicadae* fermentation products.

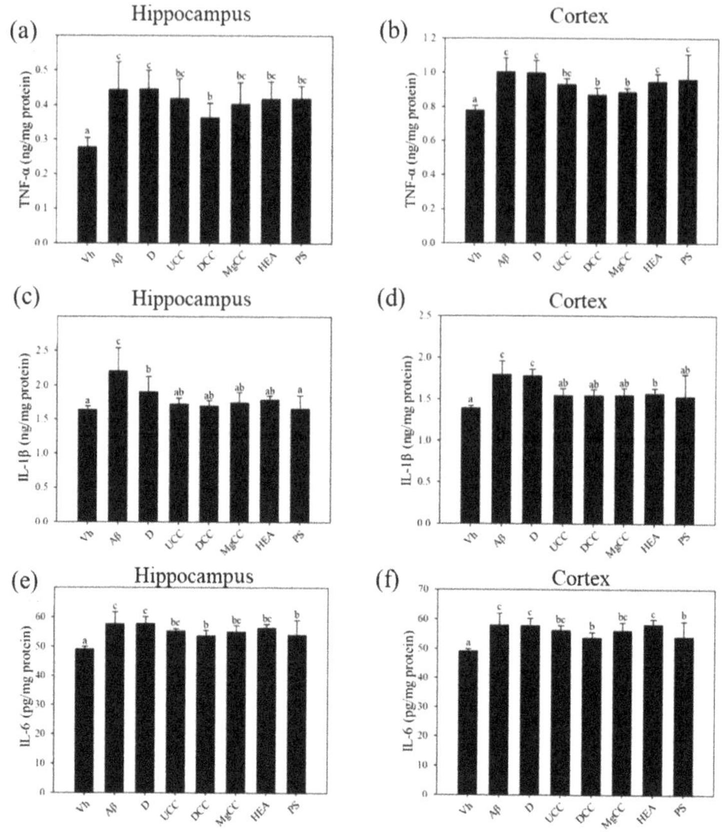

Figure 4. Effect of various *C. cicadae* NTTU 868 fermented products on TNF-α (**a**,**b**), IL-1β (**c**,**d**), and IL-6 (**e**,**f**) protein expression of rats with AD induced by Aβ40 and STZ i.c.v. infusion in the hippocampus and the cortex. Groups of rats were infused with vehicle solution (Vh group) or Aβ40 solution (Aβ group) by i.c.v. injection without administration of test materials. Other Aβ40- and STZ-infused rats were administered deep ocean water (0.0976 mL/kg/day, deep ocean water (DOW) group), ultra-pure water cultured *C. cicadae* (UCC) (200 mg/kg/day, UCC group), DOW-cultured *C. cicadae* (DCC) (220 mg/kg/day, DCC group), $MgCl_2$ solution-cultured *C. cicadae* (MgCC) (137 mg/kg/day, MgCC group), N6-(2-hydroxyethyl)-adenosine) (1.12 mg/kg/day, N(6)-(2-Hydroxyethyl) adenosine (HEA) group), or polysaccharides (1.5 mg/kg/day, PS group). The Mg^{2+} concentration of DCC and MgCC was 309.75 mg/L. Data are presented as mean ± SD; [a,b,c] Different letters indicate significantly different values according to a one-way ANOVA using Duncan's multiple test ($p < 0.05$).

3.5. *Effects of* C. cicadae *Fermentation Products on the Expression of sRAGE in the Hippocampus of Rats with Brain Infusions of Aβ40 and STZ*

An anti-inflammatory factor, sRAGE when binding with RAGE reduces the binding of Aβ and RAGE and the production of proinflammatory factors [30,31]. As demonstrated in Figure 5, after Aβ40 and STZ infusions, the expression of sRAGE was significantly lower than in the Vh group ($p < 0.05$), and feeding DOW, UCC, DCC, and MgCC can effectively increase the expression of sRAGE, with DCC having the most significant increases ($p < 0.05$), followed by MgCC ($p < 0.05$). This result indicates that the sRAGE-enhancing effect of DCC may originate from synergistic effects of the Mg^{2+} in DOW and other ions after being involved in the cocultivation of *C. cicadae* to increase sRAGE expression and reduce inflammation responses.

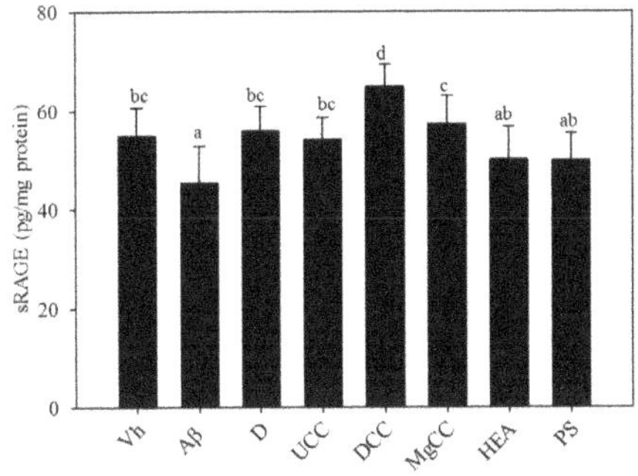

Figure 5. Effect of various *C. cicadae* NTTU 868 fermented products on sRAGE protein expression of rats with AD induced by Aβ40 and STZ i.c.v. infusion in the hippocampus. Rats were infused with vehicle solution (Vh group) or Aβ40 solution (Aβ group) by i.c.v. injection without administration of test materials. Other Aβ40- and STZ-infused rats were administered deep ocean water (0.0976 mL/kg/day, deep ocean water (DOW) group), ultra-pure water cultured *C. cicadae* (UCC) (200 mg/kg/day, UCC group), DOW-cultured *C. cicadae* (DCC) (220 mg/kg/day, DCC group), $MgCl_2$ solution-cultured *C. cicadae* (MgCC) (137 mg/kg/day, MgCC group), N6-(2-hydroxyethyl)-adenosine) (1.12 mg/kg/day, N(6)-(2-Hydroxyethyl) adenosine (HEA) group), or polysaccharides (1.5 mg/kg/day, PS group). The Mg^{2+} concentration of DCC and MgCC was 309.75 mg/L. Data are presented as mean ± SD; ^{a,b,c,d} Different letters indicate significantly different values according to a one-way ANOVA using Duncan's multiple test ($p < 0.05$).

3.6. Effects of *C. cicadae* Fermentation Products on Mg^{2+} in the Blood Serum, Hippocampus, and Cortex of Rats with Brain Infusions of Aβ40 and STZ

Studies have demonstrated Mg^{2+} imbalances in individuals with AD and lab animals with induced AD and that moderate supplementation of Mg^{2+} can effectively improve AD symptoms [32–35]. Therefore, moderate supplementation of Mg^{2+} is viewed as a key indicator in AD improvement. The results are presented in Table 2. Brain infusions of Aβ40 significantly reduced Mg^{2+} content in the blood serum, hippocampus, and cortex ($p < 0.05$). DCC significantly increases the concentration of Mg^{2+} in the hippocampus and the cortex by 16.0% and 16.0%, respectively ($p < 0.05$). MgCC also increases the concentration of Mg^{2+} in the hippocampus and the cortex by 9.7% ($p > 0.05$) and 11.2% ($p < 0.05$), respectively. This confirms that DCC can effectively treat Mg^{2+} imbalances in the brain.

Table 2. Effect of various *C. cicadae* NTTU 868 fermented products on Mg^{2+} concentration in the serum, hippocampus, and cortex of rats with AD induced by Aβ40 and STZ i.c.v. infusion.

Group	Mg^{2+} Concentration		
	Serum (mg/dL)	Hippocampus (μg/g)	Cortex (μg/g)
Vh	5.08 ± 0.44 ^{a}	2.24 ± 0.06 ^{c}	1.45 ± 0.19 ^{b}
Aβ	3.92 ± 0.50 ^{b}	1.75 ± 0.15 ^{a}	1.25 ± 0.05 ^{a}
D	4.10 ± 0.49 ^{b}	1.87 ± 0.16 ^{ab}	1.32 ± 0.07 ^{ab}
UCC	3.83 ± 0.66 ^{b}	1.87 ± 0.22 ^{ab}	1.33 ± 0.07 ^{ab}
DCC	4.02 ± 0.74 ^{b}	2.03 ± 0.18 ^{b}	1.45 ± 0.11 ^{b}
MgCC	3.73 ± 0.30 ^{b}	1.92 ± 0.14 ^{ab}	1.39 ± 0.08 ^{b}
HEA	3.90 ± 0.33 ^{b}	1.88 ± 0.09 ^{ab}	1.32 ± 0.12 ^{ab}
PS	3.87 ± 0.40 ^{b}	1.74 ± 0.11 ^{a}	1.25 ± 0.08 ^{a}

Groups of rats were infused with vehicle solution (Vh group) or Aβ40 solution (Aβ group) by i.c.v. injection without administration of test materials. Other Aβ40- and STZ-infused rats were administered deep ocean water (0.0976 mL/kg/day, deep ocean water (DOW) group), ultra-pure water cultured *C. cicadae* (UCC) (200 mg/kg/day, UCC group), DOW-cultured *C. cicadae* (DCC) (220 mg/kg/day, DCC group), $MgCl_2$ solution-cultured *C. cicadae* (MgCC) (137 mg/kg/day, MgCC group), N6-(2-hydroxyethyl)-adenosine) (1.12 mg/kg/day, N(6)-(2-Hydroxyethyl) adenosine (HEA) group), or polysaccharides (1.5 mg/kg/day, PS group). The Mg^{2+} concentration of DCC and MgCC was 309.75 mg/L. Data are presented as mean ± SD; ^{a,b,c} Different letters indicate significantly different values according to a one-way ANOVA using Duncan's multiple test ($p < 0.05$).

3.7. Effects of C. cicadae Fermentation Products on the Expression of MAGTI Protein in the Hippocampus of Rats with Brain Infusions of Aβ40 and STZ

MAGT1 protein on cell membrane carries Mg^{2+} outside the membrane into the cell, increasing Mg^{2+} content inside the cell. Increased Mg^{2+} content in brain cells reduces the expression of BACE, thus reducing the Aβ generation caused by the cleavage of APP by BACE [32,33]. Related studies have also confirmed that Mg^{2+} in brain cells can reduce hyperphosphorylation of p-tau protein by inhibiting GSK-3β expression (Gomez et al., 2006). As in Figure 6, which illustrates MAGT1 in the hippocampus of rats with AD, because brain infusions of Aβ40 and STZ reduce Mg^{2+} in the brain, MAGT1 levels drop significantly ($p < 0.05$). UCC, HEA, and PS did not increase MAGT1 expression, whereas D, DCC, and MgCC effectively increased expression of MAGT1 ($p < 0.05$). This indicates that D, DCC, and MgCC of Mg^{2+} can reduce the generation of Aβ by regulating the expression of MAGT1.

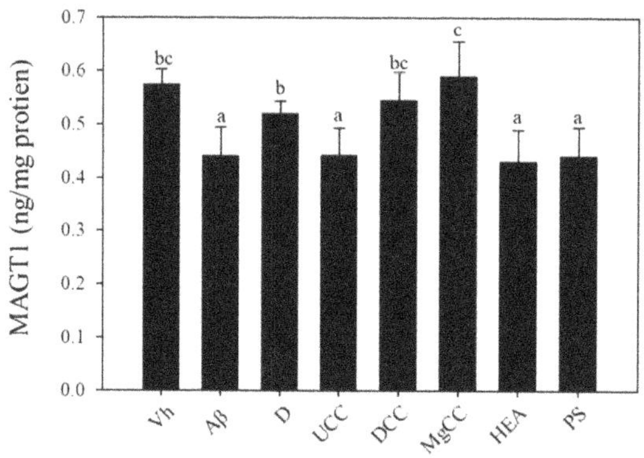

Figure 6. Effect of various *C. cicadae* NTTU 868 fermented products on MAGT1 protein expression of rats infused with Aβ40 and STZ in the hippocampus. Groups of rats were infused with vehicle solution (Vh group) or Aβ40 solution (Aβ group) by i.c.v. injection without administration of test materials. Other Aβ40- and STZ-infused rats were administered deep ocean water (0.0976 mL/kg/day, deep ocean water (DOW) group), ultra-pure water cultured *C. cicadae* (UCC) (200 mg/kg/day, UCC group), DOW-cultured *C. cicadae* (DCC) (220 mg/kg/day, DCC group), $MgCl_2$ solution-cultured *C. cicadae* (MgCC) (137 mg/kg/day, MgCC group), N6-(2-hydroxyethyl)-adenosine) (1.12 mg/kg/day, N(6)-(2-Hydroxyethyl) adenosine (HEA) group), or polysaccharides (1.5 mg/kg/day, PS group). The Mg^{2+} concentration of DCC and MgCC was 309.75 mg/L. Data are presented as mean ± SD; [a,b,c] Different letters indicate significantly different values according to a one-way ANOVA using Duncan's multiple test ($p < 0.05$).

4. Discussion

Studies have argued that DOW can promote the growth of fungal mycelium and regulate the production of functional components [15]. This study examined the cultivation of *C. cicadae* using UPW, DOW, and $MgCl_2$. Compared with using UPW, cultivating *C. cicadae* with DOW increased mycelium growth and HEA content but reduced intracellular and extracellular polysaccharide contents. Regarding the mycelium growth, *C. cicada* were similar to other fungi such as *Monascus purpureus*, *Antrodia cinnamomea*, and *Cordyceps militaris* can be significantly increased by DOW [12,14–16,36]. According to our past research, different minerals in DOW may affect the growth of *C. militaris* and the production of cordycepin [16]. $MgCl_2$ was regarded as the most important compound in the DOW. However, $MgCl_2$ and DOW performed the opposite trend on the regulation of mycelium production. $MgCl_2$ largely reduced the production of mycelium and adenosine, increased HEA content, and reduced the production of intracellular polysaccharides. The lower intracellular polysaccharides production may result from the decreased mycelium production. Although the polysaccharide content is reduced, it may also change the polysaccharide composition. The crude polysaccharides isolated from DCC still have the ability to slightly improve memory deficit and anti-inflammatory effect. It shows that the polysaccharide composition of DCC has an improvement effect on the development of AD.

Past studies have pointed out that the Cl^- in DOW would reduce the production of cordycepin and mycelium. *C. militaris* mycelium was increased by $MgNO_3$ but not by $MgCl_2$ [16]. Therefore, in this study, the main factor that $MgCl_2$ reduces the production of *C. cicadae* mycelium and adenosine may not be caused by Mg^{2+}, but by Cl^-. It is the first study to prove that DOW enhanced mycelium and HEA productions of *C. cicadae.* Although the production of polysaccharides was slightly decreased ($p > 0.05$), the co-fermentation product still performed the synergistic functional effect. This study also found that the mycelium of *C. cicadae* can absorb Mg^{2+} in DOW and increase the content of Mg^{2+} in cells. Mg^{2+} was absorbed in cells and converted into organic ions, and the absorbed Mg^{2+} should be the important factor for increasing magnesium uptake in brain (Table 2).

Aβ40 is a neurotoxic protein generated by cleaved APP and leads to immense production of inflammatory factors that cause damage to nerve cells and promote excessive phosphorylation of tau proteins. As a result, neurofibrillary tangles continually form and block the transmission of neural signals, which inhibits memory [26,27]. STZ causes necrosis of pancreatic β cells, affecting the secretion of insulin and resulting in DM; as a result, it has often been considered a drug that induces type 1 diabetes. STZ promotes the accumulation of Aβ in the brain and causes the excessive phosphorylation of tau proteins; this damages nerve cells and blocks the transmission of neural signals, affecting learning and memory. In recent years, it has been considered a drug that induces AD [37]. In this study, Aβ40 and STZ were used as drugs to co-induce AD; a mixture of Aβ40 and STZ was infused into the brains of animals for 28 days to induce AD. The results demonstrate that infusions of Aβ40 and STZ effectively impede the learning and memory of rats. An analysis of AD risk factors and inflammatory factors indicated that infusions of a Aβ40 and STZ mixture effectively increased the expression of AD risk factors Aβ, BACE, and p-tau, decreased the expression of microglia cell receptor sRAGE, and increased the expression of inflammatory factor TNF-α, IL-1β, and IL-6 proteins. This then increases the expression of BACE and causes the continuous synthesis of Aβ, leading to hyperphosphorylation of tau proteins and massive formation of neural tangles and senile plaques. This ultimately affects memory and learning, inducing AD.

The water maze involves the observation of rodents searching for a hidden platform submerged in a pool of water and recording the time and swimming trajectories of the animals locating the platform. The records are then analyzed to deduce memory and spatial awareness. The reference memory laboratory placed a hidden platform in quadrant IV and recorded the time the rats spent searching for the platform each instance; this was one of the methods employed to evaluate their learning and memory. Injections of Aβ40 and STZ into the brain reduced memory learning in the reference memory test, probe test, and working memory test. UCC had poor results in improving memory and learning, but DCC and MgCC improved memory and learning considerably, with DCC outperforming MgCC. This result may be related to the main functional components; DOW, HEA, and polysaccharides are the main functional components in this study. In the probe test, polysaccharides exhibited more favorable results in improving memory and learning than did HEA. However, although UCC had the highest polysaccharide content, it did not demonstrate significant effectiveness (Table 1). Therefore, the reason for DCC exhibiting more satisfactory results in improving memory and learning is not its higher HEA and polysaccharide content. Furthermore, MgCC also led to significant improvements, indicating that the ions in DOW may be a main factor driving DCC improvement of memory and learning capabilities. The effectiveness of DCC may be due to increases in polysaccharide and HEA content in *C. cicadae* (for HEA after DOW cultivation) and the content of Mg and other minerals in fermented *C. cicadae* mycelium. UCC contains only polysaccharides and HEA; therefore, its effects are limited.

Aβ is generated after the cleavage of APP by BACE and γ-secretase [23]. A result of increased BACE expression or enzyme activity in the brain is the generation of Aβ. Therefore, the expression of Aβ, BACE, and p-tau proteins were measured in this study to determine whether the effects of DCC in improving memory and spatial learning derive from regulating the formation of Aβ and the phosphorylation of tau proteins. After inducements with Aβ and STZ, the expression of Aβ and BACE was increased significantly ($p < 0.05$). DCC can significantly reduce the expression of Aβ and BACE when their levels have been elevated, whereas UCC, MgCC, HEA, and PS can slightly lower Aβ and BACE expression. This demonstrates that DCC can mitigate memory loss by regulating Aβ-generation pathways, and this effect may be due to ions in DOW and the synergistic effects of HEA and PS in *C. cicadae*. Phosphorylation of tau proteins is then aggravated; hyperphosphorylation of tau proteins in the brain diminishes microtubal organization and results in the formation of neural tangles. This impairs the transmission of nerve substances, resulting in some neuronal components that are typically transported between cell bodies and nerve endings (and are not rapidly degraded) accumulating between normal neuronal cells. This ultimately leads to degeneration of the neuronal cells [24,26,27]. Among the experimental substances, only MgCC significantly reduced p-tau protein expression ($p < 0.05$); no significant differences in p-tau protein expression were present between the other experimental substance groups and the Aβ group.

When Aβ is produced in massive quantities and accumulated in brain cells, the excessive Aβ binds with RAGE receptors on microglia cells, which increases the expression of inflammatory factors such as TNF-α, IL-1β, and IL-6 and induces massive release of inflammatory factors. This leads to continuous inflammatory responses and results in the apoptosis of many nerve cells [30]. The results of this study indicate that DCC can significantly reduce ($p < 0.05$) the expression of TNF-α, IL-1β, and IL-6, elevated by infusions of Aβ40 and STZ, in the hippocampus and the cortex.

As a protein that regulates RAGE, when sRAGE is synthesized with RAGE, it blocks the synthesis of intracellular Aβ and RAGE. This suppresses the production of pro-inflammatory factors and reduces the occurrence of inflammatory responses [31,38]. Feeding DCC can increase the expression of sRAGE. Among its main components, HEA and PS also slightly increase the expression of sRAGE. These results indicate that HEA and PS in DCC increase expressions of sRAGE and reduce the expression of pro-inflammatory factors TNF-α, IL-1β, and IL-6; this then reduces the production of inflammatory responses. These effects may be due to the abundant minerals in DOW as well as the Mg^{2+} fermentation product of the co-fermentation of DOW and *C. cicadae* and the effects of HEA and PS in *C. cicadae*.

According to past research, although *Cordyceps cicadae*, *C. Militaris*, and *C. sinensis* belong to *Cordyceps* species, the functional components produced are quite different. In additoin to the fungi polysaccharides, *C. militaris* is a famous for higher production of cordycepin. Cordycepin in *Cordyceps militaris* has been considered as one of the functional components with neuroprotection in past studies [39]. *Cordyceps cicadae* has been gradually discovered in recent years because its HEA was proven to perform liver protection [4], renoprotection [2], and anti-diabetic effect [39,40]. Some studies have argued that HEA in *C. cicadae* can limit NF-κB signal pathways and alleviate LPS-induced inflammatory responses [1]. *C. cicadae* polysaccharides were also proven to perform anti-oxidation and anti-inflammatory activities [2]. In AD prevention, anti-inflammatory should be an important feature against Aβ-induced neurodamage. In this study, it was found that HEA and polysaccharides of *Cordyceps cicadae* exhibit neuroprotection effects due to their anti-inflammatory effects. In the analysis results of cortex and hippocampal tissue, HEA reduced Aβ-induced IL-1β expression, and polysaccharides also significantly repressed IL1β and IL-6. Therefore, HEA and *C. cicadae* polysaccharides still performed some potentials on suppressing inflammatory responses in AD brain.

In addition, *Cordyceps militaris* and *Cordyceps cicada* have similar effects on DOW fermentation. DOW can improve both of the biomass and health benefits. However, in particular, *Cordyceps cicada* can convert DOW minerals into a form that is beneficial to biological absorption. In this study, organic Mg^{2+} of DCC was the important key factor to achieve neuroprotective effects. Mg^{2+} is among the most abundant bivalent cations in cells. It synthesizes with enzymes and related receptors in cells to regulate cell growth, metabolism, and neurotransmissions. Past studies have demonstrated deficiencies in brain magnesium content in AD animal models and feeding test subjects with experimental substances containing magnesium can effectively supplement Mg^{2+} deficiencies in the brain and improve AD symptoms [35]. Therefore, using magnesium content in the hippocampus and cortex as the study basis, our experiments investigated whether feeding the subjects with experimental substances containing magnesium can increase magnesium content in the brain and thus suggest the possibility of treating AD. Rats infused with a mixture of Aβ40 and STZ had significantly lower magnesium content in their hippocampus and cortex than rats undergoing vesicle. Being fed DCC and MgCC effectively supplemented Mg^{2+} content in the hippocampus and the cortex. Feeding DOW alone only slightly increased the Mg^{2+} content in the hippocampus and the cortex. Organic Mg^{2+} effectively mitigated the Mg^{2+} deficiency in the brain. Feeding the DOW and Mg^{2+} fermentation products in *C. cicadae* yielded considerably more satisfactory results, which can be attributed to the co-fermentation of DOW and $MgCl_2$ with *C. cicadae* producing organic state Mg^{2+} products that are more conducive to absorption and use.

MAGT1 in the brain facilitates Mg^{2+} entering neuronal cells to suppress the expression of BACE, reduces the formation of Aβ, and inhibits the expression of GSK-3β. This reduces the phosphorylation of tau proteins [34]. The results of our experiment indicated that rats infused with a mixture of Aβ40 and STZ have significantly lower expression of MAGT1 than rats undergoing sham operations ($p < 0.05$). This finding is consistent with those of past studies in which mice with induced AD had significantly lower expression of MAGT1, and their being fed experimental substances with Mg^{2+} significantly increased MAGT1 expression ($p < 0.05$); MgCC exhibited the greatest increase [32,33]. However, in our study, feeding HEA and PS did not result in significant differences ($p > 0.05$), indicating that the outcomes from DCC and MgCC were not due to the effects of HEA and PS but were possibly due to the Mg^{2+} in the DOW and $MgCl_2$ mixture. MgCC contains only single-ion magnesium, which may be more easily used by *C. cicadae* to increase the expression of MAGT1. DCC, possibly because it is restricted by other ions in DOW and its role in maintaining ion consistency in the body, resulted in lower expression of MAGT1 than that in rats fed with MgCC. These observations suggest that after DOW and $MgCl_2$ were co-fermented with *C. cicadae*, the resulting magnesium fermentation products assist with replenishing magnesium deficiency in the brain and are also conducive to increasing the expression of MAGT1 in the brain and improving AD.

UCC, DCC, and MgCC can reduce the synthesis of Aβ and RAGE by increasing the expression of sRAGE. This results in the deactivation of a large quantity of microglia cells, which lowers the expression of TNF-α, IL-1β, and IL-6 and reduces the occurrence of inflammatory responses. Due to co-fermentation with DOW and $MgCl_2$ to become fermentation products containing magnesium, DCC and MgCC can effectively regulate MAGT1. Summarizing these effects, UCC only improved inflammatory responses, and this effect is due to the functional components HEA and *C. cicadae* polysaccharides already present in *C. cicadae*. MgCC contains the inflammatory abilities of HEA and *C. cicadae* polysaccharides, but it can also regulate MAGT1 expression due to its fermentation with $MgCl_2$ producing fermentation products containing organic magnesium. As a result, compared with UCC, MgCC has more potential to improve AD. DCC contains *C. cicadae* HEA and *C. cicadae* polysaccharides and has the ability to improve inflammatory responses; it also contains organic fermentation products with DOW minerals and magnesium, enabling effective MAGT1 regulation. Combined with the synergistic effects provided by its other ions, DOW outperforms MgCC and UCC in treating AD.

This study investigated the effects of *C. cicadae* fermentation products on absorbing ions and treating AD induced by Aβ 1–40 and STZ. *C. cicadae* absorbs Mg^{2+} in DOW at a high level, which increases its growth and the production of its active components as well as the ion-absorption rate of organisms. The mechanisms were shown in Figure 7. Studies on models of animals with AD have demonstrated that *C. cicadae* cultivated with DOW contains the anti-inflammatory substances HEA and *C. cicadae* polysaccharides. Furthermore, due to effects from its co-cultivation with DOW, *C. cicadae* can increase the expression of MAGT1 and increase Mg^{2+} content in the cortex as well as effectively increase the expression of sRAGE and inhibit the release of inflammatory factors by microglia cells. This effectively reduces the occurrence of inflammatory responses, which lowers the expression of BACE and reduces the generation of Aβ. In summary, *C. cicadae* cultivated with DOW can treat AD through the effective supplementation of Mg^{2+} required by the human body and the inhibition of inflammatory responses.

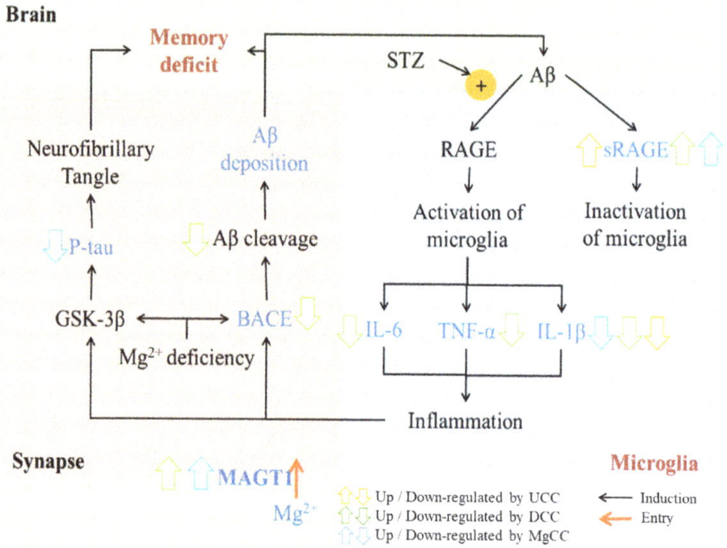

Figure 7. Mechanisms of various *C. cicadae* NTTU 868 fermented products in the regulation of the formation of Aβ and pro-inflammatory response in STZ- and Aβ40-induced Alzheimer's disease.

Author Contributions: Y.-Z.W.: experiments on fermentation, functional compound analysis, and animal test; C.-L.L.: experiments, experiment design, funding application, experiment discussion, paper writing, and submission. All authors have read and agreed to the published version of the manuscript.

Funding: This research was funded by Ministry of Economic Affairs of Taiwan, grant number MOST 105-2628-B-143-003-MY3.

Institutional Review Board Statement: The study was conducted according to the guidelines of the Declaration of Helsinki, and approved by the Institutional Animal Care and Use Committee (IACUC) of the National Taitung University (protocol code 1050321 and 2016/03/21 of approval).

Informed Consent Statement: Not applicable.

Data Availability Statement: All data included in this study are available upon request by contacting the corresponding author.

Conflicts of Interest: The authors declare no conflict of interest.

References

1. Lu, M.Y.; Chen, C.C.; Lee, L.Y.; Lin, T.W.; Kuo, C.F. N(6)-(2-Hydroxyethyl)adenosine in the Medicinal Mushroom Cordyceps cicadae Attenuates Lipopolysaccharide-Stimulated Pro-inflammatory Responses by Suppressing TLR4-Mediated NF-kappaB Signaling Pathways. *J. Nat. Prod.* **2015**, *78*, 2452–2460. [CrossRef]
2. Zheng, R.; Zhu, R.; Li, X.; Li, X.; Shen, L.; Chen, Y.; Zhong, Y.; Deng, Y. N6-(2-Hydroxyethyl) Adenosine From Cordyceps cicadae Ameliorates Renal Interstitial Fibrosis and Prevents Inflammation via TGF-beta1/Smad and NF-kappaB Signaling Pathway. *Front. Physiol.* **2018**, *9*, 1229. [CrossRef] [PubMed]
3. Huang, Y.S.; Wang, X.; Feng, Z.; Cui, H.; Zhu, Z.; Xia, C.; Han, X.; Liu, W.J.; Liu, Y.N. Cordyceps cicadae Prevents Renal Tubular Epithelial Cell Apoptosis by Regulating the SIRT1/p53 Pathway in Hypertensive Renal Injury. *Evid. Based Complement. Altern. Med.* **2020**, *2020*, 7202519. [CrossRef]
4. Ke, B.J.; Lee, C.L. *Cordyceps Cicadae* NTTU 868 Mycelium Prevents CCl4-induced Hepatic Fibrosis in BALB/c Mice Via Inhibiting the Expression of Pro-inflammatory and Pro-fibrotic Cytokines. *J. Funct. Foods* **2018**, *43*, 214–223. [CrossRef]
5. Olatunji, O.J.; Feng, Y.; Olatunji, O.O.; Tang, J.; Ouyang, Z.; Su, Z.; Wang, D.; Yu, X. Neuroprotective effects of adenosine isolated from Cordyceps cicadae against oxidative and ER stress damages induced by glutamate in PC12 cells. *Environ. Toxicol. Pharmacol.* **2016**, *44*, 53–61. [CrossRef] [PubMed]
6. Ke, B.J.; Lee, C.L. Using submerged fermentation to fast increase N6-(2-hydroxyethyl)-adenosine, adenosine and polysaccharide productions of Cordyceps cicadae NTTU 868. *AMB Express* **2019**, *9*, 198. [CrossRef] [PubMed]
7. Chu, Z.B.; Chang, J.; Zhu, Y.; Sun, X. Chemical Constituents of Cordyceps cicadae. *Nat. Prod. Commun.* **2015**, *10*, 2145–2146. [CrossRef] [PubMed]
8. Zhu, Y.; Yu, X.; Ge, Q.; Li, J.; Wang, D.; Wei, Y.; Ouyang, Z. Antioxidant and anti-aging activities of polysaccharides from Cordyceps cicadae. *Int. J. Biol. Macromol.* **2020**, *157*, 394–400. [CrossRef]
9. Hwang, H.S.; Kim, H.A.; Lee, S.H.; Yun, J.W. Anti-obesity and antidiabetic effects of deep sea water on ob/ob mice. *Mar. Biotechnol.* **2009**, *11*, 531–539. [CrossRef] [PubMed]
10. Katsuda, S.; Yasukawa, T.; Nakagawa, K.; Miyake, M.; Yamasaki, M.; Katahira, K.; Mohri, M.; Shimizu, T.; Hazama, A. Deep-sea water improves cardiovascular hemodynamics in Kurosawa and Kusanagi-Hypercholesterolemic (KHC) rabbits. *Biol. Pharm. Bull.* **2008**, *31*, 38–44. [CrossRef] [PubMed]
11. Yoshioka, S.; Hamada, A.; Cui, T.; Yokota, J.; Yamamoto, S.; Kusunose, M.; Miyamura, M.; Kyotani, S.; Kaneda, R.; Tsutsui, Y.; et al. Pharmacological activity of deep-sea water: Examination of hyperlipemia prevention and medical treatment effect. *Biol. Pharm. Bull.* **2003**, *26*, 1552–1559. [CrossRef]
12. Wang, L.C.; Kuo, I.U.; Tsai, T.Y.; Lee, C.L. Antrodia camphorata-fermented product cultured in deep ocean water has more liver protection against thioacetamide-induced fibrosis. *Appl. Microbiol. Biotechnol.* **2013**, *97*, 9955–9967. [CrossRef]
13. Lung, T.Y.; Liao, L.Y.; Wang, J.J.; Wei, B.L.; Huang, P.Y.; Lee, C.L. Metals of Deep Ocean Water Increase the Anti-Adipogenesis Effect of Monascus-Fermented Product via Modulating the Monascin and Ankaflavin Production. *Mar. Drugs* **2016**, *14*, 106. [CrossRef]
14. Hung, Y.P.; Lee, C.L. Higher Anti-Liver Fibrosis Effect of Cordyceps militaris-Fermented Product Cultured with Deep Ocean Water via Inhibiting Proinflammatory Factors and Fibrosis-Related Factors Expressions. *Mar. Drugs* **2017**, *15*, 168. [CrossRef] [PubMed]
15. Lee, C.L. The advantages of deep ocean water for the development of functional fermentation food. *Appl. Microbiol. Biotechnol.* **2015**, *99*, 2523–2531. [CrossRef]
16. Hung, Y.P.; Wang, J.J.; Wei, B.L.; Lee, C.L. Effect of the salts of deep ocean water on the production of cordycepin and adenosine of Cordyceps militaris-fermented product. *AMB Express* **2015**, *5*, 140. [CrossRef] [PubMed]
17. Dubois, M.; Gilles, K.; Hamilton, J.K.; Rebers, P.A.; Smith, F. A colorimetric method for the determination of sugars. *Nature* **1951**, *168*, 167. [CrossRef]
18. Lee, C.L.; Tsai, T.Y.; Wang, J.J.; Pan, T.M. In vivo hypolipidemic effects and safety of low dosage Monascus powder in a hamster model of hyperlipidemia. *Appl. Microbiol. Biotechnol.* **2006**, *70*, 533–540. [CrossRef] [PubMed]
19. Boyd, E. *The Growth of the Surface Area of Human Body*; University of Minnesota Press: Minneapolis, MN, USA, 1935.
20. Paxinos, G.; Watson, C. *The Rat Brain in Stereotaxic Coordinates: The New Coronal Set*; Elsevier Academic: London, UK, 2005.
21. Lee, C.L.; Kuo, T.F.; Wang, J.J.; Pan, T.M. Red mold rice ameliorates impairment of memory and learning ability in intracerebroventricular amyloid beta-infused rat by repressing amyloid beta accumulation. *J. Neurosci. Res.* **2007**, *85*, 3171–3182. [CrossRef] [PubMed]
22. Lee, C.L.; Kung, Y.H.; Wang, J.J.; Lung, T.Y.; Pan, T.M. Enhanced hypolipidemic effect and safety of red mold dioscorea cultured in deep ocean water. *J. Agric. Food Chem.* **2011**, *59*, 8199–8207. [CrossRef] [PubMed]
23. Haass, C.; Schlossmacher, M.G.; Hung, A.Y.; Vigo-Pelfrey, C.; Mellon, A.; Ostaszewski, B.L.; Lieberburg, I.; Koo, E.H.; Schenk, D.; Teplow, D.B.; et al. Amyloid beta-peptide is produced by cultured cells during normal metabolism. *Nature* **1992**, *359*, 322–325. [CrossRef]
24. Jick, H.; Zornberg, G.L.; Jick, S.S.; Seshadri, S.; Drachman, D.A. Statins and the risk of dementia. *Lancet* **2000**, *356*, 1627–1631. [CrossRef]

25. Llufriu-Daben, G.; Carrete, A.; Chierto, E.; Mailleux, J.; Camand, E.; Simon, A.; Vanmierlo, T.; Rose, C.; Allinquant, B.; Hendriks, J.J.A.; et al. Targeting demyelination via alpha-secretases promoting sAPPalpha release to enhance remyelination in central nervous system. *Neurobiol. Dis.* **2018**, *109*, 11–24. [CrossRef] [PubMed]
26. Brion, J.P.; Anderton, B.H.; Authelet, M.; Dayanandan, R.; Leroy, K.; Lovestone, S.; Octave, J.N.; Pradier, L.; Touchet, N.; Tremp, G. Neurofibrillary tangles and tau phosphorylation. *Biochem. Soc. Symp.* **2001**, 81–88. [CrossRef]
27. Geschwind, D.H. Tau phosphorylation, tangles, and neurodegeneration: The chicken or the egg? *Neuron* **2003**, *40*, 457–460. [CrossRef]
28. Liu, Z.; Zhang, B.; Xia, S.; Fang, L.; Gou, S. ROS-responsive and multifunctional anti-Alzheimer prodrugs: Tacrine-ibuprofen hybrids via a phenyl boronate linker. *Eur. J. Med. Chem.* **2020**, 112997. [CrossRef]
29. Shibata, N.; Kobayashi, M. The role for oxidative stress in neurodegenerative diseases. *Brain Nerve* **2008**, *60*, 157–170.
30. Deane, R.; Du Yan, S.; Submamaryan, R.K.; LaRue, B.; Jovanovic, S.; Hogg, E.; Welch, D.; Manness, L.; Lin, C.; Yu, J.; et al. RAGE mediates amyloid-beta peptide transport across the blood-brain barrier and accumulation in brain. *Nat. Med.* **2003**, *9*, 907–913. [CrossRef]
31. Oh, S.; Son, M.; Choi, J.; Lee, S.; Byun, K. sRAGE prolonged stem cell survival and suppressed RAGE-related inflammatory cell and T lymphocyte accumulations in an Alzheimer's disease model. *Biochem. Biophys. Res. Commun.* **2018**, *495*, 807–813. [CrossRef]
32. Li, W.; Yu, J.; Liu, Y.; Huang, X.; Abumaria, N.; Zhu, Y.; Huang, X.; Xiong, W.; Ren, C.; Liu, X.G.; et al. Elevation of brain magnesium prevents and reverses cognitive deficits and synaptic loss in Alzheimer's disease mouse model. *J. Neurosci.* **2013**, *33*, 8423–8441. [CrossRef] [PubMed]
33. Li, W.; Yu, J.; Liu, Y.; Huang, X.; Abumaria, N.; Zhu, Y.; Huang, X.; Xiong, W.; Ren, C.; Liu, X.G.; et al. Elevation of brain magnesium prevents synaptic loss and reverses cognitive deficits in Alzheimer's disease mouse model. *Mol. Brain* **2014**, *7*, 65. [CrossRef] [PubMed]
34. Li, X.; Han, X.; Yang, J.; Bao, J.; Di, X.; Zhang, G.; Liu, H. Magnesium Sulfate Provides Neuroprotection in Eclampsia-Like Seizure Model by Ameliorating Neuroinflammation and Brain Edema. *Mol. Neurobiol.* **2017**, *54*, 7938–7948. [CrossRef]
35. Xu, Z.P.; Li, L.; Bao, J.; Wang, Z.H.; Zeng, J.; Liu, E.J.; Li, X.G.; Huang, R.X.; Gao, D.; Li, M.Z.; et al. Magnesium protects cognitive functions and synaptic plasticity in streptozotocin-induced sporadic Alzheimer's model. *PLoS ONE* **2014**, *9*, e108645. [CrossRef] [PubMed]
36. Shi, Y.; Yang, S.; Lee, D.Y.; Lee, C. Increasing anti-Abeta-induced neurotoxicity ability of Antrodia camphorata-fermented product with deep ocean water supplementary. *J. Sci. Food Agric.* **2016**, *96*, 4690–4701. [CrossRef]
37. Ravelli, K.G.; Rosario, B.D.; Camarini, R.; Hernandes, M.S.; Britto, L.R. Intracerebroventricular Streptozotocin as a Model of Alzheimer's Disease: Neurochemical and Behavioral Characterization in Mice. *Neurotox Res.* **2017**, *31*, 327–333. [CrossRef]
38. Yan, S.D.; Chen, X.; Fu, J.; Chen, M.; Zhu, H.; Roher, A.; Slattery, T.; Zhao, L.; Nagashima, M.; Morser, J.; et al. RAGE and amyloid-beta peptide neurotoxicity in Alzheimer's disease. *Nature* **1996**, *382*, 685–691. [CrossRef] [PubMed]
39. Nxumalo, W.; Elateeq, A.A.; Sun, Y. Can Cordyceps cicadae be used as an alternative to Cordyceps militaris and Cordyceps sinensis?—A review. *J. Ethnopharmacol.* **2020**, *257*, 112879. [CrossRef] [PubMed]
40. Li, I.C.; Lin, S.; Tsai, Y.T.; Hsu, J.H.; Chen, Y.L.; Lin, W.H.; Chen, C.C. Cordyceps cicadae mycelia and its active compound HEA exert beneficial effects on blood glucose in type 2 diabetic db/db mice. *J. Sci. Food Agric.* **2019**, *99*, 606–612. [CrossRef]

fermentation

Article

The Effect of Fermented Kefir as Functional Feed Additive in Post-Weaned Pigs

Woosik Choi [1,†], Dang Bao Son [1,†], Jeongpyo Hong [1], Dabeen Jeong [1], Hee-Chang Kim [2,*], Hanki Lee [1,*] and Joo-Won Suh [1,*]

1 Center for Nutraceutical and Pharmaceutical Materials, Myongji University, Yongin, Gyeonggi-do 17058, Korea; woos5051@gmail.com (W.C.); baoson.dang@gmail.com (D.B.S.); hongjp05@gmail.com (J.H.); jdabin16@gmail.com (D.J.)
2 TMC Co., Ltd., Anyang, Gyeonggi-do 14057, Korea
* Correspondence: jeromekim@tmc.co.kr (H.-C.K.); hklee95@mju.ac.kr (H.L.); jwsuh@mju.ac.kr (J.-W.S.); Tel.: +82-2-2663-2773 (H.-C.K.); +82-31-330-6799 (H.L.); +82-31-330-6881 (J.-W.S.)
† These authors contributed equally to this study.

Abstract: The control of the immune system of pigs after weaning is important in pig farming because productivity depends on the survival of the post-weaned pigs. Previously, antibiotics would have been administered in the case of infectious diseases to increase the survival rate of post-weaned pigs, but now, the use of antibiotics is strictly restricted in order to prevent other problems such as the occurrence of antibiotic-resistant pathogens. In this study, the effect of fermented kefir as a functional feed additive as a replacement to antibiotics was evaluated in terms of the microbial profile in fecal samples, immunological factors in the blood of pigs, growth performance measured as average daily gain (ADG) and the feed conversion rate (FCR) of post-weaned pigs. In the kefir-treated group, the number of lactic acid bacteria and *Bacillus* spp. in the fecal samples of the pigs increased with the kefir treatments. Interestingly, the number of coliform groups as opportunistic pathogens was reduced in the fecal samples of pigs treated with kefir. We found out that treatment with kefir enhanced the innate immunity of post-weaned pigs though the reduction of IL-6 as a proinflammatory cytokine and an increase in IgG as an immunoglobulin, enhancing immunological defense against pathogens. Finally, after treatment with kefir, we observed that the ADG of post-weaned pigs increased to 135.6% but FCR decreased to 92.2%. Therefore, this study shows that fermented kefir can be used as a functional feed additive and an antibiotic alternative in order to improve both the innate immune system and growth performance of post-weaned pigs.

Keywords: fermented kefir; post-weaned pig; innate immune system; growth performance; functional feed additive

Citation: Choi, W.; Son, D.B.; Hong, J.; Jeong, D.; Kim, H.-C.; Lee, H.; Suh, J.-W. The Effect of Fermented Kefir as Functional Feed Additive in Post-Weaned Pigs. *Fermentation* **2021**, 7, 23. https://doi.org/10.3390/fermentation7010023

Academic Editor: Maurizio Ciani

Received: 30 December 2020
Accepted: 14 February 2021
Published: 16 February 2021

1. Introduction

In pig farming, it is important that the health of pigs is maintained and supported, especially in the post-weaning period, when they are more susceptible to environmental stresses and infection by pathogenic microorganisms [1]. Pathogenic microorganisms can be lethal to piglets after weaning and treatment with antibiotics is usually increased [2]. However, the use of antibiotics in livestock accelerates the occurrence of antibiotic-resistant pathogens, which is a major concern, like the emergence of antibiotic-resistant microorgansims [3]. Due to these concerns, the use of antibiotics is now strictly prohibited in pig farming for the prevention of infection by pathogenic microorganisms [3]. Therefore, it is now very important to find alternatives that can replace antibiotics.

Kefir is known to be milk fermented using a kefir grain with microbial complexes containing kefiran as the biologically active exopolysaccharide [4]. These microbial complexes in kefir grain possess probiotic characteristics [4]. Many microorganisms isolated from kefir show high survival against the low pH and bile salts in the gastrointestinal tract, thereby adhering to intestinal mucus. In addition, these microorganisms can produce antagonistic materials like organic acids and antimicrobial peptides to interfere with the adherence of pathogenic bacteria to the intestinal mucus.

Kefir grain has been shown to confer diverse biological activities such as improved digestion and tolerance to lactose, antibacterial effects, hypocholesterolaemic effects, anti-hypertensive effects, anti-inflammatory effects, antioxidant activity, anti-carcinogenic activity and anti-allergenic activity. It is thought that these diverse activities result from the microbial complexity of the kefir grain's composition [4,5]. In this study, we examined whether fermented kefir could play a role as a functional feed additive for improvement of the growth performance measures average daily gain (ADG) and feed conversion rate (FCR) and the immune condition of pigs through a field trial.

2. Materials and Methods

The experimental protocol (TMCco-2019-01) describing the management and care of animals was reviewed and approved by the Animal Care and Use Committee of Chungnam National University, Daejeon, South Korea.

2.1. Preparation of Fermented Kefir

Based on the method previously reported, we carried out the preparation of fermented kefir [6]. The kefir grain was inoculated in 4% (*w/v*) of whole fat milk medium and cultivated at 30 °C for 2 d without agitation. For main fermentation of kefir, the medium used in this study was developed based on De Man, Rogosa and Sharpe (MRS, KisanBio, Seoul, Korea), yeast extract-peptone-dextrose (YPD, KisanBio, Seoul, Korea) and nutrient broth (NB, KisanBio, Seoul, Korea) media. In order to determine the optimum conditions for the main fermentation of kefir, we modified the composition of glucose, whey protein and dipotassium phosphate and the inoculation size and chose the best compositions of each factors to increase number of viable lactic acid bacteria, *Bacillus* spp. and yeast. The main fermentation of kefir was carried out in 300 L of working volume of a 500-L fermenter (JUNGHYUN PLANT, Hwaseong, Korea) at 30 °C for 1 d. For fermentation, sterilized air was supplied at 2 vvm (volume/volume/minute) in the fermenter and mixing rate was kept at 200 rpm through an impeller (JUNGHYUN PLANT, Hwaseong, Korea). After fermentation, the total cells were harvested by continuous centrifugation (HANIL SCIENCE MEDICAL, Daejeon, Korea) at 8000 rpm and the cell pellet was mixed with 20% (*w/v*) of sterilized skim milk solution. After that, this mixture was lyophilized for 3 d.

2.2. Determination of Vible Colonies in Fermented Kefir

For determination of the number of lactic acid bacteria, *Bacillus* spp. and yeast in the lyophilized kefir, viable colony counting was carried out. Briefly, the sample was diluted by serial dilution to 0.85% with a sterilized saline solution and 100 μL of the diluted sample was spread onto MRS agar for lactic acid bacteria, NB agar for *Bacillus* spp. and YPD agar for yeast. After incubation at 30 °C for 1 d, colonies were counted.

2.3. Experimental Design of Animals

Ninety pigs regardless if sex (4 w of age, 7.66 ± 0.38 kg) were used in this experiment. The weight of each pig at the start and end of this experiment and the amount of feed consumed for 4 weeks was measured. To calculate average daily gain (ADG) and feed conversion rate (FCR), we monitored the weight of each pig and feed consumed for 4 weeks. The groups for this field trial were divided into three groups: the control, experimental I and experimental II group. Each group consisted of 30 pigs. The control group was fed the basic feed only. Experimental I and II groups were fed the basic feed containing 0.1% (*w/w*) and 0.5% (*w/w*) of lyophilized kefir of the weight of basic feed, respectively. The 0.1% and 0.5% of lyophilized kefir contained 1×10^7 and 5×10^7 CFU/mL of lactic acid bacteria, respectively.

2.4. Analysis of Microorganisms in Fecal Sample of Pigs

For analysis of microorganisms (lactic acid bacteria, *Bacillus* spp. yeast and coliform groups) in the fecal samples of pigs, fecal samples were taken from all of pigs at 0, and the second and fourth weeks of the trial. After that, the serial dilution of samples was carried out with sterilized 0.75% NaCl solution and 100 μL of diluted sample was spread onto MRS agar for lactic acid bacteria, NB agar for *Bacillus* spp., OGYE agar for yeast and MacConkey agar for the coliform group. Plates were incubated for 1 d at 30 °C and colonies were then counted.

2.5. Analysis of Porcine TNF-α, IL-6, IgA and IgG in Serum of Pigs

Blood samples were collected from the jugular vein of each pig at the start and end of the trial. Serum was collected using a vacuum tube (Becton Dickinson, Franklin Lakes, NJ, USA) and centrifuged at 4000 rpm for 10 min at 4 °C before being stored at −80 °C until immunological factors were quantified. Determination of TNF-α and IL-6 in the serum of pigs were carried out using the porcine TNF-α DuoSet ELISA (enzyme-linked immunosorbent assay, R&D Systems, Minneapolis, MN, USA) and porcine IL-6 DuoSet ELISA (R&D Systems, Minneapolis, MN, USA) according to the manufacturer's instructions, respectively. Determination of IgA and IgG in the serum of pigs was carried out using the Pig IgA ELISA Quantitation kit (Bethyl Laboratories, INC, Montgomery, TX, USA) and Pig IgG ELISA Quantitation kit (Bethyl Laboratories, INC, Montgomery, TX, USA) according to the manufacturer's instructions, respectively. The concentrations of porcine TNF-α, IL-6, pig IgA and IgG were then calculated from the standard curves.

2.6. Statistical Analysis

Statistical analyses were accomplished using SPSS 22.0 [7]. The obtained data were analyzed using the paired *t*-test for evaluating the association and significance between variables. A *p*-value <0.05 was considered significant.

3. Results

3.1. The Optimization of Culture Conditions for Kefir Fermentation

For the main fermentation of kefir, seed culture was carried out in sterilized 10% whole fat milk medium for 2 days at 30 °C without agitation. Thereafter, we optimized the culture conditions for kefir fermentation according to the concentrations of glucose as a carbon source, whey protein as a nitrogen source and dipotassium phosphate as a phosphate source. We optimized the innoculation size of the seed culture. As a result, the number of lactic acid bacteria, *Bacillus* spp. and yeast increased by 123.6%, 115.4% and 113.4%, respectively, when we used 2% of glucose. At 1% of whey protein, the number of lactic acid bacteria, *Bacillus* spp. and yeast increased by 122.7%, 120.2% and 111%, respectively. At 0.02% and 0.1% of dipotassium phosphate, the number of lactic acid bacteria, *Bacillus* spp. and yeast was not reduced, compared to other concentrations of dipotassium phosphate (Figure 1a–c) [6]. The number of lactic acid bacteria, *Bacillus* spp and yeast reached its maximum when 2% seed culture of kefir was inoculated, rather than 10% (Figure 1d) [6]. Based on these results, we finally confirmed the optimum concentration of glucose, whey protein and dipotassium phosphate, and the size of inoculum (Table 1) [6].

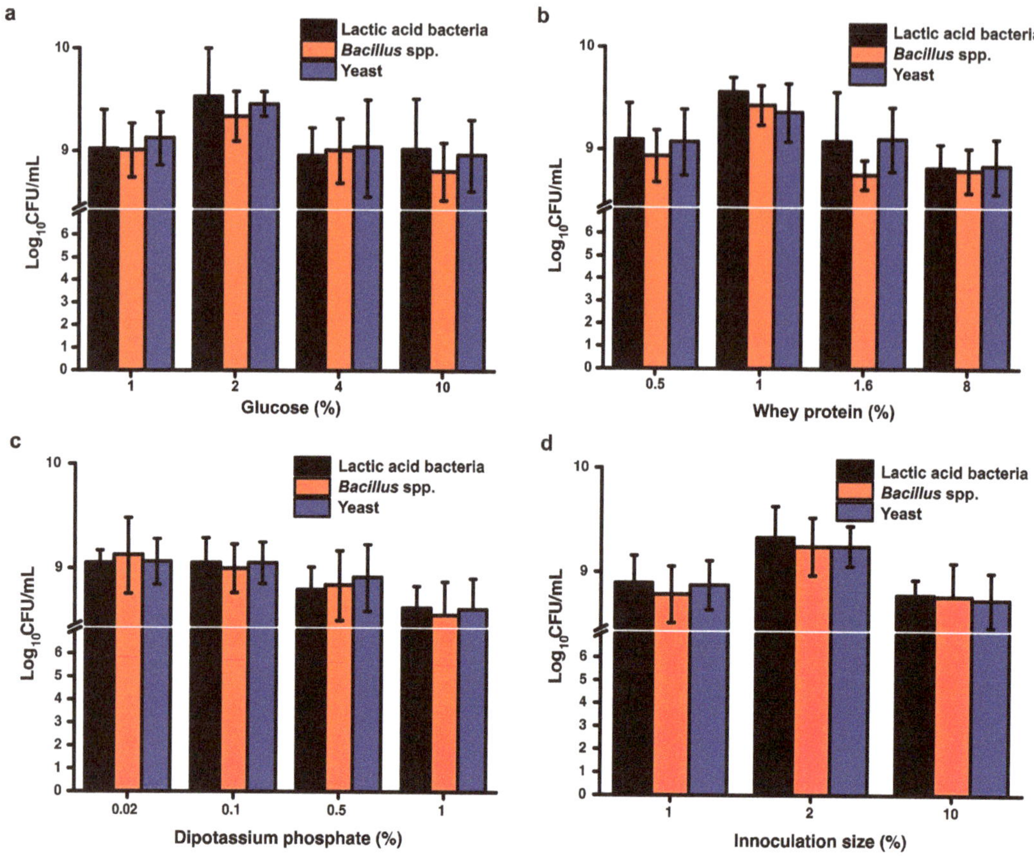

Figure 1. Proportion of lactic acid bacteria, *Bacillus* spp. and yeast in kefir according to culture conditions. (**a**) The proportion of lactic acid bacteria, *Bacillus* ssp. and yeast in kefir by glucose content. (**b**) The proportion of lactic acid bacteria, *Bacillus* spp. and yeast in kefir by whey protein content. (**c**) The proportion of lactic acid bacteria, *Bacillus* spp. and yeast in kefir by dipotassium phosphate content. (**d**) The proportion of lactic acid bacteria, *Bacillus* ssp. and yeast in kefir by inoculation size.

Table 1. The composition of optimized culture medium for kefir fermentation.

Component	Composition (%, *w*/*v*)
Glucose	2
Whey protein	1
Dipotassium phosphate	0.02
Yeast extract	2
Ammonium sulfate	0.1
$MgSO_4$	0.01
$MnSO_4$	0.05
Inoculation size	2

3.2. The Effect of Fermented Kefir on Microflora in the Fecal Samples of Post-Weaned Pigs

To examine whether treatment with kefir can alter microorganisms in the gut of pigs, we investigated the number of lactic acid bacteria, *Bacillus* spp., yeast and coliform groups in the fecal samples before and after two and four weeks of treatment with kefir. An increase in the number of lactic acid bacteria in the fecal sample was observed at the second week in all groups supplied with kefir and kept on it by end of this field test, compared to the control group (Figure 2a). In addition, the number of lactic acid bacteria in the fecal sample of the control group decreased by the end of this field test. This result indicates that direct supplementation of lactic acid bacteria can lead to lactic acid bacteria-enriched microflora. The number of *Bacillus* spp. in the fecal sample also increased in all groups supplied with kefir, similar to the increase in the number of lactic acid bacteria (Figure 2b). However, the kefir supplement did not affect the amount of yeast in the fecal sample (Figure 2c). Interestingly, the kefir supplement reduced the number of coliform groups in the fecal sample at end of this field test with statistical significance (Figure 2d). This result shows the possibility of kefir as antibiotic alternative, because the supplementation of kefir directly inhibited the growth of coliform groups. Therefore, these results indicate that supplementation with kefir was able to directly make the microflora conditions beneficial to the health of post-weaned pigs.

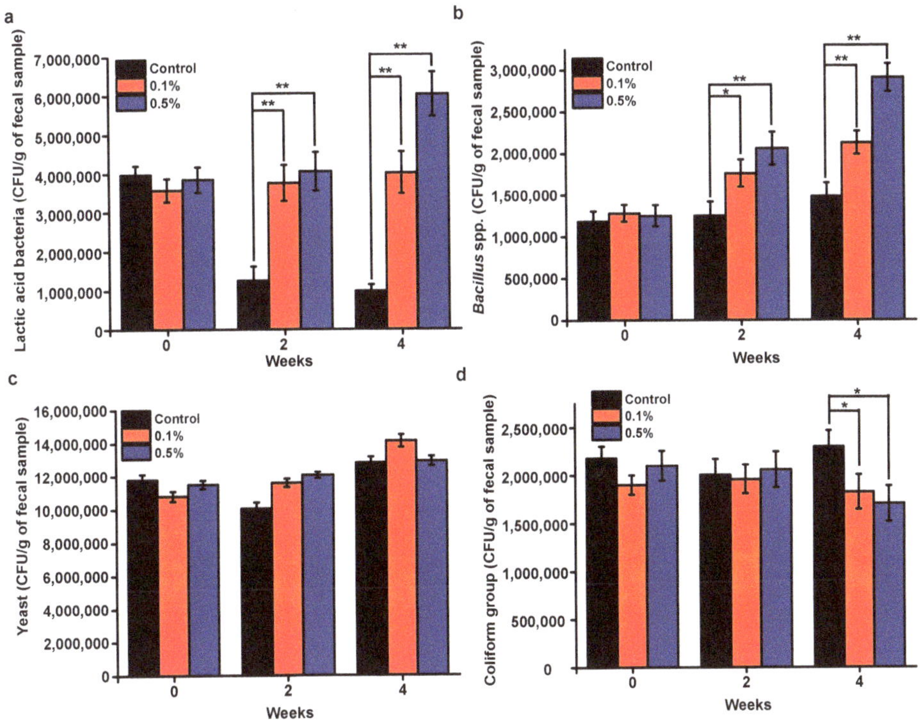

Figure 2. Monitoring microorganisms in the fecal sample of pigs fed fermented kefir. (**a**) Change in the number of lactic acid bacteria. (**b**) Change in the number of *Bacillus* spp. (**c**) Change in the amount of yeast. (**d**) Change in the number of coliform groups. Fecal samples were collected before feeding of kefir, and on the second and fourth weeks. * $p < 0.05$ and ** $p < 0.01$, determined using a paired *t*-test.

3.3. The Effect of Fermented Kefir on Innate Immunity of Post-Weaned Pigs

It is known that post-weaned pigs are more susceptible to infection by pathogenic microorganisms, compared to adult pigs [2]. One of the strategies to avoid the use of antibiotics and minimize the damage by infection of pathogenic microorganisms in pig farming is the enhancement of the innate immunity system of these pigs. The levels of TNF-α and IgA in the blood serum were not changed after supplementation with kefir (Figure 3a,c). However, the supplementation of 0.5% kefir dramatically reduced the level of IL-6 as proinflammatory cytokine after the end of this field test (Figure 3b). The level of IgG as immunoglobulin, participating in the defense system against pathogens, was increased in all groups treated with kefir (Figure 3d). This result shows that supplementation of kefir can contribute to the enhancement of the innate immunity of post-weaned pigs.

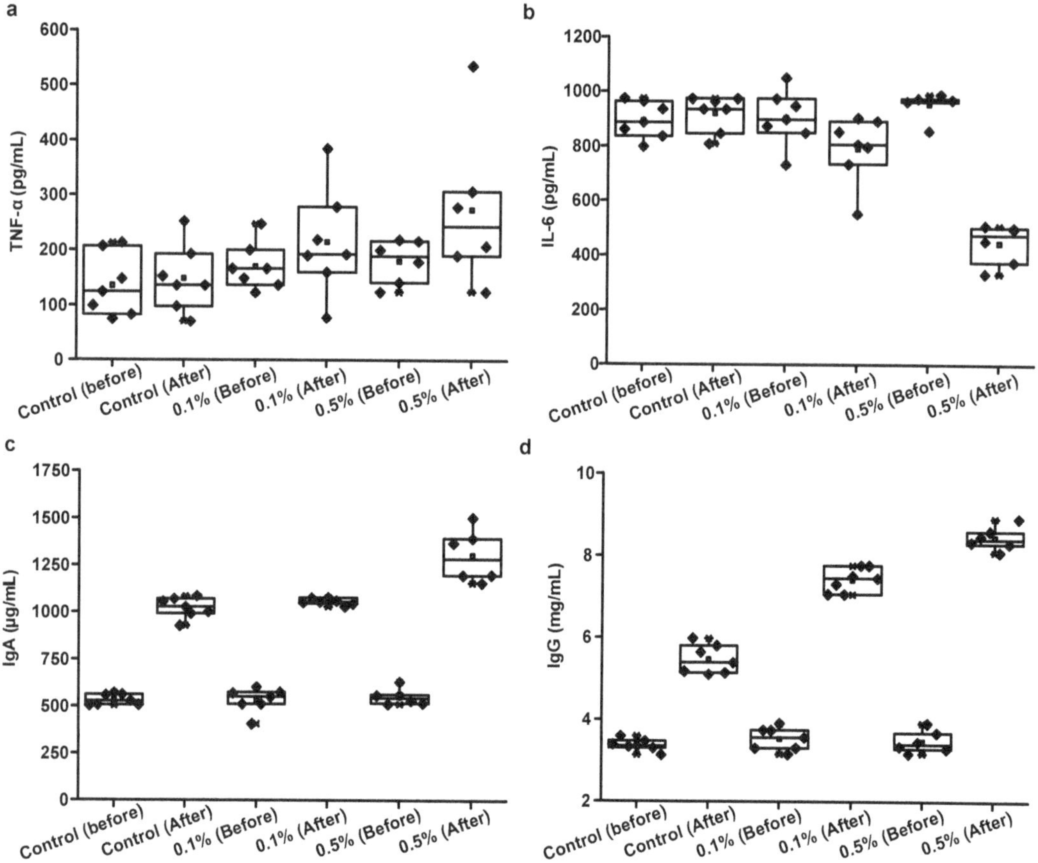

Figure 3. Monitoring cytokines and antibodies participating in innate immunity in the blood of the post-weaned pigs fed fermented kefir. (**a**) The concentration of TNF-α. (**b**) The concentration of IL-6, (**c**) The concentration of IgA. (**d**) The concentration of IgG. Each group consisted of 30 pigs. Blood was collected before and on the fourth week.

3.4. The Effect of Fermented Kefir on ADG and FCR as Growth Performance Measures in Post-Weaned Pigs

We also determined the effect of fermented kefir on ADG and FCR, as well as the innate immunity of post-weaned pigs. The ADG in the group supplied with 0.1% kefir was not changed but increased in the group supplied with 0.5% kefir with statistical significance, compared to the control group (Figure 4a). In the case of FCR, 0.1% kefir supplementation had little effect for 4 weeks. However, 0.5% kefir supplementation induced a reduction in FCR, compared to the control group (Figure 4b). Improvement of both ADG and FCR as growth performances by 0.5% kefir supplementation implies that kefir supplementation can contribute to the efficient conversion of nutrient intake to weight gain.

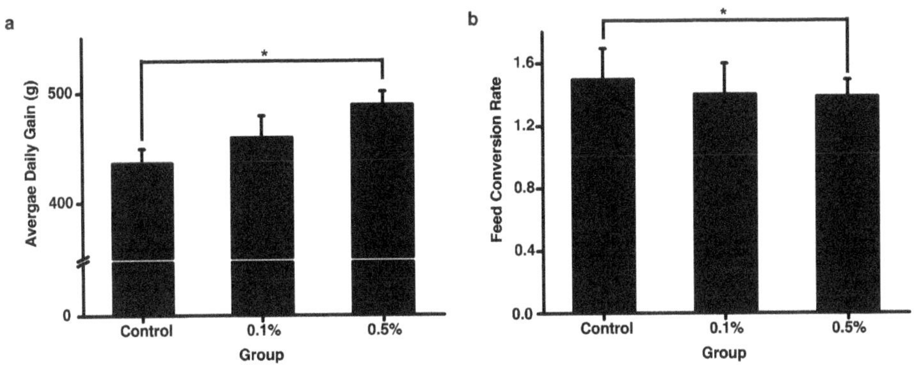

Figure 4. Improvement of average daily gain (ADG) (**a**) and feed conversion rate (FCR) (**b**) in pigs after weaning by supplementation with fermented kefir. Each group consisted of 30 pigs. * $p < 0.05$, determined using a paired *t*-test.

4. Discussion

The use of antibiotics is now restricted for the prevention of microbial infections because of antibiotic resistance; therefore, the need for functional feed additives as antibiotic alternatives in animal farming has increased [8–11]. Indeed, functional feed additives based on probiotic bacteria or natural products have been developed and used as antibiotic alternatives [12–14]. In this research, fermented kefir was evaluated as a functional feed additive for post-weaned pigs, which are relatively susceptible to infection of pathogens.

Firstly, we found that the supplementation with kefir induced an increase in the populations of lactic acid bacteria and *Bacillus* sp. but a reduction in the number of coliform groups. These coliform groups, such as opportunistic pathogens like *Citrobacter* sp., *Enterobacter* sp., *Hafnia* sp., *Klebsiella* sp. and *Escherichia* sp., often damage the intestines of pigs through toxin production, thereby leading to death [14,15]. The use of antibiotics to eliminate these groups is paradoxically one of reasons why antibiotic-resistant bacteria occur [16]. Therefore, it is important to control the number of coliform groups in the intestine to maintain the health and improve the productivity of pigs without the use of antibiotics. Previously, it was known that fermented kefir and kefiran as exopolysaccharide show antimicrobial activity through in vitro and vivo testing [17]. In this study, we showed that the antimicrobial activity of fermented kefir against coliform groups can indeed be realized in weaned pigs through supplementation with kefir.

Such microflora changes to increase the number of beneficial microorganisms can lead to modulation of the production of cytokines and antibodies, controlling the innate and acquired immunity of pigs through interactions between microflora and the intestine, as well as metabolites produced from microorganisms [18]. According to Wang et al., they observed effects of probiotics on levels of inflammatory cytokines in the serum of pigs supplied with *Lactobacillus fermentum* and *Pediococcus acidilactici* [18]. In particular, the level of IL-6 as a pro-inflammatory cytokine was decreased in the probiotic groups. The reduction of IL-6 through treatment with probiotics was also studied by Zhang et al. [19]. They demonstrated that treatment with *L. rhamnosus* GG attenuated the level of IL-6 in the serum of piglets [19]. Other studies have also shown that treatment with probiotic microorganisms induces the expression of IgG and anti-inflammatory cytokines like IL-10 and suppresses the expression of proinflammatory cytokines such as IL-6, IL-8 and TNF-α [20–22]. In agreement with these studies, we found out that the level of IL-6 was reduced and that of IgG was increased in the serum of weaned pigs supplied with kefir, compared to the control group. However, supplementation with kefir did not affect the level of TNF-α and IgA in this study.

Finally, we found out that supplementation with fermented kefir improved both ADG and FCR in post-weaned pigs. Similar to this result, it was previously reported that the supplementation of beneficial microorganisms like probiotics can maintain a healthy intestine and improve conversion rate of feed to energy and nutrients, thereby improving ADG and FCR as growth performances in swine [12–14].

Therefore, we have demonstrated that fermented kefir can be utilized as a functional feed additive in order to improve ADG and FCR, as well as the immune system in pigs through the alternation of microflora inhabiting in the intestines (Figure 5).

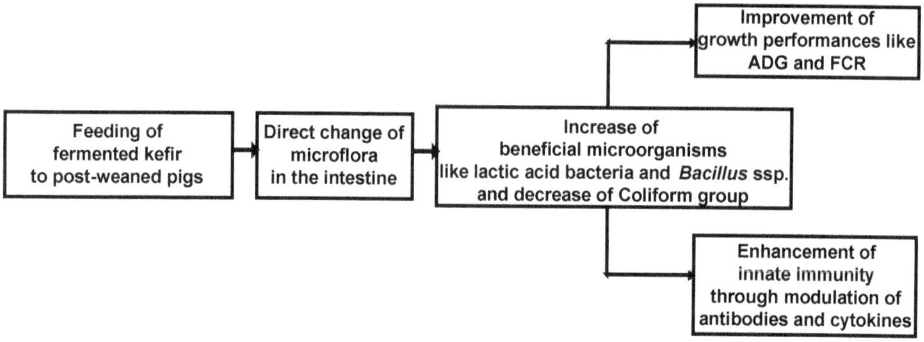

Figure 5. Proposed mode of action of fermented kefir for improving ADG, FCR and innate immunity in post-weaned pigs.

Author Contributions: H.-C.K., H.L., and J.-W.S. designed the research and conducted all experiments. W.C. and D.B.S. carried out the field trial. J.H. and D.J. carried out the fermentation of kefir, preparation of lyophilized kefir, and viable cell counting. H.L. wrote the manuscript. All authors discussed the results and commented on the manuscript. All authors have read and agreed to the published version of the manuscript.

Funding: This work was supported by the Cooperative Research Program for Agriculture Science & Technology Development (Project No. 01319103) from the Rural Development Administration, South Korea.

Institutional Review Board Statement: The study was conducted according to the guidelines of the Declaration of Helsinki, and approved by Ethics Committee of Myongji University, Yongin, Korea (TMCco-2019-01, 10 January 2019).

Data Availability Statement: The data presented in this study are available in the article.

Conflicts of Interest: The authors declare no conflict of interest.

References

1. Modesto, M.; D'Aimmo, M.R.; Stefanini, I.; Trevisi, P.; De Filippi, S.; Casini, L.; Mazzoni, M.; Bosi, P.; Biavati, B. A Novel Strategy to Select Bifidobacterium Strains and Prebiotics as Natural Growth Promoters in Newly Weaned Pigs. *Livest. Sci.* **2009**, *122*, 248–258. [CrossRef]
2. Lekagul, A.; Tangcharoensathien, V.; Yeung, S. Patterns of Antibiotic Use in Global Pig Production: A Systematic Review. *Vet. Anim. Sci.* **2019**, *7*, 100058. [CrossRef]
3. Silbergeld, E.K.; Graham, J.; Price, L.B. Industrial Food Animal Production, Antimicrobial Resistance, and Human Health. *Annu. Rev. Public Health* **2008**, *29*, 151–169. [CrossRef]
4. Leite, A.M.d.O.; Miguel, M.A.L.; Peixoto, R.S.; Rosado, A.S.; Silva, J.T.; Paschoalin, V.M.F. Microbiological, Technological and Therapeutic Properties of Kefir: A Natural Probiotic Beverage. *Braz. J. Microbiol.* **2013**, *44*, 341–349. [CrossRef]
5. Bourrie, B.C.T.; Willing, B.P.; Cotter, P.D. The Microbiota and Health Promoting Characteristics of the Fermented Beverage Kefir. *Front. Microbiol.* **2016**, *7*, 647. [CrossRef] [PubMed]
6. Choi, W.; Choi, C.-W.; Son, D.-B.; Jeong, B.-C.; Kim, H.-C.; Lee, H.; Suh, J.-W. Effects of Fermented Kefir as a Functional Feed Additive in Litopenaeus Vannamei Farming. *Fermentation* **2020**, *6*, 118. [CrossRef]
7. *IBM SPSS Statistics for Windows*, version 22.0; IBM Corp.: Armonk, NY, USA, 2020.
8. Allan, P.; Bilkei, G. Oregano Improves Reproductive Performance of Sows. *Theriogenology* **2005**, *63*, 716–721. [CrossRef] [PubMed]
9. Stein, H.H.; Kil, D.Y. Reduced Use of Antibiotic Growth Promoters in Diets Fed to Weanling Pigs: Dietary Tools, Part 2. *Anim. Biotechnol.* **2006**, *17*, 217–231. [CrossRef]
10. Windisch, W.; Schedle, K.; Plitzner, C.; Kroismayr, A. Use of Phytogenic Products as Feed Additives for Swine and Poultry. *J. Anim. Sci.* **2008**, *86*, E140–E148. [CrossRef]

11. Liu, Y.; Espinosa, C.D.; Abelilla, J.J.; Casas, G.A.; Lagos, L.V.; Lee, S.A.; Kwon, W.B.; Mathai, J.K.; Navarro, D.M.D.L.; Jaworski, N.W.; et al. Non-Antibiotic Feed Additives in Diets for Pigs: A Review. *Anim. Nutr.* **2018**, *4*, 113–125. [CrossRef]
12. Wang, J.; Ji, H.; Zhang, D.; Liu, H.; Wang, S.; Shan, D.; Wang, Y. Assessment of Probiotic Properties of Lactobacillus Plantarum ZLP001 Isolated from Gastrointestinal Tract of Weaning Pigs. *Afr. J. Biotechnol.* **2011**, *10*, 11303–11308. [CrossRef]
13. Dowarah, R.; Verma, A.K.; Agarwal, N. The Use of Lactobacillus as an Alternative of Antibiotic Growth Promoters in Pigs: A Review. *Anim. Nutr.* **2017**, *3*, 1–6. [CrossRef] [PubMed]
14. Bogere, P.; Choi, Y.J.; Heo, J. Probiotics as Alternatives to Antibiotics in Treating Post-Weaning Diarrhoea in Pigs: Review Paper. *S. Afr. J. Anim. Sci.* **2019**, *49*, 403–416. [CrossRef]
15. Hermann-Bank, M.L.; Skovgaard, K.; Stockmarr, A.; Strube, M.L.; Larsen, N.; Kongsted, H.; Ingerslev, H.C.; Mølbak, L.; Boye, M. Characterization of the Bacterial Gut Microbiota of Piglets Suffering from New Neonatal Porcine Diarrhoea. *BMC Vet. Res.* **2015**, *11*, 139. [CrossRef] [PubMed]
16. Liao, S.F.; Nyachoti, M. Using Probiotics to Improve Swine Gut Health and Nutrient Utilization. *Anim. Nutr.* **2017**, *3*, 331–343. [CrossRef]
17. Rodrigues, K.L.; Gaudino Caputo, L.R.; Tavares Carvalho, J.C.; Evangelista, J.; Schneedorf, J.M. Antimicrobial and Healing Activity of Kefir and Kefiran Extract. *Int. J. Antimicrob. Agents* **2005**, *25*, 404–408. [CrossRef]
18. Wang, S.; Yao, B.; Gao, H.; Zang, J.; Tao, S.; Zhang, S.; Huang, S.; He, B.; Wang, J. Combined Supplementation of *Lactobacillus fermentum* and *Pediococcus acidilactici* Promoted Growth Performance, Alleviated Inflammation, and Modulated Intestinal Microbiota in Weaned Pigs. *BMC Vet. Res.* **2019**, *15*, 239. [CrossRef]
19. Zhang, L.; Xu, Y.Q.; Liu, H.Y.; Lai, T.; Ma, J.L.; Wang, J.F.; Zhu, Y.H. Evaluation of *Lactobacillus rhamnosus* GG Using an *Escherichia coli* K88 Model of Piglet Diarrhoea: Effects on Diarrhoea Incidence, Faecal Microflora and Immune Responses. *Vet. Microbiol.* **2010**, *141*, 142–148. [CrossRef]
20. Frick, J.S.; Schenk, K.; Quitadamo, M.; Kahl, F.; Köberle, M.; Bohn, E.; Aepfelbacher, M.; Autenrieth, I.B. *Lactobacillus fermentum* Attenuates the proinflammatory Effect of Yersinia enterocolitica on Human Epithelial Cells. *Inflamm. Bowel Dis.* **2007**, *13*, 83–90. [CrossRef] [PubMed]
21. Candela, M.; Perna, F.; Carnevali, P.; Vitali, B.; Ciati, R.; Gionchetti, P.; Rizzello, F.; Campieri, M.; Brigidi, P. Interaction of Probiotic *Lactobacillus* and *Bifidobacterium* Strains with Human Intestinal Epithelial Cells: Adhesion Properties, Competition against Enteropathogens and Modulation of IL-8 Production. *Int. J. Food Microbiol.* **2008**, *125*, 286–292. [CrossRef]
22. Jiang, M.; Zhang, F.; Wan, C.; Xiong, Y.; Shah, N.P.; Wei, H.; Tao, X. Evaluation of Probiotic Properties of *Lactobacillus plantarum* WLPL04 Isolated from Human Breast Milk. *J. Dairy Sci.* **2016**, *99*, 1736–1746. [CrossRef] [PubMed]

fermentation

MDPI

Review

The Role of Probiotics and Synbiotics on Hirsutism

Vasiliki Lolou

Department of Applied Sciences, Northumbria University, Newcastle Upon Tyne NE1 8ST, UK; vasiliki.lolou@northumbria.ac.uk

Abstract: Probiotics and synbiotics are known to have beneficial effects on human health and disease. Hirsutism, a disorder that is characterised by the presence of coarse terminal hairs in a male-like pattern, is usually caused by elevated androgen levels in blood plasma. This disorder is usually observed in PCOS women and it is linked to insulin resistance (IR). Although idiopathic hirsutism (IH) is not shown to have excess androgen production from the ovarian and adrenal glands, increased 5α-reductase in peripheral tissues and insulin resistance are common observations. The effect of probiotics and synbiotics have been recently studied on PCOS women; androgens were also included in the hormonal groups that were investigated. Only a few studies focus on hirsutism and the potential effect of the beneficial microbes mentioned, whereas the increasing interest on insulin resistance and synbiotics indicate a potential beneficial effect on hirsutism through the management of insulin resistance.

Keywords: probiotics; synbiotics; hirsutism; androgens; insulin resistance; PCOS

Citation: Lolou, V. The Role of Probiotics and Synbiotics on Hirsutism. *Fermentation* **2021**, *7*, 10. https://doi.org/10.3390/fermentation7010010

Received: 13 December 2020
Accepted: 8 January 2021
Published: 11 January 2021

1. Introduction

Fermented food, such as yoghurt, bread, bear and wine, has been widely known since ancient times, and some have been used since then, for therapeutic purposes [1–3]. Fermented dairy products (yoghurt, kefir) were used to treat diseases like diarrhea and other infections of the intestinal tract [4,5]. Interaction between microorganisms and human health was first reported in 1907, where yoghurt microflora was described by Elie Metchnikoff [6]. Fermented food is still used as an additional treatment for disease, such as wounds [7–9], gastroenteric disorders and infections [7,10], as well as an overall support for health [7,8,10–12]. The term probiotic was introduced by Werner Kollath [13] and according to the World Health Organisation (WHO), this term refers to the microorganisms that "when administered in adequate amounts, confer a health benefit on the host" [14]. Probiotics are found to have multiple beneficial roles on human health and disease, through restoration of gut microbiota (GM) [15,16], symptom improvement on intestinal disease, such as irritable bowel syndrome (IBS) [17], modulation of hormonal profile in animals and humans [18–22] and skin infections and healing [23–25]. Along with the beneficial microorganisms, prebiotics, which are fermented compounds that support and promote modifications in the activity and composition of GM [26], have also been studied. It is shown that a controlled combination of probiotics and prebiotics, called synbiotics, can provide a more enhanced beneficial effect on human health and disease [16,27–29].

Although the effect of probiotics and synbiotics are extensively studied, there has been limited research conducted on their effect on sex hormones and sex hormone imbalances, such as hirsutism, a disorder that is mainly characterised by elevated androgen levels in women. The aim of this review is to focus on the role of probiotics and synbiotics and their metabolic process on hirsutism.

1.1. Probiotics

Lactobacillus, *Streptococcus*, *Bifidobacterium*, *Lactococci* and *Saccharomyces* are some of the probiotic species that are known [30]. Microorganisms that are mostly used as treatments are *Lactobacillus*, *Streptococcus*, *Bifidobacterium* and *Saccharomyces*, whilst most of them are naturally present in the human GM [31–37]. There has been an increase in research on the effect of probiotics on health and disease, whilst they are shown to have both beneficial effects on healthy subjects and therapeutic effect on various diseases and conditions, such as intestinal disease [38–46], skin disease [10,24,47–52] and wound healing [53–57]. Probiotics have also been used for the restoration of disturbed GM [58]. Dysbiosis of gut microbiota (DOGMA) is the alteration of gut microflora that can occur due to a variety of reasons, such as antibiotic use [59,60] and diet [61,62]. The human gut, also described as the "second brain" of the human body, plays an important role in human health and disease [63,64]. The GM is known to have a significant role in the functionality of the bowel, maintaining the gut mucosa through their role on gut homeostasis [63–65]. The probiotic *E. coli* Nissle 1917 protects and prevents against inflammatory response via TLR-4 and TLR-2-dependent pathways, whilst lower counts of *Firmicutes prausnitzii* are linked with potential inflammatory bowel disease pathogenesis due to its anti-inflammatory effects [65,66]. The impact of the metabolic processing of the gut microflora in the gut is extended outside of the bowel and can impact other functionalities of the human body, reaching the skin [10,24,67–69]. *L. casei* is shown to reduce skin inflammation through the regulation of $CD8^{+}T$ cells, which initiate inflammatory response [47]. The role of probiotics on endocrinology has also been reported and GM was shown to affect the production of hormones, such as leptin [62,70,71], stress hormones [72], insulin [73,74] and sex hormones [75,76] in the intestinal tract. Additionally, the disturbance of gut microbiota can affect the levels of endogenous hormones, such as estrogens, as well as administrated steroids, such as megestrol acetate, medroxyprogesterone acetate, norethisterone and others, suggesting that the use of antibiotics can cause hormonal imbalance and reduce the absorption of contraceptive hormones, which are used as a method of birth control and/or treatment for metabolic syndromes [75]. An altered microbiota during the early life of the diabetic mouse (Type 1) can determine sex hormones and cause metabolic changes, such as increased testosterone levels [76]. Administration of probiotics can restore the imbalanced microorganisms that live in the human intestine and improve certain conditions through their metabolic processes [58]. Additionally, DOGMA is also observed in patients suffering from inflammatory bowel disease, colitis [77], Crohn's disease, IBS [63], polycystic ovary syndrome (PCOS) [78] and other conditions.

1.2. *Synbiotics*

Synbiotics are the controlled combination of probiotics and prebiotics and their supplementation aims to provide a more enhanced health benefit on human health and disease. Known prebiotics are fructans (e.g., inulin), complex polysaccharides, oligosaccharides and sugar alcohols [16,79] and in combination with probiotics, they have been used for the treatment of conditions and disease [18,27,28,80,81]. Synbiotics containing *L. acidophilus NCC90*, oligofructose and acacia gum can have a preventive role on bone mineral loss after ovariectomy in rats [16]. It has been reported that synbiotic treatments showed beneficial effects on non-alcoholic fatty liver disease (NAFLD), reducing fibrosis and hepatic steatosis in humans [82,83], total necrosis factor α (TNF-α), total nuclear factor κ-B (TNF-κB) and other NAFLD biomarkers, such as high-sensitivity C-reactive protein [83]. Moreover, synbiotics can bring improvements on Crohn's disease through reduction in TNF-α production [84], IBS [85] and delayed Alzheimer's disease in *Drosofila melanogaster* [86]; improvement of thyroid function was observed after 8 weeks of synbiotic supplementation in hypothyroid patients, decreasing thyroid stimulating hormone (TSH) levels and increasing tri-iodothyronine (FT3), whilst overall increasing FT3/TSH ratio [18]. *Bifidobacterium Longum* in combination with inulin, reduced levels of TNF-α and IL-1α in patients suffering from ulcerative colitis [77]. Another study on patients with ulcerative colitis showed that *B. breve* and galacto-oligosaccharide (GOS)-containing beverage (Yakult) managed to clinically improve the condition of the patients, through reduction of UC markers, such as myeloperoxidase, and by lowering the faecal pH [87].

1.3. Hirsutism

Hirsutism is a condition that appears in 5–10% of women and it is recognized by the presence of coarse terminal hairs in a male-like pattern [88]. Excess hair growth that consists of terminal hairs are present in areas were women normally have thinner hair. The clinical diagnosis of hirsutism is completed based on Ferriman and Gallwey criteria (Figure 1) [88] and the areas that are scored include the face, chest, thighs, upper arms, abdomen and back. Scores (m-FG) from 1–4, with 1 describing minimal terminal hair and 4 describing frank virilization, are given to the mentioned areas whist total scores less than 8 are considered normal [88].

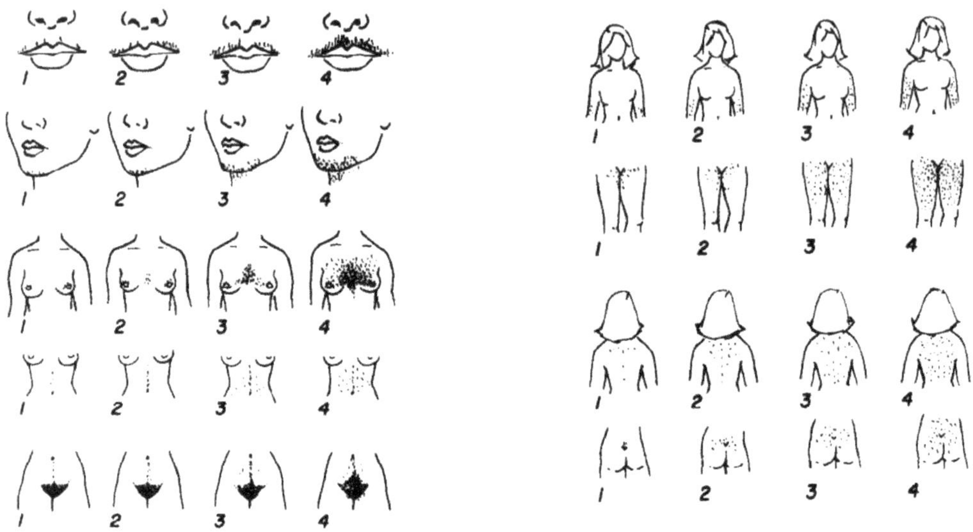

Figure 1. Hirsutism scoring system presenting scores ranking from 1—minimal hirsutism, to 4 virilization in 9 body parts. A total score of less than 8 is considered normal, whereas higher scores indicate mild to severe hirsutism [88].

Hirsutism is a quite complex condition that can occur due to other syndromes or disease, such as PCOS (Figure 2). Although the main reason for this condition is the presence of excess androgens, such as testosterone (T) and dihydrotestosterone (DHT), IH shows normal levels of androgens [89,90]. The most common reason for the excess androgen levels is PCOS and 60–80% of PCOS women suffer from hirsutism [91]. The diagnosis of this condition is conducted according to the Rotterdam criteria, by which two of the three criteria have to be met: (a) evidence of hyperandrogenism, (b) oligo- and/or anovulation and (c) polycystic ovaries [92]. Other clinical features are ovarian enlargement and IR, which is found to be very common amongst PCOS women, whilst the syndrome has also been linked with cardiovascular disease and inflammation. The heritability of this disorder has been studied and it is possible that genetic changes on the androgen receptor gene can lead to hirsutism [90]. During this study, a repeatability of the trinucleotide CAG in exon 1 was observed and more frequent repeats in the N-terminal domain of the androgen gene were linked with the development of the disorder. However, other researchers have not shown a significant role of CAG in the pathogenesis of hirsutism [93,94]. Hirsutism has also been linked with IR (Figure 3) [91,95–101], whilst excess insulin production leads to hyperinsulinemia, which increases luteinizing hormone (LH) through insulin receptor stimulation [97]. LH, along with follicle-stimulating hormone (FSH), is a hormone that regulates androgen production through the secretion of them from ovarian and adrenal glands. Moreover, high levels of insulin inhibit sex hormone-binding globulin (SHBG), a hormone that binds with plasma T and is considered its major determinant along with 17-β hydroxysteroids from plasma [102]. Consequently, both effects of IR are directly linked with hirsutism through excess androgen production. This relationship between IR and hirsutism was studied in healthy women, where adipose tissue was collected and in-vitro treatments with testosterone and/or anti-androgens were conducted on insulin stimulated adipose cells [100].

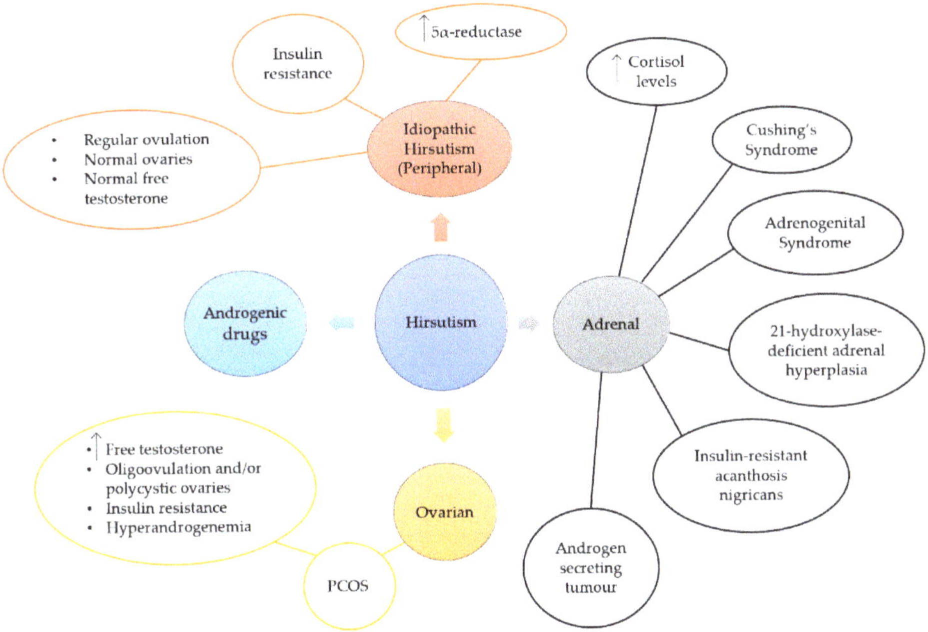

Figure 2. Hirsutism can originate from adrenal or ovarian disorders, or both. Various syndromes and disorders related to the adrenal glands, such as Cushing's syndrome, adrenogenital syndrome, 21-hydroxylase-deficient adrenal hyperplasia, insulin-resistant acanthosis nigricans and androgen secreting tumour, show high levels of cortisol hormone and hirsutism. Ovarian hirsutism is caused by PCOS that is characterised from high levels of free testosterone and/or polycystic ovaries, insulin resistance and hyperandrogenemia. Idiopathic hirsutism is caused by peripheral increase of androgens, and women that suffer from it show increased levels of 5α-reductase and insulin resistance, whilst they do not show any abnormalities in their ovarian or adrenal function.

Adipose tissue contains adipocytes, also known as fat cells and are responsible for fat energy storage. This study showed that the exposure of these cells to testosterone led to insulin resistance, suggesting there is a link between hirsutism, androgen presence and insulin resistance development.

Other disorders that are known for the increased production of androgens are hyperandrogenic insulin-resistant acanthosis nigricans syndrome [103,104], 21-hydroxylase-deficient non-classic adrenal hyperplasia [90,105], androgen-secreting tumor [106], and rarely androgenic drug intake [107]. On the other hand, the causes of IH are not well known. However, when hormonal profiles of PCOS women and IH women were compared, IR was found to be significant in both groups compared to the control group. IR and hyperinsulinemia increase the insulin-like growth factor (IGF) which affects hair follicles [95,96], and therefore it is suggested that IR is potentially a cause of IH. This disorder is also characterised as an increase in 5α-reductase in peripheral tissues, an enzyme that converts T to DHT, and is generally responsible for the metabolism of steroids [95]. Potential causes of IH include increased sensitivity of hair follicles to androgens, and androgen receptor gene polymorphism [96].

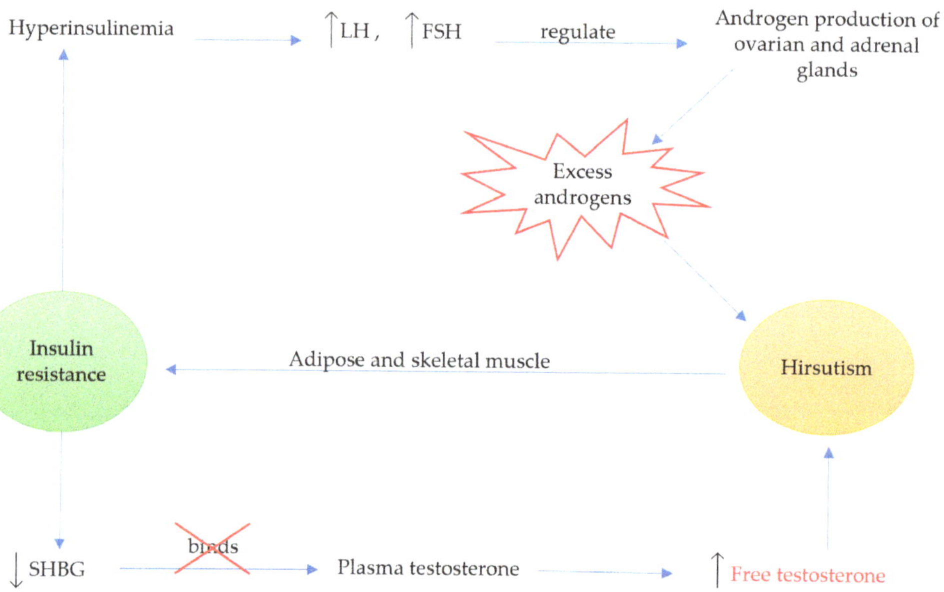

Figure 3. The relationship between insulin resistance and hirsutism. Hyperinsulinemia is caused when insulin resistance is left unmanaged, causing an increase in LH and FSH, which regulate the production of ovarian and adrenal androgens. This increase results in excess androgens, and eventually hirsutism. Moreover, insulin resistance reduces SHBG, which leaves plasma testosterone unbound leading to increased free testosterone and hirsutism. Chronic androgen exposure leads to adipose and skeletal muscle insulin resistance. The graph above presents the circular relationship between the two conditions, leading to a vicious cycle of effects.

2. Therapeutic Index

Hirsutism treatments are usually combinational treatments, and they include peripheral androgen blockage, androgen suppression and cosmetic intervention. Cosmetic and mechanical interventions are more effective and long lasting when used in combination with either androgen suppression treatment or peripheral androgen blockage. Moreover, alternative treatments have been investigated in recent years and naturally derived additional treatments have been used for the improvement of hirsutism, as well as live organisms that are shown to have a significant role in regulating and reducing androgen production on PCOS women.

2.1. *Chemically Derived Therapies*

Current drug therapies for hirsutism include androgen suppression and peripheral androgen blockage, also known as anti-androgens. For a more effective treatment these drugs are usually administered in combination [108–112].

Androgen suppression treatments include insulin sensitisers, long acting gonadotrophin releasing hormone (GnRH) analogue and estrogen-progestin oral contraceptive (OC) pills [98,112–114]. OC action can differ in composition, but it is not shown to significantly influence the effectiveness of the treatment. Their compositions can be monophasic, biphasic or triphasic and the action of the OC is through the suppression of LH and FSH and therefore a reduced androgen production from the ovaries [90]. Although this treatment is shown to reduce SHBG and hyperandrogenemia in general, it does not have an effect in low-grade inflammation and other hormonal disorders. Third generation OCs are not as effective at reducing the levels of SHBG when compared to older combined OC [115]. OCs containing sex hormones progestin and estrogen are shown to have significant effect on the hormonal profiles of hirsute women, reducing the levels of androgens, such as T, dehydroepiandrosterone sulfate (DHEAS) and DHT [116]. GnRH is a hormone produced in the hypothalamus and it stimulates the production of LH/FSH which stimulates the ovaries to produce androgens [97]. GnRH analogues are antagonists of GnRH and, when used on PCOS women, were shown to reduce the total and free T, DHEAS and androstenedione sex hormone. This treatment is usually combined with OCs. Moreover, the combination of insulin sensitisers with OC has also been studied. Insulin sensitisers, such as metformin and thiazolidenediones, are shown to reduce androgen production through the regulation of blood glucose and the increase of insulin sensitivity [108,112,117,118]. Metformin has been widely used, alone or in combination with other OCs or anti-androgens, for the treatment of PCOS as it is shown to decrease levels of LH, testosterone, and androstenedione, and improve the ovulation and decrease IR [108,109,112,113,119–122]. Metformin is shown to be equally but moderately effective on reducing the hair diameter in PCOS women when compared to anti-androgen combined with OC [112]. Although there was no decrease of androgen levels, the effect is suggested to appear due to the reduction of hyperinsulinemia. This suggests that managing insulin resistance and increasing insulin sensitivity might be a better therapy treatment and approach, than administrating anti-androgens. Overall, the combination of metformin with other drugs appears to be more effective on the treatment of hirsutism, decreasing m-FG scores, levels of androgens and further drop of free androgens through the enhancement of SHBG [109].

Peripheral androgen blockage drugs include cyproterone acetate, spironolactone, flutamide and bicalutamide. 5α-reductase inhibitors, such as finasteride, have also been used for the treatment of hirsutism [123]. These drugs bind to the androgen receptors in selective tissues and therefore reduce the synthesis of androgens [124–126]. Cyproterone acetate is shown to be effective in treating excessive hair growth and reducing androgen levels [100,110,112,123,124,127,128], although side effects, such as weight gain, tiredness, loss of libido and adrenal disfunction were reported [127–129]. The effect of this progestogenic drug is through the inhibition of gonadotropin response to reduce T levels, decrease hair growth and androgen levels [127,128,130]. Spironolactone was originally used as a diuretic drug. However, it was shown to reduce T, through cytochrome P450 inhibition, without having any effect on SHBG or free T [131]. It is shown to be an effective and tolerable drug for the treatment of hirsutism, especially when combined with OC [132–134]. Flutamide is reported to be a better choice of treatment for the disorder studied when compared to the previously mentioned drugs [134]. It is shown to have a significant effect on hirsutism, acne and male-pattern hair loss in hirsute women. However, severe side effects are reported, such as embryotoxicity, thus it is suggested to use in combination with other drugs, such as insulin sensitisers or OC [98,108,111,122,134]. Bicalutamide was shown to have similar effectiveness with the above drugs on reducing m-FG scores and lower androgen levels, although it did not have any effect on insulin resistance [135–137]. Despite the effectiveness of these drugs, many subjects are sceptical when it comes to the administration of such treatments due to the potentially severe side effects. Thus, the potential therapeutic effect of naturally derived therapies is also studied.

2.2. *Naturally Derived Therapies*

Many natural remedies have been studied and used for the management of hirsutism and acne, however, extensive research has not been conducted for many of them [138–141]. Some of the naturally derived anti-androgens are spearmint tea, green tea, licorice, Chinese peony, marjoram, and red reishi [141]. In a study where the effects of spearmint tea as an anti-androgen were investigated, forty two patients were asked to consume two cups of spearmint tea every day for thirty days [142]. Their gonadotrophin levels and androgen hormones were tested before and after the treatment and a significant reduction of total and free T was observed. Increased LH and FSH were also observed after short administration with spearmint tea [140]. Marjoram is shown to significantly reduce levels of DHEAS and improve IR, and therefore inhibit the production of androgens [143]. Diet has also been investigated on the role of hormonal imbalances especially since the production of androgens is related to IR. Studies are still at early stages, looking at the potential effect of dairy and starch on the hormonal profile of PCOS women [144–146]. Studies where women were asked to follow a low starch and low dairy diet presented mixed results and the role of these foods on hormonal profiles requires more research. In addition to plant-based remedies and diet for the treatment of hirsutism and other hormonal imbalances, live organisms have also been studied on their potential effect on the hormonal profile of hirsute women [147–151].

2.2.1. Probiotics and Hirsutism

Microbes have been widely used for the production of various chemicals in the pharmaceutical world. In a recent study, after metabolically engineering *Rhodococcus ruber Chol-4*, it was shown to produce high amounts (61%) of testosterone through the conversion of 4-androstene-3,17-dione (AD) to the mentioned hormone [152]. There is increasing literature on how probiotics can affect the hormonal profiles in animals and humans and how probiotic supplementation can be used for the management and improvement of metabolic disorders and conditions. When the GM taxa of boards and gilts was analysed, host sex hormones were shown to significantly interact with the GM, which indicates the relationship between hormones and microbes [20]. Another research study that presents the relationship between the gut microbiota and the hormonal balance, is when the androgen levels of normal GM mice were compared to germ-free mice and findings showed exceptionally low unconjugated DHT detected in the later subject's distal intestine in contrast with the former subjects [153]. Interestingly, in the same study, human male feces samples with normal GM were tested and similar findings were observed compared to the normal GM mice samples. For the mice samples, male and female subjects were included. It is therefore proposed that treatment with probiotics could potentially affect the levels of DHT and T.

As mentioned earlier, PCOS is the most common cause of Hirsutism and most women that suffer from this syndrome show symptoms of increased hair growth in a male-like pattern. There is growing literature on the role of probiotics on PCOS and how the mentioned microbes are potentially involved in the management of this condition. Tremellen and Pearce in 2012 presented a detailed hypothesis on the development of PCOS starting from dysbiosis of gut microbiota (DOGMA) and how this causes a series of effects that eventually leads to hirsutism along with other symptoms of PCOS [154]. More specifically, it is hypothesised that the disruption of the balance of GM, and in combination with a diet high-in- sugar and saturated fat, leads to increased gut mucosa permeability. This allows the gram-negative bacteria to enter the systemic circulation and stimulate the immune system causing mild chronic inflammation, leading to damage in insulin receptors that further develops into IR and hyperinsulinemia in the long term. The latter condition affects the ovaries to produce excess androgens and disrupts their normal function, causing further disturbance in the development of follicles. Therefore, it is possible to hypothesise that these conditions are less likely to occur if a well-balanced gut microbiota is maintained. The above hypothesis was further investigated a few years later, where the potential gut permeability, and the level of inflammation and GM in PCOS women was studied [155]. Even though the GM was indeed found to be altered, compared to healthy subjects, there was no significant changes in gut permeability and function suggesting that "leaky" gut is only potentially a symptom in women with PCOS. In another study where the relationship of PCOS and gut microbiota was investigated, PCOS rats' GM was compared to control groups and a decreased number of *Lactobacillus*, *Clostridium* and *Ruminococcus* were detected, whereas *Prevotella*, a gram-negative bacteria, was found in higher numbers than for the healthy subjects [156]. The rat's PCOS was induced using an aromatase inhibitor called letrozole and the subjects showed increased levels of androgens along with symptoms of abnormal hormonal levels and cycles with their ovaries morphologically altered. To further investigate this relationship, faecal microbiota and *Lactobacillus* were transplanted in the gut of PCOS rats and restoration of the gut microbiota was observed along with improvement of PCOS symptoms. Interestingly, the androgen levels were reduced, and the ovaries' morphology was improved. Such effects are in agreement with more recent studies where androgen levels and hirsutism are shown to improve after supplementation with probiotics [148,149,157]. Possibly the first investigation on hormonal profiles of PCOS women after the supplementation of probiotics, was presented in 2018 in a study where 60 women participated [157]. Probiotics *Lactobacillus* and *Bifidobacterium*, two of the most common species used in probiotic treatments, were supplied to the subjects for 12 weeks after which time significant increase in SHBG was observed, whilst m-FG scores and total testosterone were decreased. Combinational treatments of vitamin D and probiotics also showed a significant response when it comes to androgen level reduction [149]. Although this suggests that probiotics might be more efficient on managing hirsutism when combined with vitamin D, other researchers have not shown any significant effect of the vitamin alone to the referred condition [158–161], despite its significant anti-inflammatory and insulin sensitivity effect [162–164]. Therefore, it is possible that the effects of the combinational treatment are due to the improvement of IR from vitamin D, along with the probiotics effect [149]. It is also reported that PCOS-rats showed improvement in reproductive function after the treatment with probiotics [165]. Although only a few studies have looked at the effects of probiotics on hirsutism, the literature found on the role of the former on IR and sensitivity has been explored in several studies [27,165–173]. Moreover, it is now well established from a variety of studies that hirsutism is linked with IR [21,95–97,104,174–177]. As described earlier IR and hyperinsulinemia enhances the production of androgens such as T and DHT, whilst reduces SHBG levels leading to increased free testosterone levels [102].

Considering the above it is necessary to briefly present the role of probiotics on IR through which a potential improvement of hirsutism can occur. Research has been conducted on PCOS subject as well as prediabetics and healthy participants. In 2015, Shoaei et al. studied PCOS women for 8 weeks while the subjects were under probiotic supplementation, observing decreased FBS, insulin levels and IR [173]. The same observations were made in another study conducted on PCOS women where probiotics were supplemented for 12 weeks [167]. Glucose levels decreased during pregnancy and insulin sensitivity increased during the 12 month postpartum period, in a study where subjects were treated with *Lactobacillus* and *Bifidobacterium* [168]. Lower glucose levels were also observed when post-menopausal women were provided with *Lactobacillus plantarum*-containing beverages [170]. It is reported that *Lactobacillus salivarius UBL22*, when supplemented on healthy subjects, showed a significantly reduced fasting insulin compared to the control group (placebo) [27]. Another study conducted on prediabetic participants showed a reduction in fasting glucose and fasting insulin levels after treatment with probiotics (Table 1) [166].

Table 1. Probiotic treatments and the hirsutism-relevant outcomes.

Treatment	Study Population	Relevant Outcomes	Reference
Lactobacillus	PCOS [1] rats	Improvement on estrous cycles, reduction of androgen biosynthesis, normalization of ovaries, GM [2] restoration, reduction of *Prevotella*	[156]
L. acidophilus, L. casei, B. bifidum	PCOS women	Reduction of total testosterone, increase of SHBG [3], decrease of m-FG scores	[157]
L. acidophilus, L. reuteri, B. bifidum and Vitamin D	PCOS women	Reduction of total testosterone and hirsutism	[149]

Table 1. *Conts.*

Treatment	Study Population	Relevant Outcomes	Reference
E. faecali, L. reuteri, Bifidobacterium	PCOS rats	Improvement in reproductive function and GM restoration	[165]
L. acidophilus, L. casei, B. bifidum	PCOS women	Reduced fasting plasma glucose, insulin concentrations and insulin resistance. Increased insulin sensitivity	[167]
L. acidophilus, L. casei, L. rhamnosus, L. bulgaricus, L. breve, B. longum, S. thermophiles	PCOS women	Reduced FBS [4], reduced insulin levels and reduced insulin resistance	[173]
L. plantarum	Postmenopausal women	Reduced glucose levels	[170]
L. rhamnosus, B. lactis	Postpartum period women	Reduced blood glucose concentrations, improved glucose tolerance, reduced insulin concentrations and increased insulin sensitivity	[168]

[1] polycystic ovary syndrome; [2] gut microbiota; [3] sex hormone-binding globulin; [4] fasting blood sugar.

2.2.2. Synbiotics and Hirsutism

In reviewing the literature, very few studies were found on the association between synbiotics and hirsutism. A recent study looked at the effects of synbiotics on PCOS women using a mix of pomegranate juice, inulin and *Lactobacillus* [178]. Testosterone levels were significantly reduced after the treatment with synbiotics, compared to the control group. These findings were accompanied with an increase in insulin sensitivity and improved anthropometric measurements. Inulin is a prebiotic that has been shown to improve IR and has a beneficial effect on SHBG [147,179,180]. In a 12 week study, *Lactobacillus bifidum, Lactobacillus acidophilus, Lactobacillus casei* and inulin were used for the treatment of PCOS women and their hormone levels were studied along with other biomarkers, such as serum high sensitivity C-reactive protein [19]. Despite the statistical significance of the decrease of m-FG scores, the clinical picture of the subjects showed little improvement and longer treatment with prebiotic inulin was suggested. Selenium, known for its prebiotic activities [181–183] was also used in a combinational treatment with probiotics on women and the hormonal profiles were studied [150,184,185]. Although initial studies have not shown an effect of the treatment on free [184] and total [185] testosterone, a decrease on m-FG scores was observed [184]. On a 12-week study conducted two years later, 60 PCOS subjects were treated with *Lactobacilllus, Bifidobacterium* and selenium-containing synbiotics and, in contrast with the above findings, a statistically significant reduction of total testosterone and hirsutism was shown [150]. There is abundant room for further progress in determining the exact effect of synbiotics on hirsutism, whilst there is an increasing literature on the beneficial effects of synbiotics on IR on PCOS women.

Brief investigation of the role of synbiotics on IR is necessary to understand how synbiotics could potentially improve hirsutism through the improvement and management of IR. A year after Nasri et al. (2018) published their work as described earlier, Samimi et al. (2019) investigated the effects of the same synbiotics supplementation on biomarkers of IR and sensitivity along with cholesterol and triglycerides on PCOS women [180]. In this study, it was shown that *Lactobacillus bifidum, Lactobacillus acidophilus, Lactobacillus casei* and 0.8 g of inulin, significantly improved insulin metabolism markers but not FBS, after 12 weeks of treatment, suggesting that metabolic disorders can benefit from the mentioned treatment while managing symptoms. Longer treatment periods (24 weeks) were also investigated and prediabetic adults were treated with *Lactobacillus* and *Bifidobacterium* in a inulin-containing prebiotic [166]. There was no significant improvement on fasting insulin levels and plasma glucose between the two studied time points, 12 and 24 weeks. The use of a single probiotic along with inulin is found in a study that was conducted on diabetic rats and their insulin markers were observed [80]. The synbiotic containing *Lactobacillus plantarum* and inulin managed to increase insulin sensitivity. All the above studies described were conducted on subjects that were either diagnosed as prediabetics or suffered from conditions (PCOS) for which IR is known to be a symptom. However, the beneficial effect of synbiotics was also observed when healthy subjects were treated with *Lactobacillus salivarius UBL S22* and FOS and the treatment was shown to improve insulin sensitivity significantly more compared to the probiotics alone [27]. Interestingly, despite the higher variation of probiotic bacteria in the synbiotic mix (*Bifidobacterium breve, Lactobacillus casei, Lactobacillus bulgaricus, Lactobacillus rhamnosus, Bifidobacterium longum, Lactobacillus acidophilus, Streptococcus thermophilus*), in combination with the fructo-oligosaccharide inulin on PCOS women, it does not show a significant effect on the FBS and insulin sensitivity [179]. *Lactobacillus* synbiotics and 200 μg/day selenium [186] were used for the treatment of PCOS subjects for 12 weeks and it was found to improve IR and insulin levels, findings that were not observed when selenium alone was used for the treatment of diabetic participants [187]. Other studies show similar observations on fasting plasma glucose [188] and insulin levels, IR and insulin sensitivity (Table 2) [189].

Table 2. Synbiotic treatments and the hirsutism-relevant outcomes.

Treatment	Study Population	Relevant Outcomes	Reference
Lactobacillus and pomegranate juice	PCOS [1] women	Reduced testosterone levels, increased insulin resistance and improved anthropometric measurements	[178]
L. bifidum, L. acidophilus, L. casei and inulin	PCOS women	Reduced m-FG scores but did not affect the clinical picture of hirsutism	[19]
Lactobacillus, Bifidobacterium and Selenium	PCOS women	Reduced total testosterone and hirsutism	[150]
L. bifidum, L. acidophilus, L. casei and inulin	PCOS women	Improved insulin metabolism markers but did not influence FBS [2]	[180]
Lactobacillus, Bifidobacterium and inulin	Prediabetics	Decreased fasting insulin levels and fasting plasma glucose	[166]
L. plantarum and inulin	Diabetic rats	Improved insulin resistance, hypothalamic levels of insulin and leptin	[80]
L. salivarius UBL S22 and fructo-oligosaccharides	Healthy participants	Improved insulin sensitivity	[27]

Table 2. *Conts.*

Treatment	Study Population	Relevant Outcomes	Reference
L. bulgaricus, *L. acidophilus*, *L. casei*, *L. rhamnosus*, *B. longum*, *B. breve*, *S. thermophilus* and inulin	PCOS women	No effects on fasting insulin levels	[179]
L. acidophilus, *L. reuteri*, *L. fermentum*, *B. bifidum* and selenium	PCOS women	Reduced insulin levels and insulin resistance. Increased insulin sensitivity	[186]
L. acidophilus, *L. casei*, *B. bifidum* and inulin	Gestational diabetics	Decreased insulin levels and insulin resistance. Increased insulin sensitivity	[189]

[1] polycystic ovary syndrome; [2] fasting blood sugar.

3. Conclusions

Probiotics and synbiotics have been widely used for the treatment of several conditions and diseases [54,155,186,190–192]. The potentially healing properties of natural products are presented more often in recent years and people tend to turn to more natural ways of treating symptoms or even aiming towards them providing a reversing effect. On the other hand, endocrinology is a very complex area of study. Although hirsutism, a disorder characterised of excess terminal hair in a male-like pattern [88,193], has been generally studied, not much research was conducted on the effects of probiotics/synbiotics on this condition. The complexity of the disorder is also shown through its relationship with lipid metabolism [194]. Main markers for the determination of lipid metabolism, such as low-density lipoprotein (LDL), high-density lipoprotein (HDL) and triglycerides are shown to be altered in women suffering from PCOS and other related disorders [195]. Elevated levels of triglycerides and lower levels of LDL, a group of lipoproteins that are the main source of cholesterol, are observed in hirsute women. Moreover, insulin is known to increase the activity of tissue lipoprotein lipases (LPL), an enzyme that takes part in the metabolism of lipoproteins and is regulated by plasma hormones. IR, on the other hand, can lead to the elevation of triglycerides, an observation that indicates an indirect relationship of hirsutism with high levels of lipids through the increase of IR through the elevated levels of androgens and vice versa. However, this complex connection between hormones and lipids appears to be independent, as shown in a study where hirsute women were treated with GnRH and, despite the unaffected IR and lipid metabolism of the subjects, estrogens and androgens were decreased [195]. The vicious circle between increased androgens and IR, as shown in Figure 3, supports the hypothesis that hirsutism can affect the metabolic process of lipoproteins and the related enzymes, through the enhancements of IR. Therefore, treatments that provide an overall improvement to IR and steroid levels could potentially alter the metabolic process of lipids and enzymes. Although there is an extensive research on the effects of androgens on cardiovascular disease risk on men [196–199], not many studies exist on the effects of hirsutism on lipid metabolism. There is a mixed association between the effects of elevated androgen levels on women and their independent relationship with lipids and cardiovascular disease [200]. The purpose of the current study was to investigate the role of probiotics and synbiotics on hirsutism.

The limited literature found on this topic, shows that probiotics can decrease the production of androgens, and specifically total testosterone, and improve m-FG scoring on hirsute women, increasing the levels of SHBG that bind and regulate free testosterone [102,153,201]. Moreover, synbiotics are shown to have a similar effect with probiotics but more enhanced [27].

Therefore, it is suggested that a combination of probiotics and prebiotics can improve and help on the management of hirsutism and other conditions. Additionally, considering the direct relationship of hirsutism with IR [91,95,96,174], the effects of the beneficial compounds on IR was also examined. Growing literature shows a reduction of FBS, insulin levels, insulin resistance and insulin sensitivity after treatments with probiotics and/or synbiotics supports the beneficial effects on hirsutism through the mentioned mechanisms [178–180,202]. Taken together, these findings suggest a role for probiotics and synbiotics in reducing androgens and decreasing m-FG scores, directly and through managing and decreasing IR. Moreover, treatments with probiotics and/or synbiotics are shown to improve the lipid metabolism in hirsute women, reducing serum triglycerides levels [27,167,203], reducing levels of LDL and increasing HDL [27,186,203], although it is not clear whether hirsutism is independently correlated with lipoprotein elevation. The majority of studies found in the literature refer to research conducted on PCOS women, since most hirsute women also suffer from this metabolic syndrome. Current drug therapies on hirsutism including insulin sensitisers, anti-androgen monotherapies and OCPs, can have severe side effects [98,112–114] and a potential therapeutic role of probiotics along with prebiotics, can be a significant relief for women suffering from the disorder.

The effect of the studied microorganisms on the metabolic profile of women is shown to be significant and interestingly, correlated with many abnormalities. The combinational effect of them on IR, steroids and lipids profiles, shows a multifunctional character of these treatments that could potentially support the metabolic processes to such degree that conventional treatments with drugs could be replaced, especially when combined with lifestyle changes, such as exercise and diet. Although there is evidence for the beneficial role of probiotics and synbiotics on hirsutism and the related disorders, more studies need to be conducted on their direct effects, mechanisms, length of treatment and dosage.

Author Contributions: The author confirms sole responsibility for the following: study conception and design, data collection, analysis and interpretation of results, and manuscript preparation.

Funding: This research received no external funding.

Acknowledgments: The author wishes to thank Sotiris Kyriakou for his support.

Conflicts of Interest: The author declares no conflict of interest and there has been no financial support for this work that could have influenced its outcome.

References

1. McGovern, P.E.; Zhang, J.; Tang, J.; Zhang, Z.; Hall, G.R.; Moreau, R.A.; Nunez, A.; Butrym, E.D.; Richards, M.P.; Wang, C.-S.; et al. Fermented beverages of pre- and proto-historic China. *Proc. Natl. Acad. Sci. USA* **2004**, *101*, 17593–17598. [CrossRef] [PubMed]
2. Sicard, D.; Legras, J.L. Bread, beer and wine: Yeast domestication in the Saccharomyces sensu stricto complex. *Comptes Rendus -Biol.* **2011**, *334*, 229–236. [CrossRef] [PubMed]
3. Ozen, M.; Dinleyici, E.C. The history of probiotics: The untold story. *Benef. Microbes* **2015**, *6*, 159–165. [CrossRef] [PubMed]
4. Isolauri, E. Probiotics in human disease. *Am. J. Clin. Nutr.* **2001**, *73*, 1142S–1146S. [CrossRef] [PubMed]
5. Vandenplas, Y.; Zakharova, I.; Dmitrieva, Y. Oligosaccharides in infant formula: More evidence to validate the role of prebiotics. *Br. J. Nutr.* **2015**, *113*, 1339–1344. [CrossRef] [PubMed]
6. Gordon, S. Elie Metchnikoff: Father of natural immunity. *Eur. J. Immunol.* **2008**, *38*, 3257–3264. [CrossRef] [PubMed]
7. Rodrigues, K.L.; Gaudino Caputo, L.R.; Tavares Carvalho, J.C.; Evangelista, J.; Schneedorf, J.M. Antimicrobial and healing activity of kefir and kefiran extract. *Int. J. Antimicrob. Agents* **2005**, *25*, 404–408. [CrossRef]
8. Rahimzadeh, G.; Fazeli, M.R.; Mozafari, A.N.; Mesbahi, M. Evaluation of Anti-Microbial Activity and Wound Healing of Kefir | International Journal of Pharmaceutical Sciences and Research. Available online: http://ijpsr.com/bft-article/evaluation-of-anti-microbial-activity-and-wound-healing-of-kefir/?view=fulltext (accessed on 26 November 2018).
9. Atalan, G.; Demirkan, I.; Yaman, H.; Cina, M. Effect of topical kefir application on open wound healing on in vivo study. *Kafkas Univ. Vet. Fak. Dderg.* **2003**, *9*, 43–47.
10. Lolou, V.; Panayiotidis, M.I. Functional role of probiotics and prebiotics on skin health and disease. *Fermentation* **2019**, *5*, 41. [CrossRef]
11. Sharifi, M.; Moridnia, A.; Mortazavi, D.; Salehi, M.; Bagheri, M.; Sheikhi, A. Kefir: A powerful probiotics with anticancer properties. *Med. Oncol.* **2017**, *34*, 183. [CrossRef]
12. Chen, M.J.; Liu, J.R.; Sheu, J.F.; Lin, C.W.; Chuang, C.L. Study on skin care properties of milk kefir whey. *Asian-Australasian J. Anim. Sci.* **2006**, *19*, 905–908. [CrossRef]

13. Collins, M.D.; Phillips, B.A.; Paolo, Z. Deozyribonucleic acid homology studies of lactobacillus casei etc. Collins 1989. *Int. J. Syst. Bacteriol.* **1989**, *39*, 105–108. [CrossRef]
14. Report of a Joint FAO/WHO Working Group on Drafting Guidelines for the Evaluation of Probiotics in Food. Guidelines for the Evaluation of Probiotics in Food. 2002. Available online: https://www.who.int/foodsafety/fs_management/en/probiotic_guidelines.pdf (accessed on 11 January 2021).
15. Goldin, B.R.; Swenson, L.; Dwyer, J.; Sexton, M.; Gorbach, S.L. Effect of diet and Lactobacillus acidophilus supplements on human fecal bacterial enzymes. *J. Natl. Cancer Inst.* **1980**, *64*, 255–261. [CrossRef] [PubMed]
16. Scholz-Ahrens, K.E.; Adolphi, B.; Rochat, F.; Barclay, D.V.; de Vrese, M.; Açil, Y.; Schrezenmeir, J. Effects of probiotics, prebiotics, and synbiotics on mineral metabolism in ovariectomized rats - impact of bacterial mass, intestinal absorptive area and reduction of bone turn-over. *NFS J.* **2016**, *3*, 41–50. [CrossRef]
17. Milani, C.; Duranti, S.; Bottacini, F.; Casey, E.; Turroni, F.; Mahony, J.; Belzer, C.; Delgado Palacio, S.; Arboleya Montes, S.; Mancabelli, L.; et al. The First Microbial Colonizers of the Human Gut: Composition, Activities, and Health Implications of the Infant Gut Microbiota. *Microbiol. Mol. Biol. Rev.* **2017**, *81*, 1–67. [CrossRef]
18. Talebi, S.; Karimifar, M.; Heidari, Z.; Mohammadi, H.; Askari, G. The effects of synbiotic supplementation on thyroid function and inflammation in hypothyroid patients: A randomized, double-blind, placebo-controlled trial. *Complement. Ther. Med.* **2020**, *48*. [CrossRef]
19. Nasri, K.; Jamilian, M.; Rahmani, E.; Bahmani, F.; Tajabadi-Ebrahimi, M.; Asemi, Z. The effects of synbiotic supplementation on hormonal status, biomarkers of inflammation and oxidative stress in subjects with polycystic ovaryfile:///C:/Users/Vassoula/Downloads/29664663.nbib syndrome: A randomized, double-blind, placebo-controlled trial. *BMC Endocr. Disord.* **2018**, *18*, 1–8. [CrossRef]
20. He, M.; Gao, J.; Wu, J.; Zhou, Y.; Fu, H.; Ke, S.; Yang, H.; Chen, C.; Huang, L. Host gender and androgen levels regulate gut bacterial taxa in pigs leading to sex-biased serum metabolite profiles. *Front. Microbiol.* **2019**, *10*, 1–13. [CrossRef]
21. Yarak, S.; Brasil Parada, M.O.A.; Bagatin, E.; Talarico Filho, S.; Hassun, K.M. Hyperandrogenism and skin: Polycystic ovary syndrome and peripheral insulin resistance. *An. Bras. Dermatol.* **2005**, *80*, 395–410. [CrossRef]
22. Carnevali, O.; Avella, M.A.; Gioacchini, G. Effects of probiotic administration on zebrafish development and reproduction. *Gen. Comp. Endocrinol.* **2012**, *188*, 297–302. [CrossRef]
23. Sonal Sekhar, M.; Unnikrishnan, M.K.; Vijayanarayana, K.; Rodrigues, G.S.; Mukhopadhyay, C. Topical application/formulation of probiotics: Will it be a novel treatment approach for diabetic foot ulcer? *Med. Hypotheses* **2014**, *82*, 86–88. [CrossRef] [PubMed]
24. Sikorska, H.; Smoragiewicz, W. Role of probiotics in the prevention and treatment of meticillin-resistant Staphylococcus aureus infections. *Int. J. Antimicrob. Agents* **2013**, *42*, 475–481. [CrossRef] [PubMed]
25. Eggers, S.; Barker, A.; Valentine, S.; Hess, T.; Duster, M.; Safdar, N. Impact of Probiotics for Reducing Infections in Veterans (IMPROVE): Study protocol for a double-blind, randomized controlled trial to reduce carriage of Staphylococcus aureus. *Contemp. Clin. Trials* **2017**, *52*, 39–45. [CrossRef] [PubMed]
26. Frei, R.; Akdis, M.; O'mahony, L. Prebiotics, probiotics, synbiotics, and the immune system: Experimental data and clinical evidence. *Curr. Opin. Gastroenterol.* **2015**, *31*, 153–158. [CrossRef]
27. Rajkumar, H.; Kumar, M.; Das, N.; Kumar, S.N.; Challa, H.R.; Nagpal, R. Effect of Probiotic Lactobacillus salivarius UBL S22 and Prebiotic Fructo-oligosaccharid on Serum Lipids, Inflammatory Markers, Insulin Sensitivity, and Gut Bacteria in Healthy Young Volunteers: A Randomized Controlled Single-Blind Pilot Study. *J. Cardiovasc. Pharmacol. Ther.* **2015**, *20*, 289–298. [CrossRef]
28. Cozzolino, M.; Vitagliano, A.; Pellegrini, L.; Chiurazzi, M.; Andriasani, A.; Ambrosini, G.; Garrido, N. Therapy with probiotics and synbiotics for polycystic ovarian syndrome: A systematic review and meta-analysis. *Eur. J. Nutr.* **2020**, *59*, 2841–2856. [CrossRef]
29. Pandey, K.R.; Naik, S.R.; Vakil, B.V. Probiotics, prebiotics and synbiotics–A review. *J. Food Sci. Technol.* **2015**, *52*, 7577–7587. [CrossRef]
30. Azad, M.A.K.; Sarker, M.; Li, T.; Yin, J. Probiotic Species in the Modulation of Gut Microbiota: An Overview. *Biomed Res. Int.* **2018**, *2018*. [CrossRef]
31. Mori, N.; Kano, M.; Masuoka, N.; Konno, T.; Suzuki, Y.; Miyazaki, K.; Ueki, Y. Effect of probiotic and prebiotic fermented milk on skin and intestinal conditions in healthy young female students. *Biosci. Microbiota Food Health* **2016**, *35*, 105–112. [CrossRef]
32. Sánchez, B.; Delgado, S.; Blanco-Míguez, A.; Lourenço, A.; Gueimonde, M.; Margolles, A. Probiotics, gut microbiota, and their influence on host health and disease. *Mol. Nutr. Food Res.* **2017**, *61*, 1–15. [CrossRef]
33. Britti, M.; Roselli, M.; Finamore, A.; Merendino, N.; Mengheri, E. Regulation of immune response at intestinal and peripheral sites by probiotics. *Biologia (Bratisl)* **2006**, *61*, 735–740. [CrossRef]
34. Thomas, C.M.; Versalovic, J. Probiotics-host communication. *Gut Microbes* **2010**, *1*, 148–163. [CrossRef] [PubMed]
35. Bansal, S.; Mangal, M.; Sharma, S.K.; Gupta, R.K. Non-dairy Based Probiotics: A Healthy Treat for Intestine. *Crit. Rev. Food Sci. Nutr.* **2016**, *56*, 1856–1867. [CrossRef] [PubMed]
36. Wang, X.; Farnell, Y.Z.; Peebles, E.D.; Kiess, A.S.; Wamsley, K.G.S.; Zhai, W. Effects of prebiotics, probiotics, and their combination on growth performance, small intestine morphology, and resident Lactobacillus of male broilers. *Poult. Sci.* **2016**, *95*, 1332–1340. [CrossRef] [PubMed]
37. Friedrich, A.D.; Paz, M.L.; Leoni, J.; Maglio, D.H.G. Message in a bottle: Dialog between intestine and skin modulated by probiotics. *Int. J. Mol. Sci.* **2017**, *18*, 1067. [CrossRef] [PubMed]
38. White, J.S.; Hoper, M.; Parks, R.W.; Clements, W.D.B.; Diamond, T.; Bengmark, S. The probiotic bacterium Lactobacillus plantarum species 299 reduces intestinal permeability in experimental biliary obstruction. *Lett. Appl. Microbiol.* **2006**, *42*, 19–23. [CrossRef]

39. Yamada, T.; Nagata, S.; Kondo, S.; Bian, L.; Wang, C.; Asahara, T.; Ohta, T.; Nomoto, K.; Yamashiro, Y. Effect of Continuous Fermented Milk Intake Containing Lactobacillus casei Strain Shirota on Fever in Mass Infectious Gastroenteritis Rest Home Outbreak. *Kansenshogaku Zasshi* **2009**, *83*, 31–35. [CrossRef]
40. Imaoka, A.; Shima, T.; Kato, K.; Mizuno, S.; Uehara, T.; Matsumoto, S.; Setoyana, H.; Hara, T.; Umesaki, Y. Anti-inflammatory activity of probiotic Bifidobacterium: Enhancement of IL-10 production in peripheral blood mononuclear cells from ulcerative colitis patients and inhibition of IL-8 secretion in HT-29 cells. *World J. Gastroenterol.* **2008**, *14*, 2511–2516. [CrossRef]
41. Abdin, A.A.; Saeid, E.M. An experimental study on ulcerative colitis as a potential target for probiotic therapy by Lactobacillus acidophilus with or without "olsalazine". *J. Crohn's Colitis* **2008**, *2*, 296–303. [CrossRef]
42. Amara, A.A.; Shibl, A. Role of Probiotics in health improvement, infection control and disease treatment and management. *Saudi Pharm. J.* **2015**, *23*, 107–114. [CrossRef]
43. Mego, M.; Májek, J.; Končeková, R.; Ebringer, L.; Čierniková, S.; Rauko, P.; Kováč, M.; Trupl, J.; Slezák, P.; Zajac, V. Intramucosal bacteria in colon cancer and their elimination by probiotic strain Enterococcus faecium M-74 with organic selenium. *Folia Microbiol.* **2005**, *50*, 443–447. [CrossRef] [PubMed]
44. Boudeau, J.; Glasser, A.L.; Julien, S.; Colombel, J.F.; Darfeuille-Michaud, A. Inhibitory effect of probiotic Escherichia coli strain Nissle 1917 on adhesion to and invasion of intestinal epithelial cells by adherent-invasive E.coli strains isolated from patients with Crohn's disease. *Aliment. Pharmacol. Ther.* **2003**, *18*, 45–56. [CrossRef] [PubMed]
45. Anukam, K.C.; Hayes, K.; Summers, K.; Reid, G. Probiotic Lactobacillus rhamnosus GR-1 and Lactobacillus reuteri RC-14 may help downregulate TNF-alpha, IL-6, IL-8, IL-10 and IL-12 (p70) in the neurogenic bladder of spinal cord injured patient with urinary tract infections: A two-case study. *Adv. Urol.* **2009**, *2009*. [CrossRef] [PubMed]
46. Thirabunyanon, M.; Boonprasom, P.; Niamsup, P. Probiotic potential of lactic acid bacteria isolated from fermented dairy milks on antiproliferation of colon cancer cells. *Biotechnol. Lett.* **2009**, *31*, 571–576. [CrossRef] [PubMed]
47. Hacini-Rachinel, F.; Gheit, H.; Le Luduec, J.B.; Dif, F.; Nancey, S.; Kaiserlian, D. Oral probiotic control skin inflammation by acting on both effector and regulatory T cells. *PLoS ONE* **2009**, *4*, e4903. [CrossRef] [PubMed]
48. Chapat, L.; Chemin, K.; Dubois, B.; Bourdet-Sicard, R.; Kaiserlian, D. Lactobacillus casei reduces CD8+T cell-mediated skin inflammation. *Eur. J. Immunol.* **2004**, *34*, 2520–2528. [CrossRef]
49. Weise, C.; Zhu, Y.; Ernst, D.; Ku, A.A. Oral administration of Escherichia coli Nissle 1917 prevents allergen-induced dermatitis in mice. *Exp. Dermatol.* **2011**, *20*, 805–809. [CrossRef]
50. Vaiserman, A.M.; Koliada, A.K.; Marotta, F. Gut microbiota: A player in aging and a target for anti-aging intervention. *Ageing Res. Rev.* **2017**, *35*, 36–45. [CrossRef]
51. Musthaq, S.; Mazuy, A.; Jakus, J. The microbiome in dermatology. *Clin. Dermatol.* **2018**, *36*, 390–398. [CrossRef]
52. Kalliomäki, M.; Salminen, S.; Poussa, T.; Isolauri, E. *Probiotics during the First 7 Years of Life: A Cumulative Risk Reduction of Eczema in a Randomized, Placebo-Controlled Trial*; Mosby: Maryland Heights, MI, USA, 2007; Volume 119.
53. Huseini, H.F.; Rahimzadeh, G.; Fazeli, M.R.; Mehrazma, M.; Salehi, M. Evaluation of wound healing activities of kefir products. *Burns* **2012**, *38*, 719–723. [CrossRef]
54. Nasrabadi, H. Study of cutaneous wound healing in rats treated with Lactobacillus plantarum on days 1, 3, 7, 14 and 21. *Afr. J. Pharm. Pharmacol.* **2011**, *5*, 2395–2401. [CrossRef]
55. Jebur, M.S. Therapeutic efficacy of Lactobacillus acidophilus against bacterial isolates from burn wounds. *N. Am. J. Med. Sci.* **2010**, *2*, 586–591. [CrossRef] [PubMed]
56. Tsiouris, C.G.; Tsiouri, M.G. Human microflora, probiotics and wound healing. *Wound Med.* **2017**, *19*, 33–38. [CrossRef]
57. Rumbaugh, K.P.; Griswold, J.A.; Iglewski, B.H.; Hamood, A.N. Contribution of quorum sensing to the virulence of Pseudomonas aeruginosa in burn wound infections. *Infect. Immun.* **1999**, *67*, 5854–5862. [CrossRef] [PubMed]
58. de Oliveira, G.L.V.; Leite, A.Z.; Higuchi, B.S.; Gonzaga, M.I.; Mariano, V.S. Intestinal dysbiosis and probiotic applications in autoimmune diseases. *Immunology* **2017**, *152*, 1–12. [CrossRef]
59. Candon, S.; Perez-Arroyo, A.; Marquet, C.; Valette, F.; Foray, A.P.; Pelletier, B.; Milani, C.; Ventura, M.; Bach, J.F.; Chatenoud, L. Antibiotics in early life alter the gut microbiome and increase disease incidence in a spontaneous mouse model of autoimmune insulin-dependent diabetes. *PLoS ONE* **2015**, *10*, e0125448. [CrossRef]
60. Hickson, M.; D'Souza, A.L.; Muthu, N.; Rogers, T.R.; Want, S.; Rajkumar, C.; Bulpitt, C.J. Use of probiotic Lactobacillus preparation to prevent diarrhoea associated with antibiotics: Randomised double blind placebo controlled trial. *Br. Med. J.* **2007**, *335*, 80–83. [CrossRef]
61. He, C.; Cheng, D.; Peng, C.; Li, Y.; Zhu, Y.; Lu, N. High-fat diet induces dysbiosis of gastric microbiota prior to gut microbiota in association with metabolic disorders in mice. *Front. Microbiol.* **2018**, *9*, 1–9. [CrossRef]
62. Ravussin, Y.; Koren, O.; Spor, A.; Leduc, C.; Gutman, R.; Stombaugh, J.; Knight, R.; Ley, R.E.; Leibel, R.L. Responses of gut microbiota to diet composition and weight loss in lean and obese mice. *Obesity* **2012**, *20*, 738–747. [CrossRef]
63. Ochoa-Repáraz, J.; Kasper, L.H. The Second Brain: Is the Gut Microbiota a Link Between Obesity and Central Nervous System Disorders? *Curr. Obes. Rep.* **2016**, *5*, 51–64. [CrossRef]
64. Calvani, R.; Picca, A.; Lo Monaco, M.R.; Landi, F.; Bernabei, R.; Marzetti, E. Of microbes and minds: A narrative review on the second brain aging. *Front. Med.* **2018**, *5*, 1–11. [CrossRef]
65. Sokol, H.; Seksik, P.; Furet, J.P.; Firmesse, O.; Nion-Larmurier, I.; Beaugerie, L.; Cosnes, J.; Corthier, G.; Marteau, P.; Doré, J. Low counts of faecalibacterium prausnitzii in colitis microbiota. *Inflamm. Bowel Dis.* **2009**, *15*, 1183–1189. [CrossRef] [PubMed]

66. Sokol, H.; Pigneur, B.; Watterlot, L.; Lakhdari, O.; Bermudez-Humaran, L.G.; Gratadoux, J.-J.; Blugeon, S.; Bridonneau, C.; Furet, J.-P.; Corthier, G.; et al. Faecalibacterium prausnitzii is an anti-inflammatory commensal bacterium identified by gut microbiota analysis of Crohn disease patients. *J. Pathol.* **1990**, *162*, 323–327. [CrossRef] [PubMed]
67. Partlow, J.; Blikslager, A.; Matthews, C.; Law, M.; Daniels, J.; Baker, R.; Labens, R. Effect of topically applied Saccharomyces boulardii on the healing of acute porcine wounds: A preliminary study. *BMC Res. Notes* **2016**, *9*. [CrossRef] [PubMed]
68. Brouwer, M.L.; Wolt-Plompen, S.A.A.; Dubios, A.E.J.; Van Der Heide, S.; Jansen, D.F.; Hoijer, M.A.; Kauffman, H.F.; Duiverman, E.J. No effects of probiotics on atopic dermatitis in infancy: A randomized placebo-controlled trial. *Clin. Exp. Allergy* **2006**, *36*, 899–906. [CrossRef] [PubMed]
69. Sugimoto, S.; Ishii, Y.; Izawa, N.; Masuoka, N.; Kano, M.; Sone, T.; Chiba, K.; Miyazaki, K.; Ishikawa, F. Photoprotective effects of Bifidobacterium breve supplementation against skin damage induced by ultraviolet irradiation in hairless mice. *Photodermatol. Photoimmunol. Photomed.* **2012**, *28*, 312–319. [CrossRef]
70. Queipo-Ortuño, M.I.; Seoane, L.M.; Murri, M.; Pardo, M.; Gomez-Zumaquero, J.M.; Cardona, F.; Casanueva, F.; Tinahones, F.J. Gut Microbiota Composition in Male Rat Models under Different Nutritional Status and Physical Activity and Its Association with Serum Leptin and Ghrelin Levels. *PLoS ONE* **2013**, *8*. [CrossRef]
71. Naruszewicz, M.; Johansson, M.L.; Zapolska-Downar, D.; Bukowska, H. Effect of Lactobacillus plantarum 299v on cardiovascular disease risk factors in smokers. *Am. J. Clin. Nutr.* **2002**, *76*, 1249–1255. [CrossRef]
72. Messaoudi, M.; Lalonde, R.; Violle, N.; Javelot, H.; Desor, D.; Nejdi, A.; Bisson, J.F.; Rougeot, C.; Pichelin, M.; Cazaubiel, M.; et al. Assessment of psychotropic-like properties of a probiotic formulation (Lactobacillus helveticus R0052 and Bifidobacterium longum R0175) in rats and human subjects. *Br. J. Nutr.* **2011**, *105*, 755–764. [CrossRef]
73. Karlsson, F.H.; Tremaroli, V.; Nookaew, I.; Bergström, G.; Behre, C.J.; Fagerberg, B.; Nielsen, J.; Bäckhed, F. Gut metagenome in European women with normal, impaired and diabetic glucose control. *Nature* **2013**, *498*, 99–103. [CrossRef]
74. Vrieze, A.; Van Nood, E.; Holleman, F.; Salojärvi, J.; Kootte, R.S.; Bartelsman, J.F.W.M.; Dallinga-Thie, G.M.; Ackermans, M.T.; Serlie, M.J.; Oozeer, R.; et al. Transfer of intestinal microbiota from lean donors increases insulin sensitivity in individuals with metabolic syndrome. *Gastroenterology* **2012**, *143*, 913–916. [CrossRef] [PubMed]
75. Adlercreutz, H.; Pulkkinen, M.O.; Hämäläinen, E.K.; Korpela, J.T. Studies on the role of intestinal bacteria in metabolism of synthetic and natural steroid hormones. *J. Steroid Biochem.* **1984**, *20*, 217–229. [CrossRef]
76. Markle, J.G.M.; Frank, D.N.; Mortin-Toth, S.; Robertson, C.E.; Feazel, L.M.; Rolle-Kampczyk, U.; Von Bergen, M.; McCoy, K.D.; Macpherson, A.J.; Danska, J.S. Sex differences in the gut microbiome drive hormone-dependent regulation of autoimmunity. *Science* **2013**, *339*, 1084–1088. [CrossRef] [PubMed]
77. Furrie, E.; Macfarlane, S.; Kennedy, A.; Cummings, J.H.; Walsh, S.V.; O'Neil, D.A.; Macfarlane, G.T. Synbiotic therapy (Bifidobacterium longum/Synergy 1) initiates resolution of inflammation in patients with active ulcerative colitis: A randomised controlled pilot trial. *Gut* **2005**, *54*, 242–249. [CrossRef] [PubMed]
78. Liu, R.; Zhang, C.; Shi, Y.; Zhang, F.; Li, L.; Wang, X.; Ling, Y.; Fu, H.; Dong, W.; Shen, J.; et al. Dysbiosis of gut microbiota associated with clinical parameters in polycystic ovary syndrome. *Front. Microbiol.* **2017**, *8*, 1–12. [CrossRef]
79. Hutkins, R.W.; Krumbeck, J.A.; Bindels, L.B.; Cani, P.D.; Fahey, G.; Goh, Y.J.; Hamaker, B.; Martens, E.C.; Mills, D.A.; Rastal, R.A.; et al. Prebiotics: Why definitions matter. *Curr. Opin. Biotechnol.* **2016**, *37*, 1–7. [CrossRef]
80. Valenlia, K.B.; Morshedi, M.; Saghafi-Asl, M.; Shahabi, P.; Abbasi, M.M. Beneficial impacts of Lactobacillus plantarum and inulin on hypothalamic levels of insulin, leptin, and oxidative markers in diabetic rats. *J. Funct. Foods* **2018**, *46*, 529–537. [CrossRef]
81. Jiang, S.; Mohammed, A.A.; Jacobs, J.A.; Cramer, T.A.; Cheng, H.W. Effect of synbiotics on thyroid hormones, intestinal histomorphology, and heat shock protein 70 expression in broiler chickens reared under cyclic heat stress. *Poult. Sci.* **2020**, *99*, 142–150. [CrossRef]
82. Mofidi, F.; Poustchi, H.; Yari, Z.; Nourinayyer, B.; Merat, S.; Sharafkhah, M.; Malekzadeh, R.; Hekmatdoost, A. Synbiotic supplementation in lean patients with non-alcoholic fatty liver disease: A pilot, randomised, double-blind, placebo-controlled, clinical trial. *Br. J. Nutr.* **2017**, *117*, 662–668. [CrossRef]
83. Eslamparast, T.; Poustchi, H.; Zamani, F.; Sharafkhah, M.; Malekzadeh, R.; Hekmatdoost, A. Synbiotic supplementation in nonalcoholic fatty liver disease: A randomized, double-blind, placebo-controlled pilot study. *Am. J. Clin. Nutr.* **2014**, *99*, 535–542. [CrossRef]
84. Steed, H.; MacFarlane, G.T.; Blackett, K.L.; Bahrami, B.; Reynolds, N.; Walsh, S.V.; Cummings, J.H.; MacFarlane, S. Clinical trial: The microbiological and immunological effects of synbiotic consumption—A randomized double-blind placebo-controlled study in active Crohn's disease. *Aliment. Pharmacol. Ther.* **2010**, *32*, 872–883. [CrossRef] [PubMed]
85. Sivamaruthi, B.S. A comprehensive review on clinical outcome of probiotic and synbiotic therapy for inflammatory bowel diseases. *Asian Pac. J. Trop. Biomed.* **2018**, *8*, 179–186. [CrossRef]
86. Westfall, S.; Lomis, N.; Prakash, S. A novel synbiotic delays Alzheimer's disease onset via combinatorial gut-brain-axis signaling in Drosophila melanogaster. *PLoS ONE* **2019**, *14*, e0214985. [CrossRef] [PubMed]
87. Ishikawa, H.; Matsumoto, S.; Ohashi, Y.; Imaoka, A.; Setoyama, H.; Umesaki, Y.; Tanaka, R.; Otani, T. Beneficial effects of probiotic Bifidobacterium and galacto-oligosaccharide in patients with ulcerative colitis: A randomized controlled study. *Digestion* **2011**, *84*, 128–133. [CrossRef] [PubMed]
88. Hatch, R.; Rosenfield, R.L.; Kim, M.H.; Tredway, D. Hirsutism: Implications, etiology, and management. *Am. J. Obstet. Gynecol.* **1981**, *140*, 815–830. [CrossRef]
89. Carmina, E. Prevalence of idiopathic hirsutism. *Eur. J. Endocrinol.* **1998**, *139*, 421–423. [CrossRef]

90. Azziz, R. The evaluation and management of hirsutism. *Obstet. Gynecol.* **2003**, *101*, 995–1007.
91. Archer, J.S.; Chang, R.J. Hirsutism and acne in polycystic ovary syndrome. *Best Pract. Res. Clin. Obstet. Gynaecol.* **2004**, *18*, 737–754. [CrossRef]
92. Franks, S. Diagnosis of polycystic ovarian syndrome: In defense of the Rotterdam criteria. *J. Clin. Endocrinol. Metab.* **2006**, *91*, 786–789. [CrossRef]
93. Calvo, R.M.; Asunción, M.; Sancho, J.; San Millán, J.L.; Escobar-Morreale, H.F. The role of the CAG repeat polymorphism in the androgen receptor gene and of skewed X-chromosome inactivation, in the pathogenesis of hirsutism. *J. Clin. Endocrinol. Metab.* **2000**, *85*, 1735–1740. [CrossRef]
94. Vottero, A.; Stratakis, C.A.; Ghizzoni, L.; Longui, C.A.; Karl, M.; Chrousos, G.P. Androgen receptor-mediated hypersensitivity to androgens in women with nonhyperandrogenic hirsutism: Skewing of X-chromosome inactivation. *J. Clin. Endocrinol. Metab.* **1999**, *84*, 1091–1095. [CrossRef]
95. Ünlühizarci, K.; Karababa, Y.; Bayram, F.; Kelestimur, F. The investigation of insulin resistance in patients with idiopathic hirsutism. *J. Clin. Endocrinol. Metab.* **2004**, *89*, 2741–2744. [CrossRef] [PubMed]
96. Talaei, A.; Adgi, Z.; Mohamadi Kelishadi, M. Idiopathic hirsutism and insulin resistance. *Int. J. Endocrinol.* **2013**, *2013*. [CrossRef]
97. Barbieri, R.L.; Smith, S.; Ryan, K.J. The role of hyperinsulinemia in the pathogenesis of ovarian hyperandrogenism. *Fertil. Steril.* **1988**, *50*, 197–212. [CrossRef] [PubMed]
98. Barrionuevo, P.; Nabhan, M.; Altayar, O.; Wang, Z.; Erwin, P.J.; Asi, N.; Martin, K.A.; Hassan Murad, M. Treatment Options for Hirsutism: A Systematic Review and Network Meta-Analysis. *J. Clin. Endocrinol. Metab.* **2018**, *103*, 1258–1264. [CrossRef] [PubMed]
99. Moghetti, P.; Tosi, F.; Castello, R.; Magnani, C.M.; Negri, C.; Brun, E.; Furlani, L.; Caputo, M.; Muggeo, M. The insulin resistance in women with hyperandrogenism is partially reversed by antiandrogen treatment: Evidence that androgens impair insulin action in women. *J. Clin. Endocrinol. Metab.* **1996**, *81*, 952–960. [CrossRef]
100. Corbould, A. Chronic testosterone treatment induces selective insulin resistance in subcutaneous adipocytes of women. *J. Endocrinol.* **2007**, *192*, 585–594. [CrossRef]
101. Smith, S.; Ravnikar, V.A.; Barbieri, R.L. Androgen and insulin response to an oral glucose challenge in hyperandrogenic women. *Fertil. Steril.* **1987**, *48*, 72–77. [CrossRef]
102. Mathur, R.S.; Moody, L.O.; Landgrebe, S.; Williamson, H.O. Plasma androgens and sex hormone-binding globulin in the evaluation of hirsute females. *Fertil. Steril.* **1981**, *35*, 29–35. [CrossRef]
103. Dunaif, A.; Green, G.; Phelps, R.G.; Lebwohl, M.; Futterweit, W.; Lewy, L. Acanthosis Nigricans, Insulin Action, and Hyperandrogenism: Clinical, Histological, and Biochemical Findings*. *J. Clin. Endocrinol. Metab.* **1991**, *73*, 590–595. [CrossRef]
104. Dunaif, A.; Graf, M.; Mandeli, J.; Laumas, V.; Dobrjansky, A. Characterization of groups of hyperaiidrogenic women with acanthosis nigricans, impaired glucose tolerance, and/or hyperinsulinemia. *J. Clin. Endocrinol. Metab.* **1987**, *65*, 499–507. [CrossRef]
105. Azziz, R.; Carmina, E.; Sawaya, M.E. Idiopathic Hirsutism*. *Endocr. Rev.* **2000**, *21*, 347–362. [CrossRef] [PubMed]
106. Escobar-Morreale, H.F.; Carmina, E.; Dewailly, D.; Gambineri, A.; Kelestimur, F.; Moghetti, P.; Pugeat, M.; Qiao, J.; Wijeyaratne, C.N.; Witchel, S.F.; et al. Epidemiology, diagnosis and management of hirsutism: A consensus statement by the androgen excess and polycystic ovary syndrome society. *Hum. Reprod. Update* **2012**, *18*, 146–170. [CrossRef] [PubMed]
107. Irons, E.E. The clinical evaluation of drugs. *J. Am. Med. Assoc.* **1929**, *93*, 1523–1524. [CrossRef]
108. Ibáñez, L.; de Zegher, F. Low-dose flutamide-metformin therapy for hyperinsulinemic hyperandrogenism in non-obese adolescents and women. *Hum. Reprod. Update* **2006**, *12*, 243–252. [CrossRef] [PubMed]
109. Ibáñez, L.; De Zegher, F. Low-dose combination of flutamide, metformin and an oral contraceptive for non-obese, young women with polycystic ovary syndrome. *Hum. Reprod.* **2003**, *18*, 57–60. [CrossRef]
110. Rautio, K.; Tapanainen, J.S.; Ruokonen, A.; Morin-Papunen, L.C. Effects of metformin and ethinyl estradiol-cyproterone acetate on lipid levels in obese and non-obese women with polycystic ovary syndrome. *Eur. J. Endocrinol.* **2005**, *152*, 269–275. [CrossRef] [PubMed]
111. Ibáñez, L.; De Zegher, F. Ethinylestradiol-drospirenone, flutamide-metformin, or both for adolescents and women with hyperinsulinemic hyperandrogenism: Opposite effects on adipocytokines and body adiposity. *J. Clin. Endocrinol. Metab.* **2004**, *89*, 1592–1597. [CrossRef]
112. Harborne, L.; Fleming, R.; Lyall, H.; Sattar, N.; Norman, J. Metformin or antiandrogen in the treatment of hirsutism in polycystic ovary syndrome. *J. Clin. Endocrinol. Metab.* **2003**, *88*, 4116–4123. [CrossRef]
113. Diamanti-Kandarakis, E.; Kouli, C.; Tsianateli, T.; Bergiele, A. Therapeutic effects of metformin on insulin resistance and hyperandrogenism in polycystic ovary syndrome. *Eur. J. Endocrinol.* **1998**, *138*, 269–274. [CrossRef]
114. Haas, J.; Bentov, Y. Should metformin be included in fertility treatment of PCOS patients? *Med. Hypotheses* **2017**, *100*, 54–58. [CrossRef] [PubMed]
115. Dewis, P.; Petsos, P.; Newman, M.; Anderson, D.C. the Treatment of Hirsutism With a Combination of Desogestrel and Ethinyl Oestradiol. *Clin. Endocrinol.* **1985**, *22*, 29–36. [CrossRef] [PubMed]
116. Mastorakos, G.; Koliopoulos, C.; Creatsas, G. Androgen and lipid profiles in adolescents with polycystic ovary syndrome who were treated with two forms of combined oral contraceptives. *Fertil. Steril.* **2002**, *77*, 919–927. [CrossRef]
117. Zeng, X.; Xie, Y.J.; Liu, Y.T.; Long, S.L.; Mo, Z. cheng Polycystic ovarian syndrome: Correlation between hyperandrogenism, insulin resistance and obesity. *Clin. Chim. Acta* **2020**, *502*, 214–221. [CrossRef] [PubMed]

118. Acbay, O.; Gundogdu, S. Can metformin reduce insulin resistance in polycystic ovary syndrome? *Fertil. Steril.* **1996**, *65*, 946–949. [CrossRef]
119. Moghetti, P.; Castello, R.; Negri, C.; Tosi, F.; Perrone, F.; Caputo, M.; Zanolin, E.; Muggeo, M. Metformin effects on clinical features, endocrine and metabolic profiles, and insulin sensitivity in polycystic ovary syndrome: A randomized, double-blind, placebo-controlled 6-month trial, followed by open, long-term clinical evaluation. *J. Clin. Endocrinol. Metab.* **2000**, *85*, 139–146. [CrossRef]
120. Kolodziejczyk, B.; Duleba, A.J.; Spaczynski, R.Z.; Pawelczyk, L. Metformin therapy decreases hyperandrogenism and hyperinsulinemia in women with polycystic ovary syndrome. *Fertil. Steril.* **2000**, *73*, 1149–1154. [CrossRef]
121. Morin-Papunen, L.C.; Koivunen, R.M.; Ruokonen, A.; Martikainen, H.K. Metformin therapy improves the menstrual pattern with minimal endocrine and metabolic effects in women with polycystic ovary syndrome. *Fertil. Steril.* **1998**, *69*, 691–696. [CrossRef]
122. Ibáñez, L.; De Zegher, F. Flutamide-Metformin Therapy to Reduce Fat Mass in Hyperinsulinemic Ovarian Hyperandrogenism: Effects in Adolescents and in Women on Third-Generation Oral Contraception. *J. Clin. Endocrinol. Metab.* **2003**, *88*, 4720–4724. [CrossRef]
123. Carmina, E. Anti-androgens for the treatment of hirsutism. *Expert Opin. Investig. Drugs* **2002**, *11*, 357–363. [CrossRef]
124. Mowszowicz, I.; Wright, F.; Vincens, M.; Rigaud, C.; Nahoul, K.; Mavier, P.; Guillemant, S.; Kuttenn, F.; Mauvais-Jarvis, P. Androgen metabolism in hirsute patients treated with cyproterone acetate. *J. Steroid Biochem.* **1984**, *20*, 757–761. [CrossRef]
125. Corvol, P.; Michaud, A.; Menard, J.; Freifeld, M.; Mahoudeau, J. Antiandrogenic effect of spirolactones: Mechanism of action. *Endocrinology* **1975**, *97*, 52–58. [CrossRef] [PubMed]
126. Gobinet, J.; Poujol, N.; Sultan, C. Molecular action of androgens. *Mol. Cell. Endocrinol.* **2002**, *198*, 15–24. [CrossRef]
127. Dewhurst, C.J.; Underhill, R.; Goldman, S.; Mansfield, M. the Treatment of Hirsutism With Cyproterone Acetate (an Anti-Androgen). *BJOG An Int. J. Obstet. Gynaecol.* **1977**, *84*, 119–123. [CrossRef] [PubMed]
128. Kuttenn, F.; Rigaud, C.; Wright, F.; Mauvais-Jarvis, P. Treatment of hirsutism by oral cyproterone acetate and percutaneous estradiol. *J. Clin. Endocrinol. Metab.* **1980**, *51*, 1107–1111. [CrossRef]
129. Shapiro, G.; Evron, S. A novel use of spironolactone: Treatment of hirsutism. *J. Clin. Endocrinol. Metab.* **1980**, *51*, 429–432. [CrossRef]
130. Jones, D.B.; Ibraham, I.; Edwards, C.R.W. Hair growth and androgen responses in hirsute women treated with continuous cyproterone acetate and cyclical ethinyl oestradiol. *Acta Endocrinol.* **1987**, *116*, 497–501. [CrossRef]
131. Lobo, R.A.; Shoupe, D.; Serafini, P.; Brinton, D.; Horton, R. The effects of two doses of spironolactone on serum androgens and anagen hair in hirsute women. *Fertil. Steril.* **1985**, *43*, 200–205. [CrossRef]
132. Pittaway, D.E.; Maxson, W.S.; Wentz, A.C. Spironolactone in combination drug therapy for unresponsive hirsutism. *Fertil. Steril.* **1985**, *43*, 878–882. [CrossRef]
133. O'brien, R.C.; Cooper, M.E.; Murray, R.M.L.; Seeman, E.; Thomas, A.K.; Jerums, G. Comparison of sequential cyproterone acetate/estrogen versus spironolactone/oral contraceptive in the treatment of hirsutism. *J. Clin. Endocrinol. Metab.* **1991**, *72*, 1008–1013. [CrossRef]
134. Cusan, L.; Dupont, A.; Gomez, J.L.; Tremblay, R.R.; Labrie, F. Comparison of flutamide and spironolactone in the treatment of hirsutism: A randomized controlled trial. *Fertil. Steril.* **1994**, *61*, 281–287. [CrossRef]
135. Müderris, I.I.; Bayram, F.; Özçelik, B.; Güven, M. New alternative treatment in hirsutism: Bicalutamide 25 mg/day. *Gynecol. Endocrinol.* **2002**, *16*, 63–66. [CrossRef] [PubMed]
136. Moretti, C.; Guccione, L.; Giacinto, P. Di; Simonelli, I.; Exacoustos, C.; Toscano, V.; Motta, C.; De Leo, V.; Petraglia, F.; Lenzi, A. Combined oral contraception and bicalutamide in polycystic ovary syndrome and severe hirsutism: A double-blind randomized controlled trial. *J. Clin. Endocrinol. Metab.* **2018**, *103*, 824–838. [CrossRef] [PubMed]
137. Bahceci, M.; Tuzcu, A.; Canoruc, N.; Tuzun, Y.; Kidir, V.; Aslan, C. Serum C-reactive protein (CRP) levels and insulin resistance in non-obese women with polycystic ovarian syndrome, and effect of bicalutamide on hirsutism, CRP levels and insulin resistance. *Horm. Res.* **2004**, *62*, 283–287. [CrossRef] [PubMed]
138. 138. Goswami, P.K.; Khale, A.; Ogale, S. O. Natural remedies for polycystic ovarian syndrome (PCOS): A review. *Int. J. Pharm. Phytopharm. Res.* **2012**, *1*, 396–402.
139. Kolivand, M.; Keramat, A.; Khosravi, A. The Effect of Herbal Teas on Management of Polycystic Ovary Syndrome: A Systematic Review. *J. Midwifery Reprod. Heal.* **2017**, *5*, 1098–1106. [CrossRef]
140. Akdoğan, M.; Tamer, M.N.; Cüre, E.; Cüre, M.C.; Köroğlu, B.K.; Delibaş, N. Effect of spearmint (Mentha spicata Labiatae) teas on androgen levels in women with hirsutism. *Phyther. Res.* **2007**, *21*, 444–447. [CrossRef]
141. Grant, P.; Ramasamy, S. An update on plant derived anti-androgens. *Int. J. Endocrinol. Metab.* **2012**, *10*, 497–502. [CrossRef]
142. Grant, P. Spearmint Herbal Tea has Signifi cant Anti-androgen Effects in Polycystic Ovarian Syndrome. A Randomized Controlled Trial. *Phyther. Res.* **2010**, *24*, 186–188. [CrossRef]
143. Haj-Husein, I.; Tukan, S.; Alkazaleh, F. The effect of marjoram (Origanum majorana) tea on the hormonal profile of women with polycystic ovary syndrome: A randomised controlled pilot study. *J. Hum. Nutr. Diet.* **2016**, *29*, 105–111. [CrossRef]
144. Phy, J.L.; Pohlmeier, A.M.; Cooper, J.A.; Watkins, P.; Spallholz, J.; Harris, K.S.; Berenson, A.B.; Boylan, M. Low Starch/Low Dairy Diet Results in Successful Treatment of Obesity and Co- Morbidities Linked to Polycystic Ovary Syndrome (PCOS). *J. Obes. Weight Loss Ther.* **2015**, *5*, 1–12. [CrossRef]
145. Pohlmeier, A.M.; Phy, J.L.; Watkins, P.; Boylan, M.; Spallholz, J.; Harris, K.S.; Cooper, J.A. Effect of a low-starch/low-dairy diet on fat oxidation in overweight and obese women with polycystic ovary syndrome. *Appl. Physiol. Nutr. Metab.* **2014**, *39*, 1237–1244. [CrossRef] [PubMed]

146. Janiszewska, J.; Ostrowska, J.; Szostak-Węgierek, D. Milk and Dairy Products and Their Impact on Carbohydrate Metabolism and Fertility-A Potential Role in the Diet of Women with Polycystic Ovary Syndrome. *Nutrients* **2020**, *12*. [CrossRef] [PubMed]
147. Shamasbi, S.G.; Ghanbari-Homayi, S.; Mirghafourvand, M. The effect of probiotics, prebiotics, and synbiotics on hormonal and inflammatory indices in women with polycystic ovary syndrome: A systematic review and meta-analysis. *Eur. J. Nutr.* **2020**, *59*, 433–450. [CrossRef] [PubMed]
148. Tabrizi, R.; Ostadmohammadi, V.; Akbari, M.; Lankarani, K.B.; Vakili, S.; Peymani, P.; Karamali, M.; Kolahdooz, F.; Asemi, Z. The Effects of Probiotic Supplementation on Clinical Symptom, Weight Loss, Glycemic Control, Lipid and Hormonal Profiles, Biomarkers of Inflammation, and Oxidative Stress in Women with Polycystic Ovary Syndrome: A Systematic Review and Meta-analysis of Ra. *Probiotics Antimicrob. Proteins* **2019**. [CrossRef] [PubMed]
149. Ostadmohammadi, V.; Jamilian, M.; Bahmani, F.; Asemi, Z. Vitamin D and probiotic co-supplementation affects mental health, hormonal, inflammatory and oxidative stress parameters in women with polycystic ovary syndrome. *J. Ovarian Res.* **2019**, *12*, 1–8. [CrossRef]
150. Jamilian, M.; Mansury, S.; Bahmani, F.; Heidar, Z.; Amirani, E.; Asemi, Z. The effects of probiotic and selenium co-supplementation on parameters of mental health, hormonal profiles, and biomarkers of inflammation and oxidative stress in women with polycystic ovary syndrome 11 Medical and Health Sciences 1103 Clinical Sciences. *J. Ovarian Res.* **2018**, *11*, 1–7. [CrossRef]
151. He, F.F.; Li, Y.M. Role of gut microbiota in the development of insulin resistance and the mechanism underlying polycystic ovary syndrome: A review. *J. Ovarian Res.* **2020**, *13*, 1–13. [CrossRef]
152. Guevara, G.; Flores, Y.O.; De Las Heras, L.F.; Perera, J.; Navarro Llorens, J.M. Metabolic engineering of Rhodococcus ruber Chol-4: A cell factory for testosterone production. *PLoS ONE* **2019**, *14*. [CrossRef]
153. Colldén, H.; Landin, A.; Wallenius, V.; Elebring, E.; Fändriks, L.; Nilsson, M.E.; Ryberg, H.; Poutanen, M.; Sjögren, K.; Vandenput, L.; et al. The gut microbiota is a major regulator of androgen metabolism in intestinal contents. *Am. J. Physiol. Endocrinol. Metab.* **2019**, *317*, E1182–E1192. [CrossRef]
154. Tremellen, K.; Pearce, K. Dysbiosis of Gut Microbiota (DOGMA)—A novel theory for the development of Polycystic Ovarian Syndrome. *Med. Hypotheses* **2012**, *79*, 104–112. [CrossRef] [PubMed]
155. Lindheim, L.; Bashir, M.; Münzker, J.; Trummer, C.; Zachhuber, V.; Leber, B.; Horvath, A.; Pieber, T.R.; Gorkiewicz, G.; Stadlbauer, V.; et al. Alterations in gut microbiome composition and barrier function are associated with reproductive and metabolic defects in women with polycystic ovary syndrome (PCOS): A pilot study. *PLoS ONE* **2017**, *12*, 1–20. [CrossRef] [PubMed]
156. Guo, Y.; Qi, Y.; Yang, X.; Zhao, L.; Wen, S.; Liu, Y.; Tang, L. Association between polycystic ovary syndrome and gut microbiota. *PLoS ONE* **2016**, *11*, e0153196. [CrossRef] [PubMed]
157. Karamali, M.; Eghbalpour, S.; Rajabi, S.; Jamilian, M.; Bahmani, F.; Tajabadi-Ebrahimi, M.; Keneshlou, F.; Mirhashemi, S.M.; Chamani, M.; Hashem Gelougerdi, S.; et al. Effects of Probiotic Supplementation on Hormonal Profiles, Biomarkers of Inflammation and Oxidative Stress in Women With Polycystic Ovary Syndrome: A Randomized, Double-Blind, Placebo-Controlled Trial. *Arch. Iran. Med.* **2018**, *21*, 1–7. [PubMed]
158. Maktabi, M.; Chamani, M.; Asemi, Z. The Effects of Vitamin D Supplementation on Metabolic Status of Patients with Polycystic Ovary Syndrome: A Randomized, Double-Blind, Placebo-Controlled Trial. *Horm. Metab. Res.* **2017**, *49*, 493–498. [CrossRef]
159. Raja-Khan, N.; Shah, J.; Stetter, C.M.; Lott, M.E.J.; Kunselman, A.R.; Dodson, W.C.; Legro, R.S. High-dose vitamin D supplementation and measures of insulin sensitivity in polycystic ovary syndrome: A randomized, controlled pilot trial. *Fertil. Steril.* **2014**, *101*, 1740–1746. [CrossRef] [PubMed]
160. Xue, Y.; Xu, P.; Xue, K.; Duan, X.; Cao, J.; Luan, T.; Li, Q.; Gu, L. Effect of vitamin D on biochemical parameters in polycystic ovary syndrome women: A meta-analysis. *Arch. Gynecol. Obstet.* **2017**, *295*, 487–496. [CrossRef]
161. Liao, D.; Zhong, C.; Li, C.; Mo, L.; Liu, Y. Meta-analysis of the effects of probiotic supplementation on glycemia, lipidic profiles, weight loss and C-reactive protein in women with polycystic ovarian syndrome. *Minerva Med.* **2018**, *109*, 479–487. [CrossRef]
162. Laganà, A.S.; Vitale, S.G.; Ban Frangež, H.; Vrtačnik-Bokal, E.; D'Anna, R. Vitamin D in human reproduction: The more, the better? An evidence-based critical appraisal. *Eur. Rev. Med. Pharmacol. Sci.* **2017**, *21*, 4243–4251.
163. Monastra, G.; De Grazia, S.; De Luca, L.; Vittorio, S.; Unfer, V. Vitamin D: A steroid hormone with progesterone-like activity. *Eur. Rev. Med. Pharmacol. Sci.* **2018**, *22*, 2502–2512. [CrossRef]
164. Akbari, M.; Ostadmohammadi, V.; Lankarani, K.B.; Tabrizi, R.; Kolahdooz, F.; Heydari, S.T.; Kavari, S.H.; Mirhosseini, N.; Mafi, A.; Dastorani, M.; et al. The Effects of Vitamin D Supplementation on Biomarkers of Inflammation and Oxidative Stress among Women with Polycystic Ovary Syndrome: A Systematic Review and Meta-Analysis of Randomized Controlled Trials. *Horm. Metab. Res.* **2018**, *50*, 271–279. [CrossRef] [PubMed]
165. Zhang, F.; Ma, T.; Cui, P.; Tamadon, A.; He, S.; Huo, C.; Yierfulati, G.; Xu, X.; Hu, W.; Li, X.; et al. Diversity of the gut microbiota in dihydrotestosterone-induced PCOS rats and the pharmacologic effects of diane-35, probiotics, and berberine. *Front. Microbiol.* **2019**, *10*, 175. [CrossRef] [PubMed]
166. Kassaian, N.; Feizi, A.; Aminorroaya, A.; Jafari, P.; Ebrahimi, M.T.; Amini, M. The effects of probiotics and synbiotic supplementation on glucose and insulin metabolism in adults with prediabetes: A double-blind randomized clinical trial. *Acta Diabetol.* **2018**, *55*, 1019–1028. [CrossRef] [PubMed]
167. Ahmadi, S.; Jamilian, M.; Karamali, M.; Tajabadi-Ebrahimi, M.; Jafari, P.; Taghizadeh, M.; Memarzadeh, M.R.; Asemi, Z. Probiotic supplementation and the effects on weight loss, glycaemia and lipid profiles in women with polycystic ovary syndrome: A randomized, double-blind, placebo-controlled trial. *Hum. Fertil.* **2017**, *20*, 254–261. [CrossRef] [PubMed]

168. Laitinen, K.; Poussa, T.; Isolauri, E. Probiotics and dietary counselling contribute to glucose regulation during and after pregnancy: A randomised controlled trial. *Br. J. Nutr.* **2009**, *101*, 1679–1687. [CrossRef] [PubMed]
169. Andreasen, A.S.; Larsen, N.; Pedersen-Skovsgaard, T.; Berg, R.M.G.; Mller, K.; Svendsen, K.D.; Jakobsen, M.; Pedersen, B.K. Effects of Lactobacillus acidophilus NCFM on insulin sensitivity and the systemic inflammatory response in human subjects. *Br. J. Nutr.* **2010**, *104*, 1831–1838. [CrossRef] [PubMed]
170. Barreto, F.M.; Colado Simão, A.N.; Morimoto, H.K.; Batisti Lozovoy, M.A.; Dichi, I.; Helena da Silva Miglioranza, L. Beneficial effects of Lactobacillus plantarum on glycemia and homocysteine levels in postmenopausal women with metabolic syndrome. *Nutrition* **2014**, *30*, 939–942. [CrossRef]
171. Kocsis, T.; Molnár, B.; Németh, D.; Hegyi, P.; Szakács, Z.; Bálint, A.; Garami, A.; Soós, A.; Márta, K.; Solymár, M. Probiotics have beneficial metabolic effects in patients with type 2 diabetes mellitus: A meta-analysis of randomized clinical trials. *Sci. Rep.* **2020**, *10*, 1–14. [CrossRef]
172. Ardeshirlarijani, E.; Tabatabaei-Malazy, O.; Mohseni, S.; Qorbani, M.; Larijani, B.; Baradar Jalili, R. Effect of probiotics supplementation on glucose and oxidative stress in type 2 diabetes mellitus: A meta-analysis of randomized trials. *DARU J. Pharm. Sci.* **2019**, *27*, 827–837. [CrossRef]
173. Shoaei, T.; Heidari-Beni, M.; Tehrani, H.G.; Feizi, A.; Esmaillzadeh, A.; Askari, G. Effects of probiotic supplementation on pancreatic β-cell function and c-reactive protein in women with polycystic ovary syndrome: A randomized double-blind placebo-controlled clinical trial. *Int. J. Prev. Med.* **2015**, *6*, 25–30. [CrossRef]
174. Azziz, R. The hyperandrogenic-insulin-resistant acanthosis nigricans syndrome: Therapeutic response. *Fertil. Steril.* **1994**, *61*, 570–572. [CrossRef]
175. Gulan, T.; Yeernuer, T.; Sui, S.; Mayinuer, N. A rat model of maternal polycystic ovary syndrome shows that exposure to androgens in utero results in dysbiosis of the intestinal microbiota and metabolic disorders of the newborn rat. *Med. Sci. Monit.* **2019**, *25*, 9377–9391. [CrossRef]
176. Falcone, T.; Finegood, D.T.; Fantus, I.G.; Morris, D. Androgen response to endogenous insulin secretion during the frequently sampled intravenous glucose tolerance test in normal and hyperandrogenic women. *J. Clin. Endocrinol. Metab.* **1990**, *71*, 1653–1657. [CrossRef] [PubMed]
177. Tosi, F.; Negri, C.; Brun, E.; Castello, R.; Faccini, G.; Bonora, E.; Muggeo, M.; Toscano, V.; Moghetti, P. Insulin enhances ACTH-stimulated androgen and glucocorticoid metabolism in hyperandrogenic women. *Eur. J. Endocrinol.* **2011**, *164*, 197–203. [CrossRef]
178. Esmaeilinezhad, Z.; Babajafari, S.; Sohrabi, Z.; Eskandari, M.H.; Amooee, S.; Barati-Boldaji, R. Effect of synbiotic pomegranate juice on glycemic, sex hormone profile and anthropometric indices in PCOS: A randomized, triple blind, controlled trial. *Nutr. Metab. Cardiovasc. Dis.* **2019**, *29*, 201–208. [CrossRef] [PubMed]
179. Karimi, E.; Moini, A.; Yaseri, M.; Shirzad, N.; Sepidarkish, M.; Hossein-Boroujerdi, M.; Hosseinzadeh-Attar, M.J. Effects of synbiotic supplementation on metabolic parameters and apelin in women with polycystic ovary syndrome: A randomised double-blind placebo-controlled trial. *Br. J. Nutr.* **2018**, *119*, 398–406. [CrossRef] [PubMed]
180. Samimi, M.; Dadkhah, A.; Haddad Kashani, H.; Tajabadi-Ebrahimi, M.; Hosseini, E.S.; Asemi, Z. The effects of Synbiotic Supplementation on Metabolic Status in Women With Polycystic Ovary Syndrome: A Randomized Double-Blind Clinical Trial. *Probiotics Antimicro. Prot.* **2019**, 1355–1361. [CrossRef]
181. Yang, J.; Huang, K.; Qin, S.; Wu, X.; Zhao, Z.; Chen, F. Antibacterial action of selenium-enriched probiotics against pathogenic escherichia coli. *Dig. Dis. Sci.* **2009**, *54*, 246–254. [CrossRef]
182. Nido, S.A.; Shituleni, S.A.; Mengistu, B.M.; Liu, Y.; Khan, A.Z.; Gan, F.; Kumbhar, S.; Huang, K. Effects of Selenium-Enriched Probiotics on Lipid Metabolism, Antioxidative Status, Histopathological Lesions, and Related Gene Expression in Mice Fed a High-Fat Diet. *Biol. Trace Elem. Res.* **2016**, *171*, 399–409. [CrossRef]
183. Molan, A.L. Antioxidant and prebiotic activities of selenium-containing green tea. *Nutrition* **2013**, *29*, 476–477. [CrossRef]
184. Razavi, M.; Jamilian, M.; Kashan, Z.F.; Heidar, Z.; Mohseni, M.; Ghandi, Y.; Bagherian, T.; Asemi, Z. Selenium Supplementation and the Effects on Reproductive Outcomes, Biomarkers of Inflammation, and Oxidative Stress in Women with Polycystic Ovary Syndrome. *Horm. Metab. Res.* **2016**, *48*, 185–190. [CrossRef] [PubMed]
185. Mohammad Hosseinzadeh, F.; Hosseinzadeh-Attar, M.J.; Yekaninejad, M.S.; Rashidi, B. Effects of selenium supplementation on glucose homeostasis and free androgen index in women with polycystic ovary syndrome: A randomized, double blinded, placebo controlled clinical trial. *J. Trace Elem. Med. Biol.* **2016**, *34*, 56–61. [CrossRef] [PubMed]
186. Shabani, A.; Noshadian, M.; Jamilian, M.; Chamani, M.; Mohammadi, S.; Asemi, Z. The effects of a novel combination of selenium and probiotic on weight loss, glycemic control and markers of cardio-metabolic risk in women with polycystic ovary syndrome. *J. Funct. Foods* **2018**, *46*, 329–334. [CrossRef]
187. Faghihi, T.; Radfar, M.; Barmal, M.; Amini, P.; Qorbani, M.; Abdollahi, M.; Larijani, B. A randomized, placebo-controlled trial of selenium supplementation in patients with type 2 diabetes: Effects on glucose homeostasis, oxidative stress, and lipid profile. *Am. J. Ther.* **2014**, *21*, 491–495. [CrossRef]
188. Asemi, Z.; Zare, Z.; Shakeri, H.; Sabihi, S.S.; Esmaillzadeh, A. Effect of multispecies probiotic supplements on metabolic profiles, hs-CRP, and oxidative stress in patients with type 2 diabetes. *Ann. Nutr. Metab.* **2013**, *63*, 1–9. [CrossRef]
189. Ahmadi, S.; Jamilian, M.; Tajabadi-Ebrahimi, M.; Jafari, P.; Asemi, Z. The effects of synbiotic supplementation on markers of insulin metabolism and lipid profiles in gestational diabetes: A randomised, double-blind, placebo-controlled trial. *Br. J. Nutr.* **2016**, *116*, 1394–1401. [CrossRef]

190. Fitzpatrick, L.R.; Small, J.; Hoerr, R.A.; Bostwick, E.F.; Maines, L.; Koltun, W.A. In vitro and in vivo effects of the probiotic Escherichia coli strain M-17: Immunomodulation and attenuation of murine colitis. *Br. J. Nutr.* **2008**, *100*, 530–541. [CrossRef]
191. Farnworth, E.R. Kefir a complex probiotic. *Food Sci. Technol. Bull. Funct. Foods* **2006**, *2*, 1–17. [CrossRef]
192. Serafini, F.; Turroni, F.; Ruas-Madiedo, P.; Lugli, G.A.; Milani, C.; Duranti, S.; Zamboni, N.; Bottacini, F.; van Sinderen, D.; Margolles, A.; et al. Kefir fermented milk and kefiran promote growth of Bifidobacterium bifidum PRL2010 and modulate its gene expression. *Int. J. Food Microbiol.* **2014**, *178*, 50–59. [CrossRef]
193. Sonino, N.; Faval, G.A.; Mani, E.; Belluardo, P.; Boscaro, M. Quality of life of hirsute women. *Postgrad. Med. J.* **1993**, *69*, 186–189. [CrossRef]
194. Wild, R.A. Lipid metabolism and hyperandrogenism. *Clin. Obstet. Gynecol.* **1991**, *34*, 864–871. [CrossRef] [PubMed]
195. Wild, R.A.; Alaupovic, P.; Parker, I.J. Lipid and apolipoprotein abnormalities in hirsute women: I. The association with insulin resistance. *Am. J. Obstet. Gynecol.* **1992**, *166*, 1191–1197. [CrossRef]
196. Laughlin, G.A.; Barrett-Connor, E.; Bergstrom, J. Low serum testosterone and mortality in older men. *J. Clin. Endocrinol. Metab.* **2008**, *93*, 68–75. [CrossRef] [PubMed]
197. Araujo, A.B.; Dixon, J.M.; Suarez, E.A.; Murad, M.H.; Guey, L.T.; Wittert, G.A. Endogenous testosterone and mortality in men: A systematic review and meta-analysis. *J. Clin. Endocrinol. Metab.* **2011**, *96*, 3007–3019. [CrossRef] [PubMed]
198. Corona, G.; Rastrelli, G.; Monami, M.; Guay, A.; Buvat, J.; Sforza, A.; Forti, G.; Mannucci, E.; Maggi, M. Hypogonadism as a risk factor for cardiovascular mortality in men: A meta-analytic study. *Eur. J. Endocrinol.* **2011**, *165*, 687–701. [CrossRef] [PubMed]
199. Ruige, J.B.; Mahmoud, A.M.; De Bacquer, D.; Kaufman, J.M. Endogenous testosterone and cardiovascular disease in healthy men: A meta-analysis. *Heart* **2011**, *97*, 870–875. [CrossRef] [PubMed]
200. Penn, C.A.; Chan, J.; Mesaros, C.; Snyder, N.W.; Rader, D.J.; Sammel, M.D.; Dokras, A. Association of serum androgens and coronary artery calcium scores in women. *Fertil. Steril.* **2019**, *112*, 586–593. [CrossRef]
201. Diamondf, M.P.; Grainger, D.A.; Laudano, A.J.; Starick-Zych, K.; Defronzo, R.A. Effect of acute physiological elevations of insulin on circulating androgen levels in nonobese women. *J. Clin. Endocrinol. Metab.* **1991**, *72*, 883–887. [CrossRef]
202. Kim, Y.A.; Keogh, J.B.; Clifton, P.M. Probiotics, prebiotics, synbiotics and insulin sensitivity. *Nutr. Res. Rev.* **2017**, *31*, 35–51. [CrossRef]
203. Raygan, F.; Ostadmohammadi, V.; Asemi, Z. The effects of probiotic and selenium co-supplementation on mental health parameters and metabolic profiles in type 2 diabetic patients with coronary heart disease: A randomized, double-blind, placebo-controlled trial. *Clin. Nutr.* **2019**, *38*, 1594–1598. [CrossRef]

fermentation

Article

Effects of Fermented Kefir as a Functional Feed Additive in *Litopenaeus vannamei* Farming

Woosik Choi [1,†], Chi-Won Choi [1,†], Dang-Bao Son [1], Byeong-Chul Jeong [2], Hee-Chang Kim [3,*], Hanki Lee [1,*] and Joo-Won Suh [1,*]

1 Center for Nutraceutical and Pharmaceutical Materials, Myongji University, Yongin 17058, Gyeonggi-do, Korea; woos5051@gmail.com (W.C.); xlzhvxv1004@gmail.com (C.-W.C.); baoson.dang@gmail.com (D.-B.S.)

2 Division of Biosciences and Bioinformatics, College of Natural Science, Myongji University, Yongin 17058, Gyeonggi-do, Korea; bcjeong@mju.ac.kr

3 TMC Co., Ltd., Anyang 14057, Gyeonggi-do, Korea

* Correspondence: jeromekim@tmc.co.kr (H.-C.K.); hklee95@mju.ac.kr (H.L.); jwsuh@mju.ac.kr (J.-W.S.); Tel.: +82-2-2663-2773 (H.-C.K.); +82-31-330-6799 (H.L.); +82-31-330-6881 (J.-W.S.)

† These authors contributed equally to this study.

Received: 18 November 2020; Accepted: 26 November 2020; Published: 27 November 2020

Abstract: *Litopenaeus vannamei*, known as whiteleg shrimp, is susceptible to infection by pathogenic microorganisms such as viruses and bacteria. Therefore, the prevention of infections in this shrimp is important to regulate the outbreaks of pathogenic microorganisms. In this study, we investigated the effects of kefir as a functional feed additive on innate immunity, survival against WSSV (White Spot Syndrome Virus) and productivity of *L. vannamei*. As a result, the treatment of kefir could upregulate six of seven genes crucial for innate immunity of *L. vannamei*. Also, the treatment of kefir directly improved the survival rate of *L. vannamei* against WSSV infection. Finally, in order to determine whether kefir can improve the productivity of shrimp, we carried out field tests in three aquaculture farms in South Korea. The weight of shrimp fed kefir was increased by 120% as well as the length, compared with that of the control group. These results demonstrate that kefir can be utilized as a functional feed additive to improve both innate immunity and productivity of *L. vannamei* in shrimp farming with no use of antibiotics.

Keywords: fermented kefir; *Litopenaeus vannamei*; innate immunity; survival rate against WSSV; productivity; functional feed additive

1. Introduction

Litopenaeus vannamei, known as whiteleg shrimp, is an important aquaculture species in the Asia–Pacific region because this region is the largest producer contributing to nearly 80% of the value and volume of the global shrimp market [1]. This shrimp can resist poor farming conditions such as enrichment of nitrogen sources, but it is susceptible to infection by viruses and bacteria, including *Vibrio* sp. [2,3]. These infections cause the sudden death of shrimp; the prevention of these infections is important to maintain the productivity of shrimp farming [2]. Antibiotics have been effective in preventing these infections, but there has been growing concern of serious problems such as the emergence of antibiotic-resistant microorganisms [4,5]. Therefore, the use of antibiotics in aquaculture is strictly regulated to minimize their negative effects on the environment and human health [4,5]. This has encouraged researchers to develop substances such as probiotics and identify medicinal herbs, as alternatives to antibiotics, to control infections in shrimp [1,6–11].

Kefir is milk fermented by kefir grains and contains beneficial microorganisms such as lactic acid bacteria, *Bacillus* spp., and yeast. These microorganisms have health-promoting and antimicrobial activities [12–14]. Furthermore, kefir as a nutraceutical has several beneficial properties such as immunological, antimicrobial, antitumor, and hypo-cholesterolemic effects in animals and humans [15]. Also, Uluköy et al. found out that juvenile rainbow trout fed with kefir increased nonspecific immune response and improved disease resistance against lactococcosis and yersiniosis [16]. However, there are few studies on the effects of fermented kefir on *L. vannamei*. In this study, we investigated the effects of kefir as a functional feed additive on innate immunity and productivity of *L. vannamei* without the use of antibiotics.

2. Materials and Methods

2.1. Preparation of Lyophilized Kefir Cell Pellet

Kefir grain used in this study was collected in a private house, South Korea. Kefir grains were inoculated in 4% (*w*/*v*) whole fat milk medium and cultivated at 30 °C for 2 days without agitation. For main fermentation, we developed the medium based on de Man, Rogosa and Sharpe (MRS); yeast extract-peptone-dextrose (YPD); and nutrient broth (NB) media. To determine the optimum conditions for kefir fermentation, we modified the composition of glucose, whey protein, and dipotassium phosphate in the medium, and inoculation size of seed culture. The main fermentation of kefir was carried out in 300 L working volume of a 500 L fermentor at 30 °C for 1 d. For fermentation, sterilized air was supplied at 2 vvm in the fermentor, and the mixing rate was maintained at 200 rpm using an impeller. After fermentation, the total cells were harvested by continuous centrifugation with 8000 rpm and the cell pellet was mixed with 20% (*w*/*v*) sterilized skim milk solution. Thereafter, this mixture was lyophilized for 3 days. To determine the number of lactic acid bacteria, *Bacillus* spp., and yeast cells in the lyophilized sample, viable colonies were counted. Briefly, the sample was serially diluted to 0.85% with sterilized saline solution, and 100 μL of the diluted sample was spread on MRS agar for lactic acid bacteria, NB agar for *Bacillus* spp., and YPD agar for yeast culture. After incubation at 30 °C for 1 day, the colonies were counted.

2.2. Monitoring the Expression Level of Genes Related to Innate Immunity in L. vannamei after Treatment with Kefir

For this experiment, we prepared eight cages controlled by recirculating aquaculture system and transferred 30 *L. vannamei* into each cage. Thereafter, feed was supplied at 3% (*w*/*w*) of the total weight of shrimps in each cage three times per day. Furthermore, lyophilized kefir was supplied at 0.2% (*w*/*w*), 0.4% (*w*/*w*), and 0.8% (*w*/*w*) of the weight of feed supplied on 1 day. Lyophilized kefir was fed one time per day. The treatment was carried out for 30 days. The number of lactic acid bacteria, *Bacillus* spp., and yeast in lyophilized kefir was 5.6×10^9, 6.9×10^9, and 4.8×10^9 CFU/mL, respectively.

After 30 days, hemolymph and hepatopancreas were isolated from all *L. vannamei*. To determine the total hemocyte count, the hemolymph sample was fixed in 4% formalin in fixation buffer (27 mM sodium citrate, 336 mM NaCl, 115 mM glucose, and 9 mM EDTA; pH 7.0). Subsequently, the total hemocytes were counted using a Neubauer chamber. To monitor the expression level of genes related to antimicrobial peptides and the immune defense system against major pathogens (Table 1), we investigated the expression level of each gene by RT-PCR previously reported [17]. Briefly, the hepatopancreas was homogenized with liquid nitrogen, and then the total RNA was extracted using the TRIzol® Plus RNA Purification System (Thermo Fisher Scientific, Carlsbad, CA, USA). cDNA was synthesized by reverse transcription with oligo dT primer, and 0.1 μg/μL cDNA, as a template, was amplified by PCR with primers, designed using those in GenBank (https://www.ncbi.nlm.nih.gov/genbank)as described below. After PCR, the intensity of each PCR product was analyzed using the Gel Documentation System (Bio-Rad, Hercules, CA, USA) after running on 1% agarose gel. The intensity of the amplified cDNA was analyzed using ImageJ software (National Institutes of Health, Betheda, MD, USA) for quantification.

Table 1. Information of target genes and primers for monitoring the expression level of genes related to antimicrobial peptides and the immune defense system against major pathogens.

Gene	Primer	Sequence	GenBank No.
βGBP	Forward Reverse	5′-CGTGAGGTTCCCCAGTATGG-3′ 5′-TTCGGTTTGGATGGCTAAAG-3′	AY249858
proPO	Forward Reverse	5′-GGAATTGTTTTACTACATGCATCAGC-3′ 5′-GGAACAAGTCATCCACGAGCTT-3′	AY723296
Crustin	Forward Reverse	5′-ATTCTGTGCGGCCTCTTTAC-3′ 5′-ATCGGTCGTTCTTCAGATGG-3′	AF430076
Penaeidin 3a	Forward Reverse	5′-AGCCTCACCTGCAGAGACCA-3′ 5′-AATCAGGATCRCAGKCTCTTCAC-3′	Y14926
Lysozyme	Forward Reverse	5′-TTCGGGAAGTGCGAATTCG-3′ 5′-AATGGAAACCCTTGGTGAC-3′	AY170126
SOD	Forward Reverse	5′-GAGAAGAAGTTGGCTGAGCT-3′ 5′-ATGTTGGGTCCAGAAGATGG-3′	AY486424
Hemocyanin	Forward Reverse	5′-AATGCAGCCTACTTCCGTCAG-3′ 5′-TTATCGGGGTACACGCCATG-3′	X82502
β-actin	Forward Reverse	5′-TGTGTGACGACGAAGTAGCC-3′ 5′-TGGTCGTGAAGGTGTAACCA-3′	AF300705

2.3. Monitoring the Effect of Kefir through WSSV (White Spot Syndrome Virus) Challenge Test

Two hundred and forty *L. vannamei* were used for this challenge test and it was confirmed that these shrimps were specifically pathogen free by National Institute of Fisheries Science, Rep. of Korea. We divided the total amount into four groups: negative, positive and two kefir-treated groups according to kefir concentration. Each group was subdivided into three subgroups. Each 15 shrimps were tested in a plastic aquarium (1 m × 1 m × 0.7 m). Average weight of shrimps for this experiment was 4.70 g. Before infection of WSSV, all shrimps were acclimated for 7 days in plastic aquarium.

To prepare the WSSV filtrate, we obtained dead shrimps by infection of WSSV and homogenized the infected organ of dead shrimp in TNE buffer (50 mM Tris-HCl, 0.1 M NaCl, and 1 mM EDTA, pH 7.5). Next, the homogenate was centrifuged at 6000 rpm and 4 °C for 15 min and the supernatant was centrifuged again at 12,000 rpm and 4 °C for 15 min for 30 min. The final supernatant was filtrated by 0.45 μm syringe filter. To determine LD50 as the dose used for this experiment, we injected 0.1 mL of 10^{-2}, 10^{-3}, 10^{-4}, and 10^{-5} diluted filtrate in the dorsal body part of one shrimp and checked that LD50 of filtrate was 10^{-4} diluted filtrate. The infection of WSSV was carried out by the injection of the 0.1 mL of 10^{-4} diluted filtrate into the dorsal body of all shrimps without a negative control group. In the negative control group, the buffer was injected instead of the filtrate. After injection, we monitored the number of dead shrimp day by day for 15 days. The feeding for this experiment was carried out like described above.

For this experiment, the temperature of all of aquariums had been kept at 28 °C and all aquariums were aerated by air blower. The range of parameters of water quality like dissolved oxygen, pH, total ammonium nitrogen, and nitrate was kept on according to Holmström (2003) [5]. Survival rate by kefir treatment was calculated by Kaplan–Meier survival analysis provided in Prism 8 (GraphPad Software, San Diego, CA, USA).

2.4. Monitoring Weight Gain in L. vannamei after Treatment with Kefir through Field Test

For the field test, three shrimp farms located in Taean-Gun, Chungcheongnam-do, South Korea, were selected. The shrimp were treated with lyophilized kefir one time per week from May to October. The amount of lyophilized kefir in a tank of 1 m depth was 10 kg/ha, and the number of lactic acid bacteria, *Bacillus* spp., and yeast in lyophilized kefir was 5.6×10^9, 6.9×10^9, and 4.8×10^9 CFU/mL, respectively. To monitor the length and weight of the shrimp, we collected 100 shrimps before transfer to the main farm, at July, and at the end of farming. We measured the length and weight of the shrimps collected. The occurrence of significant differences in the average weight by treatment of lyophilized kefir was tested using Student's *t*-test.

2.5. Statistical Analysis

Statistical analyses were accomplished using SPSS 22.0 (IBM Corp., Armonk, NY, USA). The obtained data were analyzed by the paired *t*-test for evaluating the association and significance between variables. A p value < 0.05 indicated significance.

3. Results

3.1. Optimization of Culture Conditions for the Growth of Kefir

For the main fermentation of kefir, seed culture of kefir was carried out in whole fat milk medium for 2 days at 30 °C without agitation. To optimize the conditions of the main culture, we tested several parameters such as the content of glucose as a carbon source, whey protein as a nitrogen source, and dipotassium phosphate as a phosphate source, and the size of seed inoculated. The colony forming unit (CFU) of lactic acid bacteria, *Bacillus* spp., and yeast increased at 2% of glucose, 1% of whey protein and 0.02% and 0.1% of dipotassium phosphate, compared to other concentrations of nutrients (Figure 1A–C). In addition, we examined how the CFU of these microorganisms changed according to the size of inoculation of seed culture. Interestingly, the CFUs were the maximum with 2% inoculation of seed culture, rather than 10% (Figure 1D). Based on these results, we finally chose the optimal content of glucose, whey protein, and dipotassium phosphate, and size of inoculum (Table 2).

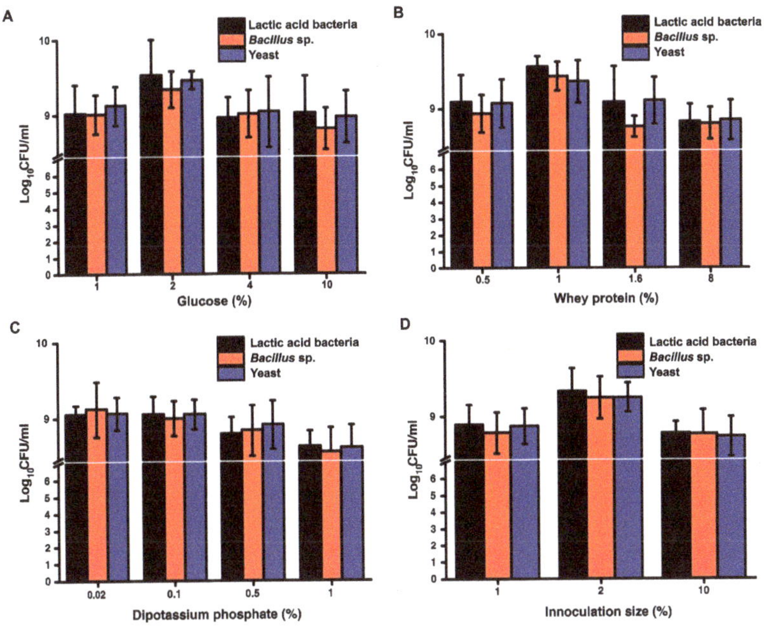

Figure 1. Proportion of lactic acid bacteria, *Bacillus* spp., and yeast in kefir according to culture conditions. (**A**) The proportion of lactic acid bacteria, *Bacillus* spp., and yeast in kefir by glucose content. (**B**) The proportion of lactic acid bacteria, *Bacillus* spp., and yeast in kefir by whey protein content. (**C**) The proportion of lactic acid bacteria, *Bacillus* spp., and yeast in kefir by dipotassium phosphate content. (**D**) The proportion of lactic acid bacteria, *Bacillus* spp., and yeast in kefir by inoculation size.

Table 2. The optimized medium condition for kefir fermentation.

Component	Composition (%, *w*/*v*)
Glucose	2
Whey protein	1
Dipotassium phosphate	0.02
Yeast extract	2
Ammonium sulfate	0.1
$MgSO_4$	0.01
$MnSO_4$	0.05
Inoculation size	2

3.2. *Effect of Kefir on the Innate Immunity of L. vannamei*

To monitor the effect of kefir on the innate immunity of *L. vannamei*, we treated lyophilized kefir with different doses. After 30 days of treatment, the total hemocytes in the hemolymph of *L. vannamei* were counted. Next, we examined the expression level of immune-related genes such as beta-glucan binding protein (βGBP), prophenoloxidase (proPO), crustin, peaneidin 3a, lysozyme, superoxide dismutase (SOD), and hemacyanin in the hepatopancreas of *L. vannamei*. The treatments of kefir did not change the number of total hemocytes (Table 3). However, the expression of immune-related genes was substantially upregulated after treatment of kefir (Figure 2). The expression of βGBP was upregulated fourfold in the group treated with 0.8% lyophilized kefir compared with that in the control group (Figure 2). Moreover, the expression of proPO, lysozyme, and SOD was substantially upregulated by lyophilized kefir treatment in a dose-dependent manner (Figure 2). The treatment of lyophilized kefir strongly induced the expression of genes encoding crustin and penaedin-3a by threefold compared with the control (Figure 2). These results indicate that kefir can contribute to the enhancement of the immune system in *L. vannamei* by upregulating the expression of immune-related genes and not cellular immunity by increasing hemocytes.

Table 3. The number of total hemocytes in *L. vannamei* by treatment of fermented kefir.

Kefir Treated (%)	Number of Total Hemocytes ($\times 10^7$/mL)
0	7.53 ± 0.30
0.20	7.46 ± 0.11
0.40	7.40 ± 0.2
0.80	7.6 ± 0.2

3.3. *Effect of Kefir in L. vannamei against WSSV Infection*

After infection of WSSV filtrate, the number of dead shrimps immediately increased and the survival rate at the end of this experiment was calculated to be 33.33% in the positive group (Figure 3). In the case of the 0.2% kefir-treated group, the death rate was retarded early, compared to the positive group. However, the survival rate was the same at the end of the experiment (Figure 3). Meanwhile, in the case of the 0.8% kefir-treated group, the survival rate was dramatically improved and calculated to be 56.8% at the end of this experiment. This result means that the enhancement of innate immunity of *L. vannamei* by treatment of kefir can increase the survival rate against WSSV infection.

3.4. *Effect of Kefir on Productivity of L. vannamei*

In the development of a functional feed additive for shrimp farming, the improvement of shrimp productivity is an important aspect. Therefore, to examine whether kefir can improve the productivity of *L. vannamei*, we treated lyophilized kefir in three farms in South Korea after transfer to the main pond for 6 months; we measured the weight of shrimp before transfer to the main pond, and at 3 and 6 months after transfer. The shrimps in the group treated with lyophilized kefir were longer than those in the control group (Figure 4A–D). Furthermore, the average weight of *L. vannamei* in the group treated with lyophilized kefir increased by up to 120% compared with that of *L. vannamei* in the control group (Figure 4E–G). The increase in both length and weight of shrimp in the group treated with lyophilized kefir was observed in all three farms. These results indicate that the supplementation of kefir can improve the utilization of feed to increase the length and weight of *L. vannamei*.

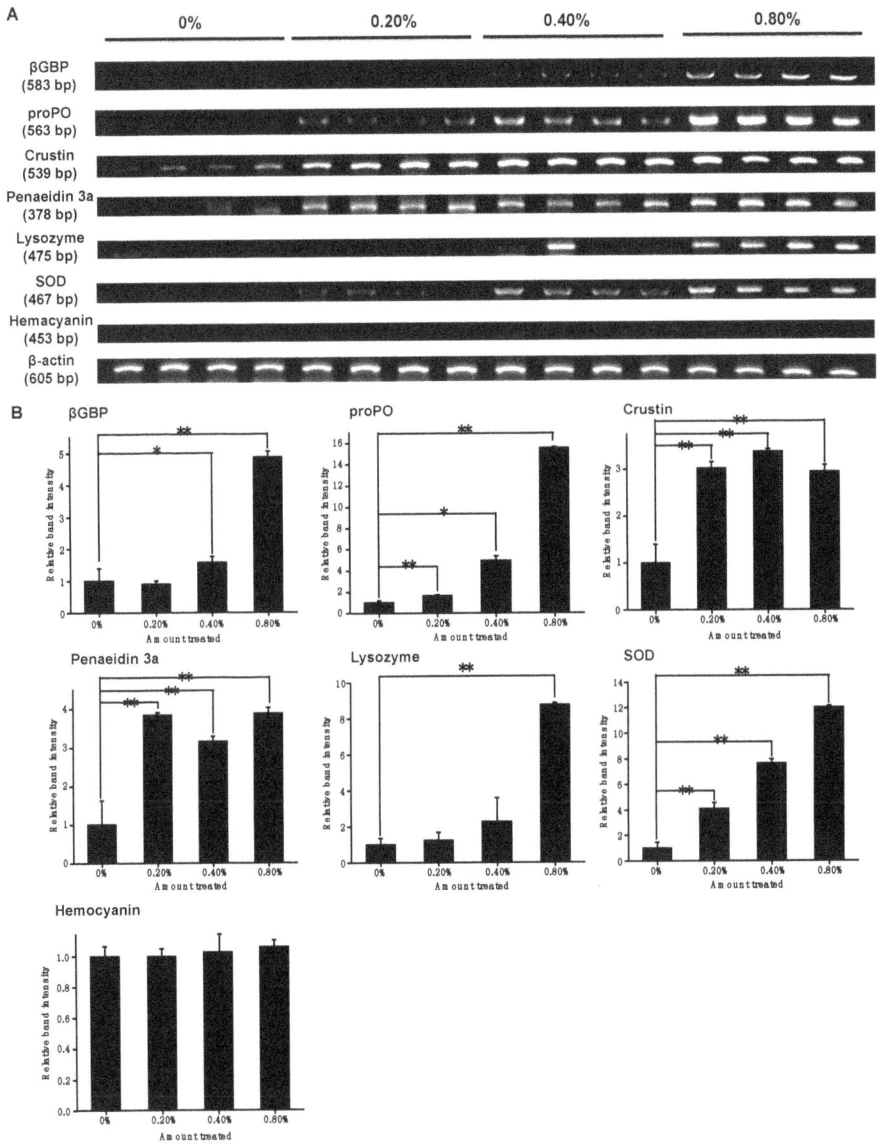

Figure 2. Transcription level of genes related to innate immunity in *L. vannamei* after treatment with kefir. (**A**) The amplified cDNA genes related to innate immunity in *L. vannamei* after treatment with kefir. (**B**) Quantification of the amplified cDNA genes related to innate immunity in *L. vannamei* using ImageJ. For quantification of the amplified cDNA genes, the intensity of the amplified cDNA in four wells was averaged and normalized by dividing with amplified cDNA intensity of the non-treated group. * $p < 0.05$ and ** $p < 0.01$, accessed using the paired t test.

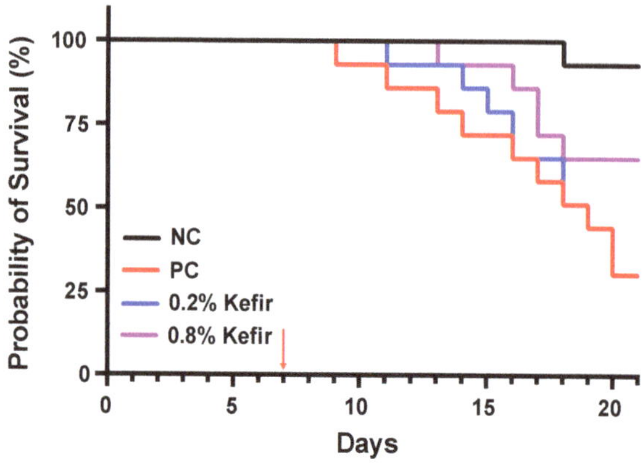

Figure 3. Survival rate of *L. vannamei* against WSSV infection according to treatment of kefir. NC and PC indicate negative and positive control group, respectively. Red arrow indicates the day when WSSV filtrate was injected except for the negative control group.

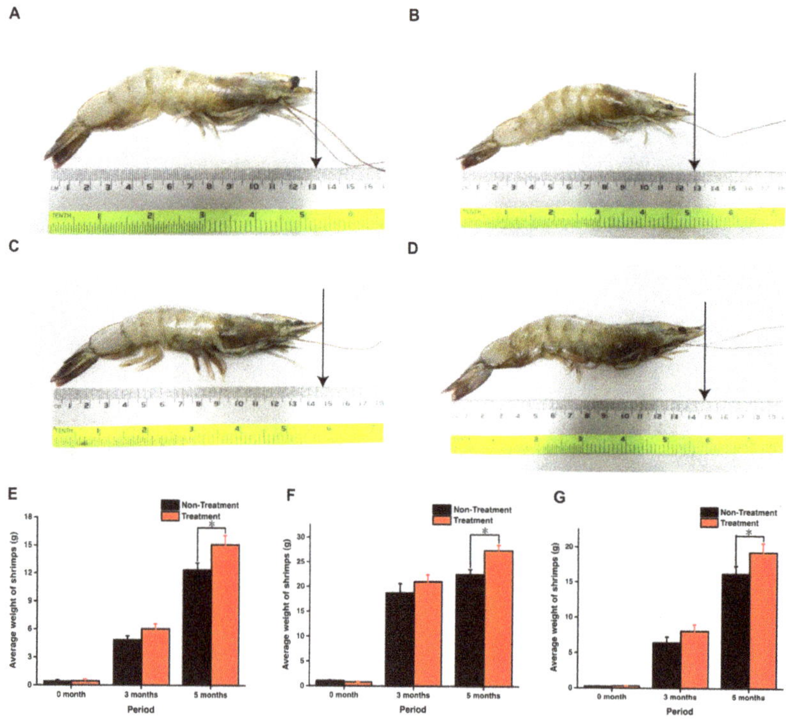

Figure 4. The growth of *L. vannamei* after treatment with kefir. (**A**,**B**) Representative images of the length of *L. vannamei* in the non-treated group. (**C**,**D**) Representative images of the length of *L. vannamei* in the kefir-treated group. Black arrows indicate the end of head of *L. vannamei*. (**E**) The average weight of *L. vannamei* after treatment with kefir in shrimp farm #01. (**F**) The average weight of *L. vannamei* after treatment with kefir in shrimp farm #02. (**G**) The average weight of *L. vannamei* after treatment with kefir in shrimp farm #03. The average weight of *L. vannamei* was calculated by the average weight of 100 *L. vannamei* individuals. * $p < 0.05$, accessed using the paired t test.

4. Discussion

The production of this shrimp is approximately 50 ton/ha/crop, which exceeds the production of Penaeus monodon, another popular shrimp species [18]. It has been reported that *L. vannamei* is highly tolerant to a wide range of salinity levels and requires a relatively low protein feed [18]. Furthermore, this shrimp can grow with high stocking densities and at low temperatures of up to 15 °C [19]. Even with these advantages, *L. vannamei* is prone to be infected by viruses such as Taura syndrome virus (TSV), white spot syndrome virus (WSSV), and yellow head virus (YHV), and pathogenic bacteria such as Vibrio parahaemolyticus [19]. Moreover, high-density farming of this shrimp that produces yields of 20,000 to 100,000 kg/ha/year results in the outbreak of infection by viruses and pathogenic bacteria because of eutrophication by unconsumed feeds and lots of shrimp feces [4]. Treatment with antibiotics is one of the approaches used to prevent these infections, but their use is strictly restricted because antibiotics can cause the development of antibiotic resistance among pathogens [4,5]. Thus, there is a need for alternatives to antibiotics in shrimp farming.

It has been reported that kefir consists of lactic acid bacteria, *Bacillus* spp., and yeast that play a role as probiotics [20], and thus, it has various probiotic properties [21]. Furthermore, the effect of probiotics as an alternative to antibiotics has been studied in shrimp farming [1,6,7]. Therefore, in this study, we explored the effect of kefir as a functional feed additive in shrimp farming to improve the immune system and productivity of *L. vannamei* with no use of antibiotics. We treated lyophilized kefir and determined the number of hemocyte and expression level of genes related to the innate immune system of *L. vannamei*. In *L. vannamei*, hemocytes play pivotal roles in pattern-recognition system, phagocytosis, proPO-activating system, encapsulation, nodule formation, antimicrobial peptide release, and lysozyme activity [2]. In this study, lyophilized kefir did not directly affect the count of hemocytes, but it substantially upregulated the expression of the βGBP, proPO, crustin, peaneidin 3a, lysozyme, and SOD genes. βGBP is a pattern-recognition protein and proPO participates in the proPO cascade [22]. Furthermore, crustin and peaneidin 3a are antimicrobial peptides [23]. Lysozyme and SOD are known to participate in the post-phagocytosis process to remove toxic materials during phagocytosis [2]. Also, in WSSV challenge experiment, we observed that the survival rate of *L. vannamei*-treated 0.8% kefir was obviously improved, compared to positive control and the 0.2% kefir-treated group. So, it is shown that the enhancement of innate immunity of *L. vannamei* directly can improve the survival of these shrimp against WSSV infection.

In terms of economic status, the use of functional feed additive might increase the cost of management; therefore, an improvement in shrimp productivity is one of the important factors when choosing a functional feed additive [7]. In this study, we carried out a field test in three farms in South Korea. During farming, we monitored the weight of shrimps before the transfer of shrimp to the main pond and after 3 months of feeding and harvest of shrimp. After 3 months of feeding, the average weight of shrimp was higher than that of the control group shrimp; at harvest, the average weight of shrimp treated with lyophilized kefir increased by 120% compared with that of the control group. This suggests that multiple bacteria in kefir can promote the growth of *L. vannamei*. Based on these results, we conclude that kefir might be a promising feed additive for both enhancement of the immune defense system and improvement of productivity of *L. vannamei* without treatment of antibiotics.

Author Contributions: H.-C.K., H.L., and J.-W.S. designed the research and conducted all experiments. W.C. and C.-W.C. mainly carried out field test and monitored the expression level of genes and weight of shrimps. D.-B.S. and B.-C.J. carried out kefir fermentation, lyophilized kefir preparation, and viable cell counting in the sample. H.L. wrote the manuscript. All authors discussed the results and commented on the manuscript. All authors have read and agreed to the published version of the manuscript.

Funding: This work was supported by the Cooperative Research Program for Agriculture Science & Technology Development (Project No. 01128905) from the Rural Development Administration, South Korea.

Conflicts of Interest: The authors declare no conflict of interest.

References

1. Van Hai, N.; Fotedar, R. A Review of Probiotics in Shrimp Aquaculture. *J. Appl. Aquac.* **2010**, 251–266. [CrossRef]
2. Lin, Y.C.; Chen, J.C.; Siti, S.N.; Wan, W.Z.; Awangku, A.S.; Cheng, S.Y.; Hsu, C.H. Modulation of Innate Immunity and Gene Expressions in White Shrimp Litopenaeus Vannamei Following Long-Term Starvation and Re-Feeding. *Results Immunol.* **2012**, *2*, 148–156. [CrossRef]
3. Mohney, L.L.; Lightner, D.V.; Bell, T.A. An Epizootic of Vibriosis in Ecuadorian Pond-Reared Penaeus Vannamei Boone (Crustacea: Decapoda). *J. World Aquac. Soc.* **1994**, *25*, 116–125. [CrossRef]
4. Bermdez-Almada, M.C.; Espinosa-Plascenci, A. The Use of Antibiotics in Shrimp Farming. In *Health and Environment in Aquaculture*; IntechOpen Limited: London, UK, 2012. [CrossRef]
5. Holmström, K.; Gräslund, S.; Wahlström, A.; Poungshompoo, S.; Bengtsson, B.E.; Kautsky, N. Antibiotic Use in Shrimp Farming and Implications for Environmental Impacts and Human Health. *Int. J. Food Sci. Technol.* **2003**, *38*, 255–266. [CrossRef]
6. Farzanfar, A. The Use of Probiotics in Shrimp Aquaculture. *FEMS Immunol. Med. Microbiol.* **2006**, 149–158. [CrossRef] [PubMed]
7. Ninawe, A.S.; Selvin, J. Probiotics in Shrimp Aquaculture: Avenues and Challenges. *Crit. Rev. Microbiol.* **2009**, *35*, 43–66. [CrossRef] [PubMed]
8. Kumar, V.; Roy, S.; Meena, D.K.; Sarkar, U.K. Application of Probiotics in Shrimp Aquaculture: Importance, Mechanisms of Action, and Methods of Administration. *Rev. Fish. Sci. Aquac.* **2016**, 342–368. [CrossRef]
9. Sivasankar, P.; Santhiya, A.V.; Kanaga, V. A Review on Plants and Herbal Extracts against Viral Diseases in Aquaculture. *J. Med. Plants Stud.* **2015**, *3*, 75–79.
10. Immanuel, G.; Vincybai, V.C.; Sivaram, V.; Palavesam, A.; Marian, M.P. Effect of Butanolic Extracts from Terrestrial Herbs and Seaweeds on the Survival, Growth and Pathogen (Vibrio Parahaemolyticus) Load on Shrimp Penaeus Indicus Juveniles. *Aquaculture* **2004**, *236*, 53–65. [CrossRef]
11. Yogeeswaran, A.; Velmurugan, S.; Punitha, S.M.J.; Babu, M.M.; Selvaraj, T.; Kumaran, T.; Citarasu, T. Protection of Penaeus Monodon against White Spot Syndrome Virus by Inactivated Vaccine with Herbal Immunostimulants. *Fish Shellfish Immunol.* **2012**, *32*, 1058–1067. [CrossRef]
12. Bourrie, B.C.T.; Willing, B.P.; Cotter, P.D. The Microbiota and Health Promoting Characteristics of the Fermented Beverage Kefir. *Front. Microbiol.* **2016**, *7*, 647. [CrossRef] [PubMed]
13. Leite, A.M.D.O.; Miguel, M.A.L.; Peixoto, R.S.; Rosado, A.S.; Silva, J.T.; Paschoalin, V.M.F. Microbiological, Technological and Therapeutic Properties of Kefir: A Natural Probiotic Beverage. *Braz. J. Microbiol.* **2013**, 341–349. [CrossRef] [PubMed]
14. Laureys, D.; De Vuyst, L. Microbial Species Diversity, Community Dynamics, and Metabolite Kinetics of Water Kefir Fermentation. *Appl. Environ. Microbiol.* **2014**, *80*, 2564–2572. [CrossRef] [PubMed]
15. Liu, J.R.; Chen, M.J.; Lin, C.W. Antimutagenic and Antioxidant Properties of Milk-Kefir and Soymilk-Kefir. *J. Agric. Food Chem.* **2005**, *53*, 2467–2474. [CrossRef]
16. Uluköy, G.; Metin, S.; Kubilay, A.; Güney, Ş.; Yıldırım, P.; Güzel-Seydim, Z.; Kok-Tas, T.; Gümüş, E. The Effect of Kefir as a Dietary Supplement on Nonspecific Immune Response and Disease Resistance in Juvenile Rainbow Trout, Oncorhynchus Mykiss (Walbaum 1792). *J. World Aquac. Soc.* **2017**, *48*, 248–256. [CrossRef]
17. Wang, Y.C.; Chang, P.S.; Chen, H.Y. Tissue Expressions of Nine Genes Important to Immune Defence of the Pacific White Shrimp Litopenaeus Vannamei. *Fish Shellfish Immunol.* **2007**, *23*, 1161–1177. [CrossRef]
18. Thakur, K.; Patanasatienkul, T.; Laurin, E.; Vanderstichel, R.; Corsin, F.; Hammell, L. Production Characteristics of Intensive Whiteleg Shrimp (Litopenaeus Vannamei) Farming in Four Vietnam Provinces. *Aquac. Res.* **2018**, *49*, 2625–2632. [CrossRef]
19. Li, E.; Wang, X.; Chen, K.; Xu, C.; Qin, J.G.; Chen, L. Physiological Change and Nutritional Requirement of Pacific White Shrimp Litopenaeus Vannamei at Low Salinity. *Rev. Aquac.* **2017**, 57–75. [CrossRef]
20. Shi, X.; Chen, H.; Li, Y.; Huang, J.; He, Y. Effects of Kefir Grains on Fermentation and Bioactivity of Goat Milk. *Acta Univ. Cibiniensis Ser. E Food Technol.* **2018**, *22*, 43–50. [CrossRef]
21. Guzel-Seydim, Z.B.; Kok-Tas, T.; Greene, A.K.; Seydim, A.C. Review: Functional Properties of Kefir. *Crit. Rev. Food Sci. Nutr.* **2011**, 261–268. [CrossRef]

22. Amparyup, P.; Charoensapsri, W.; Tassanakajon, A. Prophenoloxidase System and Its Role in Shrimp Immune Responses against Major Pathogens. *Fish Shellfish Immunol.* **2013**, *34*, 990–1001. [CrossRef]
23. Tassanakajon, A.; Somboonwiwat, K. Antimicrobial Peptides from the Black Tiger Shrimp Penaeus Monodon—A Review. In *Dissease in Asian Aquaculture VII*; Asian Fisheries Society: Kuala Lumpur, Malaysia, 2011; pp. 229–240.

Publisher's Note: MDPI stays neutral with regard to jurisdictional claims in published maps and institutional affiliations.

MDPI AG
Grosspeteranlage 5
4052 Basel
Switzerland
Tel.: +41 61 683 77 34

Fermentation Editorial Office
E-mail: fermentation@mdpi.com
www.mdpi.com/journal/fermentation

www.ingramcontent.com/pod-product-compliance
Lightning Source LLC
Chambersburg PA
CBHW040757250926
51850CB00007BA/1021
* 9 7 8 3 7 2 5 8 2 2 5 4 6 *